REGULATING AGRICULTURAL BIOTECHNOLOGY:
ECONOMICS AND POLICY

NATURAL RESOURCE MANAGEMENT AND POLICY

Editor:
David Zilberman
Dept. of Agricultural and Resource Economics
University of California, Berkeley
Berkeley, CA 94720

EDITORIAL STATEMENT

There is a growing awareness to the role that natural resources such as water, land, forests and environmental amenities play in our lives. There are many competing uses for natural resources, and society is challenged to manage them for improving social well being. Furthermore, there may be dire consequences to natural resources mismanagement. Renewable resources such as water, land and the environment are linked, and decisions made with regard to one may affect the others. Policy and management of natural resources now require interdisciplinary approach including natural and social sciences to correctly address our society preferences.

This series provides a collection of works containing most recent findings on economics, management and policy of renewable biological resources such as water, land, crop protection, sustainable agriculture, technology, and environmental health. It incorporates modern thinking and techniques of economics and management. Books in this series will incorporate knowledge and models of natural phenomena with economics and managerial decision frameworks to assess alternative options for managing natural resources and environment.

The   Series   Editor

Recently Published Books in the Series
Haddadin, Munther J.
Diplomacy on the Jordan: International Conflict and Negotiated Resolution
Renzetti, Steven
The Economics of Water Demands
Just, Richard E. and Pope, Rulon D.
A Comprehensive Assessment of the Role of Risk in U.S. Agriculture
Dinar, Ariel and Zilberman, David
Economics of Water Resources: The Contributions of Dan Yaron
Ünver,.I.H. Olcay, Gupta, Rajiv K. IAS, and Kibaroğlu, Ayşegül
Water Development and Poverty Reduction
d'Estrée, Tamra Pearson and Colby, Bonnie G.
Braving the Currents: Evaluating Environmental Conflict Resolution in the River Basins of the American West
Cooper, Joseph, Lipper, Leslie Marie and Zilberman, David
Agricultural Biodiversity and Biotechnology in Economic Development
Babu, Suresh Chandra and Djalalov, Sandjar
Policy Reforms and Agriculture Development in Central Asia
Goetz, Renan-Ulrich and Berga, Dolors
Frontiers in Water Resource Economics

REGULATING AGRICULTURAL BIOTECHNOLOGY: ECONOMICS AND POLICY

Edited by

RICHARD E. JUST
University of Maryland

JULIAN M. ALSTON
University of California, Davis

DAVID ZILBERMAN
University of California, Berkeley

Liesl Koch, Technical Editor

 Springer

Library of Congress Control Number: 2006929864

ISBN:10: 0-387-36952-X   (Printed on acid-free paper)   e-ISBN-10: 0-387-36953-8
ISBN-13: 978-0387-36952-5   e-ISBN-13: 978-0387-36953-2

Printed in the United States of America.

9 8 7 6 5 4 3 2 1

springer.com

Contents

PART III
CASE STUDIES ON THE ECONOMICS OF REGULATING
AGRICULTURAL BIOTECHNOLOGY

Section III.1 – International Evidence

Section III.2 – Refuge Policy and Regulatory Compliance at the Farm Level

Section III.3 – Crop-Specific Issues in Biotechnology Regulation

CONCLUSIONS

Introduction

REGULATING AGRICULTURAL BIOTECHNOLOGY: ECONOMICS AND POLICY

Chapter 1

REGULATING AGRICULTURAL BIOTECHNOLOGY: INTRODUCTION AND OVERVIEW

Richard E. Just,* David Zilberman,† and Julian M. Alston ‡
University of Maryland, *University of California, Berkeley,*†
University of California, Davis‡

Abstract: This chapter introduces the topic of this book, drawing upon the content of its chapters. An overview is provided on the economics of technological regulation as applied to agricultural crop biotechnologies. Key elements of agricultural biotechnology regulation in the United States are summarized.

Key words: agricultural biotechnologies, causes and consequences of technological regulation, overview and synthesis

1. INTRODUCTION

Agricultural biotechnologies apply modern knowledge in molecular and cell biology to produce new varieties and similar genetic materials. The use of genetically modified (GM) crop varieties has grown dramatically since they were introduced in 1995, and large portions of the land allocated to corn, soybeans, and cotton are grown with these varieties. The evidence from the United States, Canada, China, India, Brazil, and Argentina suggests that these applications of biotechnology in agriculture increase yield, reduce the use of pesticides, and save production costs.

Many expect agricultural biotechnologies to play a crucial role in meeting growing food demands to accommodate population and income growth in the future and, at the same time, to contribute to containing the environmental footprint of agriculture and to provide new sources of biofuels. Yet, others view agricultural biotechnologies as inadequately tested, dangerous technologies that pose unforeseen risks and that must be handled with the utmost of care. These differences in perspective have contributed to policy debates on the regulation of agricultural biotechnologies and to regulatory differences across locations.

Even when they provide net economic benefits, new technologies almost always generate gainers and losers where some of the negative consequences may involve external effects on human health or the environment. Under classical theory, whether externalities are actual or perceived, regulation is typically justified as a means of correcting such market distortions and enhancing net social benefits from production and consumption, regardless of distributional impacts. Under interest group theory (Becker 1983), individuals and groups support political activities that are in their best interest. Regulations are viewed as means of redistribution and are the consequence of influence by politically powerful interest groups who benefit as a result, often with net social loss. Understanding the political economy of current regulations likely calls for combining both approaches.

Whether primarily for efficiency or distributional reasons, the development, release, adoption, and application of agricultural biotechnologies are increasingly subject to public scrutiny and regulation. Compliance with these regulations adds considerably to the time lags and costs borne in bringing new biotechnology products to market. Because of differences in interest groups, distribution, and local circumstances, it is not surprising that biotechnology regulations differ among countries, among states within countries, and among biotechnologies. The regulations modify the rate and form of technological change and the distribution of benefits and costs. Without doubt, the economic consequences are significant, although the full consequences of technological regulation in agriculture are not well understood.

The rapid evolution of agricultural biotechnologies has led to the emergence of significant bodies of research on various aspects of the economics of crop biotechnologies, including the adoption and impact, consumer and producer attitudes, and the management of intellectual property rights. However, comparatively little research has been conducted on the economics of the regulation. This book aims to fill this void and provide a foundation for further research on the economics of regulation of agricultural biotechnologies. The chapters of the book are based on a three-day conference held in Arlington, Virginia, on March 10–12, 2005, that presented and discussed methods and current issues in the "Economics of Regulation of Agricultural Biotechnologies," with an emphasis on drawing together the collective state of wisdom on forces shaping regulation of agricultural biotechnologies and the consequences for U.S. agriculture and the food system.

The objective is to increase general understanding of the issues associated with regulation of agricultural biotechnologies. Positive and normative perspectives are presented on how and why societies do and could manage these technologies, and the actual and potential consequences in terms of benefits and costs to consumers, producers, innovators, and the environment.

The book is written mostly by economists, but is aimed at the wider audience of educated people interested in the policy debate on the future of crop biotechnology.

A unique feature of this book is that we integrate and build upon bodies of literature from disciplines both within and outside economics. We build upon the vast literature on the economics of agricultural research and technical change, but we also provide perspectives on the problems and potential of agricultural biotechnology and its health implications from a contributing plant biologist, public health scholars, and a policymaker. Several lines of economic study are integrated: (i) new methods of environmental economics, in particular, the economics of pest control and resistance management; (ii) new methods of evaluating consumer preferences and willingness to pay for environmental amenities and product quality; (iii) public economics for policy design; and (iv) political economy to assess policy viability within a political structure.

This chapter provides an introduction to the rest of the book in several ways. First, it reports on the conference that provided the genesis of the work. Second, it presents an "economic way of thinking" about the regulation of agricultural biotechnology, which provides some organizing principles for the ideas in the book and its structure. Third, it provides a summary description of the main elements of regulation of U.S. agricultural biotechnology, which serves as the context for many of the chapters concerned with U.S. agriculture, and as a contrast to regulations in other countries reported in other chapters. Finally, it provides an overview of the chapters that make up the rest of the book, and a brief synthesis and synopsis of what it all means.

2. THE ECONOMIC STATUS OF CROP BIOTECHNOLOGY

Biotechnology has transformed the production systems of major field crops, including soybeans, corn, cotton, and canola. Since their first large-scale introduction in 1996, the global area planted to biotech crops grew to 200 million acres by 2004 (James 2004). Almost 60 percent of this acreage was in the United States, where biotech varieties represented 85 percent of the soybeans, 76 percent of the cotton, and 45 percent of the corn acreage (National Agricultural Statistics Service 2004). These high rates of adoption reflect farmer benefits associated with these crops, which have been almost exclusively targeted toward providing herbicide tolerance, insect resistance, or both. Such benefits come from increased yields, lower risk, reduced use of chemical pesticides, gains from reduced tillage and other modified production practices, and savings in management, labor, and capital equipment (Kalaitzandonakes 2003).

Notably, however, the substantial adoption of agricultural biotechnology to date has been concentrated in a small number of countries and confined to a small number of traits in a small number of crops—specifically, pest resistance and herbicide tolerance in feed grains, oil seeds, and cotton. Biotech food products emphasizing output traits of value to consumers (e.g., long shelf-life tomatoes) or input traits for lightly processed crops (e.g., Bt potatoes or sweet corn) have been ignored or disadopted by food manufacturers or retailers in the face of perceived market resistance or political opposition. The fact that adoption of available biotech products has been limited to a small number of countries reflects a combination of market resistance, legal barriers to adoption, and trade barriers against importation of biotech crop products. The same barriers also have reduced incentives for biotech companies to invest in the development of new biotech products. These same factors may have contributed to the erection of regulatory barriers to development and adoption of biotech crops, which themselves provide a further disincentive for biotech companies.

3. RATIONALIZING REGULATION

The conventional economic argument for government intervention in the economy is based on the idea of market failure—that the unfettered working of the free-market mechanism has given rise to an inefficient allocation of resources or an unsatisfactory distribution of income—and that government intervention can make things better. Various types of market failures can and do arise in agriculture, often associated with the use of particular technologies, giving rise to arguments for government intervention. Examples include (i) various kinds of pollution externalities (such as pollution of air or groundwater associated with the use of agricultural chemicals); (ii) incomplete, ill-defined, or ill-enforced property rights to assets such as irrigation water or other natural resource stocks, or to intellectual property including plant varieties or other inventions; (iii) incomplete or asymmetric information about product characteristics including how a product was produced and whether it is safe to consume; and (iv) market distortions arising from the exercise of market power by agribusiness firms in the supply of inputs or technology, or in the marketing of agricultural products.

Government regulations to address these concerns are pervasive, and largely taken for granted, but the regulations evolve as knowledge, institutions, and market characteristics change. For example, various agricultural chemicals have been banned (such as DDT in U.S. agriculture); selected uses of others have been eliminated by regulation (such as use of pesticides near surface water); and environmental and occupational health and safety regulations limit how they may be applied. Similarly, the laws and rules governing rights to natural resources and to intellectual property are constantly evolving as circumstances, knowledge, and institutions change. In particular, expanded intellectual property rights applied to plant varieties have contributed importantly to development of the agricultural biotechnology industry as a predominantly private enterprise in the United States. And with rising affluence, and in the wake of various food scares, increasing attention has focused on public provision of information and food-safety assurance, leading to an attendant rise in food-safety regulation.

In contemplating the economics of regulation of agricultural biotechnologies, one must consider both policy design and policy impact. Research on policy design is needed to optimize the parameters of the regulatory systems. For example, a major challenge is to identify which biotechnology innovations should be pre-tested by government, and to identify the correct testing protocol (Zilberman 2006). Research on policy impact is needed to quantify the benefits and costs of regulations and their distribution. To measure these correctly, counterfactual circumstances must be identified, and when possible, impacts on treatment and control groups should be compared. Government intervention to correct one distortion may create another, making the full effects more difficult to discern. For instance, the provision of intellectual property rights to inventors of modified crop varieties has two somewhat offsetting effects: they enhance incentives to invest in research and development, while allowing firms to charge monopoly prices for their inventions, resulting in sub-optimal adoption rates and loss of consumer welfare because of high prices. Evaluating policy design and policy impact must consider heterogeneity of economic and political environments. This may help to explain international differences in the regulation of agricultural biotechnologies, as well as commodity trade policies.

4. CAUSES AND CONSEQUENCES

Regulations affecting the development and adoption of biotech crops, and their causes and consequences, are the focus of this book. To understand the genesis and consequences of these regulations requires some understanding of the nature of biotech crops and attitudes toward them, and the role of these attitudes in shaping regulations on laboratory development, field trials, commercial farm use, international commodity trade, and final consumer markets for food.

4.1. Lack of Consumer Acceptance and Labeling Issues

Some consumers believe that GM foods are unsafe to eat, or that the processes used to produce them are environmentally unsafe. Consequently, they may favor a labeling requirement or a ban on biotechnology (Huffman and Rousu 2006, McCluskey, Grimsrud, and Wahl 2006). The coalition that supports regulation and restriction of biotechnology includes certain consumer groups and environmentalists. Pressures by these groups and others have led to segregation and labeling requirements (and even bans) of GM crops by the European Union (EU) and other entities. This, in turn, has slowed adoption and development of new crop biotechnologies.

4.2. Substitution for Conventional Pest Control

An important potential benefit of agricultural biotechnology is the possibility of displacing conventional resource-intensive agricultural technologies by introducing pest-resistant genes and herbicide-tolerant genes. These new technologies will allow a substantial reduction in the consumption of chemicals, labor, and energy, and also reduce the environmental burden of chemical pesticides. As with conventional chemical technologies, the development of resistant pests or herbicide-tolerant weeds is an important potential consequence of the adoption of biotech crops. To slow the buildup of resistance, the U.S. government has imposed refuge requirements as part of its regulatory approval process for biotechnologies, but the policing of refuge requirements has been left largely in the hands of biotech companies and the resulting compliance appears to be very low. The design of these regulations, and their impact on adoption and economic welfare, are the subject of ongoing research (Mitchell and Hurley 2006, Langrock and Hurley 2006). There is also some evidence that unchecked adoption without refuges can potentially lead to more serious economic effects of secondary pests due to eradication of primary pests that serve as predators (Wang, Just, and Pinstrup-Andersen 2006).

4.3. Regulatory Costs

Different sets of regulations govern the research and development process, commercial release, and the commercial use of new agricultural biotechnologies depending on the type of biotechnology. For example, in the United States, prior to the development and release of a new GM crop variety, a biotech company must satisfy separately the regulations for registration and approval from the EPA, the FDA, and the USDA. Obtaining these approvals imposes substantial costs and delays in the development process (Kalaitzandonakes, Alston, and Bradford 2006). Some argue (for instance, Miller and Conko 2004) that the requirements on the development of biotech crops are more onerous than the corresponding requirements on development of crop varieties by conventional techniques. Additionally, U.S. biotech firms may seek regulatory approval abroad. The cost of compliance with international regulations can detract from potential total profits associated with product development and thus determine whether a product is introduced domestically. Analysis of these issues is crucial to the assessment of existing regulations and proposals for reform.

4.4. Noncompetitive Market Implications

Patent laws aim to achieve societal benefits from product development through a balance between producer benefits from monopoly profits under patent, and consumer benefits from post-patent competition. Imposing substantial regulatory costs on private companies can affect this tradeoff. One possibility is to lengthen the patent period to offset regulatory costs and time delays incurred by product developers. In the United States, such action has been taken by Congress for pharmaceutical products but rejected for agricultural chemicals and certain agricultural biotechnologies. Current regulation of these items forces post-patent generic competitors to share in the regulatory costs necessary to ensure product safety. But because generic entrants typically gain only a small share of competitive profits in a declining market stage, imposing a substantial share of regulatory costs on generic firms can be a major deterrent to generic entry and post-patent competition. The result can be preservation of market power for product developers and a postponing or elimination of the benefits of post-patent competition for farmers and consumers (Just 2006).

4.5. Large-Crop, Large-Country Bias

Regulatory intervention and the substantial regulatory cost it imposes have especially impeded development and adoption of food crops, minor crops, and crops grown in small countries. The reason is that biotech firms require a large potential market and a high rate of adoption to justify the large overhead costs of regulatory compliance in addition to research and development (Alston 2004, Bradford, Alston, and Kalaitzandonakes 2006). The fact that various countries and market groups have different interests in such outcomes may offer a partial explanation for differences in agricultural biotechnology policies among countries.

4.6. Comparative Advantage and Implications for International Trade

The introduction of agricultural biotechnology can favor one group of businesses over another. Groups that may lose from the technology may use their political influence to support policies that will stall the spread of the technology. Anderson (2006) suggests that farmers in Europe as a whole would be worse off if they had to compete in a world in which farmers worldwide were free to adopt, compared with a world without biotech crops. Hence, European farmers might naturally oppose the development of biotech crops generally. But Anderson also shows that European farmers can be even better off if the adoption of biotech crops in other countries combined with opposition to it in Europe leads to the erection of new regulatory barriers on imports by the EU that amount to trade protection against competition from both conventional and biotech crop producers. Graff and Zilberman (2004) speculate that agricultural technology firms in Europe have a comparative advantage in chemical technologies whereas agricultural technology firms in the United States have a comparative advantage in biotechnology. Hence, firms in Europe (perhaps in a coalition with European farmers) would oppose biotech and influence their governments to regulate accordingly, whereas firms (and farmers) in the United States would do the opposite.

A possibly contradictory view is that regulatory compliance is a barrier to entry, and that successful biotech firms in the United States have a comparative advantage in meeting the requirements (Heisey and Schimmelpfennig 2006). The implication is that incumbent U.S. biotech firms may have encouraged the introduction of more stringent and costly regulations so as to preserve their market power. These questions become more complex upon considering that the major firms are involved in both chemical technologies and biotechnologies, that they are integrated with non-agricultural applications of biotechnology, and that they are multinational.

5. REGULATION OF U.S. AGRICULTURAL BIOTECHNOLOGIES

As the potential for genetically engineered products began to take shape, the U.S. government chose in 1986 to use existing health and safety laws to regulate agricultural biotechnology under the Coordinated Framework for Regulation of Biotechnology. As a result, federal regulation of agricultural biotechnology in the United States today is scattered across three agencies in three different departments with roles that are partially complementary and in some cases overlapping.

The U.S. Department of Agriculture's Animal and Plant Health Inspection Service (APHIS) has jurisdiction over the planting of genetically engineered plants and veterinary biologics; the U.S. Environmental Protection Agency (EPA) has jurisdiction over pesticides engineered into plants, microbial pesticides, and novel microorganisms; and the Department of Health and Human Services' Food and Drug Administration (FDA) has jurisdiction over food and feed uses of biotechnology.

The laws under which agricultural biotechnology is regulated are the Plant Protection Act (PPA), originally enacted in 1930, the Federal Food, Drug, and Cosmetic Act (FFDCA), originally enacted in 1938, the Federal

Insecticide, Fungicide, and Rodenticide Act (FIFRA), originally enacted in 1947, and the Toxic Substances Control Act (TSCA), originally enacted in 1976. These laws have been modified by numerous amendments including the Food Quality Protection Act (FQPA) of 1996. New regulations, rules, and guidelines have been developed under each of these statutes in piecemeal fashion by administering agencies to address issues for genetically engineered products as they have arisen.

Under the authority of the PPA, APHIS regulations provide procedures for obtaining a permit prior to developing or importing organisms altered or produced through genetic engineering that are potential plant pests. The FDA regulates foods and feed derived from new plant varieties and enforces pesticide tolerances on foods under the authority of pre-existing food law in the FFDCA, and requires that genetically engineered foods meet the same rigorous safety standards as required of all other foods. If substances added to food through genetic engineering are significantly different from substances currently found in food, then they are treated as food additives. However, many food crops currently being developed using biotechnology do not contain substances significantly different from those already in the diet, and thus do not require pre-market approval.

The EPA regulates the distribution, sale, use, and testing of pesticidal substances, including plant-incorporated protectants such as Bt, just as for chemical pesticides. The EPA uses the authority of FIFRA to regulate the distribution, sale, use, and testing of plants and microbes producing pesticidal substances; the authority of the FFDCA to set tolerance limits for substances used as pesticides on and in food and feed (enforced by the USDA on meat, poultry, and eggs, and by the FDA on other foods); and the authority of TSCA to regulate GM microbial pesticides (microorganisms such as bacteria, fungi, viruses, protozoa, or algae) intended for commercial use. In the case of herbicide-tolerant crops, the EPA regulates the herbicide and APHIS regulates the crop.

These various regulations clearly have dual purposes as evidenced by Congressional records (see, for example, Just 2006). The purpose is not only to protect human health and the environment by facilitating regulatory test data generation and conditioning registration on each substance's health, safety, and environmental effects. The purpose is also to administer those regulations in a way that promotes social well-being and minimizes social waste and disruption to an otherwise well-functioning market economy.

This book considers some of the potential differential effects of regulations imposed by these laws and related agency rules on the competitive efficiency of regulated markets. In some cases, regulations appear to operate relatively efficiently while in others they appear to lead to inefficiency. In any case, the variety of laws and agencies regulating agricultural biotechnologies certainly leads to different types of regulations and regulations at different levels of development depending on which agency and law is applicable. While agricultural biotechnology is likely still in its infant stage, the variation in regulations and the scattering of their administration among agencies appears to be partly explained by the motivations of private interests in biotechnologies and related lobbies, but also partly the result of simply extending the regulations of old policies when the Coordinated Framework for Regulation of Biotechnology was adopted. Thus, the impact and distributional consequences of current regulations on industry structure appear to be partly a result of political interests and partly the indirect consequences of expedient approaches to law-making under uncertainty in an evolving policy setting.

6. OVERVIEW OF THE BOOK, SYNTHESIS, AND SYNOPSIS

The remainder of the book is presented in three main parts, followed by a concluding chapter. Part I is entitled *Agricultural Biotechnology in the Context a Regulated Agricultural Sector*, and it comprises a total of 10 chapters begins with four chapters that describe the consequences of regulation In Chapter 2, David Widawsky outlines the key elements of the regulatory framework administered by the EPA for agricultural biotechnologies in the United States. In Chapter 3, Nicholas Kalaitzandonakes, Julian Alston, and Kent Bradford describe the process of regulatory compliance for a new biotech crop variety from the perspective of the biotech firm, and present preliminary estimates of the costs of compliance. In Chapter 4, Bruce Gardner describes the indirect and perhaps unintended incentives that traditional farm programs have provided for technology adoption generally and suggests that other technology policies including those for biotechnology, including their questionable welfare implications, can be best understood in the overall context of commodity policy interests. In Chapter 5, Stuart Smyth, Peter Phillips, and William Kerr examine, as an alternative to the regulatory approach, the prospect of ensuring safety through imposing strict liability rules that induce firms to take socially appropriate precautions.

The next four chapters present results on measures of the benefits from the adoption of biotech crops. In Chapter 6, Robert Evenson reports estimates of annual benefits from the current level of adoption of biotech crops, and the prospective benefits from full adoption of existing biotech crop varieties, country by country and for the world as a whole. In Chapter 7, Kym Anderson presents a detailed and more formal analysis of the benefits of biotechnology for various countries based on a market model of international trade, in which he shows the implications of various trade barriers and other policies for the total benefits and their distribution. Evenson and Anderson both show that the EU may have little to gain from worldwide adoption of GM crops, but that EU non-acceptance of GM crops has adverse implications for poor, food-deficient countries.

While Evenson and Anderson consider conventional measures of benefits for the main crops, they do not consider various "hidden" benefits accruing to farmers or consumers. In Chapter 8, Michele Marra and Nicholas Piggott consider one type of hidden benefit—non-pecuniary benefits to farmers associated with greater convenience, farmer and worker safety, and environmental advantages from biotech crops—and they discuss some methodological issues associated with measuring these non-pecuniary benefits. In Chapter 9, Felicia Wu addresses a different type of hidden benefit—a lower incidence of mycotoxins in biotech crops, which means a lower rate of human or animal health problems associated with ingestion of mycotoxins—and presents empirical results on the importance of these benefits for the United States and for developing countries. Wu's contribution is one of the few to date that evaluates a genetically engineered attribute that primarily benefits consumers. To date, agricultural biotechnologies have primarily lowered producer costs, but the problem of consumer acceptance suggests that future technologies may need to focus more on attributes of value to consumers.

The last two papers in this part discuss consumer and market acceptance. In Chapter 10, Wallace Huffman and Matt Rousu present a review of the empirical literature on labeling and consumer acceptance issues. In Chapter 11, Jill McCluskey, Kristine Grimsrud, and Thomas Wahl report on several recent empirical studies showing that consumer acceptance depends on cultural, religious, and political factors. They note the critical role of the media in shaping consumer acceptance negatively, and discuss the issue of consumer sovereignty (the right of the consumer to know) versus scientific sovereignty (where science determines the safety standards).

The second main part of the book, Part II, is entitled *Conceptual Issues in Regulating Agricultural Biotechnology*, and comprises nine chapters. It begins with two chapters on the causes of regulations and their impacts. In Chapter 12, David Zilberman discusses the role of various market distortions—including monopoly power of biotech firms and positive externalities from biotech products that lead to reduced use of pesticides—and the role of distorted incentives for bureaucrats, both of which may contribute to causing current regulations to be stricter than optimal. In Chapter 13, GianCarlo Moschini and Harvey Lapan present a conceptual framework incorporating farm production efficiency gains and consumer opposition with heterogeneous consumers, to demonstrate an approach for evaluating the important issues of GM labeling and potential market segregation.

Much of the rhetoric about biotechnology relates in some way to environmental risk and the regulatory response to it. These issues are the subject of the next three chapters. In Chapter 14, Erik Lichtenberg presents a framework for assessing the risk of biotechnologies, which he uses to explain the importance of assessing not only risk (on average) but uncertainty about risk. He also discusses the balance between prevention at the stage of pre-market testing versus post-commercialization monitoring. In Chapter 15, David Ervin and Rick Welsh discuss improving statistical standards for testing biotechnologies, pointing out that risk will likely increase as virus-resistant and more novel crops are developed. In Chapter 16, Sara Scatasta, Justus Wesseler, and Matty Demont discuss the precautionary principle as it affects EU policy related to biotechnology regulation.

The remaining four chapters in this section relate to imperfect competition in the markets for agricultural technologies and its implications for the regulation of agricultural biotechnologies, drawing on evidence mainly from the agricultural chemical industry. In Chapter 17, Richard Just presents empirical evidence from pesticide regulation, suggesting regulatory issues that should be considered for biotechnology policy. He shows how post-patent competition can lead to 20–50 percent price reductions, which largely transfer surplus from monopolistic developers (under patent protection) to farmers and consumers (upon generic entry). But loopholes regarding the sharing of regulatory testing costs between market developers and generic entrants under FIFRA allow manipulation by original entrants in a way that discourages or prevents generic entry. Thus, farmers and consumers receive only a share of the benefits from innovation to which they are entitled under patent policy. In Chapter 18, Vincent Smith presents survey results from cross-border comparisons of pesticide prices in Montana and Alberta that confirm significant non-competitive pricing as suggested by Just's framework. In Chapter 19, Paul Heisey and David Schimmelpfennig identify economies of scope in regulation as the primary contributing factor for the emergent dominance of the biotechnology industry by large firms, and note the role of campaign contributions by large biotechnology firms as an explanation of statutes that permit economies of scope in regulation. Finally, in Chapter 20 James Oehmke presents a simple framework to show that going off patent is more important for social welfare than product innovation, and that a major role for the public sector is to improve technology availability in the post-patent stage—a reinforcement to the argument by Just.

The third main part of the book, Part III, is entitled *Case Studies on the Economics of Regulating Agricultural Biotechnology*, and comprises 12 chapters. The first three chapters present international evidence on the regulation of agricultural biotechnology either at the stage of innovation or in the final market for the products. In Chapter 21, Colin Carter and Guillaume Gruère give an account of the status of biotechnology regulation and innovation in other developed countries, including Canada, Australia, and Japan, with an emphasis on the role of segregation and labeling requirements. In Chapter 22, Carl Pray and his co-authors discuss biotechnology regulation and innovation in India and China, including a discussion of the regulatory processes and some estimates of the costs of regulatory compliance. In Chapter 23, José Falck Zepeda and Joel Cohen provide similar results for other developing countries, including various countries in Africa, Asia, and Latin America.

The next five chapters relate to refuge policy and regulatory compliance at the farm level. In Chapter 24, George Frisvold develops a conceptual model to illustrate major issues in the choice of parameters of refuge policy, the tradeoff of EPA refuge requirements versus resistance buildup, and the dependence of tradeoffs on local circumstances. In Chapter 25, Silvia Secchi, Terrance Hurley, Bruce Babcock, and Richard Hellmich discuss the benefits and costs of resistance management and refuge requirements in the case of the European corn borer. In Chapter 26, Ines Langrock and Terrance Hurley present results showing that farmers' demand for Bt corn depends critically on refuge requirements (to the extent they are enforced). In Chapter 27, Paul Mitchell and Terrance Hurley present empirical results showing that noncompliance with refuge requirements is widespread, and develop a conceptual model that explains lack of proper incentives for biotech firms to monitor compliance with refuge requirements effectively. In Chapter 28, Shenghui Wang, David Just, and Per Pinstrup-Andersen present evidence from Chinese agriculture showing that high levels of adoption without refuge requirements can lead to secondary pest problems that are economically more serious than the primary pest due to eradication of a natural predator pest.

The final four chapters in this part discuss the regulation of agricultural biotechnology from the perspective of specific types of crops. Chapter 29 by Richard Perrin covers field crops, which because of their predominance in agricultural biotechnologies introduced to date, are also discussed in many of the previous chapters in this volume. In Chapter 30, Greg Graff discusses aspects of biotechnology regulation related to non-food crops. In Chapter 31, Roger Sedjo discusses regulation of biotechnology as it applies to forestry. And in Chapter 32, Kent Bradford, Julian Alston, and Nicholas Kalaitzandonakes discuss horticultural biotechnology regulation. These last three chapters contain a number of common threads concerning the limited incentives of biotech companies to develop products for niche markets (e.g., horticultural crops) or markets for which aggregate revenues are low (e.g., staple crops in developing countries). They also highlight the importance of the biology of the plants as a factor that is not well reflected in the regulations (including the fact that backcrossing is not an option for some species and that, compared with annuals, trees and other perennial crops raise different issues).

Following Part III, we include a short chapter of conclusions by the editors. It summarizes and synthesizes the main points and draws implications for policy and further work in the area.

REFERENCES

Alston, J.M. 2004. "Horticultural Biotechnology Faces Significant Economic and Market Barriers." *California Agriculture* 58(2): 80–88.

Anderson, K. 2006. "Interactions Between Trade Policies and GM Food Regulations." In R.E. Just, J.M. Alston, and D. Zilberman, eds., *Regulating Agricultural Biotechnology: Economics and Policy.* New York: Springer.

Becker, G.S. 1983. "A Theory of Competition Among Pressure Groups for Political Influence." *The Quarterly Journal of Economics* 98 (3): 371–400.

Bradford, K.J., J.M. Alston, and N. Kalaitzandonakes. 2006. "Regulation of Biotechnology for Specialty Crops." In R.E. Just, J.M. Alston, and D. Zilberman, eds., *Regulating Agricultural Biotechnology: Economics and Policy.* New York: Springer.

Graff, G.D., and D. Zilberman. 2004. "Explaining Europe's Resistance to Agricultural Biotechnology." *Agricultural and Resource Economics Update* 7(5): 1–4.

Heisey, P., and D. Schimmelpfennig. 2006. "Regulation and the Structure of Biotechnology Industries." In R.E. Just, J.M. Alston, and D. Zilberman, eds., *Regulating Agricultural Biotechnology: Economics and Policy.* New York: Springer.

Huffman, W.E., and M. Rousu. 2006. "Consumer Attitudes and Market Resistance to Biotech Products." In R.E. Just, J.M. Alston, and D. Zilberman, eds., *Regulating Agricultural Biotechnology: Economics and Policy.* New York: Springer.

James, C. 2004. "Preview: Global Status of Commercialized Biotech/GM Crops: 2004." ISAAA Briefs No. 32, International Service for the Acquisition of Agri-Biotech Applications, New York. Available online at www.isaaa.org (accessed February 1, 2005).

Just, R.E. 2006. "Anticompetitive Impacts of Laws That Regulate Commercial Use of Agricultural Biotechnologies in the United States." In R.E. Just, J.M. Alston, and D. Zilberman, eds., *Regulating Agricultural Biotechnology: Economics and Policy.* New York: Springer.

Kalaitzandonakes, N. 2003. *Economic and Environmental Impacts of Agbiotech: A Global Perspective.* New York: Kluwer-Plenum Academic Publishers.

Kalaitzandonakes, N., J.M. Alston, and K.J. Bradford. 2006. "Compliance Costs for Regulatory Approval of New Biotech Crops." In R.E. Just, J.M. Alston, and D. Zilberman, eds., *Regulating Agricultural Biotechnology: Economics and Policy.* New York: Springer.

Langrock, I., and T.M. Hurley. 2006. "Farmer Demand for Corn Rootworm Bt Corn: Do Insect Resistance Management Guidelines Really Matter?" In R.E. Just, J.M. Alston, and D. Zilberman, eds., *Regulating Agricultural Biotechnology: Economics and Policy.* New York: Springer.

McCluskey, J.J., K.M. Grimsrud, and T.I. Wahl. 2006. "Comparison of Consumer Responses to Genetically Modified Foods in Asia, North America, and Europe." In R.E. Just, J.M. Alston, and D. Zilberman, eds., *Regulating Agricultural Biotechnology: Economics and Policy.* New York: Springer.

Miller, H.I., and G. Conko. 2004. *The Frankenfood Myth: How Protest and Politics Threaten the Biotech Revolution.* Westport, CT: Praeger Publishers.

Mitchell, P.D., and T.M. Hurley. 2006. "Adverse Selection, Moral Hazard, and Grower Compliance with Bt Corn Refuge." In R.E. Just, J.M. Alston, and D. Zilberman, eds., *Regulating Agricultural Biotechnology: Economics and Policy.* New York: Springer.

National Agricultural Statistics Service. 2004. "Statistical Information." U.S. Department of Agriculture, Washington, D.C. Available online at www.usda.gov/nass/ (accessed February 1, 2005).

Wang, S., D.R. Just, and P. Pinstrup-Andersen. 2006. "Damage from Secondary Pests and the Need for Refuge in China." In R.E. Just, J.M. Alston, and D. Zilberman, eds., *Regulating Agricultural Biotechnology: Economics and Policy.* New York: Springer.

Zilberman, D. 2006. "The Economics of Biotechnology Regulation." In R.E. Just, J.M. Alston, and D. Zilberman, eds., *Regulating Agricultural Biotechnology: Economics and Policy.* New York: Springer.

Part I

AGRICULTURAL BIOTECHNOLOGY IN THE CONTEXT OF A REGULATED AGRICULTURAL SECTOR

Chapter 2

ECONOMIC ANALYSIS AND REGULATING PESTICIDE BIOTECHNOLOGY AT THE U.S. ENVIRONMENTAL PROTECTION AGENCY

Derek Berwald, Sharlene Matten, and David Widawsky
U.S. Environmental Protection Agency

Abstract: This chapter discusses the role that economic analysis plays in pesticide regulation for plant-incorporated protectants and compares that to how economic
analysis is used in conventional pesticide regulatory decisions. The goal is to
provide a description, for research economists, of what makes economic research on agricultural biotechnology relevant to regulatory decision makers. It
is our hope that in providing this perspective, economists will be able to develop a stronger sense of what types of research questions and approaches could
actually inform policy. This enhanced understanding would serve the interests
of those researchers seeking to make a policy contribution and could provide useful, independent analysis to help policymakers in making regulatory decisions.

Key words: EPA, biotechnology, transgenic crops, regulation, pesticides

1. INTRODUCTION

Many widely grown crops have varieties that have been genetically modified
to protect them against insect pests, such as the cotton bollworm, pink bollworm, and tobacco budworm in cotton, and the corn rootworm and corn
borer in corn. The U.S. Environmental Protection Agency (EPA) has regulatory oversight over agricultural pesticides, which include crops with
"plant-incorporated protectants" (PIPs). "Plant-incorporated protectant" is
the EPA's term for pesticidal substances produced by plants and the genetic
material necessary for the plant to produce such substances, made possible
through the use of biotechnology. EPA's regulatory responsibility for plant
incorporated protectants is governed primarily[1] by three statutes: FIFRA,
FFDCA, and FQPA (all explained later); the same legal authorities by which

---

[1] The Migratory Bird Act and the Endangered Species Act also affect pesticide regulation.

EPA also regulates "conventional pesticides."[2] To date, with one exception (Bt potato Cry 3A), all PIP registrations for commercial production have been time-limited conditional registrations. Each conditional registration under FIFRA 3(c)7(C) must be shown to be in the public interest. EPA uses certain criteria set forth in 51 Fed. Reg. 7628 (*Conditional Registration of New Pesticides*, March 5, 1986) to make this determination. Part of a determination of public interest is an analysis of the economic benefits associated with such a registration.[3] The benefits assessments are, to some degree, unique to PIPs, but also share common features with other economic analyses that are conducted as part of the pesticide regulatory program.

This chapter will discuss the role that economic analysis plays in pesticide regulation for plant-incorporated protectants and compare that to how economic analysis is used in other pesticide regulatory decisions. The purpose of this chapter is to provide a description, for research economists, of what makes economic research on agricultural biotechnology less (or more) relevant to regulatory decision makers. It is our hope that in providing this perspective, the practitioners of policy economics will be able to develop a stronger sense of what types of research questions and approaches could actually inform policy. This enhanced understanding would serve the interests of those researchers seeking to make a policy contribution and could provide useful, independent analysis to help policymakers in making regulatory decisions.

This chapter has three essential messages to research economists. The first is that for economists seeking to conduct policy-relevant research on regulating agricultural biotechnology, it is extremely important to align the questions and testable hypotheses with the issues and questions that arise in making actual decisions in regulatory agencies. The second message is that for research to be relevant to policy-making, the models used in such research must be empirically tractable and robust, employing data that are feasible to obtain and verifiable. Lastly, economic policy research on agricultural biotechnology must be communicated effectively to non-economists if the research is expected to inform policy formation and/or regulatory decisions.

These messages are important because, in spite of the potential for valuable insights, external economic research (from academic economists, for example) does not typically have much influence on the regulation of conventional pesticides, although there are exceptions. For plant protectant traits

---

[2] Information on the regulatory framework for PIPs can be found at http://www.epa.gov/pesticides/biopesticides/pips/index.htm.
Regulations regarding registration of PIPs can be found at http://www.epa.gov/pesticides/biopesticides/pips/pip_rule.pdf.
[3] For an example of an analysis of the benefits of PIPs, see http://www.epa.gov/pesticides/biopesticides/pips/bt_brad2/5-benefits.pdf.

in genetically modified plants, however, there is a wealth of research by academic agricultural economists that could be useful to regulators. The overlap between important regulatory issues and areas of research that are interesting to economists has valuable spillover effects for those with regulatory responsibility.

The next section of this chapter provides a brief overview of pesticide regulation at the EPA. That section is followed by a section describing the role of economic analysis in regulating conventional pesticides. That role is then contrasted with the need for economic analysis to support regulatory efforts related to plant-incorporated protectants. The chapter concludes with a discussion of policy-relevant topics that may be of interest to academic researchers.

2. STATUTORY FRAMEWORK FOR PESTICIDE REGULATION

There are two main laws that give the EPA the authority to regulate pesticides in the United States. Broadly speaking, the Federal Insecticide, Fungicide, and Rodenticide Act (FIFRA) and the Federal Food, Drug and Cosmetic Act (FFDCA) provide frameworks for registering pesticides and establishing tolerances,[4] respectively. Both statutes were amended by the Food Quality Protection Act (FQPA) in 1996. Together, these statutes provide the framework for regulating pesticides, including plant-incorporated protectants.

In 1947, FIFRA established the Federal role in regulating pesticides. FIFRA has been updated several times since 1947 and was most recently amended by FQPA, as noted earlier. Under FIFRA and FQPA, pesticides must be registered or granted an exemption from registration by EPA before they can be sold, and they must be periodically reviewed to ensure that they continue to meet the requirements of registration. Pesticide registration may be granted after a review of the human health and environmental risks posed by a pesticide (or pesticide product). In some cases, pesticides may be granted conditional registrations (i.e., time-conditional restrictions are imposed on the registration) if they meet certain criteria, including being found to be in the public interest. In these cases, economic assessments of public interest may play a role in the regulatory decision and have been particularly

---

[4] A pesticide cannot be sold or used without a registration, and the registration specifies the ingredients of the pesticide, the particular site or crop on which it is to be used, the amount, frequency, and timing of its use, and storage and disposal practices. A tolerance is the maximum permissible level for pesticide residues allowed in or on commodities for human food and animal feed.

important in registration decisions for plant-incorporated protectants.[5] In all cases, the goal is to prevent any "unreasonable[6] adverse effects on the environment" (FIFRA Sec. 3 [136a]).

Under FFDCA, EPA establishes tolerances for pesticide residues in food. Tolerances are based on assessment of health risks from exposure to a given pesticide or class of pesticides. Under FFDCA, the standard for setting a tolerance is strictly a health-based standard: "a reasonable certainty that no harm" will result from exposure to the pesticide [FFDCA section 408 [6a] (b) (2) (A) (ii)]. This is a narrower standard than under FIFRA, and it precludes the balancing of benefits and costs of a pesticide in setting tolerances, except in extremely narrow circumstances (e.g., preventing public health risks or disruptions in the food supply). Either a tolerance or a tolerance exemption must be granted before a pesticide can be registered for use on a food crop.

Understanding the role of economic analysis in pesticide regulation within this statutory mandate, therefore, is key for those interested in conducting policy-relevant economic research on regulating agricultural biotechnology related to plant-incorporated protectants. There are opportunities for economics to inform the regulatory process, and the next section provides a general overview of these opportunities.

3. ECONOMIC ANALYSIS AND PESTICIDE REGULATION

There are several well-defined roles for economic analysis in pesticide regulation. In some cases, the role may be fairly narrow, such as in making decisions to balance risks and benefits for pesticide registration and reregistration where dietary risk is not of concern. In other cases, the role of economics may be broader, particularly under rulemaking, which is the process by which regulatory frameworks are developed and implemented and which require a thorough analysis of costs and benefits. Although a detailed description of economic analysis in the Office of Pesticide Programs is beyond the scope of this chapter, we provide a brief overview to aid in understand-

---

[5] Plant-incorporated protectants are regulated under FIFRA and FFDCA, but herbicide-tolerant genes are not, because these genes do not have direct pest control properties. Herbicide tolerance, where introduced into the plant genome, is regulated by the U.S. Department of Agriculture under statutes other than FIFRA or FFDCA. A list of these statutes can be found at http://www.aphis.usda.gov/brs/usregs.html#usdalaw.

[6] The term "unreasonable adverse effects on the environment" means (i) any unreasonable risk to man or the environment, taking into account the economic, social, and environmental costs and benefits of the use of any pesticide, or (ii) a human dietary risk from residues that result from a use of a pesticide in or on any food inconsistent with the standard under section 408 of the Federal Food, Drug, and Cosmetic Act (FIFRA Sec. 2 [136(bb)]).

ing what type of analysis may be important to regulating plant-incorporated protectants.

In making regulatory decisions on individual pesticides, the broadest role of economic analysis over the last 10 years has been in pesticide rereg- istration and tolerance reassessment. Under FQPA, EPA is required to reas- sess all pesticide tolerances over a ten-year span, ending in 2006. In the fu- ture, there will also be ongoing reassessments every 15 years under what is called registration review. Registration review is expected to begin in the latter part of 2006. Under both these programs, EPA recognizes that existing pesticides are productive substances that perform an important role, but also may potentially have adverse effects on human health and the environment. Under FIFRA, EPA is required to balance the risks from pesticide use with the benefits from having particular pest control options available.

There are limits, though, to how economic analysis informs findings and decisions pertaining to pesticide regulation. For example, when considering dietary risks under FFDCA, the Agency is required to make a finding of "a reasonable certainty of no harm" before allowing a particular use (or uses) to continue. This finding is made independent of economic analyses.

Although the "no harm" finding limits economic considerations, it does not eliminate them. Economic analysis can be very important in determining the least-cost way to achieve an acceptable "risk cup" under FQPA. The "risk cup" is a term that the EPA uses when describing setting the tolerances allowing for exposures from multiple dietary sources. If the risk cup is over- flowing, then tolerances must be set to reduce exposure from some uses, and EPA seeks to accomplish that in a least-cost way. Because exposure is often the result of a pesticide being used on a number of different food sources (crops), and these crops often have diverse pest control issues, pesticides have different marginal productivities for different crops depending on fac- tors such as pest damage issues, pattern and timing of pesticide use, potential pest control alternatives, and crop value. Economic analysis can be quite influential in determining the set of use restrictions that meet the risk cup constraint while minimizing economic loss from these restrictions. This type of analysis is grounded in agricultural production economics.

Under FIFRA, risks to both human health and the environment are evaluated and regulatory decisions are based on the FIFRA standard of "no unreasonable adverse effects on human health or the environment." Eco- nomic analysis of pesticide benefits is a factor that may influence whether a pesticide will be registered or be found eligible for continued registration. For conventional chemical pesticides, analysts from the Office of Pesticide Programs' Biological and Economic Analysis Division analyze the eco- nomic impacts of new uses of pesticides, registration of new pesticidal active ingredients, and potential restrictions on continued use of a particular pesti- cide. At the same time, the Environmental Fate and Effects Division evaluates the environmental risks from different pesticide use scenarios, and the Health Effects Division evaluates the possible occupational risk from various use scenarios. These analyses are all taken into account by risk managers in proposing final regulatory decisions. These same types of analyses are also performed for biological pesticides evaluated by the Biopesticide and Pollution Prevention Division and for antimicrobial pesticides evaluated by the Antimicrobials Division.

Under the existing reregistration system under FIFRA, the role of detailed economic analysis has been particularly important when the reregistration decision poses particular challenges: pesticides that have high risks and high benefits. In cases where risks are low and benefits of pesticide use are also low, there may be little need for significant regulatory action. In cases where risks are low, but benefits of pesticide use are high, risk management is much less likely to lead to restrictions on use. Conversely, when risks are high, but benefits are low, risk management is likely to favor mitigation that reduces this risk. It is only in those cases where both risk and benefits are high that some sort of tradeoff is likely to occur, and for which economic analysis may be an important factor in determining the ultimate regulatory decision as to what pesticide uses should be found eligible for reregistration. The Office of Pesticide Programs is expected to complete its existing reregistration program in 2006, after which the registration review process will begin.

Another regulatory area where economic analysis informs regulatory decisions is for emergency exemption requests for temporary registration of unregistered uses of pesticides (section 18 of FIFRA). The state lead agencies or another federal agency must petition EPA for these temporary registrations when emergency pest damage situations arise. Section 18 of FIFRA authorizes exemptions to the registration process under emergency conditions. The applications are usually submitted by state lead agencies that identify a pest situation that cannot be controlled by a registered pesticide. If the risks of the pesticide are sufficiently low, and the EPA finds the situation to be "urgent and non-routine," an emergency exemption can be granted if failure to grant the temporary registration would lead to significant economic loss.[7]

These exemptions are often important when there are emerging pest problems, and for small crops for which few chemicals are registered. One of the criteria for a section 18 exemption is that the emergency will cause a "significant economic loss" in the absence of the requested chemical, while using the next most effective registered alternative. Although economists do not grant an emergency exemption, the exemption is rarely granted without a finding of a significant economic loss. An exemption will also not be

---

[7] Past emergency exemptions can be found at http://cfpubl.epa.gov/oppref/section18/search. cfm.

granted if the dietary or environmental risks are too high, even if the economic analysis shows the situation to be severe.

3.1. External Economics Research and Pesticide (Re)Registration

Because of the way FIFRA, FFDCA, and FQPA are written, and the way economic analysis of pesticides is practiced at the EPA, external economic research plays a limited role in the day-to-day economic analysis required for registering and reregistering conventional pesticides. For reregistration of existing conventional pesticides, the economic questions are typically quite narrow, focusing on the impact of mitigating specific risks from individual pesticides through changes in use patterns with crop-specific or location-specific measures. In order to be relevant for these day-to-day decisions, economic research would have to estimate potential damage from marginal changes in use patterns for specific crops, specific regions, and specific pesticides, and evaluate the benefits of crop risk mitigation relative to the next best alternative for that situation.

With analysis that is narrowly defined by pesticide, crop, and pest, there are thousands of combinations one might analyze, all with specific data requirements and market knowledge. Academic research, therefore, could speak to either specific pesticide cases or develop models that are flexible enough, and for which there are sufficient available data, to tackle these case-specific regulatory analyses in a relatively short time frame. Unfortunately, this is fairly specialized research which appears to have limited appeal to the academic community given the way research is conducted in academia (longer time frames, limited data, directional vs. nominal results, etc.).

For economic research that does address pesticide topics, models that are developed in these studies rarely model the marginal policy decisions that may be instrumental in regulatory decisions. Typically, research results are general, or aggregated across pesticides (for example, considering the impact of total pest control expenditures on a farm, or in a region) rather than analyzing marginal policy decisions that are important to regulators (such as the value of a new pesticide compared to the next best alternative).

In addition to informing these marginal decisions, though, there is other external research that could be very valuable to the EPA for conventional pesticide regulation. Such research might include estimates of price (cost) elasticity for new pesticide registrations, or estimates of the value of additional information on human health or ecological risk, which would facilitate refinement of risk estimates. In cases where exposure-specific data allow one to depart from default assumptions about risk parameters, and lead to lower values for estimated risk, the need to mitigate risk may decline. Therefore, understanding the tradeoff between the cost of obtaining additional risk information and the cost of mitigation in the absence of such refining information could help inform the regulatory process. Additionally, a framework for being able to analyze the costs and benefits to society of the pesticide regulatory program would be a valuable contribution in an era of increasing quantitative accountability.

This does not imply that external economic research has not been useful to EPA; it has. Particularly helpful are partial equilibrium models of agricultural markets and research that can help estimate the consumer and producer surplus effects of policy changes. For example, EPA economists have devoted substantial effort working on issues surrounding the Montreal Protocol, which phases out methyl bromide (an ozone-depleting pesticide fumigant), but allows for continuing use in special cases where alternatives to methyl bromide are not technically or economically feasible. External economic research in this area has also been quite helpful, because it tends to focus very closely on the issues surrounding policy decisions. Examples of recent work that will be helpful to the EPA in future methyl bromide work are Carter et al. (2005) and Goodhue, Fennimore, and Ajwa (2005). External economic research has also been particularly helpful for the regulation of biotechnology products, which we discuss below.

3.2. Rulemaking

Another important area for economic analysis in regulating pesticides is rulemaking, the process by which legislative mandates are implemented into specific actions and protocols. Because rulemaking has the potential for imposing regulatory burdens on the regulated community and society at large, these regulatory activities have engendered a set of requirements for economic analysis, both by statute and by executive branch requirement (Presidential executive orders, Office of Management and Budget directives). These economic analyses are subject to public comment and are reviewed by the Office of Management and Budget (OMB).

For the rulemaking process, economic analyses must consider multiple policy options and contain quantitative and qualitative evaluation of the benefits and costs of the proposed regulations. A regulatory analysis will also contain a justification of the regulatory action, an analysis linking the proposed regulation to the desired outcome, an identification of second order costs and benefits, the distribution of benefits and costs, and the impact on small business. EPA is currently proposing a number of rules related to pesticide regulation, including those dealing with pesticide registration data requirements, amendments to procedures for emergency exemptions (including determination of a significant economic loss), procedures for continuing review of registered pesticides (called registration review), and third-party submission of data generated with human subjects.[8]

3.3. Conditional Registration

Another important role for economic analysis is for a Public Interest Finding (PIF). A PIF provides information in support of a conditional registration under FIFRA 3(c)7(C), rather than an unconditional registration of a pesticide under FIFRA 3(c)5. In order to conditionally register a pesticide under FIFRA 3(c)7(C), EPA must make a finding that the conditional registration is in the public interest. A PIF will include some level of economic analysis.

EPA can conditionally register a pesticide or product under several sets of circumstances described in 51 Fed. Reg. 7628 (*Conditional Registration of New Pesticides*, March 5, 1986). These include when there is a need that is not met by currently registered pesticides, when the new pesticide poses less risk to health or the environment than registered alternatives, or when the benefits of the new pesticide exceed those of alternative means of control, both with registered pesticides and non-chemical techniques. The last of these criteria provides one entry point for economic analysis.

Historically, for conventional pesticides there has been a limited amount of EPA-initiated economic analysis for PIFs, because other conditions are sufficient for finding that a conditional registration is in the public interest (i.e., the pesticide meets the criteria for a reduced risk pesticide). All of the PIPs (pesticides produced in genetically modified plants) have had PIFs prior to the Agency granting a conditional registration. Compared to PIFs for conventional pesticides, these PIFs generally include a much more comprehensive economic analysis, and are generally combined with a benefits assessment; they are described in more detail below.[9]

4. ECONOMIC ANALYSIS AND BIOTECHNOLOGY REGULATION

Almost all of the registered PIPs to date have been for Cry (crystalline) proteins isolated from different species of the soil bacterium *Bacillus thur-*

---

[8] The economic analysis for emergency exemptions can be found at http://docket.epa.gov/ edkfed/do/EDKStaffAttachDownloadPDF?objectId=090007d48031dbdd. The economic analysis for reregistration review can be found at http://docket.epa.gov/edkfed/do/EDKStaffAttach[-] DownloadPDF?objectId=090007d48081e7b3. The economic analysis for registration data requirements rule can be found at http://docket.epa.gov/edkfed/do/EDKStaffAttach Down[-] loadPDF?objectId=090007d48065b8d7.

[9] The benefits assessment and PIF for Cry2Ab2 Bollgard II cotton can be found in the Registration Action Document, http://www.epa.gov/pesticides/biopesticides/ingredients/tech_[-] docs/brad_006487.pdf.

ingiensis (Bt), and their genes have been genetically engineered into corn, potato, and cotton. These proteins provide protection against different classes of insects depending on the Cry protein. Other plants that are the result of biotechnology, such as soybeans genetically modified to provide resistance to the herbicide glyphosate, are not regulated as pesticides because the engineered trait does not fit the definition of a pesticide; these traits allow the glyphosate, for example, to be metabolized by the plant so that it does not affect the crop. This means that weeds can be controlled by glyphosate, but the plant remains unaffected. EPA does regulate the herbicide, but not the genetically modified plant that is resistant to it, because the plant does not control the weeds that are pests, so the genetically modified plant is not a PIP. The Food and Drug Administration and the U.S. Department of Agriculture do regulate crops that are genetically modified to be herbicide-tolerant.

Most Bt PIP registrations have been time-limited conditional registrations for full commercial use. These registrations must be reviewed prior to the Agency making a decision to allow continued use of Bt PIP. EPA reassessed all of the risks and benefits of the Bt (Cry1Ab and Cry1F) corn PIPs and cotton (Cry1Ac) PIPs in 2001 (see EPA, 2001). During this reassessment, the tolerances for Cry1Ab and Cry1F in corn and Cry1Ac in cotton were reassessed as required under FQPA, and the EPA determined that there was a reasonable certainty of no harm from dietary exposure to these PIPs. Under FIFRA, EPA performed an economic analysis of the benefits of these PIPs from the date on which they were first registered in 1995 through 2001. The benefits of these PIPs and their risks were both important in allowing these PIPs to be conditionally registered for another limited period of time. [10] Unlike the recent history of regulation for conventional pesticides, external economic research by academic economists has played an important role in the registration decisions for Bt and is expected to continue to do so in the future.

For a benefits assessment for a PIP, some economic issues are similar to those for conventional pesticides, and some are unique to this type of agricultural biotechnology. As for conventional pesticides, EPA is interested in estimating the change in profits at the farm and industry level due to the adoption of a PIP, which directly influences the propensity to adopt the pesticide product (in this case, a PIP) and informs the degree of exposure and/or risk. Any change in the grower's ability to manage risk or the quality of the crop is also important in the adoption decision. A typical analysis would also consider other possible benefits, such as changes in current patterns of pesticide use. In the case of PIPs, an important consideration is the degree to

---

[10] A benefits assessment for Bt corn can be found at http://www.epa.gov/pesticides/biopesti[-]cides/pips/bt_brad2/5-benefits.pdf, or at http://www.epa.gov/pesticides/biopesticides/ingre[-]dients/tech_docs/cry3bb1/2_e_cry3bb1_benefits.pdf.

which a PIP can displace use of conventional pesticides and reduce human health and environmental exposure from these pesticides.

Several economic issues are unique to regulation of PIPs. One example is the economic consequence of different types of resistance management, including refugia design. Because Bt, in particular, is considered an important resource to some agricultural production systems (both with the conventional production system and the organic agricultural production system—the Agency is interested in maintaining the sustainability of Bt in all of its forms), there is substantial policy interest in maintaining the productivity of this resource. One regulatory policy that attempts to maintain productivity of Bt is the institution of specific insect resistance management (IRM) requirements. The refuge requirement for non-Bt crops that is intended to maintain a pest population susceptible to the action of Bt has been an important part of the IRM requirements. An understanding of the economic consequences of different types of refuge design, and the costs of maintaining different levels of pest susceptibility through these refugia, is expected to be critical to the decision process as EPA revisits these conditional registrations in the coming years.

External economic research is particularly relevant in this area, due in large part to the limited decision space for analysis. For Bt technology, there are only 3 crops currently on the market (field corn, sweet corn, and cotton), and the Bt crops are targeted mainly at only five or six pests (there are other pests in which Bt has suppressive effects or even control effects compared to registered pesticide alternatives). Unlike the vast number of pest/crop combinations germane to regulation for conventional pesticides and the difficulty for an academic researcher in choosing which combinations might be of policy interest, Bt presents a fairly compact and predictable set of policy-relevant production scenarios to explore. The models of pest control have a few dimensions that can be calibrated with realistic data, and there are a finite number of choices to consider in the analysis. For example, in analyzing refugia, the farm-level choices may include the share of land planted to Bt, the share planted to refuge, and the type of refuge to adopt (e.g., level of pest control in the refuge, internal vs. external refugia). Equally important are the incentives to growers and industry: will compliance with refuge requirements be compatible with grower interests such as yield and profitability, or will they appear to be restrictive, viewed as a prohibitive cost rather than a benefit to growers?

Agricultural biotechnology is a fairly new field and it has generated substantial interest among economists, providing opportunities for innovative research and peer recognition. This has been driven in part because EPA has mandated specific IRM requirements as conditions of registration. Because of the refuge requirements, agricultural economists have been asked by a number of stakeholders to determine or predict the economic impacts of these requirements. No one wants regulatory requirements to be burden-some. The area of IRM requirements has stimulated much interest among academic researchers, government, and industry—especially growers. These requirements have focused interest among external economic researchers on their impact on agriculture and society. The Biopesticide and Pollution Prevention Division has worked with a number of these agricultural economists in the past 10 years and has used their research in its analyses of the impact of IRM requirements [for example, Hurley, Babcock, and Hellmich (2001), Livingston, Carlson, and Fackler (2004), Mitchell et al. (2002), Frisvold and Tronstad (2002), and Hurley, Mitchell, and Rice (2004)].

For external economists wishing to contribute research that could inform biotechnology policy, this is a fertile ground for research, with some caveats. First, there are a number of emerging innovative approaches for exploring the economics of refugia choice and resistance management, but in applying innovative models, there is an attraction to simplifying other parts of the production system to make a given model tractable. Given the importance of previous research in providing insight into pest control economics, particularly the damage abatement approach (such as Lichtenberg and Zilberman 1986), it might be shortsighted to overly simplify production models of crop production solely in pursuit of resistance management results.

Additionally, direct applications of resistance management models are critical, which may favor some degree of modeling parsimony, and it is also important that models be verified or calibrated to actual situations with real-world data. This makes it easier for economists and biologists at EPA to understand and use the models, and more importantly, makes it possible for the models and their results to be explained to policymakers. Models and results need to focus on the policy choices that actually face a policymaker, with special consideration toward the practical fact that policy formation favors relatively simple and straightforward instruments and/or mechanisms. This is especially important to remember when policy complexity generates only negligible improvements in measuring welfare.

Finally, it is important for external researchers to appreciate that economic analysis supporting regulatory work usually must take an *ex ante* view, considering what will happen in the event of a new registration or regulation. Research that is solely backward-looking has limited relevance to a policymaker. On the other hand, *ex post* research can be very valuable in simulating or inferring the potential consequence of future regulatory options. Much of the current research on the economics of Bt crops is *ex post* but provides information about several important issues surrounding the benefits of Bt: adoption behavior by growers, the impact on profitability, the value placed on the technology by growers, the extent to which that value is risk premium (or discount) when biotechnology changes the risk that farmers face, and changes in conventional pesticide use by farmers—all have been studied by economists. To the extent that this type of result can be used to generate insights into the possible economic consequences of future policy choices, such research could be influential in informing policy decisions.

5. THE FUTURE

For the motivated research and/or policy economist working in this area, one natural question is: what are the opportunities for policy-relevant external research in the near future? Among the several agricultural biotechnology platforms, Bt crops are still the most important sector for EPA: they combine two areas of interest to agricultural and resource economists (biotechnology and pesticides); there is a clear regulatory schedule; and a significant portion of large-acreage crops are planted to Bt varieties.[11] The conditional registrations for Bt PIPs expire in the near future, with some Bt cotton registrations expiring in 2006 and some Bt corn registrations expiring in 2008.[12]

As EPA considers renewing these registrations, benefit reassessments by the EPA will continue to favor products that can decrease health and environmental risks and reduce the use of conventional pesticides, and economic analyses will help inform these decisions. Moreover, new PIP technology targeted at the same crop and pest situations as existing Bt products will require a nuanced economic analysis because the conditions of a conditional registration will be harder to meet when there are already effective Bt products available and the expected marginal benefit of additional Bt registrations may be more subtle than attended the original registrations. For example, more attention might be focused on location-specific models.

To provide appropriate regulatory oversight and to ensure that the effective Bt products remain effective, EPA values policy-relevant economic research on resistance management, monitoring, and refuge requirements, topics for which economic analyses are still evolving and where more research is needed. Recent research on grower attitudes to resistance management is particularly helpful, and bioeconomic models of resistance can be very data-intensive, but valuable. Since EPA considers pest susceptibility to Bt a common property resource, where a policy goal is to avoid depletion of this resource, then one area of possibly useful research could be exploration

---

[11] Bt corn is planted on about 26 percent of the corn acreage, with another 9 percent planted to stacked gene varieties that control insects; and about 18 percent of cotton acreage is planted to Bt cotton, and 34 percent to stacked gene varieties that control insects (USDA 2005). Estimates of acreage planted to biotechnology varieties can be found in the USDA/ NASS document *Crop Production–Acreage–Supplement*, available at http://usda.mannlib. cornell.edu/reports/nassr/field/pcp-bba/acrg0605.pdf.

[12] A full list of Bt registrations and expiration dates can be found at http://www.epa.gov/ pesticides/biopesticides/pips/pip_list.htm.

of whether there is a market mechanism that leads to a cost-effective and sustainable resistance management plan. For example, how effective are contracts designed by the registrants of Bt crops in promoting resistance management strategies that are incentive-compatible to growers of Bt crops?

Research on resistance management is most likely to help inform regulatory policies if it contains several elements of importance to EPA. Research that explores refuge requirements for Bt crop/pest combinations, specifically looking at cost-effective and sustainable refuge choices in a dynamic way, could be particularly helpful. This type of analysis could help EPA refine refuge requirements that are both feasible and efficient. To that end, it is important for bioeconomic models of resistance to be workable and applicable to different situations, such as crops with single and multiple pests, crops with single and multiple pesticides, and areas or fields with single or multiple crops. Location- and crop-specific analysis is most likely to be influential in informing future regulatory decisions for PIPs.

There is a challenge here for academic economists and for regulators—a challenge to strengthen lines of communication. How can economists interested in relevant policy work on agricultural biotechnology provide useful and timely information to EPA? How can EPA communicate to policy economists which issues are directly relevant to regulatory decisions for PIPs? Where PIPs are concerned, there is potentially overlap in research areas of interest to academic economists and the information that regulators in the Office of Pesticide Programs seek to help inform future decisions. It is likely that this shared interest will maintain policy relevance for the next several years, and EPA is hopeful that strengthening communication among researchers and regulators will generate work and collaborations that are productive and useful to each.

REFERENCES

Carter, C., J. Chalfant, R. Goodhue, F. Han, and M DeSantis. 2005. "The Methyl Bromide Ban: Economic Impacts on the California Strawberry Industry." *Review of Agricultural Economics* 27(2): 181–197.

EPA [*see* U.S. Environmental Protection Agency].

Frisvold, G., and R. Tronstad. 2002. "Economic Effects of Bt Cotton Adoption and the Impact of Government Programs." In N. Kalaitsandonakes, ed., *Economic and Environmental Impacts of Agbiotech: A Global Perspective*. New York: Kluwer-Plenum Academic Publishers.

Goodhue, R., S. Fennimore, and H. Ajwa. 2005. "The Economic Importance of Methyl Bromide: Does the California Strawberry Industry Qualify for a Critical Use Exemption from the Methyl Bromide Ban?" *Review of Agricultural Economics* 27(2): 198–211.

Hurley, T.M., B.A. Babcock, and R.L. Hellmich. 2001. "Bt Corn and Insect Resistance: An Economic Assessment of Refuges." *Journal of Agricultural and Resource Economics* 26(1): 176–194.

Hurley, T.M., P.D. Mitchell, and M.E. Rice. 2004. "Risk and the Value of Bt Corn." *American Journal of Agricultural Economics* 86(2): 345–358.

Lichtenberg, E., and D. Zilberman. 1986. "The Econometrics of Damage Control: Why Specification Matters." *American Journal of Agricultural Economics* 68(2): 261–273.

Livingston, M.J., G.A. Carlson, and P.L. Fackler. 2004. "Managing Resistance Evolution in Two Pests to Two Toxins with Refugia." *American Journal of Agricultural Economics* 86(1): 1–13.

Mitchell, P.M., T.M. Hurley, B.A. Babcock, and R.L. Hellmich. 2002. "Insuring the Stewardship of Bt Corn: A Carrot versus A Stick." *Journal of Agricultural and Resource Economics* 27(2): 390–405.

U.S. Department of Agriculture. 2005. "Crop Production–Acreage–Supplement." National Agricultural Statistics Service, U.S. Department of Agriculture, Washington, D.C. Available at http://usda.mannlib.cornell.edu/reports/nassr/field/pcp-bba/acrg0605.pdf.

U. S. Environmental Protection Agency. 2001. "Biopesticides Registration Action Document: *Bacillus thuringiensis* Plant-Incorporated Protectants." U.S. Environmental Protection Agency, Washington, D.C. Available at http://www.epa.gov/pesticides/biopesticides/pips/ Bt_brad.htm.

Chapter 3

COMPLIANCE COSTS FOR REGULATORY APPROVAL OF NEW BIOTECH CROPS

Nicholas Kalaitzandonakes,[*] Julian M. Alston,[†] and Kent J. Bradford[†]
University of Missouri-Columbia and *University of California, Davis*[†]

Abstract: The regulatory approval process for new biotech crop varieties is said to be unduly slow and expensive, presenting important barriers to the development of new cropping technologies. To date, however, the private and social costs have not been analyzed or measured, let alone compared with alternatives. This chapter reports initial findings from our continuing project on the costs of regulatory compliance for biotech crops. In this chapter we describe and document the regulatory requirements, and we provide estimates of representative compliance costs for key biotechnologies based on confidential data supplied to us by several major biotech companies.

Key words: biotechnology, regulatory approval, compliance costs, cost estimates

1. INTRODUCTION

The regulatory approval process for new biotech crop varieties is said to be slow and expensive, presenting important barriers to the development and commercialization of new cropping technologies. For some crops these barriers may be prohibitive, resulting in technological orphans (Alston 2004, Bradford et al. 2004). Alternative approaches to regulating new crop biotechnologies could be less expensive, but to date the private and social costs of the current regulatory system have not been analyzed or measured, let alone compared with alternatives. In fact, estimates of the compliance costs for the full deregulation of a biotech crop do not exist, as such information has been closely guarded by biotechnology developers.

In this chapter we make three contributions to knowledge concerning the private costs of complying with the current regulatory approval system for agrifood biotechnologies.[1] First, we briefly review the regulatory require-

---

[1] Regulatory costs have multiple dimensions in addition to the private compliance costs,

ments in the United States and elsewhere and we explain how they translate into relevant costs of compliance. Second, we characterize the structure of such compliance costs and identify the key dimensions of their variation. Third, we provide estimates of representative compliance costs for selected crop biotechnologies. Our estimates are based on reviews and analyses of dossiers submitted to regulatory agencies by major biotech firms, and firm-level data on associated expenses.[2]

2. ARE HIGH REGULATORY COSTS IMPEDING BIOTECH INNOVATION?

The techniques of biotechnology (recombinant DNA methods) have been available for almost 30 years and have been used to enhance crop perform-ance and quality for more than 20 years. Since 1996, over 950 million cu-mulative acres (385 million hectares) of bioengineered soybeans, maize, cot-ton, and canola have been grown around the world (James 2004). Econo-mists have estimated the social benefits from biotech crop varieties to be in the billions of dollars, with the benefits shared among consumers, agricul-tural producers, and the biotechnology innovators that have developed the new crop varieties.[3]

In spite of this apparent success, however, many observers have been disappointed at the rate of development and commercialization of new bio-tech crops (e.g., Jaffe 2005). Indeed, the accumulating evidence suggests that agrobiotechnology innovation and product development have recently slowed down, and high compliance costs for regulatory approval have been cited as a key culprit (e.g., Bradford et al. 2005, Jaffe 2005, McElroy 2003, Miller and Conko 2005). Several key indicators of this innovation slowdown have been highlighted in these and other studies.

First, the rate of new biotech trait deregulations has slowed down. Since the initial deregulation and subsequent commercialization of such products as Roundup Ready soybeans, Bt corn, and Bt cotton in the mid-1990s, the

---

which are the subject of this chapter. They include bureaucratic monitoring and enforcement costs as well as a range of opportunity costs associated with potential structural impacts from regulation, potential inefficiencies in implied market structures, foregone productivity gains from unrealized innovation, and costs from loss of trade.

[2] A number of major biotech firms cooperated and provided confidential information on com-pliance costs for this project, including Bayer CropSciences, DuPont, Monsanto, and Syn-genta. These four firms and their subsidiaries own a large majority of all deregulated biotech traits, and hence our estimates of compliance costs should be representative of the whole agro-biotechnology industry.

[3] For instance, see Alston et al. (2002), Falck-Zepeda, Traxler, and Nelson (2000), Gianessi et al. (2002), Huang et al. (2002), Marra, Pardey, and Alston (2002), and several studies in Kalaitzandonakes (2003).

annual number of new deregulations has decreased drastically (Figure 1). Moreover, a large portion of the new deregulations are variants of the earlier technology (e.g., stacked traits, new Bt varieties).

Second, there has been a clear downward trend in the research and development (R&D) intensity for crops with lower market potential. In recent years, large and small biotech firms have repeatedly rationalized their R&D portfolios by focusing on a few large-acre crops and traits. Experimentation and product development in specialty and small market crops in the United States have decreased since 1998 despite continuing technological advances (Figure 2). Development of second generation products for major crops has also slowed from earlier predictions, as the pace of commercialization has been delayed due to market resistance and regulatory hurdles in some countries. However, R&D in small acreage crop biotechnologies has diminished disproportionately (Bradford, Alston, and Kalaitzandonakes 2006). As Figure 3 illustrates, the number of U.S. field trials for quality traits has been stagnant since the mid-1990s, relative to those involving agronomic phenotypes.

Third, the establishment of new biotech firms and the flow of venture capital that finances them have also slowed down in recent years (McElroy 2003). While time-to-market and market size for agrifood biotechnologies appear to be similar to those in many pharmaceutical biotechnologies, venture capital funding for agrifood biotechnology startups has been but a trickle compared with that available to pharmaceutical ones (McElroy 2003).

These trends, of course, bode ill for the future vitality of agrifood biotechnology and its product pipeline. But while it is increasingly apparent that agrobiotechnology innovation has recently slowed down, the specific

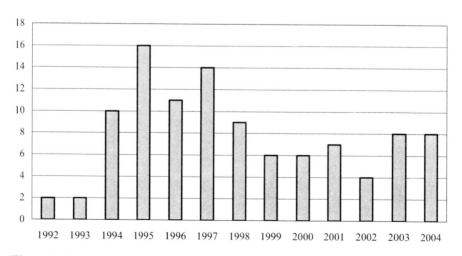

Figure 1. Agricultural Biotechnology Deregulations in the United States
Source: USDA (2005).

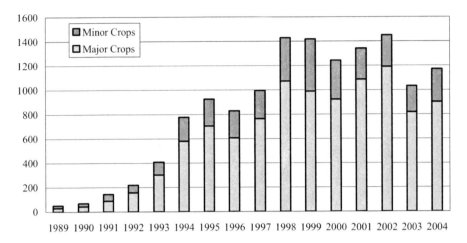

Figure 2. Field Trials in the United States for Major and Minor Crops

Notes: Major crops include corn, cotton, rapeseed, rice, soybean, and wheat. Minor crops include tomato, peanut, papaya, beet, melon, tobacco, lettuce, squash, strawberry, apple, and sunflower.

Source: USDA (2005).

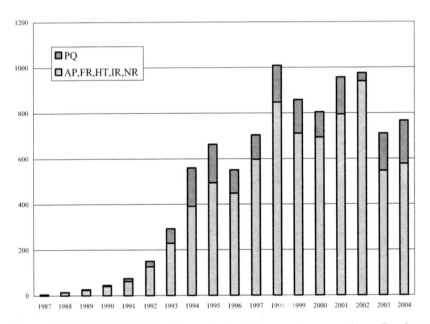

Figure 3. U.S. Field Trials by Phenotypic Category: Agronomic vs. Product Quality

Note: PQ (product quality), AP (agronomic properties), FR (fungal resistance), HT (herbicide tolerance), IR (insect resistance), NR (nematode resistance).

Source: USDA (2005).

reasons for the trends are not clear. Possible causal factors include market resistance and trade barriers as well as high and rising regulatory costs, but the relative importance of these factors is not known. In particular, there is little empirical evidence to support or refute the hypothesis that high (and possibly increasing) regulatory compliance costs are to blame. In addition, little is known about the consequences of regulatory requirements in terms of benefits to farmers, consumers, or the environment, or of costs avoided, which are relevant to questions about whether the regulations are excessive or inadequate. Indeed, the actual costs of regulatory compliance are still unclear, let alone whether they are too high, what factors shape them, and what have been the effects of specific changes in the regulatory system on the costs of compliance and the efficiency of the regulatory system. A necessary first step to answering any of these questions about the causes and consequences of the process of regulatory approval for new biotech crops is to understand the operation of the regulatory system and the size and structure of the costs of compliance. The purpose of this chapter is to describe the process of regulatory compliance in the United States, in comparison with other countries, and then to present some accounting data on the costs of compliance with U.S. regulations, supplied by four major agricultural biotechnology companies.

3. GRANTING REGULATORY APPROVALS FOR BIOTECH CROPS

All biotech crops are submitted to a battery of tests and regulatory scrutiny prior to commercialization. Analysis of the structure and size of compliance costs requires understanding of the regulatory approval process. Regulatory reviews and approvals for the cultivation and consumption of biotech crops are country-specific. Hence, at some point in the R&D cycle, biotech developers must decide in which countries they wish to seek regulatory approval for their products. In this context, they must take into account not only the countries where the cultivation of the new biotech crops could take place but also where the consumption of such crops might ultimately occur. Given the large and expanding agricultural commodity trade flows across the globe, these considerations can quickly become complex as biotech developers must balance their desire for broad regulatory approval with practical budget constraints.

Currently, nineteen countries have well-developed systems that handle submissions seeking regulatory approval for the cultivation and sale of new biotech crops, while a number of others are in the process of developing theirs. The differences and similarities of these national systems are therefore of interest. While a comprehensive review of the various national regulatory approval systems is beyond the scope of this chapter, some basic elements of interest are highlighted here. We begin with the U.S. biotech regulatory system.

3.1. Biotech Regulation in the United States

Under the auspices of the White House Office of Science and Technology Policy, the United States developed a Coordinated Framework for Regulation of Biotechnology in 1986. The Coordinated Framework places regulatory responsibility with three separate agencies: the U.S. Department of Agriculture (USDA), the U.S. Environmental Protection Agency (EPA), and the U.S. Food and Drug Administration (FDA). Each of these agencies oversees specific aspects of the technology, evaluating the safety of new biotech crops from inception to commercialization. The Coordinated Framework then focuses on protecting the safety of the agrifood system, the environment, and human health.

The USDA's Animal and Plant Health Inspection Service (APHIS) is responsible for protecting U.S. agriculture against pests and diseases and ensures that biotech products are safe for agricultural use. APHIS regulates the safe field testing of new biotech varieties. It then grants permission for commercialization after it concludes from detailed scientific data that the new varieties will not become pests and will not pose any significant risk to the environment or to wildlife.

The EPA approves new pesticides including plants expressing pesticidal proteins. In deciding whether to register a new biotech crop expressing pesticidal proteins, EPA considers the fate of the pesticidal substance in the environment, its effectiveness on the target pest, and any effects on non-target species, as well as its safety for humans. EPA grants experimental-use permits (EUPs), establishes limits (tolerances) for the amount of pesticidal proteins in foods derived from the biotech plants, and reviews all relevant environmental and toxicological studies before deciding whether to register these new biotech crops.

The FDA determines whether foods derived from new biotech crops are safe and wholesome. The FDA is also responsible for enforcing regulations to ensure that all food and feed labels, including those related to biotechnology, are truthful and not misleading.

After commercialization, all three regulatory agencies have the legal power to demand immediate removal from the marketplace of any product should new data indicate a question of safety for consumers or the environment. Such post-market surveillance has been employed in a number of commercialized agrobiotechnologies in the United States.

It is worth noting that the regulatory approach adopted within the Coordinated Framework is flexible rather than structured. Since the first dossier was submitted, the federal agencies have considered the required tests and studies on a case-by-case basis. This performance-oriented approach provides federal agencies flexibility and the ability to adapt the regulatory processes to the specific attributes of each biotechnological innovation. For example, biotechnology-derived varieties of open-pollinating canola planted in the same region as their compatible wild relatives might require more detailed studies on the impacts of gene flow than self-pollinating soybeans planted in North America, where there are no wild relatives. Hence, each new biotech crop is evaluated for both the trait and the species in which it is incorporated, and test results must establish its environmental and food safety according to relevant decision trees. Dossiers submitted to the regulatory agencies are therefore original research in their own right and, often, are uniquely differentiated.

A flexible approach to regulatory requirements has been considered by the federal agencies as desirable not only for its adaptability to the specific attributes of each individual technology but also for its ability to evolve in response to the development of new scientific knowledge. This flexibility, however, could also come at a cost. In the presence of intense political pressure and controversy, like that experienced by agrifood biotechnology across the globe, regulatory agencies could feel pressure to increase the stringency of their approval requirements, especially when the cost implications are not immediately clear. Indeed, numerous reviews of the regulatory approval process in the United States have focused on risk management (e.g., U.S. National Academy of Sciences 2000, Pew Initiative on Food and Biotechnology 2004) and some have advocated a "safety-first" approach (Kapuscinski et al. 2003), with little consideration of regulatory costs and their implications for the rate of innovation and new product development.[4] Whether political pressure has translated into increased regulatory scrutiny in the United States is less clear, however.

3.2. International Regulatory Approvals of New Biotech Traits

The U.S. system is generally simpler, more transparent, and more scientifically based than other national systems for the evaluation and approval of new biotech crops. Among nations, Canada, with its science-based coordinated approach, is the most similar to the United States. At the other extreme, the European Union (EU) has adopted a complex regulatory system

---

[4] It has been argued that the regulatory processes have become more demanding in recent years. For instance, Jaffe (2005) concluded that the time required for the deregulation of a new biotech crop in the United States has almost doubled in the past few years despite increasing familiarity with the technology. Our review of regulatory dossiers suggests that they have become substantially more complex and voluminous over the years—with recent dossiers often including thousands of pages of supporting material. However, it is unclear what factors have led to such increased complexity and what the exact cost implications have been.

where reviews and approval decisions are carried out at both the member-state and the EU-center level. Applications for field testing and for deregulation must be submitted by biotech developers to the competent regulatory authorities of some member country. The national regulatory authority can approve or reject field experimentation and can issue a favorable or unfavorable assessment for the deregulation and commercialization of a new biotech crop. In the case of a favorable assessment, the national regulatory authority informs all other member states of its assessment through the European Commission and, if there are no objections from any member state or the Commission, it moves forward with the approval. Objections set in motion a process where conciliation is initially sought. In the absence of an agreement, the decision must be taken at the EU level. In such a case, the Commission requests a scientific review from the European Food Safety Authority, and on that basis the Commission drafts a decision, which is submitted to the Regulatory Committee, composed of representatives of the member states. If the Regulatory Committee gives a favorable opinion by qualified majority, the Commission adopts the decision and submits it to the Council of Ministers for adoption or rejection. If the Council does not act within three months, the Commission moves forward with the deregulation of the new crop. Other countries, like Japan, South Korea, and Australia, have review and approval processes that generally fall between these two extremes.

Countries differ in their requirements for specific data and information for the deregulation of new biotech crops and in the speed of review and approval.[5] Obviously, having different regulatory systems in various countries, each with their own process, data requirements, and timetable, gives rise to a complex international biotech regulatory environment that biotech developers must navigate. Particularly challenging is having uneven regulatory approvals across different countries, which restricts commercialization and increases uncertainty.[6]

---

[5] The speed of review and deregulation of new biotech crops differs significantly from one country to another. For instance, the United States, Canada, Japan, and some other countries have continued to review and approve new biotech varieties, even if at variable speeds. The EU stopped considering petitions for deregulations in 2001 and began reviewing regulatory dossiers in 2004, only after mandatory labeling laws and full traceability of foods and feeds along the EU supply chain were implemented (for instance see Kalaitzandonakes 2004).

[6] Starlink corn is the best-known case of uneven regulatory approval that led to market disruptions. Starlink corn was approved in the United States for commercialization in feed but not food markets. Inability to effectively segregate the new biotech crop led to market disruptions, as small amounts found their way to U.S. and international food markets where it lacked regulatory approval. Other new biotech crops, including Roundup Ready corn, have also demonstrated the difficulties associated with the commercialization of biotech traits that have secured approval in some, but not all, possible markets.

4. SEEKING REGULATORY APPROVAL FOR BIOTECH CROPS

Despite the significant differences in the various national regulatory systems, the dossiers submitted to various regulatory agencies around the world are similar in structure and often share data and other supporting information. In most cases, the properties of the biotech crop are compared to those of the corresponding non-biotech variety with respect to various potential risk factors. Such comparative analyses include agronomic, molecular, compositional, toxicological, and nutritional assessments. A detailed agronomic assessment is essential for the identification of any unintended effects from the genetic modification at the whole-plant level, and includes examination of various morphological, phenotypic, and agronomic data. Targeted compositional analyses focused on possible changes at a metabolic level are equally important, as they quantify the expected and unexpected changes in some fifty essential crop components (e.g., amino acids, fatty acids, vitamins, minerals, antinutrients).

Similarities in the results of all these studies between the novel crop and conventional counterparts are noted, while differences trigger additional tests to examine whether any safety concerns are warranted. Of course, to conduct such comparative analyses requires well-established analytical data for the conventional crops and derivative foods and feeds that serve as benchmarks. Comparative analytical data for the novel crops and foods must also be developed through well-designed experiments and tests. Biotech developers must therefore carefully design the specific steps of their field and analytical experimentation so as to obtain relevant data to support their applications for deregulation. As the composition of plant parts is sensitive to the season of growth, the environmental or biological stresses encountered, nutritional status, and other factors, it is difficult to do comprehensive comparisons for all possible constituents of foods (Cellini et al. 2004). Thus, the FDA uses the principle of "substantial equivalence" to establish whether the composition of a biotech food falls within the range of variation expected for the conventional whole food. As these crops are being grown for testing, developers must follow the regulatory requirements for experimental testing of biotech plants established by the various national regulatory bodies of interest.

To illustrate how this general approach might be practically implemented, consider the steps biotech developers must follow while seeking regulatory approval in the United States. After the basic discovery process where biotech developers have identified a new protein of interest and inserted the gene encoding it into the host plant, the stability and efficiency of the genetic modification are tested in the greenhouse and in contained field trials. Through field trials, developers generate relevant data for their own product efficacy assessment and, ultimately, for the preparation of a regulatory dossier. To begin field trials, the biotech developers must contact APHIS and must either submit a notification or apply for a full permit for "environmental release."[7] In applying for a full permit, the developer must demonstrate that all precautionary measures will be taken to avoid the unintended or uncontrolled environmental release of the organism.

Once the developer has notified APHIS or obtained a full permit, it may proceed with field testing, provided that it complies with APHIS' performance standards. The performance standards ensure that the novel crop is sufficiently contained in transportation and in the field and that any post-harvest crop residue is destroyed completely. Upon completion of the final field tests, the developer is obligated to submit a report to APHIS documenting, in particular, any unusual occurrences in the field.

The final step to commercialization is the deregulation of the novel crop. In this stage, the developer must demonstrate to APHIS that the product, once commercialized, will pose no more a plant pest risk than its conventional counterparts. The developer must submit a petition for deregulation that includes information on the biology and agronomic performance of the novel crop, the transformation methods used, and the characteristics, stability, and expression of the introduced genetic material, the likelihood of outcrossing and gene flow, and the possibility of weediness.

The generation and compilation of the required data typically takes two to three years of field testing, or five to ten generations of plants (Belson 2000). APHIS then has 180 days to respond, during which the submitted information is posted on the Internet for public comment. Once it is deregulated, a novel biotech crop or its offspring need not undergo further review for transport or environmental release within the United States.

Insect-resistant biotech crops, like Bt corn and Bt cotton, often referred to as plant-incorporated protectants or PIPs, are also subject to EPA regulation. The registration mechanisms and criteria are similar to those required for chemical pesticides. Specifically, the EPA Office of Pesticide Programs regulates PIPs through four mechanisms: registration, experimental use permits (EUPs), labeling, and tolerance levels. All PIPs must undergo registra-

---

[7] The notification option was created as a means to expedite the regulatory process for crops of lower plant pest potential. A crop may qualify for a notification if it meets the following criteria: (i) the plant is not listed as a federally recognized noxious weed species, nor is it considered a weed in the area of release; (ii) the introduced genetic material has not been derived from a human or animal pathogen; (iii) the introduced genetic material has been stably integrated into plant's genome; (iv) the function of the introduced genetic material is known, and its expression products do not result in the creation of infectious identities such as viruses, toxic substances, plant disease, and the production of substances intended for industrial or pharmaceutical use; and (v) if the introduced genetic material is derived from a plant virus, it must not result in the creation of novel plant viruses. If the biotech crop fails to meet the criteria for the notification, the biotech firm must apply for a full permit.

tion before commercialization. In order to register a PIP, the developer must demonstrate the benefits from its safe and efficacious use.

In the case of PIPs, EUPs are necessary when testing the pesticidal substance on food or feed crops that will not be completely destroyed in the testing process, when the pesticidal substance cannot be sufficiently contained, when the field trial is conducted over an area greater than ten acres, or when a tolerance exemption for the pesticidal substance has not been established. The exact information that biotech developers must supply for the registration of PIPs is determined on a case-by-case basis but generally includes product characterization, description of pesticidal substances and their properties, assessments of impacts on non-target organisms, environmental fate assessment, assessments of impacts on human health, and an insect-resistance management (IRM) program. Once a PIP is registered, the EPA must approve its labeling, establish the conditions for its commercial use, and set a maximum acceptable residue level allowed in food.

Herbicide-tolerant crops are not regulated directly by the EPA. With such crops, the EPA is concerned only with the use of the herbicide. The developers of such crops may therefore need to petition the EPA to modify the herbicide tolerance levels or establish new conditions for their commercial use, or both.

The developers of biotech crops may also choose to undergo a voluntary FDA consultation to resolve any potential scientific, health, or legal issues that may arise in the commercialization of their crops. Most novel biotech crops do not require pre-market approval. All biotech crops developed so far, however, have undergone this voluntary consultation process, and are likely to continue to do so. It should be noted that the same requirements hold for all foods; i.e., the FDA in general does not require pre-market approval of foods but has the authority to stop sale and recall products from the market if there is evidence that they may be harmful. For both biotech and conventional foods, the responsibility is with the marketer, not with the FDA, to ensure the safety of foods offered for sale.

Consultation with the FDA begins with an initial meeting between the developer and agency representatives, after which a Biotechnology Notification File (BNF) is established by the FDA to collect all information and communication documents regarding the new product. The developer then proceeds to generate and accumulate the data necessary to ensure that the new product is safe and wholesome. Once sufficient data have been collected, the developer submits to the FDA a safety and nutritional assessment summary of the product. The summary ordinarily includes a description of the crop, information about the introduced genetic material, a compositional comparison to the parental crop, and a discussion of any possible deleterious products (such as allergens, toxicants, and anti-nutrients) that may be expressed as a result of the modification. The FDA evaluates the information provided in the summary and the BNF to ensure that all scientific and legal issues have been adequately addressed, before issuing a letter of consultation.

Taking into account all relevant overlap, one finds that biotech developers typically supply the following types of studies in order to make a case for the food safety of their new biotech crops:

- Descriptive studies of the host organism (e.g., corn, soybeans) which include information on typical uses, consumption patterns, the crop's composition, its potential allergenicity, as well as details on the presence of known toxicants and antinutrients.
- Descriptive studies of the donor organisms and the introduced genes (e.g., *Bacillus thureingiensis Cry 1Ab* gene*)*, including the genes that are used for the transformation but are not included in the genome of the new biotech crop.
- Molecular characterization studies specifying the genetic modification performed. Such studies include details on the transformation process used and the inserted genes (e.g., number of copies of DNA and sites of insertion, as well as expression levels over time and in different tissues) and analytical data on the genetic stability (e.g., expression pattern in subsequent generations) of the novel trait.
- Detailed studies on the primary and secondary gene products and evaluation of the safety of novel substances. In this context, introduced genes that are expressed are characterized; the characteristics, concentration, and localization of the expressed products are detailed (e.g., expression levels of relevant proteins in various plant tissues); potential disruptions of gene expression and their effects (e.g., potential increases of natural toxicants or allergens) are described; and when novel proteins are expressed, their identity, functionality, similarity to traditional products, and safety are assessed. Safety assessment studies include *in vitro* studies (e.g., digestion of novel proteins in simulated gastric fluids) and, in some cases, *in vivo* studies (e.g., acute mouse gavage studies).[8]
- Studies on the potential allergenicity of foods derived from the novel crops. Such studies focus on sequence homology, serum screenings, pepsin stability, and, in some cases, include additional animal studies.
- Studies on any potential effects on crop and derived food and feed composition such as changes in concentration of nutrients and

---

[8] It should be noted that the concentrations of novel proteins expressed in biotech crops are typically very low. Accordingly, the amounts of such proteins that are required for all relevant safety assessment studies are much larger than can be purified from the plant material grown in field trials. For this reason, safety studies are performed with proteins purified from bacterial expression systems, which in turn have to be demonstrated as functionally equivalent to those expressed in the novel crops.

identification of anti-nutrients. Examples of such studies include proximate analysis, amino acid composition, fatty acid composition, tocopherols, and analyses of inorganic (e.g., calcium, phosphorus) and carbohydrate components. Because of the inherent variation in nutrient levels and the possibility of unexpected effects, appropriate animal feeding studies (e.g., catfish, broiler) are often undertaken.

In addition, biotech developers supply the following types of studies in order to make a case for the environmental safety of their new biotech crops:

- Descriptive studies of the host organism, including typical agronomic practices, details on its reproductive biology, common diseases and pests, and distribution and ecology of related species.
- Genetic and phenotypic variability studies across time and field environments. Such studies include details on the behavior of the novel biotech crop in various environments (e.g., susceptibility to insects and disease, survival capacity, seed multiplication capacity, and yields).
- Stability of genetic modification from one generation to another.
- Studies on potential for gene transfer to related plants. Such studies provide details on the potential of the novel biotech crop to outcross with sexually compatible plants (e.g., biotech corn with annual teosinte but also with conventional corn), the existence of such compatible plants and suitable pollen vectors to facilitate the transfer, and the fertility and ecological fitness of the progeny.
- Studies of potential for horizontal gene transfer, that is, non-sexual exchange of genetic material with organisms of the same or different species (e.g., between plants and soil bacteria).
- Studies on the potential of the novel crop to become weedy or invasive in natural habitats. Such studies include information on seed dissemination, dormancy, and germination, as well as on the competitiveness and stress tolerance of the novel crop.
- Studies of potential adverse effects on non-target organisms. Details are obtained from field studies (e.g., through repeated non-destructive visual sampling) on numerous species of interest as well as through laboratory feeding trials on key non-target indicator species (e.g., honey bees, green lacewing, ladybird beetles, daphnia, earthworm, collembolan, and quail).
- Plans for insect resistance management that are relevant to PIPs.

Just how representative might the dossiers submitted to the U.S. agencies be of dossiers submitted to various regulatory authorities in other countries? Clearly there are meaningful differences in the various dossiers submitted to regulatory agencies outside the United States. Some of these differences stem from distinct informational requirements of regulatory agencies (e.g., the requirements for molecular characterization in Japan and the EU are arguably more demanding than in some other countries). Others arise from inherent national differences in the anticipated production and consumption patterns of the new crops and hence in the necessary supporting analytical data (e.g., goat- instead of cattle-feeding trials might be necessary). Ecosystems, non-target organisms, and sexually compatible related plants can also vary across countries and could necessitate different studies for assessing environmental fate, weediness, and non-target impacts. Despite these differences, however, the dossiers submitted to the various national regulatory authorities are more similar in structure than they are different owing to the overall common approach to environmental and food safety assessment used for all new biotech crops. This similar structure allows, then, for some general characterization of the compliance costs incurred by biotech developers.

5. REPRESENTATIVE DOSSIERS AND COMPLIANCE COSTS

How do these processes of experimentation, submission, and regulatory review then translate into relevant compliance costs? How high are these costs and how much do they vary among companies, over time, and between different types of biotech crop innovations? Significant variance in compliance costs is expected as they will tend to vary from one dossier to another with the number and type of field trials, analytical tests, bioinformatic analyses, animal studies, and other comparative safety assessments which are, principally, determined by:

- Which crop has been modified (e.g., corn, soybeans, tomato)?
- What novel trait has been introduced (e.g., insect resistance, herbicide tolerance, virus resistance, quality enhancement)?
- How many (and which) countries are petitioned for regulatory approvals?
- What kind of regulatory approvals are being pursued (e.g., production, importation)?

Despite their significant variance, some general categories of compliance costs are characteristic of all regulatory submissions. To identify these general categories of compliance costs, we first interviewed lead scientists and regulatory affairs practitioners on the basics of regulatory submissions. We also obtained and analyzed representative dossiers for various novel corn and soybean traits submitted over the past fifteen years. Using these representative dossiers and relevant cost data provided by biotech developers, we next added structure to the compliance costs by identifying aggregate categories that were characteristic across all types of dossier submissions. Last, we evaluated the degree of overlap among multiple submissions for the same technology across various national regulatory systems and the added compliance costs associated with each additional international market where regulatory approval was sought.

Following these steps, we organized private compliance costs, both variable and quasi-fixed, into the following categories, which adequately characterized all reviewed dossiers:

- preparation for handing-off of a biotech event to the regulatory process
- molecular characterization of the genetic modification performed
- compositional assessment
- animal performance and safety assessments
- protein production and characterization
- protein safety assessment
- non-target organism studies, including field and laboratory assessments
- agronomic and phenotypic assessments
- production of tissues and grain for molecular, expression, composition, animal performance, and toxicology assessments
- analytical test ELISA development and validation to detect the presence of the expressed protein(s)
- EPA expenses for PIPS (EUPs, tolerances) or herbicide residue assessments
- toxicology (90-day rat trial)—when performed
- stewardship to monitor compliance with regulatory requirements, trait efficacy, and other factors such as occurrence of insect resistance
- overhead for facilities and management

In order to provide representative figures for each of these categories, we standardized private compliance costs along certain key dimensions (trait, crop, and countries petitioned). Specifically, we evaluated, and report here, compliance costs incurred by leading biotech developers seeking deregulation of herbicide-tolerant and insect-resistant corn in ten key countries, which include the top producers (e.g., the United States, Canada, Argentina, and the EU) and the top importing countries (e.g., Japan and South Korea). Compliance costs for these two corn biotechnologies are reported in the form of ranges in Tables 1 and 2. To preserve the confidentiality of firm-level data used, we do not report any averages for the total compliance costs or the individual categories. When possible, compliance costs unique to a country because of its specific regulatory requirements are separately

Table 1. Compliance Costs for Insect-Resistant Corn (in $1000s)

Cost Categories	Range of Costs Incurred
Preparation for handoff of events into regulatory process[a]	20–50
Molecular characterization	300–1,200
Compositional assessment [b]	750–1,500
Animal performance and safety studies	300–845
Protein production and characterization [c]	162–1,725
Protein safety assessment [d]	195–853
Non-target organism studies [e]	100–600
Agronomic and phenotypic assessments [f]	130–460
Production of tissues [g]	680–2,200
ELISA development, validation, and expression analysis	415–610
EPA expenses for PIPs (EUP, tolerances, etc.)	150–715
Environmental fate studies [h]	32–800
EU import (detection methods, fees)	230–405
Canada costs	40–195
Stewardship [i]	250–1,000
Toxicology (90-day rat trial)—when done [j]	250–300
Facility and management overhead costs	600–4,500
Total[k]	**7,060–15,440**

[a] Bioinformatic evaluation of product concept, inserted gene/protein.
[b] Includes two field seasons to meet the expectations of Argentina and other countries outside the U.S.
[c] Includes fermentation, purification, characterization, mode of action, equivalence to plant-produced, development of assays, maintenance of standard.
[d] Includes bioinformatics, mouse gavage, in vitro digestibility.
[e] Includes field and laboratory, tissue and pure protein assessments.
[f] Includes field agronomic and phenotypic assessments, pollen morphology, seed germination, dormancy, etc.
[g] Production of tissues and grain for molecular, expression, composition, animal performance, and rat toxicology studies.
[h] Soil dissipation for PIP.
[i] Includes insect resistance management.
[j] Not required for the United States, Canada, or Japan.
[k] Individual cost category ranges may not include data from all participating firms and are, thus, not additive. The "total" range of costs incurred includes all available data.

reported (e.g., EU requirements for development of detection methods and other fees).

A number of observations can be readily made from the compliance costs reported above. First, there is a wide variance in the total compliance costs incurred by biotech developers. Indeed, the reported variance is much higher than expected considering that key sources of variation in compliance

Table 2. Compliance Costs for Herbicide-Tolerant Corn (in $1000s)

Cost Categories	Range of Costs Incurred
Preparation for handoff of events into regulatory process[a]	20–50
Molecular characterization	300–1,200
Compositional assessment [b]	750–1,500
Animal performance and safety studies	300–845
Protein production and characterization [c]	620–1,725
Protein safety assessment [d]	195–855
Agronomic and phenotypic assessments [e]	130–460
Production of tissues [f]	680–2,200
ELISA development, validation, and expression analysis	415–610
Herbicide residue study	105–550
EU import (detection methods, fees)	230–405
Canada costs	40–195
Stewardship [g]	165–1,000
Toxicology (90-day rat trial)—when done [h]	250–300
Facility and management overhead costs	560–4,500
Total[i]	**6,180–15,510**

[a] Bioinformatic evaluation of product concept, inserted gene/protein.
[b] Includes two field seasons to meet the expectations of Argentina and other countries outside the U.S.
[c] Includes fermentation, purification, characterization, mode of action, equivalence to plant produced, development of assays, maintenance of standard.
[d] Includes bioinformatics, mouse gavage, in vitro digestibility.
[e] Includes field agronomic and phenotypic assessments, pollen morphology, germination, dormancy, etc.
[f] Production of tissues and grain for molecular, expression, composition, animal performance, and rat toxicology studies.
[g] Includes insect resistance management.
[h] Not required for the United States, Canada, or Japan.
[i] Individual cost category ranges may not include data from all participating firms and are, thus, not additive. The "total" range of costs incurred includes all available data.

costs (e.g., the type of modified crop, the specific countries where deregulation is pursued, and the types of traits introduced) have been minimized. The variance is even larger within the individual compliance cost categories. To be sure, some firm-level differences in the individual cost categories and total compliance costs are the result of differential accounting and budgeting practices among firms. More importantly, however, these differences are also attributable to the variable strategies followed by biotech developers as they pursue deregulation of their innovations. These strategies are shaped by the (apparently distinct and often evolving) developers' expectations of the appropriate number and types of field trials and analytical tests and assessment studies that are likely to satisfy the various national regulators. For instance, some firms regularly submit toxicology (90-day rat trial) studies while others consider them irrelevant and do not include them in their dossiers. Other factors, both endogenous and exogenous, also influence the developers' individual regulatory strategies. For example, compliance costs can vary drastically depending on the number of events advanced by the developers through various regulatory stages in order to minimize uncertainty. Similarly, prior commercialization of events with similar modes of action to the one being submitted for consideration, as in the case of Bt crops, could trigger more demanding safety assessments and stewardship plans to minimize potential resistance build-up.

Second, among all variable compliance cost categories, four dominate: (i) production of tissues, (ii) compositional assessment, (iii) protein production and characterization, and (iv) molecular characterization. Indeed, these four cost categories represent almost 60 percent of all variable costs.

Third, overhead costs for facilities and management are also very significant as they represent between 10 and 25 percent of the total compliance costs for various firms. Clearly, such costs are most challenging to measure as facilities and regulatory management are shared across multiple traits and events for various crops, all being advanced in parallel at their individual development speeds; and the same facilities and some of the evaluations are used for private commercial purposes as well as regulatory compliance. Overhead costs also include regulatory outreach and other relevant activities.

While accounting and budgeting nuances make measurement of overhead costs difficult, their accurate assessment is essential for the identification of potential scale and scope economies. Our preliminary assessment indicates that large biotech firms that experiment with numerous crops and traits do not have discernible fixed cost advantages, and hence we could not detect any economies of scale and scope. This may be the result of the regulatory slowdown that has occurred in recent years, suggesting that, at least temporarily, a larger than necessary management and facility capacity is being maintained by larger biotech firms. It may also be the result of the more significant regulatory outreach efforts sustained by larger biotech firms. Finally, it might be simply the outcome of the limited variance in the firm size studied here or other data limitations.

Fourth, the gap in the compliance costs between insect-resistant and herbicide-tolerant corn is lower than expected as data requirements for both traits tend to be similar. At the same time, it appears that over time firm strategies on how to develop regulatory dossiers for those two traits have also converged, and so have the relevant compliance costs incurred.

Finally, the compliance costs incurred by biotech developers and reported here appear to be quite high, considering that they represent only part of the regulatory burden of novel biotech crops. Specifically, only direct compliance costs are reported here, counted as such by most biotech developers only after a formal assessment process with strict standards known as "good laboratory practices" has commenced. Informal pre-regulatory safety assessments of various discovered proteins and events are regularly carried out but are normally budgeted as R&D costs. Similarly, indirect private compliance costs from unnecessary and unexpected regulatory delays are not presented here. These costs include increased expenditures (e.g., for seed inventories that are carried over), foregone profits from delays in commercialization, costs for channeling and segregating biotech crops away from certain markets in cases of partial approvals, and others. Such indirect regulatory costs are likely significant but more difficult to estimate than direct ones.

6. SOME CONCLUDING COMMENTS

Assessments on whether compliance costs are "high" or "low" are arbitrary and subjective unless they are made against an appropriate benchmark. The figures reported here are, no doubt, large in an absolute sense especially since they represent costs incurred by biotech developers upfront and on top of R&D expenses, while commercial success is still an uncertain outcome. Clearly, further research is needed to assess how such costs vary from one crop to another and whether they are large enough to discourage development of biotech traits in certain crops with limited market size, leading to unrealized potential productivity gains and technological orphans.

An additional important question that needs to be addressed is whether compliance costs have increased over time. Such assessments are extremely difficult considering the relatively small number of deregulations that have been spread over a relatively long period of time. Nevertheless, some incomplete data and our cursory comparisons of dossiers that have been submitted over time indicated certain differences. Most obvious are expansions of the molecular characterization of the genetic modification studies and of the stewardship plans with parallel increases in the compliance costs. Other supportive safety assessments also appear to have become more complex and voluminous, but we do not have sufficient data to accurately measure any relevant cost changes, if any have occurred. Clearly, these last issues are important in the own right and deserve additional detailed research.

REFERENCES

Alston, J.M. 2004. "Horticultural Biotechnology Faces Significant Economic and Market Barriers." *California Agriculture* 58(2): 80–88.

Alston, J.M., J. Hyde, M.C. Marra, and P.D. Mitchell. 2002. "An Ex Ante Analysis of the Benefits from the Adoption of Corn Rootworm Resistant, Transgenic Corn Technology." *AgBioForum* 5(3): 71–84.

Belson, N.A. 2000. "U.S. Regulation of Agricultural Biotechnology: An Overview." *AgBio-Forum* 3(4): 268–280.

Bradford, K.J., J.M. Alston, and N. Kalaitzandonakes. 2006. "Regulation of Biotechnology for Specialty Crops." In R.E. Just, J.M. Alston, and D. Zilberman, eds., *Regulating Agricultural Biotechnology: Economics and Policy*. New York: Springer.

Bradford, K.J., J.M. Alston, D.A. Sumner, and P. Lemaux (eds.). 2004. "Challenges and Opportunities for Horticultural Biotechnology." *California Agriculture* 58(2): 68–71.

Bradford, K.J., A. Van Deynze, N. Gutterson, W. Parrott, and S.H. Strauss. 2005. "Regulating Transgenic Crops Sensibly: Lessons from Plant Breeding, Biotechnology and Genomics." *Nature Biotechnology* 23 (April 6): 439–444.

Cellini, F., A. Chesson, I. Colquhoun, A. Constable, H.V. Davies, K.H. Engel, A.M.R. Gatehouse, S. Kärenlampi, E.J. Kok, J.-J. Leguay, S. Lehesranta, H.P.J.M. Noteborn, J. Pedersen, and M. Smith. 2004. "Unintended Effects and Their Detection in Genetically Modified Crops." *Food and Chemical Toxicology* 42(7): 1089–1125.

Falck-Zepeda, J., G. Traxler, and R. Nelson. 2000. "Surplus Distribution from the Introduction of a Biotechnology Innovation." *American Journal of Agricultural Economics* 82(2): 360–369.

Gianessi, L.P., C.S. Silvers, S. Sankula, and J.E. Carpenter. 2002. "Plant Biotechnology: Current and Potential Impact for Improving Pest Management in U.S. Agriculture: An Analysis of 40 Case Studies." National Center for Food and Agricultural Policy, Washington, D.C. Available at http://www.ncfap.org/40CaseStudies.htm (August 10, 2005).

Huang, J., S. Rozelle, C. Pray, and Q. Wang. 2002. "Plant Biotechnology in the Developing World: The Case of China." *Science* 295(25): 674–677.

Jaffe, G. 2005. "Withering on the Vine: Will Agricultural Biotechnology's Promise Bear Fruit?" Center for Science in the Public Interest, Washington D.C. Available at http://cspinet.org/new/pdf/withering_on_the_vine.pdf (accessed July 31, 2005).

James, C. 2004. "Preview: Global Status of Commercialized Biotech/GM Crops: 2004." ISAAA Briefs No. 32, International Service for the Acquisition of Agri-Biotech Applications, Ithaca, NY. Available at http://www.isaaa.org (accessed May 7, 2005).

Kalaitzandonakes, N. (ed.). 2003. "Economic and Environmental Impacts of Agbiotech: A Global Perspective." New York: Kluwer-Plenum Academic Publishers.

Kalaitzandonakes, N. 2004. "Another Look at Biotech Regulation." *Regulation* 27(1): 44–50.

Kapuscinski, A.R., S. Goodman, L. Hann, E. Jacobs, C. Pullins, J. Johnson, R. Kinsey, A. Krall, M. LaVina, M. Mellon, and V. Ruttan. 2003. "Making Safety First a Reality for Biotechnology Products." *Nature Biotechnology* 21(6): 599–601.

Marra, M., P. Pardey, and J. Alston. 2002. "The Payoffs to Transgenic Field Crops: An Assessment of the Evidence." *AgBioForum* 5(2): 43–50.

McElroy, D. 2003. "Sustaining Agrobiotechnology Through Lean Times." *Nature Biotechnology* 21(9): 996–1002.

Miller, H., and G. Conko. 2005. "Agricultural Biotechnology: Overregulated and Underappreciated." *Issues in Science and Technology* 21(2): 76–81.

Pew Initiative on Food and Biotechnology. 2004. "Issues in the Regulation of Genetically Engineered Plants and Animals." The Pew Initiative on Food and Biotechnology, Washington, D.C. Available at http://pewagbiotech.org/research/regulation/request.php (accessed August 10, 2005).

U.S. Department of Agriculture. 2005. "Field Testing Database." Animal and Plant Health Inspection Service, U.S. Department of Agriculture, Washington, D.C. Available at http://www. aphis.usda.gov/brs/database.html (accessed May 13, 2005).

U.S. National Academy of Sciences. 2000. *Genetically Modified Pest-Protected Plants: Science and Regulation*. Washington, D.C.: National Academy Press.

Chapter 4

REGULATION OF TECHNOLOGY IN THE CONTEXT OF U.S. AGRICULTURAL POLICY

Bruce Gardner
University of Maryland

Abstract: The United States has a long history of regulation of both private-sector activity and public resources in agriculture and food. Some regulation has been aimed at creating, encouraging, or controlling agricultural technology, mostly in pursuit of public health and safety, but in some cases best understood as a political response to interests supporting or opposing technological innovations. Other policy effects on agricultural technology are unintended consequences of price and income support programs. In recent years there have been recurrent efforts to harness farm commodity support programs in service of influencing the implemented technology in U.S. farming, particularly technology promoting conservation and environmental policy. The chapter uses historical examples of policy/technology interaction to examine the prospects for joint pursuit of technological and income support goals in the future. Both the likelihood of success and the desirability of results, from a welfare economic viewpoint, are questionable in light of political experience.

Key words: commodity programs, regulation, income support, politics of agriculture

1. INTRODUCTION

This chapter considers how U.S. agricultural policy has influenced technology used on farms. Some provisions of commodity policy, mainly conservation programs, provide the examples discussed in most detail in this chapter. Other elements of commodity policy have influenced technology indirectly. Payment limits and crop insurance programs are the main ones addressed here. The chapter concludes with a discussion of the political economy and welfare economics of prospective regulatory policies.

Julian Alston provided a number of suggestions and comments that have been incorporated into the chapter.

2. HISTORICAL BACKGROUND

Regulating or influencing agricultural technology might not be expected to be on the front burner in debate about agricultural policy, given the tenor of argument among the interest groups most centrally involved. The more important issues for farmers and the general public tend to be ones of prices of farm products and incomes of producers. Economists tend to focus on the costs to taxpayers and consumers, and the overall welfare effects. Even so, some interest groups, arguably increasing in influence, do want to see agricultural policy influencing technology. Animal welfare advocates want to discourage certain practices in animal husbandry. Opponents of biotechnology have spent considerable energy opposing BST (bovine sematotropin) for dairy cows, and their arguments have carried over into the debate on dairy support policy.[1] Those interests are likely to look for ways to use commodity programs to hinder or stop other innovations derived from genetic engineering. These sentiments along with environmental interests can make a difference in agricultural price and income policies when alternative programs have different environmental consequences.

Moreover, in earlier history, before commodity programs began to dominate governmental efforts in the 1930s, attempts to influence, and generally to foster, advances in technology were at the core of U.S. agricultural policy. The Morrill Act of 1862, and subsequent legislation funding agricultural research, extension, and education, embodied objectives of increasing agricultural productivity through technological change and improved farm management. Early subsidy programs, notably of irrigation projects in the Federal Reclamation Act of 1902, were technologically focused, as were drainage and flood control activities that have been important federal programs since the 1930s. Programs to promote and finance rural electricity, telephones, and roads were also key elements of federal efforts to improve the technological status of U.S. farming throughout the first half of the twentieth century. While questions have been raised about large drainage and flood control projects of the U.S. Army Corps of Engineers in recent decades, the general assessment has been that the public investments in agricultural research, extension, and infrastructure development have had high social rates of return and have in that sense been major success stories. Empirical work by economists has provided data evidence for these high returns, and political success is indicated by the emulation of public investments of these kinds in developing countries around the world.

---

[1] A recent example is "The Corporation," a prize-winning 2004 documentary film. One of its most effective segments is a discussion of BST as outrageously harmful. It illustrates the evil-doing of Monsanto with pictures of very sick-looking cows that the viewer is led to believe are the result of BST use.

At the same time, there is also a long history of regulation of technology, in some cases the result of concerns about or hostility to particular farming practices. Nineteenth century examples typically involve health risks from farm products. Regulation of sanitary conditions and of animal diseases thought to be transmissible to humans is longstanding in many localities, notably regulation of dairy farming in order to be allowed to ship milk to urban markets. Regulation of farming to combat animal diseases transmissible to animals on other farms has long been regulated in the interests of the farming community as a whole. This has involved requirements that animals be vaccinated, and in case of outbreaks isolated and destroyed.

Regulation pursued in the interests of animal welfare also have a long history. The reaction to "distillery dairies" in which cows were fed on by-products under distressing urban conditions led the State of New York to enact "swill-milk" legislation in 1862 that defined milk coming from distillery dairies as impure and unlawful to produce.[2] Agricultural interests themselves have supported legislation mandating aspects of technology when their interests were seen as calling for this step, notably the uniformity of product size and characteristics required under many federal marketing orders. The pursuit of environmental quality has resulted in more recent regulation of agricultural technology, at the federal level most notably the Environmental Protection Agency's rules for controlling farm workers' and consumers' exposure to pesticides, and emerging regulations of manure handling by large confined animal feeding operations (CAFOs).

The idea that certain processed products, or the production processes themselves, are harmful and should be regulated goes back to the nineteenth century too. Laws regulating oleomargarine, at both state and federal levels, persisted from the 1880s until today (in the form of a few state-level taxes, although federal taxation ended in 1950). Less well known is the attempt in the 1880s to regulate glucose, which was thought to be a harmful product by some food-safety pioneers (because it was made, from corn, in a process using sulfuric acid that was thought to risk leaving dangerous residues in the product—see Young 1989, pp. 66–71).

The preceding sketches a varied menu of regulatory approaches. Economic assessments of their effectiveness have been spotty and mostly informal. Some regulation, notably of sanitary conditions in farm and food establishments, and measures to control the spread of animal diseases, are widely judged as successful (although formal benefit/cost analyses that support such findings rigorously are difficult to come by—see Golan et al. 2000 for a useful discussion of the difficulties). Other regulation, notably that under marketing orders and under laws to hamstring politically unpopular

---

[2] A good source for details on this and other regulatory efforts of the nineteenth century is Young 1989 (see especially pp. 35–39 for the distillery milk story).

products like margarine, has been found wanting of economic sense (see for example Bockstael 1984).

Federal regulation of food products evolved during the twentieth century away from early ambitions to mandate what ingredients are allowed to be sold and toward labeling requirements that allow products not demonstrably dangerous to be sold, but mandating labels that truthfully state the nature of the product (Gardner 2003 reviews this evolution). A version of this approach, which is possibly a paradigm for direct regulation of technology in the next farm bill, is organic standards legislation, enacted in the Food, Agriculture, Conservation, and Trade Act of 1990. The 12-year gestation period before the regulations to implement the standards went into effect in October 2002 testifies to the difficulty of reaching decisions on an identity standard for organic food, as do the 275,000 public comments on the draft regulations. Indeed the difficulty of reaching agreement on the standards, and the questionable net benefits of the regulations finally arrived at, constitute a cautionary note for future like efforts on other aspects of agricultural technology (see Greene 2000).

In the remainder of this chapter, the focus is on policies that address commodity markets, and so are in the mainstream of farmers' interests in agricultural policy, and also that influence agricultural technology. Some of these policies intentionally address both farm income and environmental quality objectives simultaneously. Others influence technology as a perhaps unintended by-product of commodity policy.

3. CONSERVATION PROGRAMS

The longest-standing efforts of commodity policy to influence technology are the series of laws intended to promote soil conservation. These laws go back to the very birth of commodity programs in the 1930s and have continued in various forms until the present. In 1936 the Supreme Court ruled that the federal government had no authority to administer acreage controls under New Deal farm legislation, on the Constitutional grounds "that powers not granted are prohibited. None to regulate agricultural production is given, and therefore legislation by Congress for that purpose is forbidden" (U.S. Supreme Court 1936). The Court further found that "Congress has no power to enforce its commands on the farmer to the ends sought by the Agricultural Adjustment Act. It must follow that it may not indirectly accomplish those ends by taxing and spending to purchase compliance" (U.S. Supreme Court 1936).

From today's perspective, this opinion appears even quaint, but at the time it forced Congress to seek alternative means of commodity support. The result was a marriage of prior concerns about conservation with measures to remove acreage from commodity production. This was done in the Soil Conservation and Domestic Allotment Act of 1936 principally by defining "soil-depleting" crops (the main bulk commodities) and "soil-conserving" crops (grasses and legumes), and by paying farmers to substitute the latter for the former. This stated purpose passed Constitutional muster while serving the end of production control.

Under the Bankhead-Jones Act of 1937, the federal government purchased "submarginal" lands and removed that acreage from production. By 1945, 7.2 million acres had been removed from crops under this program, 6.2 million of that total converted to grazing, and over half of it in three states (Montana and the Dakotas). (For state-level and alternative-use detail, see USDA *Agricultural Statistics* for 1946, Table 715.) The approach of long-term removal of land from production has been continued in conservation titles of agricultural legislation to the present day. The Soil Bank Program of 1956 is notable as an attempt to rationalize both short-term and long-term acreage idling in a single legislative framework. It established an "acreage reserve" with land formerly devoted to program crops eligible for idling in return for payments under annual contracts. Under this program 12 million acres were idled in 1956, 21 million in 1957, and 17 million in 1958, with payments that averaged about $43 per acre.[3] The "conservation reserve" component drew from any acreage but paid lower rental rates, averaging $9 per acre in 1956–58, for idling erosion-prone land for periods ranging from 3 to 15 years. In addition the Great Plains Conservation Program, also introduced in 1956, offered contracts for permanent conversion of cropland to grass in 10 states in the Great Plains. By 1960, the area diverted under these programs together totaled over 28 million acres, about 7 percent of U.S. cropland.

The last contracts under the Soil Bank expired in 1972, but the idea was revived in the Conservation Reserve Program, Title XII of the 1985 Food Security Act. It had a target of idling 40 to 45 million acres of highly erodible cropland, paying rents averaging about $50 per acre annually over a 10- to 15-year contract period. In subsequent legislation the criteria for enrollment were expanded to include objectives of water quality improvement. Under the Conservation Reserve Enhancement Program (CREP), higher rents were paid for land with greater cash rental value than the Great Plains semi-arid areas that dominated the first five years of enrollments in the program, with particular aim at signing up cropland bordering streams, lakes, or wellheads where contaminants could easily generate water pollution. Since the late 1990s a relatively constant level of around 36 million acres has been maintained in the program at a federal budget cost of about $1.8 billion an-

---

[3] For more details on the Soil Bank, see Cochrane and Ryan (1976). In 1959 no funds were appropriated for this program, and thereafter it ended. Annual acreage idling was resurrected in 1963 and continued as a policy tool until 1995.

nually. About $100 million of this total each year is spent on cost-sharing conservation practices such as planting trees on idled cropland.

In addition to the Conservation Reserve Program, the Food, Agriculture, Conservation, and Trade Act of 1990 established a Wetlands Reserve Program. It started as a small pilot program and expanded substantially after 1993 in response to Mississippi Valley floods. The program pays farmers for "conservation easements" either permanently or for 30 years, at the farmer's option, to restore wetlands on low-lying land that had formerly been converted to agriculture. By 2001 over a million acres had been enrolled in this program under 6,500 projects.

Policy to influence technology on land remaining in crop production was initiated in the legislation of the 1930s and received longstanding codification in the Agricultural Conservation Program. Payments were made for a wide variety of soil-conserving methods, but this program never rose much above the $200 million spending level, far below spending on the acreage idling programs. The Federal Agricultural Improvement and Reform (FAIR) Act of 1996 introduced the Environmental Quality Incentive Program, which pays farmers for developing and adopting environmentally benign practices. The 2002 Farm Act authorized spending of about $2 billion per year on this and other cost-sharing programs, but appropriated funds (necessary for this program each year) have been far less. About $100 million annually was paid to farmers during 2001–2004.

Since the 1970s a regulatory approach more direct than cost-sharing conservation practices has been debated in Congress, and in the Food Security Act of 1985 several "conservation compliance" measures were enacted. Farmers were given until 1990 to begin applying a conservation plan on highly erodible land, and until 1995 to fully implement the conservation plan in order to stay eligible for other USDA programs. In an attempt to prevent the plowing up of fragile grasslands, the Act's "sodbuster" provision regulated uses of any highly erodible field that was neither planted to an annual crop nor used as set-aside for at least one year between December 31, 1980, and December 23, 1985. Farmers who brought such land into production lost eligibility for price supports and other program benefits unless they applied an approved conservation system to control erosion on the field.

The "swampbuster" provision stipulated that farmers would lose eligibility for USDA programs if they drained wetlands after the date of the passage of the 1985 Act. Subsequent legislation has maintained these provisions but, in response to farmer complaints of overzealous enforcement, the provisions were amended in the Food, Agriculture, Conservation and Trade Act of 1990 and now appear relatively toothless. In 1992, 156 farms lost $1.66 million in benefits, 0.02 percent of the $8.2 billion paid in program benefits that year (Zinn 1994).

4. PAYMENT LIMITS

Limitations on benefits to large farms have been in place for some programs since their inception. The Soil Bank introduced a limit of $5,000 per farm in 1958. The Agricultural Act of 1970 introduced a limit of $55,000 per commodity per farm, which was reduced to $23,000 in the Agriculture and Consumer Protection Act of 1973. It is unclear how many farms have been affected by payment limits. Several different approaches have been taken to make them more effective, but at the same time to not have them become obstacles to the operation and expansion of family farms. The President's Budget for FY2006 proposed tightening of limits, including elimination of the "three-entity rule," which lets a farm operator have one whole and two half-interests in a payment-receiving entity.

4.1. Effects on Technology

It has been argued since the commodity supports were first debated in Congress in the 1920s that the programs encouraged specialization and larger-scale production of farm products. [See, for example, President Calvin Coolidge's veto message on the McNary Haugen bill of 1927, reprinted in McGovern (1967, pp. 126–134).] Payment limits could constrain incentives to increase scale by enlarging one's farm, and hence could influence agricultural technology.

The Commission on 21st Century Agriculture, created in the 2002 Farm Act, gave considerable attention to payment limits, but the literature and comments submitted to the Commission give essentially no useful quantitative information on how tightened payment limits would affect the cost of producing the program crops. To estimate this cost one needs to know how the number of farms in different size categories would change, and the differences in cost of production between size categories.

Both proponents and opponents of tightening payment limits agree that tightening would result in fewer large relative to small farms (where small is defined to cover all farms up to the sizes that are subject to limits). Is it likely that large farms, especially in cotton and rice, would produce less of those crops if a payment cap limited their returns from programs for those crops? For direct payments and counter-cyclical payments, as administered under the 2002 Act, an argument for no effect is that the payments are already independent of commodity production, so it must be the market returns rather than payments that are causing these commodities to be grown. On the other hand, market support through loan deficiency payments is output-dependent and should be expected to respond to reduced payments.

Further questions are (i) whether the land withdrawn from the payment-limited crop would be planted to another crop, idled, or transferred (leased or sold) to another farmer who had not reached the payment cap, who would then plant the same crop the original farmer would have if not payment-limited, and (ii) how large the loss is of net return for each acre whose use changes because of tightened payment limits. A principal reason for expecting a loss in net return when an acre moves from a large to a small farm is economies of size—it costs more per acre to produce on smaller farms, so the returns above the cost of purchased inputs are reduced.

While we do not have evidence on what would happen to payment-limited acreage, what seems most likely is that such acreage would be leased to another farmer, one who was eligible for payments, and that that farmer would find it most attractive to plant the same crop the original farmer would have planted. The result in that case is that instead of (to take an arbitrary example) 100 farmers in a given area growing 150,000 acres of rice, we would have 150 farmers growing that acreage, i.e., the average enterprise size would fall from 1,500 to 1,000 acres. The cost of that change to the United States as a producer of rice is the difference between the cost of growing rice on 1,500-acre farms versus 1,000-acre farms. Unfortunately, the evidence either on such cost differences or on the extent of farms getting smaller appears inconclusive. What is clear however is the substantial likelihood that there will be a cost. As long as there are significant economies of scale in growing the program crops, if average farm size falls, it will cost more to produce them.

4.2. Milk Payments

A particularly interesting issue at present is the payment limits in the dairy (Milk Income Loss Contract) program introduced in the 2002 Act. That program places an upper limit of 2.4 million pounds of milk annually that can receive payments for any producer's dairy operation. At the 2003 average annual milk production of about 18,500 pounds per cow, a producer would get payments on the production of 130 or fewer cows only. The newest and most economically promising dairy operations have more cows than that. For an example of the difference this makes, with 2003 payments averaging about $1.60 per hundredweight, a dairy operation with 520 cows would get $1.20 less for its milk than an operation with 130 cows or fewer, a penalty of about 10 percent of the average price of milk received by farmers; so the 520-cow operation would have to be able to produce milk for 10 percent less cost in order to break even at the same market price as the smaller operations. Whether this is viewed as leveling the playing field (as the smaller operators might view it) or tilting the playing field (as the larger operators might view it), this is seemingly enough of a difference to have a substantial impact on the size structure of dairy farming. So, unless the larger operations can find a way around the payment limits, we are in the midst of a real-world experiment in payment-limit tightening that could prove instructive.

4.3. Direct versus Marketing Loan Payments

Because the effects of payment limits depend crucially on farmers' reactions in their land-use and crop-choice decisions, the effects of tightening payment limitations will depend a lot on the kinds of payments limited, as mentioned above. With respect to payments under CCC loan programs, the effects are most apparent in that producers subject to the limits, if effective, actually take home less per unit sold at the margin than producers not subject to the limit.

But for direct payments and counter-cyclical payments under the 2002 Farm Act, the producer gets the same payment regardless of acreage planted or quantities sold (subject to restraints such as not switching acreage to vegetables). So subjecting a producer to a tightened payment is mainly like the government drawing a sum of money from the producer's bank account. It is difficult to estimate the consequences of that. The big incentive would be to avoid leaving money on the table by cash-renting sufficient acreage to stay within the payment limit. Then the question from the viewpoint of economic efficiency in crop production is how suboptimal cash rental is—the presumption being that there must be some sub-optimality or the land would have been cash-rented already.

4.4. What Do We Know?

Proponents and opponents of tightening the limitations both tend to emphasize structural effects, mainly that there would be less incentive for farms to keep getting larger, and that smaller farms (those under the limit) would receive higher returns per unit output than larger farms and therefore would be in an improved competitive position. Proponents also point to reduced upward pressure on land prices and a generally better situation in rural communities when there are more and smaller farms as compared to fewer and larger ones. Based on information we have to date, we know one thing with reasonable certainty: tightened payment limits that are effective will improve the competitive position and net returns of small as compared to large-scale farm operations. Beyond that, it is plausible that the structure of farming would be changed, especially in cotton and rice growing, such that there would be fewer large-scale operations and more small-scale operations. But it is not clear that these changes would be significant from the viewpoint of the functioning of U.S. agricultural commodity markets. The main reason for uncertainty is that we do not know how the most likely adjustments in structure, accomplished mainly by changes in cash renting and shifting management decisions from large-scale operators to smaller-scale operators, would work out.

Overall, our knowledge of the costs and benefits of tightening payment limitations is as follows. First, taxpayers will gain if payments to large operators are reduced and no offsetting increases to other producers occur; but if tightening is done in a budget-neutral way, there would be neither costs nor benefits to taxpayers. Second, the main social cost would be that it would cost more to produce our crops on smaller operations; but for reasons discussed it is impossible to state with confidence how large this cost would be. Third, the main benefits would be social gains that would accrue from policy giving a boost to smaller-scale farming. Virtually nothing can be said about how these structural effects would play out, and even if we knew these effects, there is no objective way to place a value on them. What can and should be done is to try to nail down the efficiency costs. Then legislators would be in a better position to judge if the structural consequences are worth it or not.

5. UNINTENDED CONSEQUENCES

In assessing agricultural policy effects on technology, indirect effects of policies not directly aimed at technology, e.g., of acreage reduction programs, payment limits, or relative price distortions on farmers' choice of technology, may be as important as directly intended effects. For example it is likely that mechanization of tobacco was held back by the small and non-transferable acreage allotments that prevailed for many years. A change in the flue-cured tobacco program in 1965 provided an experiment in the effect of changing incentives on technology. The program's production controls before then were based on acreage allotments, generating a substantial incentive to increase yields in response to higher prices. The resulting higher yields were thought to reduce quality of the leaf, and in 1965 the program added poundage quotas, so that yield increases would not earn support. Did the change make a difference? Apparently, since during the 25 years before 1965 average flue-cured tobacco yields increased 50 pounds annually per acre on average, while in the 35 years after 1965, until 2000, yield increases averaged 11 pounds annually.[4]

---

[4] Calculated using data from USDA *Agricultural Statistics*, volumes for 1952, 1972, 1992, and 2002. Julian Alston, in comments on a draft, cited this as a striking instance of effects of price policy on technology.

5.1. New Deal Programs

An example much discussed by historians is the effect of New Deal farm support programs on sharecropping and cotton-growing technology in the South. Historians have pointed to the incentives created by payments intended to be shared proportionally with crop shares as "just like the enclosures of sixteenth century England [in that they] ruthlessly displaced Southern tenants and sharecroppers from the land" (Brown 2002). The mechanism was that payments had to be shared with tenants, but not with wage laborers.

Wright (1986) goes a step further, pointing to mechanization of pre-harvest operations as being available since the 1920s but not being adopted until the New Deal Cotton Program made sharecropper and tenant labor more expensive through a range of regulations that attempted to raise returns to farm labor. This induced growers to adopt labor-saving technology, and this together with supply control under commodity programs helped to force labor off of Southern farms, even before the mechanical cotton picker and World War II off-farm opportunities accelerated the great out-migration in the 1940s. Wright's summary statement is that "mechanization in the South was induced by economic incentives, and in the 1930s, these incentives were largely created by government programs" (Wright 1986, p. 233). The most telling evidence he documents is rapid adoption of tractors in plantation areas of the South during the 1930s, pointing out that "Accelerated mechanization in a depressed region, a depressed sector, and a depressed economy is anomalous. But it makes sense in conjunction with the incentives to switch to wage labor that were offered under the federal farm programs" (Wright 1986, p. 234).

Cochrane and Ryan (1976) advanced a generalization of the idea that price supports stimulated investment and technological change, stating that price and income support programs "provided the stable prices, hence price insurance, to induce the alert and aggressive farmers to invest in new and improved technologies and capital items, and the reasonably acceptable farm incomes and asset positions to induce lenders to assume the risk of making farm production loans" (Cochrane and Ryan 1976, p. 373). Clarke (1994), addressing mainly farmers' investments in tractors in the Midwest in the 1930s, gives the story a positive spin. She concludes that "farmers' willingness to invest turned in large part on the long-term changes initiated by the New Deal farm policy" (Clarke 1994, p. 200).

However, the New Deal also introduced a variety of regulatory requirements and action-specific subsidies that arguably retarded adoption of new technology; and while market sources of instability were reduced, uncertainties associated with the policies themselves were increased. Sunding and Zilberman (2001) cite evidence that irrigation subsidies *retarded* the adoption of new water-saving technology. But their review did not uncover evidence that would either support or refute the specific Cochrane-Ryan and Clarke assertions.[5]

Moving outside commodity programs, but still within agricultural policy as opposed to regulation of technology *per se*, one can see the likelihood of more transparent effects on technology. Irrigation subsidies are an example. Land use as well as the technology used in parts of the arid West would likely be substantially different without them. Subsidized lending could also be important in investment and hence technology adoption. These policies provide a different twist because, involving public goods and missing markets, they carry a larger prospect of generating welfare gains. Other areas are where local or state as well as federal regulation influences land use in ways that affect technology. Zoning for rural areas or farmland preservation programs may be intended to promote agriculture and hence be part of agricultural policy, and these may influence technology, especially of livestock enterprises.

5.2. Crop Insurance[6]

One agricultural policy that has been quite thoroughly examined as an influence on technology is provision of subsidized insurance covering crop failure. The effects could be substantial because the government's cost of the crop insurance program now rivals that of the major price support programs.

Under the Federal Crop Insurance Reform Act of 1994, USDA's Risk Management Agency does not itself issue insurance policies but acts mainly as a regulatory and support agency for private providers of insurance and farmers who buy such insurance. The 1994 Act increased premium subsidies from about 25 percent of premiums paid before its implementation to an average of 50 percent afterwards. Because of increased sales of insurance, federal outlays for premium subsidies increased even more, from about $250 million to $900 million annually. But the 1994 Act also tightened indemnity procedures, so the ratio of indemnities to premiums declined.

Even with expanded coverage under the 1994 Act, Congress felt impelled in 1998 and 1999 to appropriate $4 billion for various forms of disas-

---

[5] A case may also be made for an alternative hypothesis, the scientific "supply side" view that the key factor was the availability of a continuing stream of better and more applicable new technology beginning in the 1930s. An acceleration in productivity growth during that decade could be explained by the acceleration of agricultural research that took place between 1910 and 1930, with long lags for developing commercially viable new technology from this research. Public spending on agricultural research tripled between the decade of 1900–1909 and the 1910s, and tripled again between the 1910s and the 1920s, and agricultural extension efforts under both federal and state support also grew rapidly (see Alston and Pardey 1996, pp. 34 and 54).

[6] This section reproduces material from Gardner (2004).

ter relief for farmers.[7] This led to another round of legislation, the Agricultural Risk Protection Act of 2000. That Act further increased subsidies on crop insurance. For example, for an insurance policy that provided indemnity payments when a farmer's yield fell below 75 percent of the established yield for the farm, under the 1980 Act the federal government paid 16.9 percent of the premium. In the 1994 Act the subsidy rate was increased to 23.5 percent, and in the 2000 Act the subsidy for that insurance policy rose to 55 percent of the premium.

Given that crop insurance is a good deal for so many farmers, it is perhaps surprising that only 56 percent of eligible field crop acreage was insured in 2000 above the "catastrophic" coverage that USDA gave away with no premium charge (but a $60 processing fee). One reason may be that in every year between 1988 and 1994, and again in 1998 and 1999, the federal government has enacted ad hoc disaster relief payments to producers in areas where yields were low. In discussing the market for commercially supplied insurance, USDA analysts have asked: "Why pay a premium for something that you would likely get for free?" (Schnepf and Heifner 1999, pp. 15–18). In response to this concern, and in order not to discourage farmers from purchasing insurance, the 1999 disaster bill made disaster payments even to farmers who had their yield losses covered by crop insurance; so these producers could well end up better off with disaster than if they had a normal crop—a sure invitation to moral hazard for producers who live in risky production areas.

Another factor in spotty demand for crop insurance is that the premium rates are much more favorable for some crops and regions than for others. In the Eastern Corn Belt rainfall is very seldom disastrously low, and while insurance premiums are low, the fact of their being low means that a 35 percent premium subsidy does not mean much in dollar terms. Moreover, because the area is the nation's breadbasket, if yields were disastrously low there would very likely be a sufficiently widespread loss of the crop that the natural hedge provided by higher crop prices would cushion the financial loss to a substantial extent. In these circumstances, many farmers are not sufficiently risk averse to buy crop insurance coverage.

The 1996 FAIR Act authorized several programs that experiment further with market-based risk management. They expand upon federal crop insurance programs, which have since 1980 been increasingly marketed through

---

[7] Legislation of 1998 provided $2.4 billion for financial assistance to farmers who had crop losses due to drought or other natural disasters and reached new levels of generosity in two respects. First, it covered losses not only in 1998 but retroactively provided assistance to producers who had crop losses in three of the preceding five years. Second, under the 1994 Act, producers who had declined to purchase subsidized crop insurance had to sign waivers indicating they would be ineligible for disaster assistance; but nonetheless the 1998 legislation made such producers eligible.

private sector insurance companies, with federal subsidies and oversight. Perhaps the most interesting of the new insurance products is "crop revenue insurance," an elaboration of the existing federal crop insurance program developed by a private company, American Agri-insurance. The farmer selects from a schedule of yield and price protection options, paying a higher premium the higher the protection. An indemnity is paid if actual yield times the harvest-time price in the producer's county falls below the insured price times yield. The government pays an average of about 30 percent of the premium cost.

In 1998, the first year large numbers of crop revenue insurance policies were sold, problems occurred that are illustrative of pitfalls of subsidized insurance. These tend to occur when overly generous policy terms are offered on near in-the-money strike prices, and producers, having better information than the rate-setters, buy these policies. For example, in 1999 crop revenue insurance policies were available on durum wheat with an unusually favorable price option (apparently because the insurance company miscalculated the basis between the North Dakota durum price and prices of the exchange-traded wheat on which they based the premium schedule). This not only resulted in large sales of the policies but seems even to have caused substantial numbers of North Dakota growers to plant durum so that they could buy the insurance. USDA's September 1999 survey of acreage harvested found that North Dakota durum acreage increased 12 percent from the previous year, despite market weakness. At the same time, other spring wheat acreage in North Dakota *declined* by 15 percent in 1999 as compared to 1998, and durum acreage in Arizona and Canada also declined substantially. Thus the pricing of insurance in this case also appears to have significantly affected farmers' planting decisions.

By 2004 the government's total cost of crop insurance had risen to $3.5 billion, which amounts to an average of $12 per insured acre. This is half the cash rental rate on unirrigated Western Plains cropland, sufficiently large to provide a significant incentive to produce crops in locations where production is sufficiently risky that producers would arguably choose to produce less risky crops or pasture the land instead of cropping it, if subsidized insurance were not available. Estimates of the effects are difficult to make with confidence, and attempts to provide such estimates have resulted in greatly varying findings both for shifts among crops and for aggregate crop acreage. Estimates in the literature imply that $3 billion in crop insurance subsidies would increase aggregate U.S. crop acreage by 0.5 to 10.0 percent, a remarkably wide range of uncertainty. (See, for example, Glauber and Collins 2002, or Young, Vandeveer, and Schnepf 2001.) The issue for the present discussion is not so much acreage *per se*, but the fact that the implemented technology on many acres is influenced by the program. The supply of riskier crops at riskier locations is increased and the demand for risk-reducing technologies is decreased.

6. POLITICS

Politically, legislation to regulate agricultural technology has had mixed results. Regulation that has succeeded has tended to provide benefits to farmers along with constraints on farm activities. Examples are soil conservation programs and the main environmental improvement legislation such as CREP, EQIP, and the 2002 Act's Conservation Security Program. These programs succeed because both commercial agriculture and environmental interests support them and opposition is slight or unorganized. A key issue for future regulation of biotechnology is whether legislation opposed by commercial agriculture can succeed politically. What examples are there where this has occurred?

Regulation aimed at protecting human and animal health from careless or dangerous farming practices is a possible example, but farm opposition has not been prevalent enough to make this a real test of political strength. Examples of regulation enacted over the objections of farm groups are found in some environmental legislation of the 1970s, notably the Clean Water Act of 1972 and related regulation in the Farm Acts of 1976, 1980, and 1985. Penalties for farming wetlands were especially resented. The result, in the Farm Acts of 1990, 1996, and 2002, was a substantial pulling back of regulatory ambitions. In the last two decades, EPA's efforts to regulate farmworker exposure to pesticides and environmentally harmful emissions from CAFOs have been political battlegrounds, more difficult for agricultural interests because of division of view among growers and the support of labor and suburban interests; nonetheless, the political strength of commercial agriculture has been apparent in this area too.

That strength is further enhanced when farmers and agribusiness have common interests. Ethanol subsidies are a good example of a policy that despite high costs and dubious benefits can be a political winner—in the 2004 Presidential campaign, none of the credible candidates of either party, in the primaries or in the general election, took a position other than endorsement of ethanol subsidies. This situation is relevant to biotechnology regulation. Despite highly motivated and articulate opponents, regulations that would stymie agricultural innovations based on genetic engineering and other worrisome approaches involving irradiation and chemical substances do not look politically promising. The more evident trend is for past precautionary regulation to be weakened, such as the now-defunct "Delaney Clause" that for decades precluded any additive causing cancer, at any dosage in laboratory animals, to be used in food manufacturing.

7. WELFARE ECONOMICS AND POLITICAL ECONOMY

Bringing the political discussion into an approach congenial to economic analysis requires considerations of efficiency—what are the costs of achieving the benefits that interest groups seek? In terms of welfare economics, the costs of gains for one group are the losses of others; we analyze trade-offs between groups and the sum of gains to all. The sum, if negative, is the deadweight loss of the policy considered. Policies that generate deadweight losses may nonetheless be viewed as optimal in a political environment in which some groups' gains are weighted more than others.

Figure 1 summarizes the information needed to choose optimally among commodity support and related environmental program options. Each policy option is represented by a point that summarizes the information contained in a cost-benefit analysis of the policy. The vertical and horizontal axes measure the level of farmer and environmental net benefits under the policy. The benefits and costs of other groups would have to be shown in added dimensions. For present purposes, these are summed; and since the main interests in this aggregated third group are those of taxpayers and consumers, this amount is expected to be a net cost.

To analyze policy options, we proceed as follows. Take, as a reference point, A in Figure 1, a policy that simply transfers funds to farmers, e.g., direct payments under the 2002 Farm Act. The net cost of the policy to achieve point A_0 is Z_0 (measured in dollars). Z_0 will be less than Y_0, the gains of farmers at point A_0, because of deadweight losses, e.g., the efficiency losses involved in raising the taxes redistributed through direct payments. Environmental gains are shown, as amount X_0, because farmers spend some small amount of their gains on environmental improvement.

Now consider another policy that also has costs Z_0. Let the policy be mandatory regulation of farmers aimed at environmental improvement. In this case the costs would largely be loss of consumers' surplus as regulations increase commodity production costs. This policy reduces farmers' incomes but generates environmental benefits, resulting in point A_1. Farmers have losses, as compared to their situation with no policies in place, because they bear costs of compliance that are not fully offset by an increase in market commodity prices.

A third policy option is one that ties environmentally improving practices to farmers' program payments (conservation compliance). Again we calibrate this policy to cost Z_0, in this case both taxpayer costs and some consumer losses, as the induced production practices generate less output (but less taxpayer costs than direct payments and less consumer costs than mandatory regulation). This policy option generates point A_2. This policy results in less farmer benefit than direct payments because farmers have to

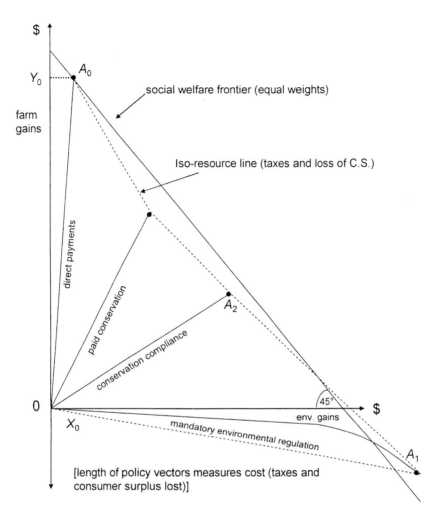

Figure 1. Benefit-Cost Trade-Offs Between Farm Support and Environmental Regulation Policies

spend some of their subsidies on environmental practices that reduce their net income (otherwise they would have adopted those practices without the program). Environmental benefits are less than under mandatory regulation (because more regulatory changes could have been achieved at less cost if farmers could have been forced to bear the burden rather than being paid to undertake the desired practices).

Which of the three policies is preferred? The dotted line segments represent a set of benefit possibilities, analogous to the utility possibility frontier in standard welfare economics, as we consider alternative policies. (There need be no continuous frontier—there may be no "in-between" policies or practical ways of blending policies.) To evaluate the alternative we need a social welfare function. Social welfare level sets are linear with a slope of -1 in the diagram if farmer gains and environmental gains are equally weighted in the social welfare function. To see if a policy is optimal, we draw the iso-welfare line through that point and see if any other policy of equal or lower cost lies outside it. As drawn in Figure 1, the optimal policy is mandatory regulation. This results from an assumption that regulation generates a net social gain—that because the policy corrects a market failure the sum of all gains including environmental benefits is positive—but the sum of all gains to direct payments is at best zero and in practice is negative (and measures the deadweight loss).

However, if one or the other interest has greater weight, the iso-welfare lines will have different slopes than the slope of -1 in the line drawn. Direct payments could be chosen over environmental regulation if the slope of the iso-welfare line were sufficiently shallow, i.e., if the relative weight on farm compared to environmental benefits were sufficiently high.

Note that, as sketched here, the "voluntary" regulatory scheme, where farmers are paid to undertake environmental improvement practices, is suboptimal for any weights on interests in the social welfare function. This is just by construction of course; the question is, should point A_2 be further from the origin? I would argue, likely not for the kinds of paid programs currently being discussed; but this hypothesis needs empirical support.

Consider conservation policy from this perspective. From the beginning it was recognized that using soil conservation programs as a mechanism for farm income support created tension in that more money could be placed in the farmer's pocket per dollar of government outlay if payments unrelated to conservation were used. Similarly, more conservation per dollar of government outlay could be generated by programs that did not just switch cropland from wheat to grass, for example. Thus conservation payments are a suboptimal measure from the viewpoints either of maximizing farm incomes or maximizing soil conservation (see Elliott 1937). However, given the political context, the 1936 Act's approach may have been optimal for achieving the maximum of each objective jointly attainable given the constraints and the political realities that required both objectives to be pursued.

Figure 2 shows federal spending on conservation-related programs along with farm income support during 1962–2005. For comparison, spending on agriculture-related research (mainly cooperative state research and extension, and federal agricultural research) is also shown. The data are the

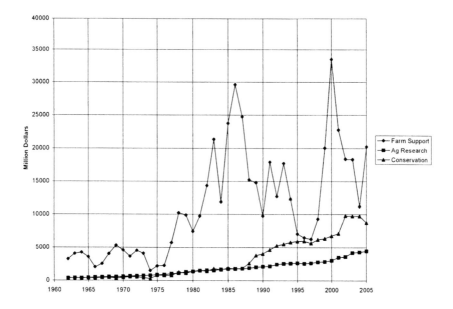

Figure 2. Spending by Fiscal Year

Source: U.S. Office of Management and Budget (2005) (historical tables).

classifications of outlays by the Office of Management and Budget by their "subfunctions" of "farm income stabilization," "agricultural research and services," and "conservation and land management." Conservation program spending accelerated sharply relative to commodity programs and research after 1985. But in more recent years the data do not show the trend toward such programs gaining political favor as compared to either farm support or agricultural research that would lead to an expectation of substantial change in orientation in the next farm bill.

These considerations do not provide evidence that I would claim substantiates my hypothesis above, but they do add to the cautionary side with respect to the idea that substituting paid environmental farm practices for traditional farm commodity supports is the way forward in the next farm bill. The New Deal conservation programs and their descendant the Conservation Reserve Program are the outstanding exceptions. They are attributable to the Constitutional constraints imposed in 1936 generating what was seen even at the time as an interior solution from the social welfare point of view. The Conservation Reserve Program as implemented in the 1990s, with its wide range of benefits, including most significantly wildlife habitat, has generated net social benefits according to a USDA analysis (U.S. Department of Agriculture 2003, Chapter 6.2), and thus would be an example of a policy that would possibly be optimal with equal welfare weights in Figure 1. It is less likely that policies aimed at regulating farmers' use of biotechnology would be either optimal in this sense or politically successful, because of the absence of documentable benefits of regulation and the likely opposition of key agricultural interests.

REFERENCES

Alston, J., and P. Pardey. 1996. *Making Science Pay: The Economics of Agricultural R&D Policy.* Washington, D.C.: American Enterprise Institute.

Bockstael, N. 1984. "The Welfare Implications of Minimum Quality Standards." *American Journal of Agricultural Economics* 66(4): 466–471.

Brown, M. 2002. "Political Culture and Antipoverty Policies in the New Deal and Great Society." Paper presented at the annual meeting of the American Political Science Association, Boston, MA, August 29–September 1.

Clarke, S.H. 1994. *Regulation and the Revolution in United States Farm Productivity.* New York: Cambridge University Press.

Cochrane, W.W., and M.E. Ryan. 1976. *American Farm Policy, 1948–1973.* Minneapolis: University of Minneapolis Press.

Elliott, F.F. 1937. "Economic Implications of the Agricultural Conservation Program." *Journal of Farm Economics* 19 (February): 16–27.

Gardner, B. 2003. "U.S. Food Quality Standards: Fix for Market Failure or Costly Anachronism?" *American Journal of Agricultural Economics* 85(3): 725–730.

_____. 2004. "U.S. Agricultural Policies and Effects on Western Hemisphere Markets." In Marcos Jank, ed., *Agricultural Trade Liberalization.* Inter-American Development Bank, Washington, D.C.

Glauber, J.W., and K.J. Collins. 2002. "Risk Management and the Role of the Federal Government." In R. Just and R. Pope, eds., *A Comprehensive Assessment of the Role of Risk in U.S. Agriculture.* Boston: Kluwer Academic Publishers.

Golan, E.H., S. Vogel, P. Frenzen, and K. Ralston. 2000. "Tracing the Costs and Benefits of Improvements in Food Safety." Agricultural Economic Report No. 791, Economic Research Service, U.S. Department of Agriculture, Washington, D.C.

Greene, C. 2000. "Organic Labeling." In E. Golan, F. Kuchler, L. Mitchell, C. Greene, and A. Jessup, eds., *Economics of Food Labeling.* Agricultural Economic Report No. 793, Economic Research Service, U.S. Department of Agriculture, Washington, D.C.

McGovern, G. (ed.). 1967. *Agricultural Thought in the Twentieth Century.* Indianapolis: Bobbs-Merrill.

Schnepf, R., and R. Heifner. 1999. "Crop and Revenue Insurance." In *Agricultural Outlook,* U.S. Department of Agriculture, Washington, D.C. (August).

Sunding, D., and D. Zilberman. 2001. "The Agricultural Innovation Process: Research and Technology Adoption in a Changing Agricultural Sector." In B. Gardner and G. Rausser, eds., *Handbook of Agricultural Economics* (Vol. I). Amsterdam: North-Holland.

U.S. Department of Agriculture. 2003. *Agricultural Resources and Environmental Indicators.* Agricultural Handbook No. 722, Economic Research Service, U.S. Department of Agriculture, Washington, D.C.

U.S. Office of Management and Budget. 2005. "Budget of the United States Government,

Fiscal Year 2006." U.S. Office of Management and Budget, Washington D.C. Available online at http://www.gpoaccess.gov/usbudget/fy06/browse.html.

U.S. Supreme Court. 1936. "Opinion of the Court." *United States vs. Butler* (decided January 6, 1936). Available online at http://supct.law.cornell.edu/supct/html/historics/USSC_CR_ 0297_0001_ZO.html (accessed May 4, 2005).

Wright, G. 1986. *Old South, New South: Revolutions in the Southern Economy Since the Civil War.* New York: Basic Books.

Young, C.E., M.L. Vandeveer, and R.D. Schnepf. 2001. "Production and Price Impacts of U.S. Crop Insurance Programs." *American Journal of Agricultural Economics* 83(5): 1196–1203.

Young, J.H. 1989. *Pure Food.* Princeton: Princeton University Press.

Zinn, J. 1994. "Wetlands and Agricultural Policy Issues in the 1995 Farm Bill." Congressional Research Service, Washington, D.C. (December).

Chapter 5

MANAGING LIABILITIES ARISING FROM AGRICULTURAL BIOTECHNOLOGY

Stuart Smyth, Peter W.B. Phillips, and W.A. Kerr
University of Saskatchewan

Abstract: While most innovations commonly enter the marketplace with little notice or fanfare, this cannot be said for products of agricultural biotechnology. The commercialization of innovative new transgenic crops and the resulting food products have resulted in some products entering the market that are not desired by some consumers. It is argued that these new products are creating a new class of socioeconomic liabilities in the marketplace. We examine three strategies for managing these liabilities: scientific, regulatory, and market strategies.

Key words: risk, liability, regulation, transgenic crops

1. INTRODUCTION

The global agri-food industry has reoriented itself in the past decade around technological change and innovation. Both farmers and the rest of the agri-food supply chain have recognized that the long-term threat to their livelihoods is other local and regional demand for land, labor, and capital. Ultimately, the sector will need to achieve productivity gains at least equal to those in other domestic sectors, which will require significant technological and institutional change. Change creates risk, which can, if not anticipated and managed, create liabilities for someone. The purpose of this chapter is to identify the context for these changes, examine the concepts of risk and liability, and review the institutional responses to these new liabilities.

While the technological imperative is not a new feature for agriculture—waves of change involving machinery (1930–60) and chemicals (1950–90) have swept through the industry in the past—the acceleration of biotechnological innovation since 1985 has fundamentally challenged the industry. In the first instance, governments have encouraged the search for new technologies and products by extending monopoly intellectual property rights (both patents and plant breeders' rights) and by offering new or different forms of government subsidy and support to develop and bring new technologies to market. Second, even though ultimate ownership is more clearly defined, the scale and complexity of using this globalized science has precipitated a wide array of collaborations between traditional competitors and between public and private research organizations, which spreads responsibility much more widely than previously. Third, the privatization of agrifood research has led many national governments to renationalize and enhance their regulatory oversight of new products (Phillips and Khachatourians 2001).

The combination of these three trends has opened a broader debate about what to do if and when one of these new technologies creates some undesirable, adverse effect. Who is ultimately responsible, how should the problem be managed, and what compensation, if any, should be paid? Ultimately, this is a debate about liability.

Liability is an evolving concept, especially as it pertains to agriculture. Historically, lawsuits in crop agriculture have been mostly about one-to-one production externalities, such as aerial spraying. Occasionally, an aerial application of a chemical would be too close to a neighboring farmer's land and would drift onto a crop belonging to another farmer. Depending on the crop, the damage could be substantial. In some instances, the farmer whose crop was adversely affected sued the commercial sprayer of the chemical for damages suffered. Another commonly cited example is the situation where a scrub bull escapes an enclosure and impregnates pure-bred cattle indiscriminately.

The genetic modification of crops has changed the nature of the liability debate and the application of the term. The commercial release of transgenic crops has created a split within agriculture, both within and between countries. In North America, there is a small organic agriculture market that is strongly opposed to further commercialization of transgenic crops due to the potential for co-mingling. The organic market's fear is that if transgenic seeds are detected in organic shipments, then domestic and export markets may be destroyed. Other producers and processors have adopted quality control systems to differentiate between GM and GM-free produce. Even in the European Union (EU), divisions are forming. Spain, for example, has produced between 45,000 and 55,000 acres of *Bacillus thuringiensis* (Bt) maize annually for the past five years (Brookes 2002), while many farm and consumer lobby groups are strongly opposed to any GM traits in the EU food system.

Internationally, there has been a split between EU countries and North America (the United States and Canada). The EU views transgenic crops as a liability and will not allow domestic production of transgenic crops for large-scale food consumption, or the importation of transgenic raw materials or processed food products. North America has approved the commercial release of a variety of transgenic food crops, which, by some estimates, are incorporated into nearly 70 percent of all processed foods. In North America, the production of transgenic crops and the consumption of the resulting food products have become the norm. In 2004, over 80 percent of all soybeans grown in the United States were transgenic, while 77 percent of all canola grown in Canada was transgenic (James 2004). Even the adoption of transgenic maize has grown rapidly, with transgenic varieties now accounting for 40 percent of all maize grown in the United States. A major challenge facing the agri-food industry is that most of the GM production is concentrated in countries that supply the bulk of the global trade in those crops. More than half and in some cases up to 90 percent of the export trade in these crops comes from the countries extensively adopting GM varieties, and those exports go to as many as 170 countries annually. In short, even though production may be concentrated in a handful of countries, any potential adverse effects are likely to spread rapidly and widely around the world through trade (Phillips 2001).

The focus of this chapter is to examine the application of liability to agriculture and more specifically to agricultural biotechnology. We are already beginning to see some effects. International trade flows have been and increasingly could be damaged when a commodity export is tested and found to contain unacceptable levels for transgenic traits. Domestic production has also been affected by the adoption of some transgenic traits: the uncontrolled spread of StarLink™ corn cost more than a billion dollars (the cost of lawsuits, product recalls, and depressed corn prices) to contain and remove, while organic canola growers in Canada assert that their market prospects have been harmed by actual or potential cross-pollination between organic and GM crop varieties. Ultimately, one overriding issue is beginning to emerge: is a liability created if a sales market is damaged by co-mingling of genetically modified seeds (or in cases where co-mingling is anticipated), and, if so, who should be liable?

2. DIFFERENTIATING BETWEEN RISK AND LIABILITY

One major challenge is to differentiate between risks and liabilities resulting from new technologies. While there are numerous options available to assess, manage, and communicate risks, there are presently no obvious structures in place to manage any resulting new liabilities. Methods to determine risk, developed over the past 50 years, now involve a number of approaches, including the Risk Analysis Framework and the Precautionary Principle. There are abundant sources of literature on risk (Shrader-Frechette 1990, Sandman 1994, Stanbury 2000, and Leiss 2001), and this chapter will not attempt to delve into the topic of risk in any great detail. Instead, we focus our efforts on defining why liability is different from risk within the agricultural biotechnology setting, and on examining how existing systems can be used to manage any resulting liabilities.

The focus of all risk analysis is to try and determine the likelihood of a risk event occurring. Risk assessment, risk management, and risk communication all involve probabilistic assessment of theoretical options and evaluation of ways to minimize the potential for the development of risks, the efficient and economical containment of risks should they occur, and strategies for informing society.

Once a risk has been realized, it is no longer a risk—it becomes a liability. Those involved in dealing with the situation are no longer dealing with probabilities, they are dealing with certainty. Once a liability is realized, it is going to cost money to rectify the situation. This is the defining difference between risk and liability. Risk is costless—liability is not. Risk is the probabilistic likelihood of an unplanned, undesired, or unwanted event actually happening. This probability may range from almost perfect certainty (i.e., 100 percent) to infinitesimally small amounts. Once a risk has been realized, however, the probability of being affected is effectively 100 percent. The challenge for those involved then is not how to assess or avoid it but how to inform people about it, how to limit the impact, and how to apportion the costs.

The issue of who pays in a case of liability is at the heart of many debates within biotechnology, especially agricultural biotechnology. In discussing a regulatory framework for liability, the New Zealand Royal Commission on Genetic Modification (2001, p. 313) advocated that:

> ... legislation regulating genetic modification should include provision for liability and compensation; there ought to be strict liability for environmental and economic damage; "liability funds" should be established; and users of genetic modification technology should be required to give bonds for cleaning up adverse environmental effects.

This proposal raises four points worth considering. First, it is quite a step to assert that liability should be formally recognized and lawsuits for compensation should be allowed. While farmer-to-farmer lawsuits are relatively common, groups of farmers have tended in many countries to be protected from civil action by outsiders through "right to farm" legislation that delimits the ability of others to control farmers. Second, the creation of liability funds would be significant, but not unique. It would be similar to the "superfund" concept for nuclear and chemical sites in the United States, where a pool of resources would be available to clean up any problems. While perhaps new to agriculture, both the legislation and liability funds are relatively straightforward extensions from other sectors.

A third consideration, the application of the concept of strict liability for environmental and economic damages resulting from the use of transgenic crops, should be a concern to all involved in the agricultural biotechnology industry. In discussing strict liability, the New Zealand Royal Commission on Genetic Modification (2001, p. 318) argued:

> The rule [of strict liability] applies to the "escape" from the defendant's land of something likely to cause damage. Liability applies even if the defendant was not at fault or took all reasonable precautions to prevent the escape; the defendant must be in possession or control of the land from which the "harm" came and be making a "non-natural" use of the land; and the possibility of escape and the consequent harm must have been foreseeable, although the manner or immediate cause of the escape need not have been foreseeable.

The possibility of being found strictly liable in spite of enacting all possible preventative measures exposes the use of transgenic crops to malicious lawsuits launched on behalf of those opposed to the concept of transgenic crops. Simply the threat of having a lawsuit brought against the uses of the technology will be enough to dramatically reduce the rate of adoption, possibly to the point of making adoption no longer commercially or economically viable. This could give birth to the largely undesirable scenario of having critics of innovation and change being in charge of commercialization of new technologies simply through their ability to threaten to launch malicious lawsuits under strict liability. The other wording that is potentially troublesome is the reference to "non-natural" use of the land. In plant agriculture it is difficult to fathom how this term can or would be applied by a court of law. Given that modern, industrial agriculture has been using modern improved crop varieties, sophisticated machinery and agronomic practices and chemicals, fertilizers, and pesticides to improve yields and quality for more than 50 years now, it is easily arguable that conventional agriculture (and possibly even organic agriculture) could now be viewed as a "non-natural" use of the land. Defining what is and what is not "non-natural" use of agricultural farm land (even deciding the body that will operationalize the definition) will be hotly contested in all societies.

A fourth consideration, that users of the technology should have to post bonds to cover environmental damages, is at the heart of assessing where the liability lies. The New Zealand Royal Commission believes that the liability lies with the actual producers that use transgenic crops. However, where national regulatory agencies have reviewed the submissions of any seed development company seeking to sell a transgenic crop and have approved the transgenic variety as safe for all forms of use and consumption, it is difficult to understand how the end users can be held liable. If the use of a transgenic crop results in an environmental problem, the technical assessment required to determine this is well beyond the ability of the local producer prior to purchasing and planting the transgenic crop. This type of environmental liability would seem to result from a lack of due diligence by the federal regulatory body, whose responsibility it is to approve new crop types safe for all uses, and, therefore, the liability should lie with the federal regulators.

Blanket statements about the application of strict liability to the production of agriculture heighten the debate but do not provide clear focus on the issues. Simply stating that strict liability should apply to all environmental and economic damage resulting from transgenic crops indicates that a serious intellectual discourse has not taken place.

3. APPROACHES TO LIMITING OR ALLOCATION LIABILITIES

The analysis of liability can be understood and assessed only in an institutional setting. Generally, institutions are mechanisms or a bundle of rules through which choices are made and conflicts resolved (Atkinson 1993). The formal structures of institutions are transparent and understandable, but it is the informal structures that present challenges. Informal structures are operating rules, norms, and cultural values that cannot be easily identified at a cursory glance.

North (1990) addresses technical and institutional change and attempts to understand the commonalities and differences between the two. North believes that institutional changes are the more complex of the two due to the multifaceted interrelationships that exist with both formal and informal constraints. North argues that stakeholders have varying degrees of vested interests in institutional change and will try to influence institutional changes towards their favor.

Particular institutions tend to be best suited to govern particular circumstances. Picciotto (1995) offers one framework to highlight the essential institutional fundamentals related to public, private, and collective sectors and their interaction. This institutional methodology provides insights into how these three stakeholders need to come together to foster successful development projects. The framework suggests that liability issues may be addressed in three discrete domains: through collective academic and industrial action, in the formal government sector, or through the marketplace. These three perspectives will be examined in the following sections.

3.1. Scientific Control of Liabilities

One means of controlling liabilities is to engineer them out of existence. While this may at one level be seen as simply a question of whether we have the science, at root it is a question of what mechanisms are economically and socially acceptable. One fundamental lesson we have learned about living organisms is that they are seldom static. They cross breed, mutate, and adapt to new circumstances. Scientists attempt to control these changes through a variety of biological control mechanisms to manage any adverse effects of change. While we have some scientific success, the key lesson seems to be that our ability to use science to control liabilities will be as much an institutional challenge as a scientific one.

In flowering plants, pollen is responsible for delivery of male gametes to the female reproductive organ (the carpel) of the same or another plant, respectively resulting in self- or cross-pollination (Sawhney 2001). The control of pollination can involve the option of genetic manipulation of its development or function. There are several loci for genetic manipulation and interference with steps of pollen development, including pre-meiotic and meiotic events in the microsporogenous tissue, the development of microspores and pollen in the anther, pollen maturation and pollen release, pollen dispersal, the attachment of pollen to the stigma of the carpel, pollen germination and tube growth, and the release of sperm cells into the female gametophytes in the ovule of an ovary. Genetic blocking of genetic functions governing these events can lead to absence of seed and fruit development (Smyth, Khachatourians, and Phillips 2002).

The 1990s brought new efforts to ensure sterility, not for processing or gustatory value, but for commercial intellectual property rights (IPR) protection. Efforts are underway to limit the diffusion of transgenes through genetic use restriction technologies (GURTs) to turn off reproduction for either transgenic varieties or traits. Some dub this approach the "terminator gene."

The use of sterile seeds *per se* is widely practiced and has not raised objections, while the use of GURTs is being criticized because it could deny poor farmers the option of saving seeds for future use. However, the crudest form of this technology was known and documented long before notions of inheritance and genetics were known. Seedless grapes were known to have existed from writings of ancient Egyptians and Greeks (ca. 3000 BC). Since the 1930s, seedless edible crop varieties produced by traditional plant breeding methods have been produced (e.g., seedless grapes in 1936 and watermelons in 1951), and they possess certain advantages. The prerequisite knowledge of flower pollination, fertilization, fruit development, genetics, and rDNA-based technologies has now enabled scientists to apply this technology to new crops.

A number of other biological options exist, depending on the crop and its attributes. Both traditional and molecular genetic methods already provide mechanisms to create hybrids, while working at a more refined molecular level offers the potential to control GM traits. Recently there has been an effort to reduce the risk of biotechnology crops by engineering foreign genes via the chloroplast instead of the nuclear genome. Such recombinants would express the new traits only in selected parts of the plant, rather than in the whole plant. Hence, any pollen drift would not include the transgenes. These and other options offer some promise.

Biological control of liabilities, either through contemporary technologies described above or those yet to be devised, is the science side of the story. The human, institutional element is the complementary other side. Ultimately, these two parts must fit together in a discussion of the relative costs (risks) and benefits of alternative options. As noted above, the costs of not managing the liabilities are potentially very high, ranging up to the US$110 million out-of-court settlement cost of the StarLink^{TM} failure. Similarly, control mechanisms are not cheap. Incomplete institutional approaches can lead to millions of dollars of losses when technologies are widely dispersed. One potential advantage of the GURTs' biological control mechanism is that while it is costly to develop, the marginal cost may be as low as US$250,000 per new variety released (Visser et al. 2001), which would add only about 10 percent to the cost of a new commercial variety. Given that many firms report that they lose at least 10 percent of their returns due to incomplete property rights, this option may be significantly more effective than other approaches. Of course, offsetting this economic argument is the social debate about the impact on subsistence farmers and the tradition of farmers' privilege.

3.2. Government Control of Liability

Governments tend to have a larger role in defining, managing, and adjudicating liability because we live in an imperfect world, where individual market actions do not lead to socially acceptable outcomes. In a world characterized by perfect, costless information, governments could simply define rights and then allow adjudication institutions to operate. If those systems did not impose any transaction costs on those who chose to enter into a transaction, the existence of the threat of liability would lead to an efficient market outcome. The existence of legal liability would act as a perfect deterrent, as the full costs of any transaction would have to be accounted for by those entering into the transaction. Liability would ensure that only economically viable transactions were entered into (i.e., after fully compensating anyone who might be negatively affected by the transaction, there is a net gain for the owner).

Of course, for liability to fulfill its deterrence/compensation role perfectly requires that the institutions that adjudicate and enforce liability claims
operate without cost. This means that the adjudication institutions would
need costless access to the same information as the parties to the transaction
and those external to the transaction. As a result, those who could be negatively impacted by the transaction would have the credible threat of litigation
and have the assurance that they would be fully compensated if the transaction were to go ahead.

The theoretical characterization of the role of liability developed above
corresponds to the role that liability plays in a transaction under the normal
assumptions of neoclassical economics (Hobbs and Kerr 1999). If the assumptions of the neoclassical model hold, there would be no need for government regulatory intervention in the economy. The neoclassical transaction, however, is only a theoretical benchmark. While the neoclassical transaction assumes that all participants can be fully informed, this is clearly not
possible in many markets, especially for markets for new transformative
technologies, such as biotechnology, where the novelty of the technology
makes information on possible externalities costly and incomplete. This is
particularly true in the case of the distant future regarding the human health
effects of consumption or potential damage to the environment. Given long
lead times and imprecise measurement, it may not be possible to identify
with any degree of certainty the source of a negative externality. This means
that liability cannot be accurately assigned and, as a result, that it cannot fulfill its deterrence/compensation role. Any adjudication system clearly imposes
considerable transaction costs on those that wish to invoke its use. Further, it
does not operate in a predictable fashion, especially in the case of international
transactions where no international private commercial law exists.

Faced with a market failure in the liability system, governments have
attempted to intervene in the markets for biotechnology through regulation.
The pace at which the transformation is taking place as a result of biotechnology has outstripped the pace at which international organizations can
adapt to accommodate the new technology. As a result, national governments have had to act independently in establishing their regulatory regimes
(Phillips 2001). Predictably, they have arrived at different conclusions regarding the nature and degree of the market failure in the liability system. In
particular, the United States and the EU have radically different views of the
extent of the failure, which shape their response to the international regulation of biotechnology.

When it is perceived that liability is prevented from effectively fulfilling
its deterrence/compensation role, the most common regulatory alternative is
to act to reduce the risks that may arise from consuming the product or from
its release into the environment. Typically, governments require that firms
wishing to sell their products in the marketplace satisfy certain regulatory

standards before the product can be sold. Often, this involves testing and providing scientific evidence for review by experts appointed by the government. This is an attempt to replace the deterrent role of liability with a science-based system for product approval. If the product clears the regulatory hurdles, then there should be no need for compensation through liability because a problem should not arise. Passing the regulatory hurdle usually means that firms are absolved from liability if all of the information provided was accurate. Firms that are caught cheating on regulatory procedures are subject both to prosecution by the government and to private liability. Problems arise, however, when firms follow all the regulatory procedures and subsequently there is found to be a problem with the product. The use of asbestos in building materials and the drug thalidomide are two obvious examples. Thus, government regulatory actions can replace the role of liability only to a partial extent.

If the technology is new, there may not be sufficient scientific information upon which to determine if the absence of liability can be offset by a set of regulatory hurdles. Faced with incomplete information, governments may wish to act with precaution. Of course, many products fail to pass the regulatory barriers and are never allowed to be released commercially. The heart of the regulatory divergence in domestic policy across the Atlantic lies in the operationalization of precaution—how lack of scientific certainty is dealt with. As a result, two quite distinct regulatory trajectories exist for biotechnology—the scientific rationality and the social rationality trajectory (Isaac 2002). The end result has been more rapid adoption of the technology in North America and very slow and tentative adoption in Europe.

Apart from the question of safety, one other group often is concerned— namely those that may lose from the changes brought by transformative technologies. These are competitors with the products arising from the transformative technology that continue to use the less efficient previous technology. Replacing the inefficient with the efficient through competition is seen as economic progress and welfare-enhancing for society. For the most part, societies do not compensate these losers. The economic model that underlies this view of technological change is based on the premise that consumers always win from the lower prices arising from increased efficiency and that their gain outweighs the losses of inefficient producers. When the gain to society is greater than the loss, it is theoretically possible for the winners to compensate those who lose and still be better off (the compensation principle). Whether or not compensation takes place is not considered as being important, only that compensation could take place.

This does not mean that inefficient producers will not seek protection from governments to prevent their loss. This is the basis of the protectionism that is the heart of trade policy and which organizations such as the WTO have been put in place to manage. It is not common in most countries, however, for winners to be liable for the economic losses suffered by other groups in society as a result of transformative technological change. Governments may, at times, be willing to step in and offer groups protection from losses that arise from the forces of competitive technological change, but these interventions need to be put in a separate category from those where governments intervene when liability would apply if there were perfect costless information and the adjudication system operated without transaction costs.

One regulatory strategy that would seemingly address this would be to adopt the tools used to deal with nonpoint-source emissions. Nonpoint-source emissions are typically diffuse and not concentrated in a specific location, which makes control very difficult. Commonly listed examples of this source of emissions are pollutants that are spread by spring run-off or pesticide residue run-offs from agricultural lands that damage watersheds. Regulatory strategies that have been used to reduce these emissions are to place a tax on the chemical that is causing the problem, thereby reducing the usage of the chemical. While this strategy might be attractive to those seeking to slow or reverse the adoption of biotechnology, it does not address the socioeconomic problem of widely disconnected causes and effects. If the issue is simply one of farm-to-farm contamination, more direct measures would be appropriate. If the concern is the creation of socioeconomic liabilities that affect only a subset of society that is often widely removed from the farm, then targeting farmers of biotechnology crops with nonspecific emissions taxes would do little or nothing to address the problem.

The EU has moved farthest and fastest to adopt more formal government rules to manage liability. The current orientation to liability within the European context is derived from three separate arrangements. First, the 1993 European Convention on Civil Liability for Damage Resulting from Activities Dangerous to the Environment (Lugano Convention) was designed to ensure that there would be adequate levels of financial compensation for any damage that was the result of activities deemed to be dangerous to the environment. Second, the 1999 Basel Protocol on Liability and Compensation for Damage Resulting from Transboundary Movements of Hazardous Wastes and their Disposal sets out very strict regulations governing the transboundary movement of hazardous waste between contracting states in Europe. Third, the 2000 European Commission White Paper on Environmental Liability offers the concept of the "polluter pays" principle and how it may be used to improve the application of the environmental principles of the European Commission Treaty.

Even though there is a great deal of research and discussion about liability in Europe, no EU-wide rules are in place. Neither the Lugano Convention nor the Basel Protocol is in force. As of December 1, 2004, the Basel Protocol had thirteen signatories and four parties, yet it requires twenty members to ratify it prior to coming into force. Similarly, the Lugano Convention had nine signatories, no parties, and requires three members to ratify it before taking effect. The member states of Denmark and Germany have gone the farthest and have enacted legislation that impose strict liability on adopters of GM crops for any subsequent financial damages to neighboring producers, but the absence of much cultivation leaves these systems largely untested.

In spite of recent efforts to enact new rules, no coherent governance system is emerging from the array of mechanisms being tried in various countries.

3.3. Market Management of Liability

Given that scientists working collectively and governments individually or collectively are unlikely to quickly and effectively delineate and manage liabilities related to new technologies, the market for the foreseeable future is going to need to manage its own affairs. In the past, industry focus has been on getting new products into the market, with adjustments to the supply chain being made along the way. Clearly, this strategy is risky. If diligence is not taken in the first delivery of products destined for the marketplace, co-mingling of product ingredients may occur, leading to inappropriate matching between product and consumers. However that happens, liabilities will develop. Industry is trying a variety of product differentiation systems to reduce potential liabilities.

The definition of product differentiation can have several nuances, depending on the justification for the differentiation. Frequently, the terms "identity preserved production and marketing" (IPPM), "segregation," and "traceability" are used interchangeably in the biotechnology and supply chain literature. This creates misconceptions about the distinct role that each of these product differentiation systems has in the supply of food products and has the potential to compound potential liabilities. Smyth and Phillips (2002) reviewed the literature and offered a more comprehensive set of definitions.

The first product differentiation system, identity-preserved production and marketing, has evolved over time in the grain and oilseed industries. Identity-preserved production and marketing systems are initiated by private firms in the grain and oilseed industry to extract premiums from those parts of the market that express a willingness to pay for an identifiable and marketable product trait or feature. An IPPM system is a "closed loop" channel that facilitates the production and delivery of an assured quality by allowing identification of a commodity from the germplasm or breeding stock to the processed product on a retail shelf (Buckwell, Brookes, and Bradley 1999, Lin 2002). These IPPM systems are predominantly voluntary, private, firm-based initiatives that range between systems that are loosely structured (e.g., malting barley) with high tolerance levels and those with rigid structures (e.g., non-GM produce for European markets) with low tolerance levels. Firms operating in the low tolerance area need to develop and adhere to strict protocols that specify production standards, provide for sampling, and ensure appropriate documentation to audit the flow of product. Most of the literature on IPPM systems relates to theoretical and operational uses of IPPM systems. Bullock, Desquilbet, and Nitsi (2000) and Bullock and Desquilbet (2001) discuss differentiation between GM and non-GM products, and Herrman, Boland, and Heishman (1999) examine the feasibility of wheat segregation. Bender et al. (1999), Bender and Hill (2000), and Good, Bender, and Hill (2000) have released a series of papers on handling specialty maize and soybean crops, with costs being the focus, not the defining of the system used to handle the specialty crop. Additionally, Miranowski et al. (1999) offer some perspectives on the economics of IPPM, while Maltsbarger and Kalaitzandonakes (2000) provide a solid theoretical model for examining the cost of identity preservation. Moss, Schmitz, and Schmitz (2004) use an empirical model in an attempt to identify the costs of identity preservation when differentiating between GM and non-GM markets. Numerous IPPM systems are operating around the world. Some extend only between the breeders and the wholesale market or processor, while others extend right up to the retailer. Their structure depends on the attribute they are trying to preserve. Some novel oils, such as low linolenic oils that are more stable in fryers, have value only at the processing level, while others, such as high oleic oils, have health attributes that can be marketed to consumers. IPPM systems are important for providing information to consumers about the provenance of a product, as those attributes are not visible or detectable in the product itself. Organic products are one of the most noticeable IPPM products in today's marketplace. Others include Cargill's IPPM system for the export of an Intermountain Canola variety that gives off virtually no odor when used to fry food to Japan, General Mills' IPPM system for a select variety of white wheat that possesses a special trait for "flake curling" when processed into breakfast cereal, and DowAgro Sciences' export program for Nexera canola to Japan, where it is sold in the specialty gift oil market.

A second product differentiation system, segregation, has frequently been confused with the grading of different classes of grains and oilseeds in order to receive a higher price for the commodity than if it were allowed to be co-mingled. Segregation systems have a formal structure and in fact can act as regulatory standards. Segregation differs from IPPM in that the focus of the system is not on capturing premiums but rather on ensuring that potentially hazardous crops are prevented from entering supply chains that have products destined for human consumption. Segregation can be defined as a regulatory tool that is required for varieties that, while approved and commercially released, could enter the general supply chain and create the potential for serious health hazards. Segregation systems are usually developed as part of a variety registration process, where government regulators use contract registration to ensure that certain novel varieties do not enter the supply channels of like varieties. The private firm seeking registration of the novel variety has to demonstrate that there is a segregation system developed to ensure the containment of the variety. Lin (2002, p. 263) defines segregation as the requirement "that crops be kept separate to avoid commingling during planting, harvesting, loading and unloading, storage and transport." Segregation systems are used when potential food safety concerns exist over the co-mingling of the segregated product and all other like products. In a recent paper, Lin and Johnson (2004) estimate the cost of segregating non-GM maize and soybeans at 12 percent of the average farm price. In short, IPPM is used to capture premiums, and segregation is used to ensure food safety.

The third product differentiation system, traceability, is commonly used in the food industry. The International Organization for Standardization (ISO) has defined traceability as the "ability to trace the history, application or location of an entity by means of recorded identifications," and the Codex Alimentarius Commission has adopted this as their working definition for all Codex standards (Codex Alimentarius Commission 2001). The EU (European Union 2001) has defined traceability quite clearly in relation to GM products. Directive 2001/18/EC (p. 2) defines traceability as "the ability to trace GMOs and products produced from GMOs at all stages of the placing on the market throughout the production and distribution chains facilitating quality control and also the possibility to withdraw products. Importantly, effective traceability provides a 'safety net' should any unforeseen adverse effects be established." Generally, retail products found with unacceptable bacteria levels, intolerable levels of pesticide or chemical residues, or inappropriate mixtures of inputs need to be quickly and completely removed from shop shelves. Traceability systems allow for retailers and the supply chain to identify the source of contamination and thereby initiate procedures to remedy the situation. The key focus of traceability is on greater acceptance and food safety. Recently, the focus for developing traceability systems for new sectors of the marketplace has shifted from food safety towards extracting premiums from the marketplace. But market premiums are unlikely to be large enough to sustain a traceability system, as traceability systems do not in and of themselves define quality, but simply trace it. If market premiums are the driver, then the developers need to use an IPPM system, as IPPM systems are the only systems properly structured to capture premiums. The economic literature from supply chain management defines traceability as the information system necessary to provide the history of a product or a process from origin to point of final sale (Wilson and Clarke 1998, Jack, Pardoe, and Ritchie 1998, Timon and O'Reilly 1998). Price,

Kuchler, and Krissoff (2004) provide a solid overview of European traceability as it relates to the American soybean industry. While Dickinson and Bailey (2001) suggest that their results from a laboratory auction market regarding features of meat traceability show that there is willingness by consumers to pay premiums for traceability, the key focus has to be on food safety. Prior to adopting traceability systems, there has to be a clear indication of specifically what aspects of food safety can be improved by the adoption. Marginal improvements in food safety would be a dubious reason for proceeding—rather, there must be a clear and evident improvement in the level of food safety. Traceability systems have been developed for beef products in many countries around the world. Traceability has been developed in conjunction with a quality assurance system to reassure export markets about the quality of meat products (Spriggs and Isaac 2001). However, it should be noted that these systems have been met with some resistance at the farm level in some countries, as producers often do not want to allow government regulators onto their farms or to provide regulators with any sensitive farm information.

A recent review of the theory and evidence underlying each of these models suggests they have widely different objectives, structures, and outcomes (Table 1).

IPPM systems, for example, are driven by revenue opportunities (Table 1, column 1). Premiums need to be available to attract participants, and the efforts of participants will be directed towards receiving a share of the premium. Participation in these systems is inevitably voluntary. The lead stakeholders in IPPM systems are private firms seeking to capture the increased value of special traits. The role of the regulatory body will be to ensure that industry standards are in place to prevent consumer fraud from occurring. The information may be asymmetric as only the product seller can know with certainty what level, if any, of cheating has occurred in the delivery of the product. Moral hazards may be present due to the presence of premiums. Effective IPPM systems that span entire supply chains must have accurate two-way information flows. This means that information about purity and quality of the product flows downstream and that information coming from consumer demand due to identified willingness to pay flows upstream. While the information flow in IPPM systems is two-way, the focus of these systems is downstream. Each participant in the system wants to ensure that they are extracting a portion of the value of the special trait, whether they are involved with the production, processing, or retailing of the product. This means that each participant will focus on the needs of the next participant in the supply chain. Market failure can result in fraud charges for mislabeling or improper labeling and can also raise questions among consumers about whether to trust brand names. Testing and auditing will usually be done by second parties acting on behalf of the brand owner or developer of the

Table 1: Comparative Management Systems for IPPM, Segregation, and Traceability

	IPPM	Segregation	Traceability
Objective	Revenue management	Liability management	Product safety
Status	Voluntary	Mandatory	Voluntary or mandatory
Lead stakeholder	Private company	Regulator	Commodity group, standards organization, or regulator
Regulatory agency involvement	Consumer fraud	Regulatory oversight	Consumer fraud
Information	Asymmetric	Full	Asymmetric
Risk	Moral hazard	None	Moral hazard
Information flow	Two way	Two way	One way
Supply chain focus	Downstream	Downstream	Upstream
Penalties for failure in product market	Consumer fraud charges; lost brand value	Criminal prosecution; mandated product recalls	Consumer fraud charges; exclusion from product category
Testing/ auditing	2nd party/brand owner	1st party/regulator	3rd party/standards organization

Source: Smyth et al. 2004.

special trait.

Segregation systems, in contrast, are focused on managing any and all liabilities that may arise through the production and processing of a commodity. Participation is not optional—any producer or firm involved with segregated products will have to comply with standards established that have been approved by the regulatory agency. The private firm will have the responsibility of developing the actual system, but the regulatory agency will be the final arbiter approving the system for field use. Information will be fully disclosed due to the importance of protecting food safety, which will result in the reduction of risks in the system. Segregation systems must have two-way information flow due to compliance with food safety standards. The focus of product delivery within a segregation supply chain will be downstream. Segregated commodities commonly have industrial value, so these products will be supplied to meet the criteria of the processor. The costs of market failure would most definitely see a complete recall of any and all products suspected of being affected. Market failure may also result in criminal prosecution in the most severe instances. Testing and auditing will be vital features of segregation systems and will be conducted by agents of, or acting on behalf of, the regulator. This process will also reinforce the level of trust with foreign export markets.

Finally, traceability systems seek to minimize resulting liabilities by facilitating product recalls if unsafe products enter the supply chain. Participation in a traceability system can be voluntary or mandatory, depending on where in the supply chain the participant is located. The closer the participant is to the start of the supply chain, the more likely it will be that participation is voluntary. The lead stakeholder may be a commodity group demanding greater clarity in or selection of food products, a standards council that is comprised of industry representatives from all sectors of the supply chain, or a regulator seeking to ensure consumer protection. Information may be asymmetric due to the voluntary nature at the start of traceability supply chains. A moral hazard may also exist due to the inability to fully test for some features of traceability. Only traceability systems will have information flows that are only one-way, as these systems are designed to react quickly to food safety concerns. If a product is found to exceed any defined tolerance level at any point in the supply chain, traceability will be used to identify the source of the problem and to locate any and all in-chain and retail products that may be affected. This results in the focus of traceability systems being upstream. Market failures can also result in consumer fraud charges in addition to permanent exclusion from selling into that supply chain. Testing and auditing will be conducted according to the standards developed by third party organizations.

Looked at in aggregate, each of these private supply chains is inextricably linked to both governmental authorities (e.g., courts to adjudicate and regulators to enforce) and collective organizations (e.g., scientific communities and associations of consumers and users). As such, these systems cannot operate independently of each other—they are interconnected.

4.   EVOLVING CHALLENGES OF MANAGING LIABILITIES

The institutions that have been developed over the past 100 years to handle risks and adjudicate liabilities in the food system are coming under increasing scrutiny as they endeavor to handle the uncertainties created by new foods risks, especially those related to GM traits. Scientists have offered far from comforting words—in many cases the results of scientific investigations have raised more questions than answers about the efficacy of our systems. Governments, both through their regulatory systems and through new legislation, have found it next to impossible to clearly define an unambiguous approach that will improve outcomes. Meanwhile, although the marketplace has come a long way in developing new mechanisms for controlling quality, various actors are discovering that many of the critical levers of power are in the hands of others (e.g., governments, the scientific community, and consumer and environmental groups). The main lesson from what has happened to date is that there are no solitudes anymore. Rather, industry, government, and collective groups are inextricably interconnected in the management of liability in the global food system.

Thus far, the situation has been worsening and not improving. Science is becoming more fractious, consumers and citizens more jaded, regulators more frustrated, and industry more anxious. If the technology is going to provide future benefits, the issue of liability will need to be addressed.

While scientists can provide some new answers, they cannot be expected to resolve all of the issues related to a transformative technology. Given the nature of transformative technologies—that is, that all of the potential uses and impacts are unknowable because of the widespread application of the technology—it is unlikely that scientists unaided could frame the appropriate questions to guide their research. Whatever science is done will need to be informed by the needs, desires, constraints, and opportunities of consumers, citizens, regulators, and industry. Basic science will likely need both more resources and new interdisciplinary models of investigation.

Government regulators undoubtedly have a new and more engaged role to play. Given the globalized science and markets, all national regulatory systems are inextricably dependent on others—scientists and their professional associations, firms, other national regulators, and international regulatory authorities—and will be able to effectively and economically undertake accurate risk analysis and enforce appropriate liability regimes only if they collaborate. While there has been some useful work undertaken through

Codex and other technical agencies (Buckingham and Phillips 2001), progress has been very slow and spotty. Governments may need to involve more actors (both scientists and industry) and, at times, relinquish control to others with more at stake.

Finally, industry is inevitably going to need to take a more interest-based approach. Most firms with biotechnologies or products of biotechnology assert that the traditional scientific risk assessment approach should be maintained at all costs. While that is a sound base for risk management, other mechanisms such as IPPM, segregation, and traceability may need to evolve to handle the liabilities related to new product introduction. Ultimately, each of these approaches pursued independently has some potential to manage one or more risk, but none alone will resolve the challenge of managing risks and liabilities of a transformative technology. Liabilities in a global system can and will be effectively managed only through joint action of science, government, industry, and other interested parties.

REFERENCES

Atkinson, M. 1993. *Governing Canada: Institutions and Public Policy.* Toronto: Harcourt Brace Jovanovich Canada Inc.

Bender, K., and L. Hill. 2000. "Producer Alternatives in Growing Specialty Corn and Soybeans." Report No. AE-4732, Department of Agricultural and Consumer Economics, University of Illinois at Urbana-Champaign.

Bender, K., L. Hill, B. Wenzel, and R. Hornbaker. 1999. "Alternative Market Channels for Specialty Corn and Soybeans." Department of Agricultural and Consumer Economics, University of Illinois at Urbana-Champaign. Available online at www.ngfa.org/specialtybk. html.

Brookes, G. 2002. "The Farm Level Impact of Using Bt Maize in Spain." Brookes West, Canterbury, UK. Available online at http://www.bioportfolio.com/pgeconomics/spain_[-] maize.htm.

Buckingham, D., and P.W.B. Phillips. 2001. "Hot Potato, Hot Potato: Regulating Products of Biotechnology by the International Community." *Journal of World Trade* 35(1): 1–31.

Buckwell, A., G. Brookes, and D. Bradley. 1999. "Economics of Identity Preservation for Genetically Modified Crops." Report prepared for the Food Biotechnology Communications Initiative, Wye, UK.

Bullock, D.S., and M. Desquilbet. 2001. "Who Pays the Costs of Non-GMO Segregation and Identity Preservation?" Proceedings of the 5th International Conference of the International Consortium on Agricultural Biotechnology Research (ICABR), Ravello, Italy, June 15–18.

Bullock, D.S., M. Desquilbet, and E. Nitsi. 2000. "The Economics of Non-GMO Segregation and Identity Preservation." Paper presented at the American Agricultural Economics Association Annual Meeting, Tampa, Florida, July 30–August 2.

Codex Alimentarius Commission. 2001. "Matters Arising from Codex Committees and Task Forces: Traceability." Available online at ftp://ftp.fao.org/codex/ccexec49/al0121ee.pdf.

Dickinson, D.L., and D. Bailey. 2001. "Meat Traceability: Are U.S. Consumers Willing to Pay for It?" Journal Paper No. 7458, Utah Agricultural Experiment Station, Logan, UT.

European Union. 2001. "Regulation of the European Parliament and of the Council Concerning Traceability and Labeling of Genetically Modified Organisms and Traceability of Food and Feed Products Produced from Genetically Modified Organisms and Amending Directive 2001/18/EC." Available online at http://europa.eu.int/comm/food/fs/biotech/biotech09_en.pdf.

Good, D., K. Bender, and L. Hill. 2000. "Marketing of Specialty Corn and Soybean Crops." Report No. AE-4733, Department of Agricultural and Consumer Economics, University of Illinois at Urbana-Champaign.

Herrman, T., M. Boland, and A. Heishman. 1999. "Economic Feasibility of Wheat Segregation at Country Elevators." Available online at www.css.orst.edu/nawg/1999/ herrman. html.

Hobbs, J.E., and W.A. Kerr. 1999. "Transaction Costs." In S. Bhagwan Dahiya, ed., *The Current State of Economic Science* (Vol. 4). Rohtak, India: Spellbound Publications PVT Ltd.

Isaac, G.E. 2002. *Agricultural Biotechnology and Transatlantic Trade: Regulatory Barriers to GM Crops.* Wallingford, UK: CABI Publishing.

Jack, D., T. Pardoe, and C. Ritchie. 1998. "Scottish Quality Cereals and Coastal Grains: Combinable Crop Assurance in Action." *Supply Chain Management* 3(3): 134–138.

James, C. 2004. "Review: Global Status of Commercialized Biotech/GM Crops: 2004." ISAAA Briefs No. 32, International Service for the Acquisition of Agri-Biotech Applications, Ithaca, NY.

Leiss, W. 2001. *Understanding Risk Controversies.* Montreal: McGill-Queen's University Press.

Lin, W. 2002. "Estimating the Costs of Segregation for Non-Biotech Maize and Soybeans." In V. Santaniello, R.E. Evenson, and D. Zilberman, eds., *Market Development for Genetically Modified Foods.* Wallingford, UK: CABI Publishing.

Lin, W.W., and D.D. Johnson. 2004. "Segregation of Non-Biotech Corn and Soybeans: Who Bears the Cost?" In V. Santaniello and R.E. Evenson, eds., *The Regulation of Agricultural Biotechnology.* Wallingford, UK: CABI Publishing.

Maltsbarger, R., and N. Kalaitzandonakes. 2000. "Direct and Hidden Costs of Identity Preservation." *AgBioForum* 3(4): 236–242.

Miranowski, J.A., G. Moschini, B. Babcock, M. Duffy, R. Wisner, J. Beghin, D. Hayes, S. Lence, C.P. Baumel, and N.E. Harl. 1999. "Economic Perspectives on GMO Market Segregation." Staff Paper No. 298, Iowa State University (available online at http://agecon.lib.umn.edu/cgibin/detailview.pl?paperid=1768).

Moss, C.B., T.G. Schmitz, and A. Schmitz. 2004. "Differentiating GMOs and Non-GMOs in a Marketing Channel." In V. Santaniello and R.E. Evenson, eds., *The Regulation of Agricultural Biotechnology.* Wallingford, UK: CABI Publishing.

New Zealand Royal Commission on Genetic Modification. 2001. *Report of the Royal Commission on Genetic Modification.* Wellington: Printlink.

North, D.C. 1990. *Institutions, Institutional Change and Economic Performance.* Cambridge: Cambridge University Press.

Phillips, P.W.B. 2001. "International Trade in Genetically Modified Agri-Food Products." In C. Moss, G. Rausser, A. Schmitz, S. Taylor, and D. Zilberman, eds., *Agricultural Globalization, Trade and the Environment.* New York: Kluwer Academic.

Phillips, P.W.B., and G.G. Khachatourians. 2001. *The Biotechnology Revolution in Global Agriculture: Invention, Innovation and Investment in the Canola Sector.* Wallingford, UK: CABI Publishing.

Picciotto, R. 1995. "Putting Institutional Economics to Work: From Participation to Governance." World Bank Discussion Paper No. 304, World Bank, Washington, D.C.

Price, G.K., F. Kuchler, and B. Krissoff. 2004. "E.U. Traceability and the U.S. Soybean Sector." In V. Santaniello and R.E. Evenson, eds., *The Regulation of Agricultural Biotechnology.* Wallingford, UK: CABI Publishing.

Sandman, P.M. 1994. "Mass Media and Environmental Risk: Seven Principles." *Risk: Health, Safety, and Environment* 5 (Summer). Available online at www.piercelaw.edu/risk/vol5/summer/sandman.htm.

Sawhney, V.K. 2001. "Pollen Biotechnology." In G. Khachatourians, A. McHughen, W.-K. Nip, R. Scorza, and Y.-H. Hui, eds., *Transgenic Plants and Crops.* New York: Marcel Dekker Inc.

Shrader-Frechette, K. 1990. "Perceived Risks Versus Actual Risks: Managing Hazards Through Negotiations." Available online at www.piercelaw.edu/Risk/Vol1/fall/ShraderF.htm.

Smyth, S., G. Khachatourians, and P. Phillips. 2002. "The Case for Institutional and Biological Mechanisms to Control GM Gene Flow." *Nature Biotechnology* 20(6): 537–541.

Smyth, S., and P.W.B. Phillips. 2002. "Product Differentiation Alternatives: Identity Preservation, Segregation and Traceability." *AgBioForum* 5(2): 30–42.

Smyth, S., P.W.B. Phillips, W.A. Kerr, and G.G. Khachatourians. 2004. *Regulating the Liabilities of Agricultural Biotechnology.* Wallingford, UK: CABI Publishing.

Spriggs, J., and G.E. Isaac. 2001. *International Competitiveness and Food Safety: The Case of Beef.* Wallingford, UK: CABI Publishing.

Stanbury, W.T. 2000. "Reforming Risk Regulation in Canada." In L. Jones, ed., *Safe Enough? Managing Risk and Regulation.* Vancouver: The Fraser Institute.

Timon, D., and S. O'Reilly. 1998. "An Evaluation of Traceability Systems Along the Irish Beef Chain." In C. Viau, ed., *Long-term Prospects for the Beef Industry.* Ivry-sur-Seine, France: Institut National de la Recherche Agronomique (INRA).

Visser, B., D. Eaton, N. Louwaars, and I. van der Meer. 2001. "Potential Impacts of Genetic Use Restriction Technologies (GURTs) on Agrobiodiversity and Agricultural Production Systems." Study carried out for Food and Agriculture Organization, Rome.

Wilson, N., and W. Clarke. 1998. "Food Safety and Traceability in the Agricultural Supply Chain: Using the Internet to Deliver Traceability." *Supply Chain Management* 3(3): 127–133.

Chapter 6

STATUS OF AGRICULTURAL BIOTECHNOLOGY: AN INTERNATIONAL PERSPECTIVE

Robert E. Evenson
Yale University

Abstract: This paper provides a review of health and environmental effects of GM crops (foods), intellectual property right developments, and regulatory developments for GM crops (foods). Estimates of realized and potential cost reductions are also reported. These estimates show that European Union countries (original EU member countries) have not taken advantage of cost reductions for GM crops to date. More importantly, estimates of potential cost reductions indicate that European Union countries have very low potential cost reduction gains. (Even with 80 percent GM crop adoption cost reductions, gains are around one percent.) However, because EU countries are net importers of many agricultural commodities, EU countries have a major impact on developing countries. Many developing countries in Africa have failed to adopt GM crops because of EU countries' advice regarding the "precautionary principle."

Key words: economic development, technology choice, invention and innovation

1. INTRODUCTION

The introduction of the first commercially successful genetically modified (GM) crop varieties occurred in 1995.[1] As we approach the tenth year of the "Gene Revolution" in crop varieties, an assessment of the developments over the decade has merit. In conducting this assessment, I have organized this paper around several topics. The first is a review of health and environment effects of GM foods. The second is developments in intellectual property rights (IPRs). The third is a review of regulatory conditions for GM crops. The fourth is an assessment of mechanisms for GM crop cost reductions.

---

[1] Earlier GM crops, the "ice-minus" crops and Calgene's "flavor-saver" tomato varieties, were not commercially successful. Bovine Somatotrophin Hormone (BST) was introduced in 1993.

The fifth is an evaluation of the actual cost reduction gains from the adoption of GM crop varieties. The sixth is the development of estimates of cost reduction gain potential for GM crops.

2. HEALTH AND ENVIRONMENTAL EFFECTS OF GM CROPS

It is now well known that an unusual degree of "political" opposition to GM crops and GM foods has emerged over the past decade. Chapter 5 of the *2003–04 State of Food and Agriculture of the Food and Agriculture Organization of the UN* (SOFA) is based on a critical review of several recent scientific assessments of health and environmental impacts of GM crops. The first is the International Council for Science (ICSU) report which itself is based on 50 scientific assessments by national science academies and other independent researchers. Other reports surveyed in the *SOFA* chapter were prepared by the Nuffield Council on Bioethics, the United Kingdom GM Science Review Panel, and the Royal Society farm-scale evaluation report. FAO expert consultations and decisions of the FAO/WHO Codex Alimentarius Commission and the International Plant Protection Convention are also reviewed in the SOFA.

The *2003–04 SOFA* concluded that the scientific evidence for food safety showed no exceptional food safety problems associated with any GM foods currently on the market. There are two main food safety concerns associated with GM foods: (i) the potential introduction of allergens and toxins and (ii) possible negative effects from the consumption of antibiotic-resistant marker genes and viral-promoter genes used in the transformation process. Levels of allergens and toxins can increase through conventional breeding as well as genetic engineering, although only the latter are routinely tested. So far no allergenic or toxic effects from the consumption of GM foods have been confirmed anywhere in the world. Any risks associated with the use of antibiotic resistant and viral genes in the development of GM foods are very small; nevertheless, their use has been discouraged and scientists have developed "clean" methods of genetic transformation that eliminate these substances. Scientists generally agree that existing GM foods are as safe as their conventional counterparts, and that new foods should be tested on the basis of their product characteristics rather than on the method used to develop them.

The environmental safety of GM crops is also considered in *SOFA 2003–04*. Scientists generally agree on the types of hazards that are associated with GM crops, although they differ regarding their likelihood and potential severity. Most of these hazards can also occur with conventionally bred crops, and scientists generally agree that GM crops should be evaluated on a case-by-case basis depending on the crop, trait, and ecosystem where it will be used. Three main environmental hazards potentially associated with GM crops are (i) outcrossing or gene flow to related crops or wild species, (ii) harm to non-target organisms, and (iii) the emergence of resistant pests.

Gene flow occurs naturally between varieties of the same crop and between crops and sexually compatible wild species. This has occurred ever since farmers began selecting seeds to improve crop performance thousands of years ago. It occurs, therefore, with GM crops as well. In purely scientific terms, such gene flow is a cause for concern only when the resulting hybrid achieves a "competitive advantage" that would enable it to invade natural ecosystems or other agricultural fields. Since a trait like herbicide tolerance confers an advantage only in the presence of a specific herbicide, it is highly unlikely that this trait would survive in nature, and in any case other herbicides could be used to control it. Furthermore, GM crops have no wild relatives in many production areas so gene flow into natural ecosystems would be impossible. Similar reasoning based on the trait, crop, and location can be used to assess the risk of gene flow associated with other GM crops.

Gene flow from GM crops could pose an economic problem, especially for crops produced and certified as organic if such "contamination" results in the loss of organic status. Since GM pollen drift is distance limited, it is expected that this problem can be handled by imposing minimum distances between GM crops and organic crops, although the cost of doing so will depend on the tolerance limits for the accidental presence of GM traits in organic products.

The second environmental concern involves the potential for GM traits or associated farming practices to harm non-target organisms. The potential for Bt corn to harm Monarch butterfly larvae or for herbicide-tolerant crops to reduce the presence of weeds in farmers' fields—and hence the availability of food for farmland birds—are examples of such concerns. So far these traits have not caused serious harm to non-target organisms; on the contrary, compared with conventional cropping practices, they can be beneficial depending on how they are managed. Nevertheless, much remains unknown about the potential ecosystem effects of GM crops, and scientists recommend continued monitoring.

A final concern is that insects may develop resistance to Bt crops, leading to the emergence of "super-pests." This has not happened to date, in part because of the use of "refugia" where some proportion of a GM crop is planted to non-GM varieties and to the development of second-generation GM crops that contain two Bt genes (which dramatically reduces the probability that resistance will develop). The emergence of pests that are resistant to Bt crops would not necessarily cause an ecological problem since other insecticides could be used against them.

Developing countries are confronted with two conflicting recommendations regarding GM crops. The recommendation from most North American universities and policy research centers is that developing countries should develop the food safety and environmental safety regulatory structures to take advantage of actual and potential cost reductions afforded by GM crops and engage in active "gene-renting" in regions where such renting makes economic sense. The recommendation from most European Union universities and policy research centers is that developing countries should take a "precautionary" approach to GM crops.[2] This means slower development of regulatory structures for the introduction of GM crops. It also has implications for the development of the skills associated with developing GM crops. This is because modern plant breeding encompasses "marker-aided selection" techniques that facilitate conventional plant breeding.[3]

3. DEVELOPMENTS IN INTELLECTUAL PROPERTY RIGHTS

As GM techniques were developed, a parallel strengthening of IPRs, particularly patent rights, took place in the United States. This strengthening took place entirely through "case law." One of the key cases was Diamond vs. Chakarbaty [447 U.S. 303 (1980)], where the Supreme Court ruled that living multicellular organisms (plants and animals) were not excluded from patent protection. This ruling was followed by cases enabling patent protection for plants [*ex parte* Hibbard 227 U.S. Pa. 443 (1985)] and animals [*ex parte* Allen 2 U.S. Pa. 2d 1495 (1986)].[4]

Another development of importance was the inclusion of IPRs in the GATT Uruguay Round, the WTO-TRIPS agreement. The WTO-TRIPS agreement has two implications for plant breeders in both the public sector and the private sector in developed and developing countries. The first is that it allows signatory countries to exclude plant varieties from patentability, provided they "provide for the protection of plant varieties either by patents or by an effective *sui generis* system or by any combination thereof." The second is that the patent system will almost certainly cover biotechnology inventions in the spirit of the case law expansion in the United States.

Developing countries generally oppose strengthening of IPRs and fought for the *sui generis* system provision. This is generally interpreted to mean some form of "breeders' rights" system. Developing countries also oppose the provisions of the case law based expansion of the U.S. system. But in-

---

[2] See Conway (2000) and Duffy (2001) for assessments.

[3] Marker-aided selection techniques have been utilized in many developing country programs, but the sophistication of use is dependent on the acquisition of science-based skills. Dreher et al. (2000) provide a case study of marker-aided selection at the International Center for Maize and Wheat Improvement (CIMMYT).

[4] After Hibbard, crop varieties were given patent rights. Animal patents were initiated by the "Harvard Mouse" patent (licensed to DuPont). The U.S. Patent and Trademark office had a moratorium on further animal patenting but in 2000 resumed granting patents on animals.

creasingly, it is becoming clear that the term "plant varieties" is unlikely to apply to GM plant varieties or to GM techniques.[5]

A third development is reflected in the Convention on Biodiversity (CBD), where a new right (supported by developing countries) was introduced. The CBD states that states have "a sovereign right to natural resources" and that "the authority to determine access to genetic resources rests with the national government and is subject to national legislation." The CBD further notes that the genetic resources subject to such legislation apply to "countries of origin" of such resources.

The CBD thus raises the prospect that a country can deny access to its own "country of origin" genetic resources. This has profound implications for plant breeders because virtually all "modern" plant varieties incorporate landraces from multiple countries (Evenson and Gollin 1997). The FAO has introduced provisions for an agreement to preserve the traditional free exchange of genetic resources between countries.

4. GM CROP APPROVAL PROCESSES

Two major GM "traits" for GM crops—herbicide tolerance and insect resistance (Bt)—are presently available to farmers. Four major crops—soybeans, maize, cotton, and canola—have these GM traits. Many more crops have been approved for commercial use in some countries.[6] Effectively, a country must go through approval processes for commercial production [P], regulatory approval [A], field study [F], and lab/greenhouse study [L].[7]

Table 1 depicts current approval stages for field crops. At present, only four crops have full approval for commercial sales, although sugar beets, rice, and flax are nearing approval in Canada and the United States.

Table 2 shows comparable data for vegetable crops and Table 3 shows data for fruits. Table 4 shows data for other crops. As these tables indicate, the approval process for commercial production varies greatly by country. A number of developing countries have approved commercial production of GM crops, notably Argentina, Mexico, China, South Africa, Brazil, Uruguay, India, Colombia, the Philippines, Paraguay, and Chile. A number of developing countries, however, are in earlier stages in the approval process. These countries include Kenya, Bangladesh, Morocco, Pakistan, Malaysia, Costa Rica, Bolivia, Zimbabwe, Venezuela, Thailand, Cuba, Belize, Honduras,

---

[5] The European Patent Office has recently developed rulings indicating that GM plant varieties are unlikely to be included in the *sui generis* provision.

[6] The United States has granted approvals for the following crops: canola, chicory, cotton, flax and linseed, maize, melon, papaya, potatoes, rice soybeans, squash, sugar beets, tobacco, and tomatoes.

[7] This section relies heavily on Runge and Ryan (2004).

Table 1. Global Biotech Activity: Field Crops – Highest Level of Biotech Development

Field crops by country	Soybean	Cotton	Maize	Canola	Sugar beet	Rice	Flax	Wheat
Canada	P	A	P	P	A	A	A	F
U.S.	P	P	P	P	A	A	A	F
Australia	A	P	A	A	A			F
West Europe (15/15)	A	F	P	A	F	F		F
Argentina	P	P	P		F			F
Mexico	A	P	F	F		F	F	F
China	F	P	F	L	L	F		L
Japan	A	A	A	A	A	F		L
South Africa	P	P	P	F				
Brazil	P	F	F			F		
East Europe (8/13)	P		A	F	L		L	F
Indonesia	F	A	F			L		
Uruguay	P		P					
Egypt		A	F	A				F
India		P		F		L		
Colombia		P						
Philippines			P			L		
Paraguay	P							
Chile	P		P					
South Korea	A		A					
Honduras			A					
Belize	F	F	F					
Cuba			L			L		
Thailand		F				F		
Venezuela						L		
Zimbabwe		F						
Bolivia	F	F						
Costa Rica			L			F		
New Zealand				F				
Malaysia						L		
Pakistan		L				L		
Morocco								L
Bangladesh						L		
Kenya			L					

cont'd.

Table 1 (cont'd.)

Field crops by country	Sugar cane	Barley	Alfalfa	Cassava	Sunflower	Clover	Safflower	Sorghum
Canada		F	F		F	F	F	
U.S.	F	F	F				F	
Australia	F	F				F		
West Europe (15/15)		F	F		F			
Argentina	L	L	F		F			
Mexico								
China		L						L
Japan								
South Africa	F							
Brazil	F	L						
East Europe (8/13)		L	F		F			
Indonesia	L			L				
Uruguay								
Egypt	F	L						
India								
Colombia				L				
Philippines								
Paraguay								
Chile								
South Korea								
Honduras								
Belize								
Cuba	F							
Thailand				L				
Venezuela	L			F				
Zimbabwe				F				
Bolivia								
Costa Rica								
New Zealand								
Malaysia								
Pakistan								
Morocco								
Bangladesh								
Kenya								

Notes: P = commercial production, A = regulatory approval, F = field study, L = lab/greenhouse.

Table 2. Global Biotech Activity: Vegetables – Highest Level of Biotech Development

Vegetables by country	Potato	Tomato	Squash	Pepper	Pea/Bean	Lettuce	Cucumber
West Europe (13/15)	F	F	F		F	F	
U.S.	A	A	P		F	F	F
Canada	A	A	A				
Australia	A	F			F	F	
Japan	A	A			F	L	F
China	F	P		P			
Mexico	F	A	F	F			
Brazil	F	F			F	L	
Egypt	F	F	F		L		F
Thailand		F		F	L		
Argentina	F	F					
East Europe (10/13)	F	L			F		
Cuba	F	L					
Zimbabwe	F						
Bolivia	F						
Peru	F						
South Africa	F						
Kenya	F						
Guatemala		F					
New Zealand							
South Korea				F			
Indonesia	L	L		L			
Malaysia				L	L		
India	L	L					
Chile	L	L					
Colombia	L	L					
Bangladesh					L		
Philippines		L					
Tunisia	L						

cont'd.

Egypt, Tunisia, and Guatemala. Thus, 11 developing countries have given approval to farmers to plant GM crops. An additional 16 countries are in earlier stages of approval. Approximately 60 developing countries with populations over one million report lab/greenhouse and field studies of GM crops.

Table 2 (cont'd.)

Vegetables by country	Cabbage	Carrot	Eggplant	Onion	Cauliflower	Broccoli	Spinach
West Europe (13/15)	F	F	F		F	F	F
U.S.				F			
Canada							
Australia							
Japan					F	F	
China	F	L					
Mexico							
Brazil		F					
Egypt							
Thailand							
Argentina							
East Europe (10/13)							
Cuba							
Zimbabwe							
Bolivia							
Peru							
South Africa							
Kenya							
Guatemala							
New Zealand				F			
South Korea							
Indonesia							
Malaysia			L				
India	L		F				
Chile							
Colombia							
Bangladesh							
Philippines							
Tunisia							

Notes: P = commercial production, A = regulatory approval, F = field study, L = lab/greenhouse.

Table 3. Global Biotech Activity: Fruits – Highest Level of Biotech Development

Fruits by country	Papaya	Melon	Banana	Pineapple	Apple	Grape	Plum	Strawberry	Watermelon	Citrus	Cherry	Cantaloupe	Kiwi	Raspberry	Mango	Coconut
U.S.	P	A	F		F		F		F							
West Europe (8/15)		F			F	F	F	F	F	F	F	F	F	F		
Australia	F			F	F	F										
Canada	A					F										
Mexico	F	F	F	F												
Cuba	F		L	L							L					
Philippines	L		F												L	L
China	F	F														
Egypt		F	L									F				
Japan	L	F							L							
East Europe (3/13)						L	F									
South Africa									F							
Brazil	F															
Malaysia	L	L	L	L												
Chile		L			L	L	L									
Venezuela	L		L												L	
Colombia			L													
Costa Rica			L													
Bangladesh	L															
Thailand	L															

Notes: P = commercial production, A = regulatory approval, F = field study, L = lab/greenhouse.

5. MECHANISMS FOR COST REDUCTIONS IN DEVELOPED AND DEVELOPING COUNTRIES

Five mechanisms are specified in this section.

5.1. GMs for Rent: Developed Country Suppliers

This mechanism entails negotiations between private agro biotech suppliers of crop GMs and farmers in developing countries. The supplier provides the GM in return for a technology fee or a seed price premium. The supplier may incorporate the GM (e.g., Bt) product into more than one crop variety (e.g., several cotton varieties). These varieties may have been developed by

Table 4. Global Biotech Activity: Other Crops – Highest Level of Biotech Development

Other crops by country	Tobacco	Chicory	Mustard	Peanut	Coffee	Lupins	Oilseed poppy	Olive	Oil palm	Cocoa	Garlic
U.S.	P	A		F	F						
West Europe (9/15)	A	A	F					F			
Australia			F			F	F				
China	F			F							
Brazil	F									L	
Canada			F								
East Europe (3/13)	F										
South Korea	F										
India	F										
Mexico	F										
Indonesia	L			L	L				L	L	
Chile	L										L
Bangladesh	L			L							
Malaysia	L								L		
Venezuela					L						
Philippines	L										
Argentina	L										
Cuba					L						
Japan	L										

Notes: P = commercial production, A = regulatory approval, F = field study, L = lab/greenhouse.

public national agricultural research systems (NARS) or international agricultural research center (IARC-NARS) programs or by private seed companies. The supplier may even provide the rDNA technical services, so that little or no rDNA technical skills are actually required in the host economy.

5.2. GMs for Rent: Developing Country Suppliers

This mechanism is similar to the first mechanism except that a private firm or public NARS program in a developing country is the GM product supplier. Public NARS suppliers may choose to set different technology fees for domestic and foreign purchasers.

5.3. GMs for Rent: International Agency Purchase

For this mechanism, an International Donor Agency negotiates with a GM product supplier to provide specific GM products to farmers in specific countries. The International Donor Agency makes payments to the GM product supplier. Farmers may then utilize the GM product without paying a technology fee.

5.4. GM Product Germplasm Conversion

Most GM products being marketed today can be converted to germplasm in the form of "breeding lines." Once the initial "transgenic" incorporation of DNA into a breeding line is made, the GM product is expressed in the variety and in most cases will be expressed in progeny varieties where the transgenic line is utilized as a parent in a conventional cross. This effectively converts the GM product into a form where "conventional" breeding methods can be utilized to replicate the GM product. This germplasm conversion could be utilized by IARC programs in much the same way that wide crossing methods were used to incorporate "wild" (i.e., uncultivated) species' DNA into breeding lines.

5.5. Quantitative Enhancement: Genomics, Proteonomics Research

This mechanism entails "quantitative" trait breeding. Some prospects for quantitative trait locus (QTL) breeding have been developed to date, but the science of genomics and proteonomics is still in its infancy. There are, however, prospects for important gains in achieving gains in photosynthetic efficiency in plants. This research is very demanding of skills and creativity.

It should be noted that at present, GM products are basically "qualitative trait" products. And qualitative trait products endow plants with specific cost advantages that vary from environment to environment, but are "static" in nature. That is, the cost advantage gains are of a "one-time" nature. They do not grow over time. It is possible to "stack" more than one GM product in a crop variety, but stacking does not produce cumulative gains.

It is sometimes said that the Gene Revolution will replace the Green Revolution. But this will not happen until and unless this mechanism enables breeders to produce "dynamic" gains in generations of varieties. Until such time, the Gene Revolution's GM products can only complement conventional Green Revolution breeding. This complementarity takes the form of installing "static" GM products on the dynamic generations of varieties produced by conventional Green Revolution methods.[8]

---

[8] The Roundup Ready product produced by Monsanto has been "installed" on approximately 1,500 soybean varieties produced by 150 seed production companies.

6. ESTIMATES OF COST REDUCTION GAINS FOR GM CROPS

A number of cost studies have now been undertaken and can form the basis for estimates of cost reduction gains when GM crops are introduced.[9],[10] Tables 5 and 6 summarize consensus estimates from these studies for the two major GM traits, herbicide tolerance and insect resistance. Note that these estimates are related to adoption levels. At present few rice GM products are available to farmers, but estimates of cost reduction are available (I will use these in computations of potential gains presented below).

Most soybean, cotton, canola, and rice varieties are actually "modern" varieties in the sense that they have had significant breeding activity over the past 40 years or more. But not all maize varieties are modern. Many are landrace selections by farmers. It is unlikely that the addition of GM traits to landrace selections by farmers has any cost reduction value.

Soybeans. Soybeans were planted on 88 million hectares worldwide in 2003/04, with global production estimated at 190 million metric tons, and the world price averaging $250 per metric ton. The top five biotech countries represented 84 percent of the land area planted to soybeans and 90 percent of production.[11] More than half (54 percent) of soybean production in the top five biotech countries is from biotech varieties (Table 7). Total biotech soybean market value in 2003/04 was $23.5 billion—the highest of any biotech crop. The United States had the largest area in soybeans and highest biotech crop value ($13.3 billion). Brazil had the next largest biotech soy area, but due to a low (official) adoption rate, generated only $1.6 billion in biotech market value. Some reports suggest that the real biotech adoption rate in Brazil is as high as 30 percent. Argentina grew $8.3 billion in biotech soybeans in 2003/04. China grew 8.7 million hectares of conventional soybeans, but had no biotech production. The Canadian soybean area is just over a million hectares, and about half was planted to biotech varieties.

Maize. Maize was grown on 140 million hectares worldwide in 2003/04, producing 614 million metric tons, at an average world price of $100 per metric ton. The top five biotech countries represent 70 percent of worldwide maize production and 49 percent of the global maize land area. Biotech varieties are grown on 19 percent of maize production land in the top five

---

[9] Herbicide tolerance estimates are from Carpenter and Gianessi (2001), Fernandez-Cornejo and McBride (2000), and Qaim and Traxler (2004).

[10] Insect resistance estimates are based on Traxler et al. (2003), Qaim and de Janvry (2003), Qaim and Zilberman (2003), Pray and Huang (2003), Gianessi et al. (2002), and Falck-Zepeda, Traxler, and Nelson (1999).

[11] These data are from Runge and Ryan (2004).

Table 5. "Estimates" of Cost Reduction for GM Crops: Herbicide Tolerance

Crop	Country	Adoption level (percent reduction in costs)				Adoption level 2003 (percent)
		0–30	20–50	50–75	75–100	
Soybeans	U.S.	15	15	12	10	81
	Argentina	15	12	10	10	99
Cotton	U.S.	10	7	5	5	59
Canola	Canada	12	10	8	8	75
Maize	U.S.	12	10	8	8	15

Table 6. "Estimates" of Cost Reduction for GM Crops: Insect Resistance (Bt)

Crop	Country	Adoption level (percent reduction in costs)			Adoption level 2003 (percent)
		0–30	20–50	50–75	
Maize	U.S.	13	10	6	29
	South Africa	10	8	8	3
Cotton	U.S.	15	13	12	41
	China	30	25	20	58
	India	30	20	10	10
	Mexico	30	20	10	20
	South Africa	30	20	10	20
Other crops					
Rice		8	8	6	0

biotech countries, which collectively produced $11.2 billion in biotech maize (Table 8).

The United States is the leading biotech maize producer, with $10.3 billion in production market value. Argentina has a modest area of land planted to maize.

Cotton. Cotton was planted on 32.6 million hectares worldwide in 2003/04. Production (lint only) is estimated at 93.5 million bales of 480 pounds each. The adjusted world price averaged 59 cents per pound. Half of the world's cotton production takes place in the top five biotech countries, and 61 percent of that is from biotech varieties (Table 9).

Table 7. Global Biotech Soybean Value: Leading Countries

Soybean 2003/04 Price = $250/MT	Crop Area[a] M Ha	Production[b] MMT	Biotech Adoption Rate	Biotech-Related Crop Value[c]	Estimated Cost Reduction
Five countries	74.2	171.8	54%	$23.5 billion	$3.52 billion
U.S.	29.2	65.8	81%	$13.3 billion	$1.86 billion
Brazil	21.3	53.5	30%	$4.0 billion	$0.60 billion
Argentina	14.0	34.0	98%	$8.3 billion	$0.97 billion
China	8.7	16.2		--	--
Canada	1.1	2.3	50%	$284 million	$0.085 billion
Rest of the World	*13.8*	*18.3*	--	--	--

Source: Runge and Ryan (2004).

[a] Area in million hectares.
[b] Production in million metric tons.
[c] Assumes world price of $250/metric ton.

Table 8. Global Biotech Maize Value: Leading Countries

Maize 2003/04 Price = $100/MT	Crop Area[a] M Ha	Production[b] MMT	Biotech Adoption Rate	Biotech-Related Crop Value[c]	Estimated Cost Reduction
Five countries	68.5	434.5	19%	$11.2 billion	$0.685 billion
U.S.	28.8	256.9	40%	$10.3 billion	$0.66 billion
China	23.5	114.0	--	--	--
Brazil	12.6	41.5	--	--	--
Argentina	2.1	12.5	40%	$500 million	$0.016 billion
Canada	1.2	9.6	40%	$384 million	$0.009 billion
Rest of the World	*71.9*	*179.5*	--	--	--

Source: Runge and Ryan (2004).

[a] Area in million hectares.
[b] Production in million metric tons.
[c] Average world price of $100/metric ton.

The global value of the biotech cotton in 2003/04 was $7.8 billion. China has the most area in cotton, the highest production and yields, and generates the most biotech cotton market value. The United States has almost as much area as China, higher adoption, lower yields, but essentially the same biotech production value. Argentina grew $75 million in biotech cotton on a relatively modest land area. This assumes a 60 percent adoption rate, although some reports suggest it may be as low as 20 percent. Brazil has more area in cotton production than Argentina, but no biotech adoption.

Table 9. Global Biotech Cotton Value: Leading Countries

Cotton 2003/04 Price = $0.59/lb.	Crop Area[a] M Ha	Production[b] MMT	Biotech Adoption Rate	Biotech-Related Crop Value[c]	Estimated Cost Reduction
Five countries	11.2	46.7	61%	$7.8 billion	$1.67 billion
U.S.	5.1	22.4	62%	$3.9 billion	$1.02 billion
China	4.9	18.3	73%	$3.8 billion	$0.53 billion
Brazil	1.0	5.7	--	--	--
Argentina	0.3	0.4	60%	$75 million	$0.02 billion
India	--	--	20%	$500 million	$0.6 billion
Rest of the World	*21.4*	*46.8*	--	--	--

Source: Runge and Ryan (2004).

[a] Area in million hectares.
[b] Production in million metric tons.
[c] Assumes world price of $0.59/pound.

Canola. Canola (or rapeseed) was planted on 26 million hectares world-wide in 2003/04, with total production estimated at 39 million metric tons, and an average world price of $285 per metric ton. The top five biotech countries account for half the worldwide land area devoted to canola and half the global production (Table 10).

Among these top five countries, 28 percent of canola was a biotech variety. In 2003/04 the worldwide market value of the biotech canola crop was $1.4 billion. China grows the most canola among the five countries, but none in biotech varieties. Canada has the next largest land area planted to canola worldwide, but the majority of this crop is biotech, generating nearly $1.3 billion in biotech market value. The United States, by contrast, has modest canola production, but still produces $1.38 million in biotech canola value. Argentina and Brazil have no meaningful canola production and no biotech varieties in use.

Other countries grow biotech varieties of soybeans, cotton, and maize, apart from the five leading nations, and James (2002) identifies 13 countries with biotech crop production, with 8 of these at meaningful levels.[12] Combined, these countries grew more than 600,000 hectares of biotech crops commercially in 2003/04, producing an additional $160 million in global biotech crop value.

---

[12] Clive James is Director of the International Service for the Acquisition of Agri-Biotech Applications (ISAAA). Annual reports are submitted. See James (2002).

Table 10. Global Biotech Canola Value: Leading Countries

Canola 2003/04 Price = $285/MT	Crop Area[a] M Ha	Production[b] MMT	Biotech Adoption Rate	Biotech-Related Crop Value[c]	Estimated Cost Reduction
Five countries	12.6	18.8	28%	$1.43 billion	$0.17 billion
China	7.5	11.4	--	--	--
Canada	4.7	6.7	68%	$1.29 billion	$0.15 billion
U.S.	.4	.7	73%	$138 billion	$.02 billion
Argentina	--	--	--	--	--
Brazil	--	--	--	--	--
Rest of the World	13.4	20.2	--	--	--

Source: Runge and Ryan (2004).

[a] Area in million hectares.
[b] Production in million metric tons.
[c] Average world price of $285/metric ton.

7. ESTIMATES OF *POTENTIAL* GAINS FROM RENTING GM CROPS

Estimates of potential gains possible from "renting" GM crops in maize, cotton, canola, soybeans, and rice are computed and reported in Table 11. Insect-resistant (BT) GM rice varieties will be available to farmers in 2005 or 2006. China is likely to release BT rice in 2005, and this will be the first publicly released GM crop variety.

The estimates reported are made for each country based on 2004 production data.[13] The estimates presume 80 percent adoption levels. That is, no gains are computed for adoption levels above 80 percent. Table 11 reports gains as a percentage of crop value. This percentage is roughly proportional to the area planted to the five crops. Estimates of annual cost reductions in millions of U.S. dollars are reported.

Table 11 reports wide diversity in the cost reduction values of GM crop-renting as a percentage of the crop value of the country. This percentage figure suggests that many countries have little stake in renting GM crop varieties. Specifically, only 5 of 15 countries in Western Europe have more than a one percent potential for cost reduction gains. For all 15 Western European countries the potential gains are only one percent of crop value.

Contrast this with the U.S. value of about 9 percent and with Latin American potential gains. Argentina, Brazil, Paraguay, Bolivia, and Costa Rica all

---

[13] The estimated cost reduction calculations were 14 percent for maize, 17 percent for cotton, 10 percent for soybeans, 10 percent for canola, and 8 percent for rice. These were applied to 80 percent of production and valued at international prices.

Table 11. Estimates of Cost Reduction Potential by Country

Country	Million $	% Crop Value	Country	Million $	% Crop Value
North America			*Eastern Europe*		
Canada	300	4.43	Albania	3	n/a
U.S.	6,000	8.95	Armenia	1	n/a
			Azerbaijan	12	n/a
Oceania			Bosnia-Hrzg	7	n/a
Australia	78	1.14	Bulgaria	17	1.06
New Zealand	2	.38	Croatia	22	--
Papua New Guinea	1	.10	Czech Rep.	23	n/a
Japan	82	1.38	Hungary	80	3.73
			Poland	44	.66
Latin America			Slovakia	9	n/a
Southern Cone			Macedonia	2	n/a
Argentina	1,055	8.43	Romania	121	3.45
Brazil	2,205	6.81	Serbia	70	n/a
Chile	13	.62	Slovenia	5	n/a
Paraguay	137	9.14			
Uruguay	15	2.60	*Africa*		
Andean			North Africa		
Bolivia	53	6.36	Algeria	1	.09
Colombia	44	1.00	Egypt	153	2.40
Ecuador	56	2.80	Morocco	3	.28
Peru	49	2.81	Tunisia	1	.17
Venezuela	26	1.93			
			East Africa		
Central America			Ethiopia	44	1.51
Costa Rica	43	9.93	Kenya	38	2.49
El Salvador	1	.35	Madagascar	24	1.93
Guatemala	2	.60	Sudan	28	2.30
Honduras	6	2.30	Uganda	51	1.83
Mexico	271	2.34			
Nicaragua	1	.48	Central Africa		
Panama	1	.20	Burundi	1	.17
			Cameroon	32	2.51
Caribbean			Chad	41	7.70
Cuba	9	1.45	Congo (Zaire)	26	2.50
Dominican Republic	1	.16	Ctrl. African Rep.	7	2.24
Haiti	5	2.74	Rwanda	3	.49
Jamaica	1	.87			
			West Africa		
Western Europe			Benin	77	11.00
Austria	23	2.25	Burkina Faso	72	10.25
Belgium	7	.60	Côte d'Ivoire	55	1.79
Denmark	8	.60	Guinea	12	2.00
Finland	3	.60	Ghana	22	.89
France	270	1.87	Mali	85	12.14
Germany	119	1.92	Mauritania	1	1.36
Greece	81	1.96	Niger	1	.20
Italy	145	1.22	Nigeria	190	1.59
Netherlands	3	.22	Senegal	15	2.45
Norway	2	.85	Sierra Leone	2	.92
Portugal	11	.97	Togo	32	2.53
Spain	74	.96			
Sweden	4	.50			
Switzerland	4	.92			
UK	37	.83			

cont'd.

Table 11 (cont'd.)

Country	Million $	% Crop Value	Country	Million $	% Crop Value
Southern Africa			*South East Asia*		
Angola	8	1.45	Cambodia	34	3.93
Botswana	1	1.42	Indonesia	526	2.11
Malawi	32	3.97	Laos	21	5.18
Mozambique	32	3.95	Malaysia	17	.37
Namibia	1	.64	Myanmar	219	n/a
South Africa	130	4.06	Philippines	152	1.48
Tanzania	85	3.80	Thailand	247	2.38
Zambia	22	1.95	Vietnam	287	3.41
Zimbabwe	72	10.38			
			East Asia		
West Asia			China	5,315	3.80
Afghanistan	1	.05	North Korea	40	Na
Iran	63	.82	South Korea	52	Na
Jordan	1	n/a	Mongolia	1	2.60
Saudi Arabia	1	.13			
Syria	32	1.43	*Former Soviet Union*		
Turkey	139	1.00	Belarus	2	n/a
Yemen	4	1.10	Estonia	2	n/a
			Georgia	6	n/a
South Asia			Kazakhstan	35	n/a
Bangladesh	285	4.27	Kyrgyzstan	11	n/a
India	2,353	2.81	Latvia	1	n/a
Nepal	51	3.21	Lithuania	3	n/a
Pakistan	471	4.57	Russian Federation	38	n/a
Sri Lanka	22	1.16	Tajikstan	41	n/a
			Turkmanistan	103	n/a
			Ukraine	89	n/a
			Uzbekistan	201	n/a

have significant cost reduction potential. For Latin America as a whole, potential cost reduction gains are roughly 5 percent (as in Canada).

By contrast, potential gains in Eastern Europe and Oceania are modest.[14] West Asia appears to have modest potential as well. But South, Southeast, and East Asian countries have significant cost reduction potential.

North African countries also have low potential gains. East and Central African countries have modest potential for GM cost reduction. Several West African and Southern African countries have significant cost reduction potential from renting GM crops.

8. POLICY IMPLICATIONS

The calculations of cost reduction potential by country on the one hand explain much of the current regulatory aspects covered in Tables 1 to 4. West-

---

[14] Data from former Soviet republics do not allow calculation of this figure.

ern European countries have little to gain from GM crop approval. The United States, Canada, Argentina, Brazil, Paraguay, Bolivia, and Costa Rica have much to gain. Several countries in South, Southeast, and East Asia do as well, and most of these countries are further along in regulatory system development.

For a number of countries in Africa it appears that cost reduction potential is high, but regulatory systems in many countries (South Africa being an exception) are not well developed. These countries may be paying a high price for following the "precautionary principle" advice of West European countries. This circumstance is further accentuated by the fact that Western European countries have very low stakes in terms of cost reduction potential. But they do have influence over both aid and trade support.

REFERENCES

Carpenter, J.E., and L.P. Gianessi. 2001. *Agricultural Biotechnology: Updated Benefits Estimates*. National Center for Food and Agricultural Policy, Washington, D.C.

Conway, G. 2000. *Crop Biotechnology: Benefits, Risks and Ownership.* Speech by the President of the Rockefeller Foundation delivered at the OECD Edinburgh Conference on the Scientific and Health Aspects of Genetically Modified Foods. Available online under "news archive" at http://www.rockfound.org (accessed March 2004).

Dreher, K., M. Morris, M. Khairallah, J.M. Ribaut, S. Pandey, and G. Srinivasan. 2000. *Is Marker-Assisted Selection Cost-Effective Compared to Conventional Plant Breeding Methods? The Case of Quality Protein Maize.* Paper presented at the 4th ICABR Conference on Economics of Agricultural Biotechnology, Ravello, Italy, August 24–28.

Duffy, M. 2001. *Who Benefits from Biotechnology?* Paper presented at the American Seed Trade Association meeting, Chicago, IL, December 5–7. Available online at http://www.leopold.iastate.edu/pubinfo/papersspeeches/biotech.html (accessed March 2004).

Evenson, R.E., and D. Gollin. 1997. "Genetic Resources, International Organizations, and Rice Varietal Improvement." *Economic Development and Cultural Change* 45(3): 471–500.

Falck-Zepeda, J.B., G. Traxler, and R.G. Nelson. 1999. *Rent Creation and Distribution from the First Three Years of Planting Bt Cotton.* ISAAA (International Service for the Acquisition of Agri-biotech Applications) Briefs No. 14, Ithaca, NY.

Fernandez-Cornejo, J., and W.D. McBride. 2000. *Genetically Engineered Crops for Pest Management in U.S. Agriculture: Farm Level Effects.* Agricultural Economic Report No. 786, Economic Research Service, U.S. Department of Agriculture, Washington, D.C.

Gianessi, L.P., C.S. Silvers, S. Sankula, and J.E. Carpenter. 2002. *Plant Biotechnology: Current and Potential Impact for Improving Pest Management in U.S. Agriculture: An Analysis of 40 Case Studies.* National Center for Food and Agricultural Policy, Washington, D.C.

James, C. 2002. *Preview: Global Status of Commercialized Transgenic Crops: 2002.* ISAAA (International Service for the Acquisition of Agri-biotech Applications) Briefs No. 27, Ithaca, NY.

Pray, C.E., and J. Huang. 2003. "The Impact of Bt Cotton in China." In N. Kalaitzandonakes, ed., *The Economic and Environmental Impacts of Agbiotech: A Global Perspective*. New York: Kluwer-Plenum Academic Publishers.

Qaim, M., and A. de Janvry. 2003. "Genetically Modified Crops, Corporate Pricing Strategies, and Farmers' Adoption: The Case of Bt Cotton in Argentina." *American Journal of Agricultural Economics* 85(4): 814–828.

Qaim, M., and G. Traxler. 2005. "Roundup Ready Soybeans in Argentina: Farm Level, Environmental, and Welfare Effect." *Agricultural Economics* 32 (1): 73–86.

Qaim, M., and D. Zilberman. 2003. "Yield Effects of Genetically Modified Crops in Developing Countries." *Science* 299(5608): 900–902.

Runge, C.F., and B. Ryan. "The Global Diffusion of Plant Biotechnology: International Adoption and Research in 2004." Report prepared for the Council on Biotechnology Information, Washington, D.C. (December 8, 2004).

Traxler, G., S. Godoy-Avila, J. Falck-Zepeda, and J. Espinoza-Arellano. 2003. "Transgenic Cotton in Mexico: Economic and Environmental Impacts." In N. Kalaitzandonakes, ed., *The Economic and Environmental Impacts of Agbiotech: A Global Perspective*. New York: Kluwer-Plenum Academic Publishers.

Chapter 7

INTERACTIONS BETWEEN TRADE POLICIES AND GM FOOD REGULATIONS

Kym Anderson
World Bank, University of Adelaide, and Centre for Economic Policy Research

Abstract: Agricultural biotechnologies, and especially transgenic crops, have the potential to offer higher incomes to biotech firms and farmers, and lower-priced and better quality food for consumers. However, the welfare effects of adoption of genetically modified (GM) food and feed crop varieties are being affected not only by some countries' strict regulations governing GM food production and consumption, but also by their choice of food trade policy instruments. Specifically, notwithstanding the ending of the European Union's GM moratorium in April 2004, the continuing use by the EU of strict labeling and liability laws and of variable trade taxes-cum-subsidies and tariff rate quotas is reducing the aggregate gains from new biotechnologies and the incentive for EU taxpayers and for life science companies to support GM food research. The use of variable levies and prohibitive out-of-quota MFN tariffs in particular is yet another reason to push for an ambitious outcome from the WTO's Doha Round of agricultural trade negotiations.

Key words: agricultural biotechnology, trade policy, regulation of GM foods

1. INTRODUCTION

There is a small but rich literature on the consequences of market distortions for the aggregate size and distribution of the benefits (positive and negative) from agricultural R&D. A neat synopsis of the key analyses and conclusions can be found in Alston and Pardey (1996, pp. 184–198). A consensus in that literature is that the aggregate size of the benefits is likely to be far less af-

The paper has benefited from the author's conversations with Julian Alston, Colin Carter, Phil Pardey, and Brian Wright, and from earlier joint work with Lee Ann Jackson, Chantal Pohl Nielsen, and Sherman Robinson. The views are his own, however, and not necessarily those of his current employer. Thanks are due to the UK's Department for International Development for supporting the author's research on this topic at the World Bank.

fected by price-distorting policies than is their distribution. While this certainly is the case if distortion rates are not altered when new farm technologies appear, it turns out to be less so if those rates are endogenous to technological change at home or to terms of trade changes following adoption of new technology abroad. And despite the efforts of the Uruguay Round's agricultural and sanitary and phytosanitary agreements to discipline agricultural protection, much scope evidently remains for World Trade Organization (WTO) members to vary their price distortions. With respect to products that may contain genetically modified organisms (GMOs), one way this is being done is to limit imports produced with the new biotechnology (via bans or strict labeling regulations), on the grounds that they may harm the environment or be a risk to human health. Another is by having support measures that vary import tariff/export subsidy rates (or the extent of direct domestic producer price support) so as to maintain a constant domestic price in the wake of new technology being adopted abroad or at home.

The emergence in the 1990s of transgenic crop varieties initially offered hope that the private sector might boost public funding of agricultural research. But concerns soon arose to dampen that optimism. A key one was that Europeans and others would reject the technology on environmental and food safety grounds, thereby thwarting export market prospects for adopters of the transgenic crops. That concern was vindicated when the European Union imposed in 1998 a *de facto* moratorium on the production and importation of food products that may contain GMOs. As a result, widespread adoption of new food crop varieties from the fledgling "gene revolution" has been limited to date to just three products (maize, soybean, and canola) in three countries: Argentina, Canada, and the United States (James 2004). True, the European Union (EU) replaced its moratorium on May 1, 2004, with new legislation, but it involves strict GM labeling regulations and liability laws that demand the implementation of expensive segregation and identity preservation systems that—especially for developing countries—may be as restrictive of exports of GM products as was the moratorium. With a number of other countries also imposing strict labeling regulations on GM foods and no harmonization of those standards (Carter and Gruère 2006), biotech firms are increasingly diverting their R&D investments away from food. Many public agricultural research systems also have remained shy about investing heavily in this technology, including the CGIAR's international agricultural research centers, which depend largely on rich-country grants. The legality of the EU's restrictions on imports of GM products has begun to be tested by the WTO's Dispute Settlement Body, but the issue will take years to resolve (Anderson and Jackson 2005c).

Varying import taxes/export subsidies/domestic price supports is legally possible under WTO law for any member while ever its applied tariff or producer or export subsidy is below the member's bound commitment for the product and measure in question. It also happens for a country automati-cally—and again without contravening WTO commitments—when the in-ternational price falls following adoption of new technology abroad and a tariff rate quota (TRQ) applies and is filled and the out-of-quota MFN tariff is prohibitive. Since TRQs are prevalent in the EU and more than 30 other countries, and there is a great deal of "binding overhang" even in cases where MFN tariffs are not prohibitive, many WTO members still have scope to vary their agricultural protection rates. In the case of import tariffs, the bound rate for all agricultural products averages more than twice the applied rate for both developed and developing countries (Table 1). The binding overhang is even greater for the crop products of most relevance to the new agricultural biotechnologies, although somewhat less so for the European

Table 1. Agricultural Weighted Average Import Tariffs in 2001, by Region (percent, *ad valorem* equivalent, weights based on imports)

	Bound tariff	MFN applied tariff	Actual applied tariff[a]
Developed countries	27	22	14
Developing countries	48	27	21
of which LDCs	78	14	13
WORLD	37	24	17

Source: Jean, Laborde, and Martin (2006).

[a] Includes preferences and in-quota TRQ rates where relevant, as well as the *ad valorem* equivalent of specific tariffs. Developed countries include Europe's transition economies that joined the EU in April 2004. The "developing countries" definition used here is that adopted by the WTO and so includes East Asia's four newly industrialized tiger economies.

Table 2. Weighted Average Import Tariffs, EU and All Developed Countries, Selected Crop Products, 2001 (percent, *ad valorem* equivalent, weights based on imports)

	European Union		All developed countries	
	Bound tariff	MFN applied tariff[a]	Bound tariff	MFN applied tariff[a]
Maize	69	51	138	53
Oilseeds	0	0	27	9
Wheat	62	58	119	36
Rice	106	87	241	25
Other cereals	70	69	66	16

Source: Martin and Wang (2004).

[a] When account is taken for in-quota tariffs of items subject to tariff rate quotas, and non-reciprocal pref-erences to developing countries, the actual applied tariff averages are even lower, as indicated in Table 1.

Union than for other developed countries (Table 2). It is TRQs plus this gap, together with a similarly large binding overhang in agricultural domestic support commitments, that has allowed the aggregate levels of producer subsidy equivalent estimates to remain almost as high today as they were when the Uruguay Round negotiations began in the latter 1980s (Figure 1). Nor is the Doha Development Agenda likely to lead to a rapid closing of that gap unless WTO trade negotiators become far more ambitious over the next year or two (Anderson and Martin 2005), so this prospect for variable protection rates will be with us for the foreseeable future.

This chapter begins by using standard partial equilibrium theory to show the ways in which the adoption of GM crop varieties by some countries affects other countries in the absence of trade policy distortions, and then how those two types of endogenous policy responses to the new biotechnology (a virtual import ban, or the maintenance of domestic prices via variable levies/subsidies or TRQs) can alter not just the distribution but also the aggregate size of the economic benefits from that R&D. It then draws on recent computable general equilibrium simulation analyses to show how important

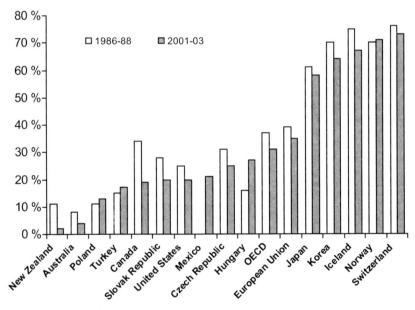

Source: PSE estimates from the OECD's database (see www.oecd.org).

Notes: Czech Republic, Hungary, Poland, and the Slovac Republic data are for 1991–93 in the first period. Austria, Finland, and Sweden are included in the OECD average for both periods, but also in the EU average for the latter period.

Figure 1. Agricultural Producer Support in High-Income Countries, by Country, 1986–1988 and 2001–2003 (percentage of total farm receipts from support policy measures)

those possibilities may be empirically. The final section draws out some policy implications of the results, particularly for the Doha Round of multilateral trade negotiations.

2. SIMPLE ECONOMICS OF GM CROP TECHNOLOGY IN A WORLD WITH VARIABLE TRADE DISTORTIONS

Leaving aside the biotechnology research industry, this section considers the market for a single crop such as maize for which a new variety is genetically engineered and made available in the form of purchasable seed by a biotech firm or public research institution. To set the scene, we first look at the effects in a small country of adoption of this cost-reducing technology at home or abroad in the absence of any trade policy responses. This is done both without and with an existing import tariff/export subsidy in place, to show the (un)importance of such measures on the results when the rate of protection is unchanged. The large-country case is then similarly considered, but in the case of adoption abroad we also explore the impact of it responding with either a variable levy to maintain the domestic price or a ban on imports from GM-adopting countries (or having a TRQ in place whose delivered protection rises when the international price falls). The final case analyzes the effects on the international market, so as to be able to see also the impacts on GM-adopting and third countries.

2.1. Effects of GM Adoption on a Small Country Imposing an Import Tariff/Export Subsidy

Consider a small country importing this product but unable to influence its international price, P_w. The effects domestically of its GM adoption are shown in Figure 2, assuming that domestic consumers are indifferent to whether the product may contain GMOs (to be relaxed later). The cost-reducing technology causes a downward shift in the supply curve from S to S', which is assumed for convenience to be parallel. In the absence of trade policy distortions, this increases producer and national economic welfare by area *abcd*.

If there was a tariff in place that had raised the domestic price from P_w to P_t, GM adoption would raise producer welfare by area *aefd* (> area *abcd*), but decrease government tariff revenue by area *efhg* (= area *bcfe*), so national welfare would increase by only area *abcd*—the same as in the absence of the tariff.

Were producers supported even more by an export subsidy (and an accompanying tariff to prevent imports for subsidized re-export) that held the domestic price at P_s, producers would gain—and taxpayers would lose—

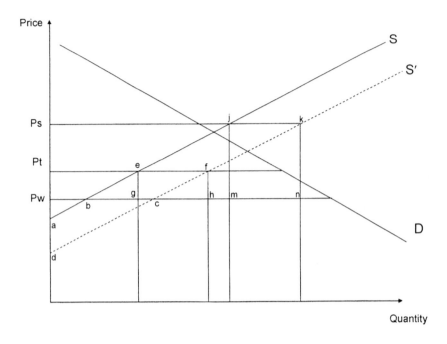

Figure 2. Effects of GM Crop Adoption at Home by a Small Distorted Economy

even more than with just the lower tariff: the producer surplus following adoption would be raised by area *ajkd* (> area *aefd*), while government reve-nue would be lowered by the increase in the export subsidy payment of area *jknm* (= area *bckj*), so net national welfare is again increased by only area *abcd* in Figure 2.

Both cases provide the standard conclusion for a small country that an unchanged price-distorting policy alters the distribution of welfare but not the aggregate national welfare gain from a new farm technology. Presuma-bly that protection policy thus raises the producers' demand for the new technology, while lowering the taxpayers'/finance ministry's incentive to subsidize such research.

What about the effect on this economy of GM adoption abroad? Adop-tion by enough producers to lower the international price from P_w to P_w' in Figure 3 would benefit this importing economy provided its tariff was not and did not remain prohibitive, but that gain is greater the smaller the tariff. With no tariff the gain in consumer welfare net of the loss to producers is area *qbwx* in Figure 3—the maximum gain from improved terms of interna-tional trade. If instead a non-prohibitive specific tariff of $P_t - P_w$ applied, the gain would be smaller but still positive, namely area *yevu*, following the in-ternational price fall. That is, even with an unchanged tariff, the importing country gains less from adoption abroad the larger is that tariff. And if that

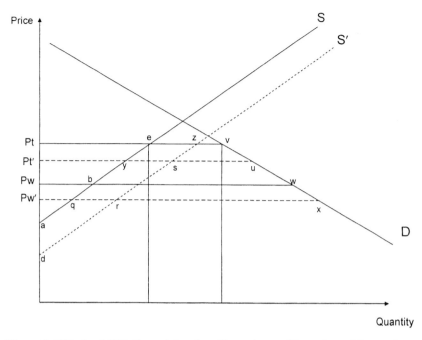

Figure 3. Effects of GM Crop Adoption Abroad or at Home for a Distorted Economy

country chose to raise its tariff to offset the effect of the fall in P_w on the domestic price, then it would gain even less from GM adoption abroad—and nothing at all if the tariff increase fully offset the international price. The latter would also be the case if a tariff rate quota operated and the out-of-quota tariff was prohibitive and the quota itself was filled.

2.2. Effects of GM Adoption on a Large Country Imposing an Import Restriction

What about when the importing country is large enough to influence the international market for this product? The analysis of Figure 3 in the previous paragraph applies equally to the large importing country enjoying a terms of trade improvement from GM adoption abroad (where again its consumers are assumed for the moment to be indifferent to GM food), assuming no change in this country's tariff. The only difference is that P_w would have fallen less than in the small-country case following GM adoption abroad, because of the greater quantity demanded by this large country in the international market (to be shown later using Figure 4).

When the assumption of no trade policy response is relaxed, qualifica-
tions are again needed. For example, if the importing country's tariff is
raised following the international price fall so as to keep the domestic price
at P_t, or if that is achieved by having a tariff rate quota that is filled and an
out-of-quota tariff that is prohibitive, the national gain from GM adoption
abroad again would be foregone. But unlike in the small country case, in this
case the larger the importing country's imports prior to adoption abroad, the
larger would be the fall in P_w with this variable levy response. That is, the
variable levy not only eliminates the gain to that country from GM adoption
abroad, but it also (i) reduces the gain from this new biotechnology to the
adopting countries, and (ii) raises the loss to all other countries that, as net
exporters of the conventional (non-GM) variant of this product, suffer a
terms of trade loss. The same would apply with a variable export subsidy.

Alternatively, if the importing country were to ban imports of this product from GM-adopting countries, that would push the domestic price above
P_t' to a level that could be more or less than P_t, depending on how large the
GM-adopters are in international markets. As shown in the next sub-section,
this reduces the gain from this new biotechnology to the adopting countries
but raises the loss to other countries exporting non-GM varieties of this crop.

Were this large importing country to allow GM adoption domestically,
its supply curve would shift out from S to S' (again assumed for convenience
to be a parallel shift) in Figure 3. In the presence of a zero or fixed import
tariff this would depress the international price (for diagrammatic convenience to, say, P_w') following the decline in the country's import demand by
quantity $bw–rx$ (or $ev–su$ in the case of a positive tariff). That would reduce
the prospects of GM adopters abroad gaining and would lead to gains to domestic producers only if area $daqr$ exceeds area qbP_wP_w' in the case of no
tariff (or if area $adsy$ exceeds area yeP_tP_t' in the case of a positive tariff).
However, if the tariff is raised after GM adoption so as to maintain the pre-
adoption domestic price P_t, producer welfare in this country would un-
equivocally improve (by area $daez$), while that of overseas producers would
be depressed by further downward pressure on P_w.

2.3. International Market Effects of Sub-Global GM Adoption

Adoption of that new GM maize variety in some maize-exporting countries
provides importing countries with the option of continuing to buy from those
exporters a crop that now may contain GMOs, or buying a GM-free product
from non-adopting countries. The former international market is shown in
the top panel of Figure 4, the latter in the bottom panel, in both cases assuming no price-distorting policies are introduced. In the top panel, ES_g and ED_g
are the excess supply and excess demand curves for the GM-adopting countries' surplus prior to adoption; and in the bottom panel, ES_n and ED_n

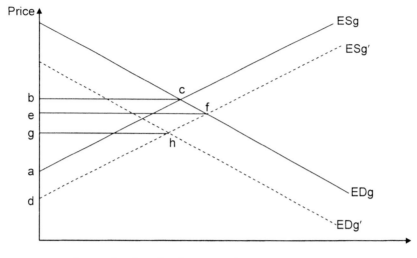

International market for a crop that may contain GMOs

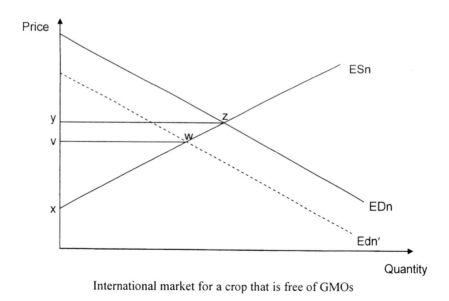

International market for a crop that is free of GMOs

Figure 4. Effects of GM Crop Adoption by Some Exporting Countries

are the excess supply and excess demand curves for the exporting countries choosing not to adopt GM varieties yet. Adoption of a GM variety by some producers in the first group of exporting countries is assumed to lower production costs there such that ES_g shifts down to ES_g'. If consumers in importing countries (as is assumed throughout to be the case in exporting countries) were indifferent about whether the crop may contain GMOs, ED_g would remain unchanged and the unit value of that bilateral trade would fall from b to e. Net economic welfare would increase in the importing countries by area $bcfe$ (the gain to their consumers would exceed the loss to their domestic producers), while net economic welfare in the GM-adopting countries would change by the area def minus area abc. A sufficient condition for that to be positive is that ES_g is parallel to ES_g'.[1]

What if some consumers in some of the importing countries consider the GM-free variety to be superior and therefore prefer it over supplies that may contain GMOs? That would cause the ED_g curve in the top panel of Figure 4 to shift leftwards to ED_g'. It would depress the unit value of that bilateral trade even more, to g, and may even lead to the export volume being smaller than before adoption (if h is to the left of c). Consumers of the GM variety benefit even more from its low price, but producers in the adopting countries (who are assumed not to segregate GM and non-GM varieties) are more likely to be worse off.

How does all this affect the GM-free exporting countries? On the one hand, if consumers consider the GM and non-GM varieties as perfect substitutes, then the increased volume and lower price of exports from GM-adopting countries would shift ED_n to ED_n' in Figure 4's bottom panel, lowering export revenue and net economic welfare in non-adopting exporting countries where producer losses would exceed gains to consumers in those countries by area $vwzy$. On the other hand, if some consumers in some importing countries prefer the GM-free product, the leftward shift in ED_n would be smaller or that curve may even end up to the right rather than the left of its original position. If that substitution effect is strong enough to place ED_n' to the right of ED_n, producers in the GM-free exporting countries would be better rather than worse off following GM adoption abroad, and conversely for consumers in those exporting countries.

---

[1] If, however, that supply shift is wedge-shaped such that points a and d are closer together or coincide, we know from the theory of immizerizing growth (Bhagwati 1958) that the GM-adopting countries could be worse off if the excess demand curve they face is sufficiently inelastic. Lindner and Jarrett (1978) show that, even with a completely inelastic demand curve, a parallel shift (but not a pivotal shift) downwards in the supply curve will not reduce the exporting countries' surplus. Were the supply curve to shift downwards only to the left of point c in the top panel of Figure 4—that is, if only lower-cost infra-marginal producers were able to adopt the new biotechnology—all the benefits would stay with the GM-adopting producers (but be shared with the biotech firm that engineered that new GM variety, not modeled above) and none would be shared with consumers abroad or at home.

Should some of the importing countries choose to ban imports of this product from countries adopting the GM variety (in the assumed absence of segregation and an identity preservation system), the leftward shift of ED_g in Figure 4's top panel would be greater and that of ED_n in the bottom panel would be less (or be more likely to be a rightward shift). In this case the government's import ban is effectively forcing all of its consumers to buy only GM-free varieties, which makes them higher-priced, so consumer welfare in such importing countries—and producer welfare in GM-adopting countries—will be less than if consumers were free to choose whether to avoid imports that may contain GMOs.[2] However, producers in the countries imposing the ban, and in GM-free exporting countries, will be better off with such a ban in place than under free trade while ever they are denied the right to adopt the new technology. Hence it is an empirical question as to whether they would be better off in that situation or in being able to adopt the technically more productive GM variety if the latter required (to avoid violating WTO national treatment rules) the removal of the GM import ban.

Were that import ban to be replaced by strict labeling and associated liability laws, as happened in the EU in April 2004, importers would be even less inclined to buy from GM-adopting countries—including in cases where the GM crop varieties grown there had been approved by the EU prior to 1998 (as with some GM soybean varieties)—because of the now-higher costs of compliance with the new labeling laws that require segregation and identity preservation back down the value chain to the farmer. The more GM-adopting countries there are that cannot provide credible certification to importers, the larger would be the further leftward shift in the ED_g curve in Figure 4's top panel.

It is clear from the above sample of situations that not even some of the signs, let alone the sizes, of the welfare effects of GM adoption by a subset of countries can be determined *a priori* when there are trade policy responses by other countries. Hence the need for quantitative analysis of global markets for the relevant products. Such analysis needs to go beyond the above one-crop model so as also to take account of products that are close substitutes or complements in production and/or consumption or that are inputs into other activities. Of particular importance in this case is the livestock sector, since maize and soybean (the first two GM food crops developed) are major inputs into the intensive segment of livestock production but not the extensive segment that still relies on grazing pastures. The most comprehensive way to meet these needs is to use a global economy-wide model of trading nations.

---

[2] The presumption here is that the deadweight welfare cost of segregation and identity preservation through the value chain is modest, as it seems to be for segregating different qualities of each traditional grain at present. For a model in which such costs are non-trivial and so the welfare costs are less clear, see Moschini and Lapan (2005).

3. EMPIRICAL ESTIMATES OF THE EFFECTS OF GM ADOPTION AND TRADE POLICY RESPONSES

Estimating the welfare consequences of actual and prospective GM crop adoption by some countries and of policy responses by others has been the focus of numerous recent studies. Here there is room only to provide summaries of three sets of simulation results that pertain specifically to the analytical issues raised in the previous section. All employ the same well-received CGE model of the world economy known as GTAP (described in Hertel 1997). The first set assumes that the world's price-distorting trade policies are all non-varying exogenous taxes/subsidies, apart from the EU imposing a ban on imports from GM-adopting countries of products that may contain GMOs. The second set specifically examines the impact of the EU's existing trade taxes/subsidies on the size of the estimated welfare effects of GM adoption abroad without versus with a ban on imports from GM-adopting countries. By way of contrast, the third set assumes that the EU's Common Agricultural Policy insulates domestic producers somewhat from the fall in international prices that follows the adoption of GM crop varieties abroad.

3.1. Effects of Sub-Global GM Adoption Without and With the EU Moratorium

A recent study by Anderson and Jackson (2005a) begins with GM adoption for just coarse grains and oilseeds but then adds rice and wheat, to get a feel for the relative economic importance to different regions and the world as a whole of current versus prospective GM crop technologies. That study modified the GTAP model (with its Version 5 database) so that it could capture the effects of productivity increases of GM crops, some consumer aversion to products containing GMOs, and substitutability between GM and non-GM crop varieties as intermediate inputs into final consumable food. The impacts of GM adoption by just the United States, Canada, and Argentina are considered first, without and then with an import ban by the EU. The EU is then added to the list of adopters to explore the tradeoffs for the EU between productivity growth via GM adoption and the non-pecuniary benefits of remaining GM-free given the prior move to adopt in the Americas. A change of policy in the EU to allow the adoption and sale of GM crop products would reduce the reticence of the rest of the world to adopt GM crop varieties, so the effects of all other countries then adopting is explored as well.

Specifically, the base case in the GTAP model, which is calibrated to 1997 just prior to the EU moratorium being imposed, is compared with an alternative set of simulations whereby the effects of adoption of currently available GM varieties of maize, soybean, and canola by the first adopters

(Argentina, Canada, and the United States) is explored without and then with the EU *de facto* moratorium on GMOs in place.[3] Plausible assumptions about the farm productivity effects of these new varieties and the likely percentage of each crop area that converts to GM varieties are taken from the latest literature including Marra, Pardey, and Alston (2002), Qaim and Zilberman (2003), and Huang et al. (2004).

The global benefits of GM adoption by the United States, Canada, and Argentina are estimated to be US$2.3 billion per year net of the gains to the biotech firms (which are ignored in all that follows) if there were no adverse reactions elsewhere. About one-quarter of that is shared with the major importing regions of the EU and Northeast Asia; 60 percent goes to the three GM-adopting countries; and Brazil, Australia, New Zealand, and sub-Saharan Africa other than South Africa lose very slightly because of an adverse change in their terms of trade. But when account is taken of the EU moratorium, which is similar to an increase in farm protection there, the gain to the three GM-adopting countries is reduced by one-third. The diversion of their exports to other countries lowers international prices, so welfare for the food-importing regions of the rest of the world improves—but only very slightly. Meanwhile, the EU is worse off by $3.1 billion per year minus whatever value EU consumers place on having avoided consuming GMO products. If the EU instead were to allow adoption and importation of GM varieties, it would benefit because of its own productivity gains, and so too would net importers of these products elsewhere in the world, while countries that are net exporters of coarse grains and oilseeds (both GM adopters and non-adopters) would be slightly worse off (only slightly because coarse grains and oilseeds are minor crops in the EU compared with North America).

However, if by adopting that opposite stance in the EU the rest of the world also became uninhibited about adopting GM varieties of these crops, global welfare would increase by nearly twice as much as it would when just North America and Argentina adopt, and almost all of the extra global gains would be enjoyed by developing countries. If one believes the EU's policy stance is determining the rest of the world's reluctance to adopt GM varieties of these crops, then the cost of the EU's moratorium to people outside the EU-15 has been up to $0.4 billion per year for the three GM-adopting countries and $1.1 billion per year for other developing countries.

On the one hand, those estimates overstate the global welfare cost of the EU moratorium to the extent that without it, the EU would have used its variable import levy capability (or TRQ) to reduce/stop any decline in EU

---

[3] This has to be done in a slightly inflating way in that the GTAP model is not disaggregated below "coarse grains" and "oilseeds." However, in the current adopting countries (Argentina, Canada, and the United States), maize, soybean, and canola *are* the dominant coarse grains and oilseed crops.

domestic prices for these products (see sub-section 3.3 below). But on the other hand, the above estimates understate the global welfare cost of the EU's moratorium in at least three respects. First, the fact that the EU's stance has induced some other countries to also impose similar moratoria has not been taken into account.[4] Second, these are comparative static simulations that ignore the fact that GM food R&D is ongoing and that investment in this area has been reduced considerably because of the EU's extreme policy stance, with biotech firms redirecting their investments towards pharmaceuticals and industrial crops instead of food crops. And third, the above results refer to GM adoption of coarse grains and oilseeds only. The world's other two major food crops are rice and wheat, for which GM varieties have been developed and are close to being ready for commercial release.

How have EU farm households been affected by the EU moratorium? The above simulation study finds that their real incomes are somewhat higher with the EU moratorium than they would be if there were no moratorium and they were allowed to adopt GM varieties of maize, soybean, and canola (and the EU did not use its variable levy capability or TRQs to prevent domestic prices of these products from falling). This is because the price decline for those products would more than fully offset the productivity gain from adopting GM varieties. It is therefore not surprising that EU farmers were not lobbying for a pro-GM policy stance.

A second set of simulations from the Anderson and Jackson (2005a) study involves a repeat of the first set except that China and India are assumed to join America in adopting existing GM crop varieties, and GM rice and wheat varieties are assumed to be made available to the GM adopting countries' farmers, again without and then with an EU import moratorium. In this case it is simply assumed that total factor productivity in GM rice and wheat production would be 5 percent greater than with current non-GM varieties. If China were to decide to approve the release of GM rice and wheat varieties, India would probably follow soon after. China and India account for 55 percent of the world's rice market and 30 percent of the wheat market, being close to self sufficient in both. They therefore do not have to worry greatly about market access abroad. If that led to enough other non-EU countries accepting GM varieties of rice and wheat, this could well lead North American and Argentina also to adopt them.

Allowing China and India to join the GM-adopters' group, and adding rice and wheat to coarse grain and oilseeds, almost doubles the potential global gains from this biotechnology. The global economic welfare gain if there were no moratoria by the EU or others is estimated to be $3.9 billion

---

[4] Sri Lanka was perhaps the first developing country to ban the production and importation of GM foods. In 2001 China did the same (with some relaxation in 2002), having been denied access to the EU for some soy sauce exports because they may have been produced using GM soybeans imported by China from the United States.

with just rice added, or $4.3 billion if wheat were also added, instead of the $2.3 billion per year when just the original three countries and commodities are involved. North America gains only a little more from the addition of GM rice and wheat, which might seem surprising given the importance to it of wheat, but it is because its productivity gain is almost offset by a worsening of its terms of trade as a consequence of its and the other adopters' additional productivity. Two-thirds of the extra $2.0 billion per year from adding rice and wheat would accrue to China and India, with other developing countries, as a net grain-importing group, enjoying most of the residual via lower-priced imports. When the EU moratorium is in place, the cost to the EU of its moratorium would rise from $3.4 to $5.5 billion per year (again not counting the benefit to EU consumers of knowing they are not consuming GMOs), while for the rest of the world (again assuming the EU policy is discouraging GM adoption elsewhere) it rises from $1.5 billion to $2.9 billion per year. The adding of further crops to the GM family would continue to multiply that latter estimate.[5]

3.2. How Much Do the Estimated Welfare Effects of the EU Moratorium in Response to GM Adoption Abroad Depend on the EU's Existing Trade Policy Distortions?

To test the proposition in Section 2 above that at least the net welfare effects of technology adoption are not affected greatly by existing trade policies, Anderson and Nielsen (2004) conducted a similar set of experiments to the one above (but with the earlier Version 4 of the GTAP model and database). Their study also is simpler in that it assumes all of each crop in adopting countries uses GM varieties, in contrast to the above study which assumes that only a subset of the crop (the percentage varying by country) is planted to GM varieties in adopting countries. In that respect it overstates the likely gains from adoption. But it understates the cost of the EU moratorium in the sense that it compares that scenario with one in which the EU simply enjoys the terms of trade improvement, rather than also adopting GM technology domestically [as assumed in the above Anderson and Jackson study (2005a)]. These differences are unimportant for the present purpose though, which is to ask how different the estimates of the effect of the EU moratorium would be if Western Europe's agricultural protectionism were not there (and hence if its farm sector were considerably smaller and its market more import-dependent).

The Anderson and Nielsen (2004) simulations were run first with 1997 farm policies in place and then without any agricultural protectionism in West-

---

[5] For a closely related study that focuses on what the EU moratorium means for Africa, see Anderson and Jackson (2005b).

ern Europe, so as to get two different base cases to compare with the alternative of the EU moratorium being imposed. The results suggest that, without those protectionist policies, an EU import ban would cost the EU only $0.4 billion per year less than it has cost with those policies in place. This is consistent with the expectation, from the theory in Alston, Edwards, and Freebairn (1988) and Alston and Pardey (1996), and from Figure 3 above, of the aggregate domestic effect even for a large economy being only slightly smaller in the presence of protectionist policies. However, without that protectionism the EU moratorium would have hurt GM-adopting countries by an estimated $1.05 billion per year less, and would have helped non-adopters outside the EU by $2.8 billion less. Those signs are as predicted from Figure 4.

3.3. How Much Difference Does It Make If the EU Simply Maintains Domestic Prices with a Variable Levy or TRQ?

To address the question of how much bias is introduced if the reality of the EU's variable levy policy and TRQs is ignored, van Meijl and van Tongeren (2004) use the GTAP model with the Version 5 database to do a similar but slightly more complicated analysis than that of Anderson and Nielsen (2004). Variable import levies and variable export subsidies are allowed for cereals, and they are triggered by an endogenous price transmission equation involving the EU's grain trade position.[6] That is, the domestic price is not kept completely constant, but its downward movement is heavily constrained. As a result, the EU is shielded from the international price falls following GM adoption abroad.

Van Meijl and van Tongeren (2004) estimate that this price-insulation mechanism halves the welfare gains that the EU would otherwise enjoy from a terms of trade improvement following the productivity gains from GM adoption abroad, while reducing the welfare of North American adopters (who have to find markets elsewhere for their exports)—although only very slightly, and much less so than when an EU ban is imposed. These results are thus consistent with the finding in Section 2 above. They imply that empirical studies that do not incorporate the EU's variable levy and TRQ policies may overstate the cost to GM adopters of EU discrimination against GM-adopting suppliers, although only slightly, but may overstate somewhat more the cost to the EU itself (depending on the extent to which the EU would have reduced the decline in domestic prices of these products). This problem of overstatement may be greater in the case of other countries with

---

[6] The EU's oilseed levies are not varied because they have been bound in the GATT/WTO at zero (see Table 2). It is the tariff/export subsidy not only on maize but also on other cereals that is endogenously adjusted when international prices for all these substitute products change.

larger tariff binding overhangs, should they also be using such variable levy schemes or have TRQs also on oilseeds.

4. POLICY IMPLICATIONS

Clearly, the welfare effects of adoption of genetically modified (GM) food and feed crop varieties are being affected not only by some countries' strict regulations governing GM food production and consumption, but also by their exogenous and endogenous choices of food trade policy instruments. Most notable have been the bans on imports of food products that may contain GMOs, particularly by the EU. Even with the ending of the European Union's GM moratorium in April 2004, the EU regulations replacing it— and those of numerous other countries—demand costly segregation and identity preservation systems that may be just as restrictive as a ban on exports from GM-adopting regions (especially developing countries). Indeed they may be even more restrictive than was the moratorium, because at least the latter allowed approved GM varieties to be shipped without being subject to such strict labeling and liability laws. WTO dispute settlement procedures provide an avenue to try to reduce these barriers, but no quick and easy resolution is expected (Anderson and Jackson 2005a).

Meanwhile, variable trade taxes-cum-subsidies by the EU and others will continue to reduce the aggregate gains from new biotechnologies and the incentive for EU taxpayers and life science companies to support GM food research. Because the EU—like many other countries—has WTO-bound tariffs well above applied rates, its use of variable levies is not inconsistent with its WTO obligations. Nor is its TRQ regime. Moreover, the July 2004 framework agreement for the WTO's current Doha Round of negotiations (World Trade Organization 2004) includes provision for a special safeguard mechanism which could allow developing countries even more scope to use variable levies, as suggested for example by Foster and Valdes (2005). The theory and results presented above show that using such a capability to insulate an economy from the international price-reducing effects of biotechnology adoption abroad will reduce the welfare gains from that new technology (i) for adopters abroad and (ii) for consumers in the insulating countries. They also dampen the incentive for biotech research providers to invest in GM food crops. Given the persistently high rates of return to agricultural R&D (Alston et al. 2000), this is yet another reason, on top of the standard ones,[7] as to why it is important to seek major reductions in bound agricultural tariffs in the current Doha Round of multilateral trade negotiations.

---

[7] See, for example, Anderson and Martin (2005).

REFERENCES

Alston, J.M., G.W. Edwards, and J.W. Freebairn. 1988. "Market Distortions and Benefits from Research." *American Journal of Agricultural Economics* 70(2): 281–288.

Alston, J.M., M.C. Marra, P.G. Pardey, and T.J. Wyatt. 2000. "A Meta Analysis of Rates of Return to Agricultural R&D: Ex Pede Herculum?" IFPRI Research Report No. 113, International Food Policy Research Institute, Washington, D.C.

Alston, J.M., and P.G. Pardey. 1996. *Making Science Pay: The Economics of Agricultural R&D Policy*. Washington, D.C.: AEI Press.

Anderson, K., and L.A. Jackson. 2005a. "Standards, Trade and Protection: The Case of GMOs." Paper prepared for the 41st Panel Meeting of *Economic Policy* in Luxembourg, April 15–16.

_____. 2005b. "Some Implications of GM Food Technology Policies for Sub-Saharan Africa." *Journal of African Economies* 14(3): 385–410.

_____. 2005c. "What's Behind GMO Disputes?" *World Trade Review* 4(2): 203–228.

Anderson, K., and W. Martin. 2005. "Agricultural Trade Reform and the Doha Development Agenda." *The World Economy* 28(9): 1301–1327.

Anderson, K., and C.P. Nielsen. 2004. "Economic Effects of Agricultural Biotechnology Research in the Presence of Price-Distorting Policies." *Journal of Economic Integration* 19(2): 374–394.

Bhagwati, J.N. 1958. "Immizerizing Growth: A Geometric Note." *Review of Economic Studies* 25(2): 201–205.

Carter, C.A., and G.P. Gruère. 2006. "International Approval and Labeling Regulations of Genetically Modified Food in Major Trading Countries." In R.E. Just, J.M. Alston, and D. Zilberman, eds., *Regulating Agricultural Biotechnology: Economics and Policy*. New York: Springer.

Foster, W., and A. Valdes. 2005. "Variable Tariffs as Special Safeguards: The Merits of a Price Floor Mechanism Under Doha for Developing Countries." Paper presented at the Workshop on Managing Food Price Risks and Instability, World Bank, Washington D.C., February 28–March 1.

Hertel, T.W. (ed.). 1997. *Global Trade Analysis: Modeling and Applications*. Cambridge and New York: Cambridge University Press.

Huang, J., R. Hu, H. van Meijl, and F. van Tongeren. 2004. "Biotechnology Boosts to Crop Productivity in China: Trade and Welfare Implications." *Journal of Development Economics* 75(1): 27–54.

James, C. 2004. *Global Review of Commercialized Transgenic Crops: 2003*. International Service for the Acquisition of Agri-biotech Applications, Ithaca, NY.

Jean, S., D. Laborde, and W. Martin. 2006. "Consequences of Alternative Formulas for Agricultural Tariff Cuts." In K. Anderson and W. Martin, eds., *Agricultural Trade Reform and the Doha Development Agenda*. London: Palgrave Macmillan. Co-published with the World Bank, Washington, D.C.

Lindner, R.J., and F.G. Jarrett. 1978. "Supply Shifts and the Size of Research Benefits." *American Journal of Agricultural Economics* 60(1): 48–58.

Marra, M., P. Pardey, and J. Alston. 2002. "The Payoffs to Agricultural Biotechnology: An Assessment of the Evidence." *AgBioForum* 5(2): 43–50 (available online at http://www.agbioforum.org/v5n2/v5n2a02-marra.pdf).

Martin, W., and Z. Wang. 2004. "The Landscape of World Agricultural Protection." Mimeo, World Bank, Washington, D.C.

Moschini, G., and H. Lapan. 2005. "Labeling Regulations and Segregation of First- and Second-Generation GM Products: Innovation Incentives and Welfare Effects." In R.E. Just, J.M. Alston, and D. Zilberman, eds., *Regulating Agricultural Biotechnology: Economics and Policy*. New York: Springer.

Qaim, M., and D. Zilberman. 2003. "Yield Effects of Genetically Modified Crops in Developing Countries." *Science* 299: 900–902.

van Meijl, H., and F. van Tongeren. 2004. "International Diffusion of Gains from Biotechnology and the European Union's Common Agricultural Policy." *Agricultural Economics* 31(2): 307–316.

World Trade Organization. 2004. "Decision Adopted by the General Council on 1 August 2004." Report No. WT/L/579 (the July Framework Agreement), World Trade Organization, Geneva.

Chapter 8

THE VALUE OF NON-PECUNIARY CHARACTERISTICS OF CROP BIOTECHNOLOGIES: A NEW LOOK AT THE EVIDENCE

Michele C. Marra and Nicholas E. Piggott
North Carolina State University

Abstract: In this chapter we examine the non-pecuniary aspects of the earliest crop bio-technologies. We analyze the stated values of the non-pecuniary aspects, taken from three farm-level surveys. We focus particularly on the phenomenon of part-whole bias, which is the empirical finding that the sum of the stated part-worths (the value of each non-pecuniary characteristic) is greater than the stated total value of all the non-pecuniary characteristics. We analyze the empirical evidence of part-whole bias in the surveys, while decomposing it to further understand the phenomenon and to rescale the stated values of the non-pecuniary characteristics in the surveys. We find for all three surveys that the degree to which part-worths should be rescaled is about 60 percent.

Key words: crop biotechnology, non-pecuniary characteristics, stated preference, part-whole bias, rescaling

1. INTRODUCTION

Crop biotechnologies have been available commercially in the United States for ten years and by most accounts have been a great success. A glance at rates of adoption and diffusion attests to the fact that farmers find most of these new technologies to be preferable to their conventional counterparts. Even though enough time has passed for the price of substitutes to adjust so that many of the early crop biotechnologies, such as Roundup Ready®

Senior authorship is not assigned. The authors wish to thank Kerry Smith and Mitch Renkow, as well as participants at the NC-1003 "Economics of Regulation of Agricultural Biotechnologies Conference" held on March 10–12, 2005, in Washington, D.C., especially Giancarlo Moschini, for helpful comments. All remaining errors are our own.

soybeans, appear to have lost most of their advantage in terms of accounting profitability, farmers are still planting them on a high percentage of total crop acres. Corn biotechnology crops were planted on 40 percent of U.S. corn acreage in 2003 and on 45 percent of acreage in 2004 (USDA 2004). Over three-quarters of cotton and soybean acreage in the United States is planted with biotech varieties of these crops, with herbicide-tolerant soybeans reaching the 85 percent level in 2004 (USDA 2004). When asked why they continue to plant these crops, farmers often will mention their convenience value, including their simplicity, or their relative environmental safety. That is, the decision to continue planting these crops is based, in part, on their *non-pecuniary* characteristics relative to other alternatives.

The term *non-pecuniary* comes originally from the law. It means a loss or a gain that cannot be expressed in terms of an amount of money. Non-pecuniary damages are those things that detract from one's utility, that are not traded in markets, and so that do not have a market price with which damages can be calculated. A classic example is "pain and suffering." In cases where environmental damages are contested, the total non-pecuniary loss in value may be quite large. When Exxon Corporation was sued by the state of Alaska, the U.S. Environmental Protection Agency, and others for the Valdez oil spill in the early 1980s, the value to society of the wildlife losses alone was calculated to be in the tens of millions of dollars (Exxon Valdez Oil Spill Trustee Council 2005).

The purpose of this chapter is to examine the value of the non-pecuniary characteristics of the earliest crop biotechnologies—those with enhanced agronomic properties embodied in a pest control system. First, we discuss measurement issues related to valuation of these characteristics, paying particular attention to the importance of precise measurement for pricing decisions and for guiding future research and development. Second, we focus on the phenomenon of part-whole bias and its implications for valuation. Then we present and discuss a new decomposition of the part-whole bias found when goods with non-pecuniary characteristics are valued using stated preference techniques. Finally, we present some evidence of empirical measures of part-whole bias, its decomposition, and the implied values of the characteristics of these crop biotechnologies that have some non-pecuniary aspects, using information taken from several recent farm-level surveys.

2. THE FARMER'S CHOICE PROBLEM

The farmer has available two choices of technology to employ in the production of a final output. The first uses a conventional seed variety and the second, a biotech seed variety. In the short run the farmer has a fixed amount of acres, A, for which he decides the number of acres to be allocated to each of the seed varieties—A_B for biotech acres and A_C for conventional acres—based on the potential (private) pecuniary and non-pecuniary costs and benefits from their use. Because of the potential for non-pecuniary benefits, the non-separable agricultural household production model is appropriate.

Let the household utility function, U, be defined over consumption of a market good x and M non-pecuniary amenities, q_m comprising $\mathbf{q}$.[1] Suppose further that the level of each non-pecuniary amenity is determined by the choice of A_B acres planted using the biotechnology [i.e., $\mathbf{q}(A_B) = q_1(A_B)$, ..., $q_M(A_B)$]. The utility function is given by $U[x, \mathbf{q}(A_B)]$, where the marginal utility derived from an additional acre of biotechnology is the sum of the product of the marginal utility of the non-pecuniary amenity and the marginal change in the amenity from a change in A_B, i.e.,

$$\frac{\partial U}{\partial A_B} = \sum_{m=1}^{M} \frac{\partial U}{\partial q_m} \frac{\partial q_m}{\partial A_B}.$$

Technology in the production of the final output is given by $f[A_B, A_C, z_B, z_C]$, where z_B and z_C denote the quantities of other inputs associated with the two technologies. The farm household's maximizing problem can be expressed as

$$\max_{x, A_B, z_B, z_C} U[x, \mathbf{q}(A_B)] \tag{1}$$

subject to

$$p^o f[A_B, A_C, z_B, z_C] - r(A_B + A_C) - w_B z_B - w_C z_C \geq p_x x - e,$$

where p^o is output price, r is the rental rate for land, w_B and w_C are the prices of other inputs associated with each technology, p_x is the price of the market good, and e is endowment income. Hence, the optimization problem can be restated as the Lagrangean:

$$\max_{x, A_B, z_B, z_C} L = U[x, \mathbf{q}(A_B)] + \tag{2}$$
$$\lambda \{ e + p^o f[A_B, A_C, z_B, z_C] - r(A_B + A_C) - w_B z_B - w_C z_C - p_x x \},$$

where λ is the Lagrange multiplier. The solution to this problem consists of optimal levels of the market good, x, the technology adoption decision (A_B),

---

[1] We use one market good for ease of exposition since the focus of this chapter is on the non-pecuniary amenities. It is straightforward to generalize the problem to N market goods.

the variable inputs employed, z_B and z_C, and the marginal utility of an additional dollar of profit or endowment, λ.

The value of the non-pecuniary amenities associated with choice of A_B is similar, but not strictly analogous, to the willingness to pay (WTP) amount elicited from contingent valuation (CV) studies. The difference is that farmers' choices of A_B affect their total expected income, whereas the CV paradigm assumes income is exogenous. In concept, it is more akin to the marginal value of a characteristic in a hedonic model, although ours is a stated, rather than revealed, value. Because we elicit the farmers' valuation of the non-pecuniary amenities through survey questions, similar to CV except we elicit marginal values rather than WTP, we call the farmers' responses "stated marginal values" (SMV).

3. NON-PECUNIARY CHARACTERISTICS
OF CROP BIOTECHNOLOGIES

The non-pecuniary characteristics we consider are associated with two types of genetically engineered agronomic properties. First are the crops that have the naturally occurring insecticide *Bacillus thuringiensis* (Bt) genetically engineered into the plant, so that when chewing insects susceptible to the particular strain of Bt begin feeding on the plant's vegetative parts, they are killed. One example is Monsanto's Yieldgard Rootworm® corn, introduced in 2003, which has in it both the original strain of Bt, geared toward controlling the European corn borer and the tobacco budworm and bollworm, and a second strain that kills corn rootworm larvae as they feed on the corn plant's roots.

The other biotech crop type considered in this study is tolerant to externally applied, broad spectrum herbicides that when applied as post-emergent, over-the-top sprays, kill the weeds, but leave the growing crop unaffected. An example of this type is the set of crops tolerant to glyphosate, in particular Roundup Ready® (RR) corn, soybeans, and cotton.

Both types of biotech crops have some degree of the three, potentially important, non-pecuniary amenities mentioned by farmers as having value to them. The non-pecuniary aspects are sometimes embedded within a characteristic that has both pecuniary and non-pecuniary components. The first of these is increased human (farmer and worker) safety. The Bt crops require no handling of pesticides, at least for the insects susceptible to the Bt strains engineered into the seed. The Bt also replaces insecticides that are more toxic so they provide an increased level of safety to the farmer and his workers who no longer need to handle the more toxic pesticides. The RR crops replace currently used conventional herbicides with the generally less toxic herbicide glyphosate, again giving the farmer more peace of mind about his own safety and the safety of his workers. So the added human safety characteristic is comprised of the expected cost reductions for protective gear and less liability, a pecuniary value, and an overall increase in human safety, a non-pecuniary value.

The second characteristic is environmental improvement. Both Bt and RR pest control systems are safer for the environment. The Bt crops eliminate external application of insecticides to control the insects for which the Bt strain is effective, so that additional contaminants are not introduced into the ambient air, wells, or waterways downstream from the field. Also, the Bt strains are so safe that they are approved for use in spray form for organic farmers and gardeners. Glyphosate is one of the most environmentally benign herbicides on the market today.

> Glyphosate has a half-life in the environment of 47 days, compared with 60–90 days for the herbicides it commonly replaces. The herbicides that glyphosate replaces are 3.4 to 16.8 times more toxic, according to a chronic risk indicator based on the EPA reference dose for humans [USDA 2000, p. 17].

It also binds to the soil very quickly so that leaching into groundwater is minimized. The RR crops minimize herbicide run-off, as well as further enhance surface water quality as compared to other alternative weed control systems. This characteristic could contain some pecuniary benefits in the form of reduced cost of averting behavior, but the non-pecuniary amenities associated with a cleaner environment are also present.

A third characteristic these seed types have in common is their relative convenience as compared to their alternatives. Most farmers report convenience as an important factor in adopting these technologies. Convenience includes such pecuniary characteristics as time and equipment cost savings if the number of pest control applications is reduced, but it also includes such non-pecuniary aspects as "ease and simplicity" and less worry about precise timing of pest control applications because of a wider window of opportunity for pest control.[2] These crops also may be viewed as less risky than their conventional counterparts. Even though the precise levels of the non-pecuniary characteristics embodied in each biotech seed would be nearly impossible to quantify from secondary sources, producers can judge for themselves the additional value of these characteristics based on their preferences and their assessment of the relevant alternatives, as modeled above. The remainder of the chapter focuses on measuring their value.

---

[2] In fact, some farmers have told us that RR soybeans are so simple to grow, even "ivy-covered academicians" could grow them.

4. MEASUREMENT ISSUES

Several measurement issues have arisen regarding stated preference approaches, such as CV. Among them are potential biases due to the hypothetical nature of CV, the payment vehicle used, the framing of the CV question,
the problem of passive use value, the difference between WTP and willingness to accept (WTA) measures of value, the problem of scope, and various
forms of part-whole bias (Mitchell and Carson 1989). While all of these
measurement issues are important to CV researchers and public goods
values, we focus on one common to both WTP and SMVs generated from
the more transparent trade-offs farmers make in adopting a new technology.
This is the type of part-whole bias that can occur when the value of the
characteristics of a good are elicited separately and/or in addition to the total
value of the same set of characteristics. This phenomenon has not yet been
studied in the context of stated preferences in technology adoption, but has
been examined somewhat in the CV literature.

This type of part-whole bias can be defined as

$$\sum_{m=1}^{M} \phi(q_{jm}) > \phi(\sum_{m=1}^{M} q_{jm}), \tag{3}$$

where ϕ is the function that maps the mth non-pecuniary characteristic into
the value placed on it by individual j. The sum of the value of each of the
parts is greater than the value of the whole. Numerous reasons have been
proposed in the literature for part-whole bias. Mitchell and Carson (1989)
lay out a typology of the potential biases that can occur with the CV method.
They place part-whole bias as a subset of a broader class of biases, which
they term "scenario misspecification." One such category of this is "amenity
misspecification bias," which is defined as "where the perceived good being
valued differs from the intended good." Part-whole bias has two potential
sources within the misspecification bias category: (i) "Where a respondent
values a larger (or a smaller entity) than the researcher's intended good," or
(ii) "benefit part-whole bias," as "where a respondent includes a broader or a
narrower range of benefits in valuing a good than intended by the researcher" (Mitchell and Carson 1989, p. 235).

Morrison (2000) defines part-whole bias as the observation that "people
are often WTP close to the same amount for a basket of goods as for separate
components of that basket, or WTP the same amount for very different quantities of the same good" (p. 1). She also terms this type of part-whole bias as
the "embedding" effect, as do others (e.g., Boyle et al. 1994).

Within their typology, Mitchell and Carson (1989) also list "incentives
to misrepresent responses" as another broad class of potential biases. Two
sub-categories within this classification will be shown to be of interest later

in the chapter because they too can be reasons for observing part-whole bias. They are (i) "strategic bias," which is said to be caused by "a respondent giving a WTP amount that differs from his or her true WTP amount (conditional on the perceived information) in an attempt to influence the provision of the good and/or the respondent's level of payment for the good," and (ii) under the sub-category of "compliance bias," "Sponsor bias, where a respondent gives a WTP amount that differs from his or her true WTP amount in an attempt to comply with the presumed expectations of the sponsor (or assumed sponsor)." (Mitchell and Carson 1989, p. 235) Some have considered the case where a respondent values the whole and the sum of the parts equally as the "adding up" property and have even proposed it as a test of internal consistency of WTP results (Diamond 1996). Others have shown that adding up would be internally consistent only if the goods (or a good's characteristics) were perfect substitutes. Otherwise, the whole should be expected to be less than the sum of the individual parts, in general, for goods (or a good's characteristics) that are imperfect substitutes for each other (Kopp and Smith 1997).

Following on from (3), we define v_j as the ratio of the stated value of the total to the sum of the stated values of the m parts:[3]

$$v_j = \frac{\phi(\sum_{m=1}^{M} q_{jm})}{\sum_{m=1}^{M} \phi(q_{jm})} = \frac{t_j}{\sum_{m=1}^{M} p_{jm}} = \frac{t_j}{s_j} , \qquad (4)$$

where t_j is the stated value of the total bundle of M characteristics, and s_j is the sum of the separately valued parts p_{jm} for the jth respondent. There may be a set of t_j and s_j values one might find in an empirical distribution of ratios from the sample of respondents in a stated preference study. Each category may contain behavioral information useful for an innovator attempting to price a product or a policymaker trying to value a set of public projects. Further, the stated value of each component part *relative* to the values of the other parts in the bundle conveys information useful to both types of decision makers.

Consider the set of values for t_j, s_j, and v_j in Table 1. Category 1 (c^1) is comprised of the responses where t_j, s_j, and v_j are all equal to zero. This is most likely the set of respondents who are giving protest zeros, a kind of *strategic bias*. The only other explanation is that the respondents place no

---

[3] Notice this measure is the inverse ratio, or the whole/part ratio. We introduce this bit of confusion because the whole/part ratio better serves to illustrate the points we want to make in the sections that follow.

Table 1. Respondent Valuation Categories

Category	t_j value	s_j value	v_j	Behavioral Implications/Source of Bias
c^1	0	0	0	1. A protest zero response *(strategic bias)*. 2. The person places no value on the non-pecuniary characteristics (unlikely).
c^2	0	> 0	0	Person thinks of the parts as having value, but places a protest zero on the total for *strategic* reasons.
c^3	> 0	> 0 and $t_j < s_j$	< 1	Person displays diminishing marginal utility in the characteristics and the characteristics are substitutes in valuation. *The most representative case.*
c^4	> 0	> 0 and $t_j = s_j$	$= 1$	Person displays *sponsor bias* in that he wants to appear to be consistent to the evaluator or the sponsor.
c^5	> 0	> 0 and $t_j > s_j$	> 1	Person is valuing more characteristics in the whole than he was asked about separately in the parts. *Amenity misspecification bias/benefit part-whole bias.*
c^6	> 0	0	undefined	Person places no value on the characteristics asked about, but places some value on other characteristic(s). *Amenity misspecification bias/benefit part-whole bias.*

value on any of the characteristics of the good, which is highly unlikely. The respondents in category 2 (c^2) value some of the characteristics in the bundle, but place zero value on the total. This is evidence of a *strategic* bias in that they most likely relate the total value to the price they may have to pay for the product and wish to *misrepresent* the true value they place on the product. In category 3 (c^3), both t_j and s_j are positive but t_j is less than s_j, which implies $v_j < 1$. This should be the dominant category found in a set of respondents asked to value non-pecuniary characteristics of a good. This dominance is supported by empirical evidence in the following sections. Bateman et al. (1997) found this case to be pervasive in a set of controlled experiments designed to test for the presence of part-whole bias. In category 4 (c^4), t_j and s_j are positive and equal, which implies $v_j = 1$. This response is a type of *sponsor bias* where the respondent wants to appear internally consistent to the enumerator or to the sponsor of the survey, reasoning that the sum of the parts should equal the total and so adjusting his responses accordingly. Responses in c^5 imply that the total is greater than the sum of the parts, ($v_j > 1$). This could be the *benefit part-whole bias* form of *amenity misspecification bias* in Mitchell and Carson's (1989) parlance. In category 6 (c^6), t_j is positive, but all of the parts, s_j, are stated to have a zero value, which implies that v_j is undefined. This could also be an extreme form of *benefit part-whole bias* where none of the characteristics asked about in the

CV study have value to the respondent, but one or more left-out characteristics have some value.[4]

The respondents in c^3 conform to the predictions of utility theory, as laid out by Hoehn (1991), and conform to the utility maximization problem laid out earlier in this chapter. Strategic misrepresentations or other anomalies are not expected to be present in c^3. Therefore, it should be viewed as the *category that is most representative of the underlying value*, to which all other categories are compared, and from which calculated rescaling factors are more likely to be better representations of the true part-worths. We now turn to our empirical findings.

5. SURVEY EVIDENCE OF NON-PECUNIARY BENEFITS

5.1. The Surveys

The following empirical evidence is based on three, computer-aided telephone surveys conducted by Doane's Market Research over the past 4 years. The purpose of the first survey, conducted in early 2001 and comprised of 601 responses ($J=601$), was to elicit U.S. corn farmers' opinions and evaluations of the new Yieldgard Rootworm® technology, which was introduced commercially the following year. The sample was stratified by corn acres in each USDA region where there is significant corn acreage and by two additional sub-regions where a corn rootworm variant has evolved to overcome a corn/soybean rotation (Figure 1). One variant, the SB variant, is able to lay its eggs and have the larvae feed on the soybean crop, and the other, the Extended Diapause (EDP) variant, has evolved to be able to skip a growing season to lay its eggs and have the larvae feed in the next corn crop. Each respondent was asked to place a value per acre, if any, on the additional time savings and equipment savings, the additional farmer and worker safety, the additional environmental benefits, and more consistent corn rootworm control (a reduction in production risk) that they thought would be provided by this technology relative to the existing corn rootworm control technologies (Alston et al. 2002).

The second survey was taken in 2002 and is comprised of 610 responses ($J=610$). It was designed to elicit U.S. soybean farmers' valuation of Roundup Ready (RR) soybeans, which were introduced in 1996. This sample was stratified by soybean acreage in the northern and southern regions of the United States. The same questions were asked about the additional safety

---

[4] In all of the studies considered here, the "total" valuation question was asked specifically and exclusively about the set of characteristics the respondents were asked to value individually.

Figure 1. Map of Regions for the Corn Rootworm Survey

and environmental impact of the technology, plus additional total convenience of RR soybeans relative to conventional soybeans (Marra, Piggott, and Carlson 2004).

The third survey of North Carolina crop farmers is comprised of 293 responses ($J=293$) and was conducted in 2003 to elicit their valuation of herbicide-tolerant crops (mostly RR). The main crops sampled were corn, cotton, and soybeans. The same valuation questions were asked of the respondents as in the national soybean survey (Marra, Piggott, and Sydorovych 2005).

All valuation questions in the surveys were asked as open-ended questions. This question format was chosen because U.S. farmers are familiar with the basic technologies and implicitly value the trade-offs between the biotech and conventional technologies frequently. An example of the series of valuation questions used in the surveys can be found in the appendix. The major alternative question type, dichotomous choice, has been demonstrated to produce higher values than the open-ended or the choice modeling approaches in many studies (Brown et al. 1996; Ready, Buzby, and Hu 1996), though rebuttal evidence exists (e.g., Huang and Smith 1998). Open-ended CV questions also have been shown to be as reliable as dichotomous choice questions and easier to interpret (Loomis 1990).

5.2. Values of Individual Characteristics

Tables 2, 3, and 4 show the unscaled valuation results from each survey.[5] Each entry is the sample mean of the responses in that cell. These results are presented only so that the reader can see the differences in the values by geography and by the degree of adoption of the technology in question. In the case of the corn rootworm study, which is an *ex ante* study, the likelihood that the new technology will be adopted is the only measure of the degree of adoption available.

It is clear from Table 2 that the production risk reduction provided by more consistent control of the corn rootworm is valued more highly by these respondents than either the operator and worker safety or the environmental safety characteristic. Likely adopters generally place a higher value on each characteristic than do unlikely adopters. Even though the characteristic values differ across regions, there seems to be no discernable geographic pattern in the differences.

Additional operator and worker safety, environmental safety, and total convenience characteristics were valued by respondents in the national soybean survey. These results are shown in Table 3. In this survey, total convenience was valued more highly than operator and worker safety or environmental safety (generally about twice as much). Full adopters placed a higher value on each characteristic, with partial adopters placing the next highest value, and non-adopters placing the lowest value on each. Growers in the South who had adopted RR soybeans placed a higher value on convenience than their northern counterparts, possibly because of more diverse crop mixes in the South or less contiguous land per farm.

The characteristic values from the North Carolina (NC) herbicide-tolerant crop survey are presented in Table 4. Total convenience was valued highest by a factor of 2 to 3 compared to operator and worker safety and environmental safety by the North Carolina respondents who were non-adopters or partial adopters. The full adopters regarded total convenience to be five to six times as valuable as either operator and worker safety or environmental safety. Environmental safety was valued slightly less than operator and worker safety except in the full adopter category, where environmental safety was valued higher than operator and worker safety on a per acre per year basis. Partial adopters seemed to value total convenience about half as much as the non-adopters or the full adopters. This makes some sense because the partial adopters are planting both types and are not taking full advantage of the convenience inherent in planting just one type, although

---

[5] In the empirical evidence presented in the next section, the sample size of each survey was less than the total number of survey responses because one or more of the important variables used in the analysis were missing for some observations.

Table 2. Unscaled SMVs for the Corn Rootworm Survey

Region	Adoption Intention	J	Time Savings	Equipment Savings	Operator and Worker Safety	Environmental Safety	More Consistent Control (Risk Reduction)
					$/acre/year		
Heartland	likely	68	2.77	2.23	2.11	1.62	4.23
	unlikely	14	2.73	1.51	1.50	0.59	3.20
Northern Crescent	likely	63	2.64	1.78	2.20	1.46	4.06
	unlikely	14	1.86	1.44	0.97	0.16	2.96
N. Great Plains	likely	59	2.79	1.73	2.76	2.47	4.77
	unlikely	18	2.68	0.98	2.22	1.61	2.00
Other	likely	61	3.65	2.79	1.55	1.73	5.38
	unlikely	12	1.84	1.48	1.85	0.39	3.65
Prairie Gateway	likely	70	3.50	2.07	2.50	1.46	3.94
	unlikely	13	2.13	1.21	1.08	0.60	1.69
SB variant	likely	42	3.40	2.86	2.94	2.24	3.24
	unlikely[a]	--	--	--	--	--	--
EDP variant	likely	35	3.32	1.83	2.50	2.03	4.42
	unlikely	7	1.44	0.81	1.43	0.46	2.57
All	likely	398	3.07	2.17	2.33	1.81	4.32
	unlikely	80	2.08	1.26	1.55	0.71	2.61

[a] Not reported to avoid revealing individual respondents' information.

Notes: Estimates are sample means. The values in this table are not rescaled. Their absolute magnitudes should not be used for policy or pricing purposes.

Table 3. Unscaled SMVs from the National Soybean Survey

Characteristic	Region (J)	All Farms (304)	Adopter Category (J)		
			Non Adopters (59)	Partial Adopters (78)	Full Adopters (167)
			$/acre/year		
Operator and worker safety	North (273)	2.21	1.40	1.53	2.82
	South (31)	1.29	0.00	1.22	1.61
Environmental safety	North (273)	2.54	1.35	2.36	3.07
	South (31)	1.65	0.00	0.11	2.78
Total convenience	North (273)	5.10	3.10	4.12	6.30
	South (31)	6.13	0.00	4.89	8.11

Note. Estimates are sample means. The values in this table are not rescaled. Their absolute magnitudes should not be used for policy or pricing purposes.

Table 4. Unscaled SMVs from the NC Herbicide-Tolerant Crops Survey

Characteristic	Adopter Category (J)			
	All Farms (71)	Non Adopters (49)	Partial Adopters (10)	Full Adopters (12)
Operator and worker safety	7.10	8.51	4.40	3.59
Environmental safety	6.14	7.04	4.20	4.08
Total convenience	16.90	17.69	8.90	20.33

Note: Estimates are sample means. The values in this table are not rescaled. Their absolute magnitudes should not be used for policy or pricing purposes.

this result could just be a sparse data problem. The full adopters value total convenience higher than do the non-adopters, but value operator and worker safety and environmental safety about half as much.

It is clear from all three of these studies that farmers generally value these non-pecuniary characteristics and that the relative contributions of each to the total value placed on the technology are not the same. *The unscaled values reflect the relative differences in the characteristic values only and do not represent the "true" characteristic values*, as we demonstrate below. The next section takes up the matter of part-whole bias in measurement of these non-pecuniary values and introduces a method of correcting the problem, while leaving the relationships among the individual values unchanged.

6. THE RATIO OF THE WHOLE TO THE SUM OF THE PARTS

Thus far, researchers have reported the overall mean of the empirical part-whole bias in their work (Bateman et al. 1997 or Boyle et al. 1994, for example). Morrison (2000) decomposes the answers to open-ended WTP questions about a car safety feature that would reduce risk of injury by varying amounts. The first question asked about a risk reduction of 12 in 1,000, or a 1.2 percent reduction in the chance of a permanent, non-fatal injury. The second asked about a risk reduction of 4 in 1,000, or a 0.4 percent reduction in the chance of the same injury. This implies that the 1.2 percent risk reduction is a 300 percent greater risk reduction than the 0.4 percent reduction in the chance of injury. The respondents were only willing to pay an extra 20 percent for the 300 percent reduction in risk of injury. Morrison categorizes the responses according to whether they are deemed to be internally consistent or not, based on their responses to the above questions. This is the same approach we take to decomposing v_j into the categories described in Table 1, although we are dealing with valuing different characteristics embodied in a good, while Morrison was dealing with different levels of the same good, or the embedding effect. Further, we are decomposing the degree of "*bias*" into its component parts, while Morrison is comparing the individual *values*. In the next section we consider the distributional properties of v_j for all three surveys.

7. DECOMPOSITION OF THE MEAN PART-WHOLE BIAS

7.1. Methodology

The three surveys provide three different samples of the ratio of the stated total value to the sum of the values of the parts, or what we refer to as v_j. A given v_j represents the factor by which each of the M parts for individual j would need to be rescaled so that the sum of the parts would equal the stated total values. This estimate of v_j also provides some measure of the magnitude of the part-whole bias as other researchers have defined it.[6] A relatively small value of v_j, say $v_j = 0.1$, would imply a large difference between the value of the whole and the sum of the value of the parts, with the whole representing only 10 percent of the sum of the parts. A larger value of v_j, say $v_j = 0.95$, would imply a small degree of part-whole bias, with the total value representing 95 percent of the sum of the value of the parts.

We are interested in several distributions and measures of central tendency of v_j, over various categories and sets of categories. First, we are interested in the distribution of v_j *over all categories*, for which v_j is defined $(c^1 - c^5)$, as a measure of dispersion of the ratio of the total values to the sum of the values of the parts in general.[7] The distribution and central tendency of v_j *across all individuals* provides an indication of the part-whole bias in the entire sample across all categories. Second, we are interested in the distribution of v_j for c^3 respondents. This distribution characterizes the rescaling factor used in each survey for the jth individual. It also characterizes

---

[6] Another way to estimate the magnitude of the part-whole bias would be to consider the difference between the stated total value (t_j) and the sum of the parts (s_j) (bias $= s_j - t_j$). Since the difference approach is subject to the magnitudes involved, we chose the ratio approach instead $(v_j = t_j / s_j)$ as a unitless measure of the bias, representing the rescaling factor by which each of the parts would need to be multiplied such that their sum would equal the stated total value.

[7] A slight disadvantage of using (t_j / s_j) rather than $(s_j - t_j)$ as a measure of the part-whole bias is that this precludes the individuals in c^6 to be included in the calculation of v_j over all the categories since the ratio is undefined when $s_j = 0$. Since we use only respondents belonging to c^3 $(0 < v_j < 1)$, when we turn to the task of rescaling the parts so that they sum to the stated total value, this disadvantage is minimal for our purposes.

the amount of part-whole "bias" in this category. A measure of central tendency (say, the mean or median) of the v_j of all categories ($c^1 - c^5$) might be larger or smaller than that of v_j for c^3 respondents alone, depending on the relative number of individuals in c^1 and c^2 compared with c^4 and c^5, and also the measure of central tendency of v_j in c^5. The fact that individuals belong to different categories within each sample complicates the task of identifying an appropriate measure of the degree of part-whole bias for a given sample.

To illustrate the continuous distributions in categories 3 and 5, we estimated kernel densities using the Analytical Methods Committee's statistical add-in to Excel V1.0e. This add-in fits a kernel density to the sample data (Royal Society of Chemistry 2001). It fits the density curve according to

$$\hat{f}(v;h) = \frac{1}{jh}\sum_{j=1}^{J}\Phi(\frac{v-v_j}{h}), \qquad (5)$$

where v is a point on the x-axis, v_j is a data point from the sample of J individual points, $\Phi(\ )$ is the standard normal density, and h is standard deviation designated by h^{Opt}.[8] This procedure fits an estimated probability density function by generating a series of R pairs $(v_r, \hat{f}_r(v;h))$, which reflect the properties of the sample for v_j. The sum of all the probability density estimates $\hat{f}_r(v;h)$, multiplied by the step size δ (the distance between each v_r in the series $\delta = v_r - v_{r-1}$), is approximately equal to

$$\sum_{r}^{R} \delta \cdot \hat{f}_r(v;h) = 1.$$

By construction, a small number of these series of pairs (which we will denote T) will have a v_r, which falls outside the feasible range of values for v_j for the probability density. To correct and account for this, the offending v_r's probability density estimates are summed to create an adjustment factor,

$$ADJ \ [ADJ = \sum_{r=1}^{T} \hat{f}_r(v;h)],$$

---

[8] In fitting these curves we used the default value of kernel width h which is denoted h^{Opt} and is calculated as

$$h^{Opt} = 0.9 \cdot \min(std, IQR/1.34)*j^{-0.2},$$

where std is the standard deviation for the survey, IQR is the inter-quartile range, and j is the number of data points.

and the remaining $R-T$ probability density estimates, which are based on v_r's within the feasible range, are re-scaled by ADJ according to

$$\tilde{\tilde{f}}_r(v;h) = \hat{f}_r(v;h) \cdot (1 + ADJ),$$

such that the series of pairs within the feasible range satisfy the property that

$$\sum_{r}^{R-T} \delta \cdot \tilde{\tilde{f}}_r^{J}(v;h) = 1.$$

Figures 2–4 depict the results of this procedure for each survey. In the top panel of each figure is the kernel density estimated for the respondents in c^3. The middle panel shows the kernel density function for those in c^5. The bottom panel shows the cumulative distribution function of the mixture distribution for each survey. Notice that the probability density functions are positively skewed for c^3 and c^5 respondents in *each* survey. This has implications for which of the measures of central tendency is the most appropriate. In the case of non-symmetric distributions, the median is the preferable measure.

The cumulative density function in each figure reveals the discontinuities imposed by $v_j = 0$ and $v_j = 1$ respondents. These discrete distributions result in a stepped CDF function that begins above the origin, reflecting probability mass at $v_j = 0$ (c^1 and c^2), and also a step at the value of $v_j = 1$, reflecting probability mass at $v_j = 1$ (c^4).

7.2. Empirical Properties of v_j

The sample properties of v_j are shown in Table 5. First, notice that the proportion of respondents falling in c^3 is the highest of all the categories in each survey, although the proportion varies. The national soybean survey has the smallest proportion of respondents in c^3, with about 39 percent, while almost 77 percent of the respondents in the corn rootworm survey fall into c^3. This implies that the other, less representative, categories account for 23 percent to as much as 60 percent of the respondents in these surveys. That is, there appears to be strong empirical support, at least from the surveys considered in this analysis, for a significant number of observations in categories of respondents who, for one reason or another, appear to be giving biased estimates of the values according to some strategic stance or other reason as spelled out in Table 1.

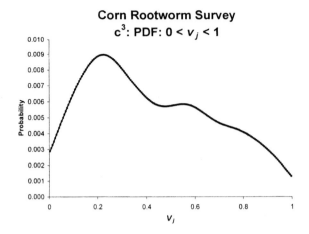

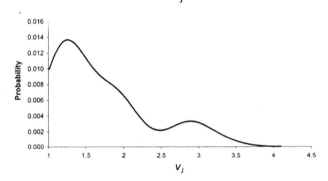

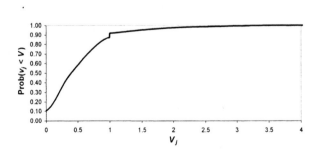

Figure 2. PDFs and CDF for v_j for the Corn Rootworm Survey

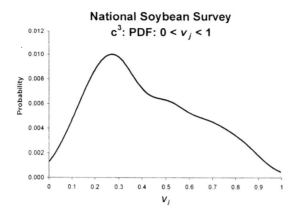

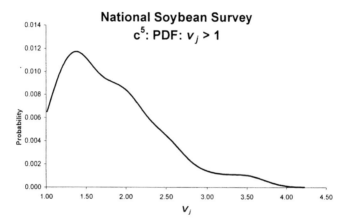

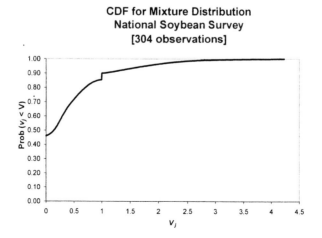

Figure 3. PDFs and CDF for v_j for the National Soybean Survey

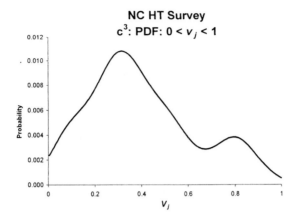

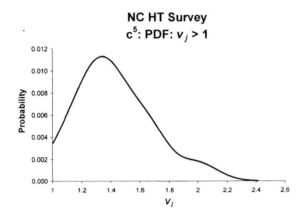

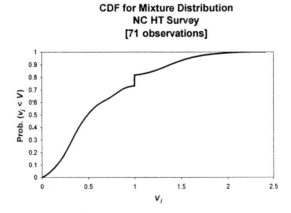

Figure 4. PDFs and CDF for v_j for the NC HT Survey

Table 5. Sample Properties of the Ratio of Total Value and Sum of the Parts (v_j)

Categories	J	w_j	Mean	Std. Dev.	Skewness	Median
			Corn Rootworm Survey			
$c^1 : t_j = 0; \ s_j = 0; \ v_j = 0$	0	0	0	0	--	0
$c^2 : t_j = 0; \ s_j > 0; \ v_j = 0$	51	0.107	0	0	--	0
$c^3 : t_j > 0; \ s_j > 0; \ v_j < 1$	**367**	**0.768**	**0.422**	**0.260**	**0.377**	**0.383**
$c^4 : t_j > 0; \ s_j > 0; \ v_j = 1$	21	0.044	1	0	--	1
$c^5 : t_j > 0; \ s_j > 0; \ v_j > 1$	39	0.082	1.672	0.638	1.143	1.429
$[c^1 - c^5]$	**478**	**1**	**0.504**	**0.489**	**2.150**	**0.385**
$c^3 / [c^1 - c^5]$			0.837	0.532	0.176	0.996
			National Soybean Survey			
$c^1 : t_j = 0; \ s_j = 0; \ v_j = 0$	97	0.34	0	0	--	0
$c^2 : t_j = 0; \ s_j > 0; \ v_j = 0$	35	0.122	0	0	--	0
$c^3 : t_j > 0; \ s_j > 0; \ v_j < 1$	**113**	**0.394**	**0.417**	**0.222**	**0.483**	**0.357**
$c^4 : t_j > 0; \ s_j > 0; \ v_j = 1$	13	0.045	1	0	--	1
$c^5 : t_j > 0; \ s_j > 0; \ v_j > 1$	29	0.101	1.795	0.601	1.055	1.667
$[c^1 - c^5]$	**287**	**1.00**	**0.391**	**0.585**	**2.198**	**0.167**
$c^3 / [c^1 - c^5]$			1.066	0.380	0.220	2.143
			North Carolina Herbicide-Tolerant Survey			
$c^1 : t_j = 0; \ s_j = 0; \ v_j = 0$	0	0	0	0	--	0
$c^2 : t_j = 0; \ s_j > 0; \ v_j = 0$	0	0	0	0	--	0
$c^3 : t_j > 0; \ s_j > 0; \ v_j < 1$	**52**	**0.732**	**0.401**	**0.228**	**0.591**	**0.333**
$c^4 : t_j > 0; \ s_j > 0; \ v_j = 1$	6	0.085	1	0	--	1
$c^5 : t_j > 0; \ s_j > 0; \ v_j > 1$	13	0.085	1.426	0.255	0.781	1.380
$[c^1 - c^5]$	**71**	**1**	**0.639**	**0.466**	**0.919**	**0.500**
$c^3 / [c^1 - c^5]$			0.628	0.489	0.643	0.666

Notes: t_j = stated total value, s_j = sum of stated characteristic values, $v_j = t_j / s_j$, and w_j = the proportional weight of the individual category in the total.

7.2.1. All Categories $c^1 - c^5$

Turning attention to the distribution of v_j over all categories for which it is defined, $c^1 - c^5$, Table 5 reveals that the mean estimate of v_j ranges from 0.391 for the national soybean survey to 0.639 for the NC herbicide-tolerant survey. Thus, there appears to be a significant amount of variation in the mean of v_j across surveys when all categories are included. Each of the surveys exhibits a significant amount of positive skewness over all categories, with skewness coefficients ranging from 0.919 for the NC herbicide-tolerant survey to 2.198 for the national soybean survey. This degree of skewness suggests that the median would be more appropriate as a measure of central tendency than the mean. The range of the median reveals that the central tendencies of the overall distributions of v_j in the surveys are even more diverse than that which the means reveal, ranging from 0.167 for the national soybean survey to 0.500 for the NC herbicide-tolerant survey. These comparisons of the distributions and central tendencies of the v_j's over all categories for the three surveys reveal several important findings and mask another.

Adopting the approach of simply calculating the mean of v_j over j individuals as a measure of central tendency of the part-whole bias, while ignoring the idea that there may be strategic responses in the sample that should be categorized and treated differently, could mask other important factors. Both the mean and median calculated over all individuals (in all categories) might be significantly impacted either downward or upward by the relative frequencies of responses known to be biased in categories 1, 2, 4, and 5. This point can be seen by comparing the ratios of the medians for c^3 and $c^1 - c^5$ in Table 5. This ratio, $c^3/(c^1 - c^5)$, ranges from 0.666 in the NC herbicide-tolerant survey to 2.143 in the national soybean survey, which indicates the influence of the observations in the categories considered to contain biased responses. This suggests that it might be more appropriate to limit attention to the most representative (or non-strategic) category c^3, and it provides another reason to consider the median rather than the mean as the best representation of the central tendency of the distribution of the sample part-whole bias. This is a departure from previous literature, which tends to report only the sample mean part-whole bias over all observations.

7.2.2. Category c^3

Consideration of only the most representative estimates (c^3) reveals remarkably similar properties of the distributions of the part-whole biases across surveys as measured by the first three moments (mean, standard deviation, and skewness) of each sample, as well as by the median. The mean v_j is about the same for c^3 respondents in each survey, ranging from 0.401 in the NC herbicide-tolerant survey to 0.422 in the corn rootworm survey, and all

c^3 means have similar standard deviations falling in the range of around 0.222–0.260. All three surveys exhibit positive skewness for the respondents falling in the c^3 category, ranging from 0.377 for the corn rootworm survey to 0.591 for the NC herbicide-tolerant survey. The medians for v_j were 0.383, 0.357, and 0.333 for the corn rootworm survey, the national soybean survey, and the NC herbicide-tolerant survey, respectively. These estimates of the median indicate that the sum of the parts for the majority of the respondents should be rescaled downward *by more than 60 percent*.

Tables 6 and 7 show how the categories of v_j are dispersed across two important cuts of the data. The regional dispersion of v_j is shown in Table 6 on the left-hand side, along with the overall median v_j in each region (and for all observations) and the medians within each region for the two categories of v_j where the distributions of observations are continuous: $0 < v_j < 1$ and $v_j > 1$. There seems to be little regional influence in the dispersion of the categories for the surveys (where regions are relevant). In particular, the percentage of observations in the categories with likely misrepresentations of one form or another do not seem to differ very much across regions. In the corn rootworm survey the exception might be in the SB variant region, where over 86 percent of the observations are in the category where the respondents' answers are expected to be most representative. This may be due to the intensity of the corn rootworm problem in that region and the respondents' desire to demonstrate the true value to them of the new technology. In the national soybean survey 43 percent of the observed v_j's are zero, but they are dispersed fairly evenly across regions on a percentage basis, with 41.8 percent in the North and 58.0 percent in the South. The North region has the only responses that fall into c^6; but with only 17 observations overall in that category, it is difficult to draw any inference or speculate about a reason. The NC herbicide-tolerant survey has no observations in categories 1, 2, or 6.

The overall median v_j's range from 0.313 in the Prairie Gateway region to 0.528 in the Northern Crescent in the corn rootworm survey. The median v_j's in c^3 range from 0.382 in the SB Variant region to 0.496 in the EDP Variant region. In the category where $v_j > 1$ (c^5) the range is larger, from 1.214 in the EDP Variant region to 2.119 in Northern Great Plains region. The median values for v_j in the national soybean survey are lower by about 20 percentage points overall when compared with the corn rootworm survey. In the corn rootworm and the NC herbicide-tolerant surveys, the observations falling into c^3 dominate, as expected, in all regions. The national soybean survey shows more observations in the categories where $v_j = 0$ (c^1 and c^2) combined, but c^3 has the most respondents overall.

Overall comparison of the medians of c^1–c^5 and c^3 shows the difference in variability of the two measures across regions and surveys. The median value for the c^1–c^5 responses ranges from 0.000 (58 percent of observations in c^1 and c^2 in the South region in the national soybean survey) to 0.508 in

Table 6. Properties of v_j Across Categories and Region

| Region | Percent of Respondents in Category and Region | | | | | | All (%) | All (# of resp.) | $0 < v_j < v_j^{max}$ ($c^1 - c^5$) | | | $0 < v_j < 1$ (c^3) | | $v_j > 1$ (c^5) | |
	c^1	c^2	c^3	c^4	c^5	c^6			J	v_j^{med}	v_j^{max}	J	v_j^{med}	J	v_j^{med}
Corn Rootworm Survey															
Heartland	0.0%	8.5%	80.5%	6.1%	4.9%	0.0%		82	82	0.343	1.805	66	0.400	4	1.360
N. Crescent	0.0%	10.4%	72.7%	2.6%	14.3%	0.0%		77	77	0.508	2.182	56	0.451	11	1.548
N. Great Plains	0.0%	11.7%	75.3%	2.6%	10.4%	0.0%		77	77	0.385	3.333	58	0.432	8	2.119
Other	0.0%	8.2%	76.7%	5.5%	9.6%	0.0%		73	73	0.444	2.857	56	0.419	7	1.636
Prairie Gateway	0.0%	15.7%	74.7%	3.6%	6.0%	0.0%		83	83	0.313	2.941	62	0.401	5	1.868
SB variant	0.0%	4.5%	86.4%	4.5%	4.5%	0.0%		44	44	0.323	1.500	38	0.382	2	1.283
EDP variant	0.0%	14.3%	73.8%	7.1%	4.8%	0.0%		42	42	0.523	1.429	31	0.496	2	1.214
Total (# of resp.)	0	51	367	21	39	0		478	478	0.385	3.333	367	0.422	39	1.672
National Soybean Survey															
North	30.8%	11.0%	38.1%	4.4%	9.5%	6.2%		273	256	0.182	3.000	104	0.357	26	1.633
South	41.9%	16.1%	29.0%	3.2%	9.7%	0.0%		31	31	0.000	3.500	9	0.333	3	2.000
Total (# of resp.)	97	35	113	13	29	17		304	287	0.167	3.500	113	0.357	29	1.667
North Carolina Herbicide-Tolerant Survey								71							
Total (# of resp.)	0	0	52	6	13	0		71	71	0.500	2.000	52	0.333	6	1.389

Notes: v_j = the ratio of total value and the sum of the parts; v_j^{med} is the sample median; v_j^{max} is the maximum value in the sample.
North region: Iowa, Illinois, Indiana, Kansas, Michigan, Minnesota, Missouri, Nebraska, Ohio, South Dakota, Wisconsin, and Kentucky.
South region: Alabama, Arkansas, Louisiana, Mississippi, North Carolina, Tennessee, and South Carolina.

Table 7. Properties of v_j Across Categories and Adopter Category

Adopter Category	% of Respondents in Category and Adoption Category						All (# of resp.)	$0 < v_j < v_j^{max}$ ($c^1 - c^5$)			$0 < v_j < 1$ (c^3)		$v_j > 1$ (c^5)	
	c^1	c^2	c^3	c^4	c^5	c^6		J	v_j^{med}	v_j^{max}	J	v_j^{med}	J	v_j^{med}
Corn Rootworm Survey														
Likely	0.0%	7.8%	79.2%	4.5%	8.5%	0.0%	398	398	0.412	3.330	315	0.385	34	1.373
Unlikely	0.0%	25.0%	65.0%	3.8%	6.3%	0.0%	80	80	0.261	3.000	52	0.333	5	1.923
Total (# of resp.)	0	51	367	21	39	0	478	478	0.385	3.333	367	0.383	39	1.429
National Soybean Survey														
Non-adopter	40.2%	11.4%	10.6%	0.0%	3.5%	17.7%	56	56	0.000	1.250	12	0.388	1	1.250
Partial adopter	27.8%	28.6%	23.0%	23.1%	24.1%	29.4%	73	73	0.000	2.400	26	0.500	7	1.400
Full adopter	32.0%	60.0%	66.4%	76.9%	72.4%	52.9%	158	158	0.250	3.500	75	0.333	21	1.667
Total (# of resp.)	97	35	113	13	29	17	304	287	0.167	3.500	113	0.357	29	1.667
North Carolina Herbicide-Tolerant Survey														
Non-adopter	0.0%	0.0%	71.2%	11.5%	17.3%	0.0%	49	49	0.455	2.000	37	0.376	9	1.429
Partial adopter	0.0%	0.0%	60.0%	20.0%	20.0%	0.0%	10	10	0.625	1.364	6	0.345	1	1.363
Full adopter	0.0%	0.0%	66.7%	11.1%	22.2%	0.0%	12	12	0.700	1.429	9	0.444	3	1.389
Total (# of resp.)	0	0	52	6	13	0	71	71			52	0.333	13	1.389

the Northern Crescent in the corn rootworm survey. Looking across the medians for the c^3 category shows significantly less variability, from 0.333 in the South region in the national soybean survey and in the NC herbicide-tolerant survey to 0.496 in the EDP Variant region in the corn rootworm survey. This supports the hypothesis that the category of the responses we have shown to be most representative presents a more reliable measure of central tendency than that taken from all the categories for which v_j is defined.

Table 7 contains the properties of v_j across categories of adopter for the three surveys. No observations in categories 1 or 6 appear in the corn root-worm survey. The observations are more concentrated in c^3 among the likely adopters compared with the unlikely adopters in the corn rootworm survey. It could be that the growers unlikely to adopt the Yieldgard Rootworm technology are from regions where corn rootworm is not as prevalent and so are less familiar with the problem.

The national soybean survey has over 62 percent of the observations in categories 1, 2, 4, 5, and 6, with over 40 percent in categories 1 and 2. The highest proportion of respondents in c^1 is in the non-adopter category (over 40 percent). All adopter categories in this survey have observations in c^6. There are no observations in c^6 in the other two surveys.

The NC herbicide-tolerant survey observations are found only in categories 3, 4, and 5 and are fairly well dispersed across adopter categories within those categories of v_j. The exception is that, similar to the national soybean survey, there are relatively few observations in c^4 for the non-adopters. Recall that c^4 contains observations where $t_j = s_j$ and $v_j = 1$, which demonstrates sponsor bias. This result may be because these surveys are both based on *ex post* surveys where the non-adopters reveal their actual choice. They may not feel any incentive to "comply with" or "please" the sponsor in this case by trying to make their responses "internally consistent."

The range of median values of v_j based on all the categories (c^1–c^5) is from zero for the non-adopters and partial adopters in the national soybean survey to 0.700 for the full adopters in the NC herbicide-tolerant survey. Looking at the median values of v_j for c^3 responses across surveys shows a range similar in magnitude to the c^3 responses across regions. The range is from 0.333 for unlikely adopters in the corn rootworm survey to 0.500 for partial adopters in the national soybean survey, and is about one-fourth the range based on all categories of responses.

The median values for all responses in each survey show a similar pattern. The smallest overall median is 0.167 and the largest is 0.500, for the distribution of v_j based on all categories, while the range of median v_j's based on c^3 responses only is from 0.333 to 0.387. These results attest to the greater reliability in the measure of part-whole bias when only the most representative category of responses is included.

In sum, we do not find a discernable pattern in the observations in the different categories of v_j where the data are categorized by timing of the survey (*ex post* or *ex ante*), geographic criteria, or adoption intensity criteria. We can conclude tentatively that observations in the different categories of v_j might be found in responses to any valuation survey that attempts to elicit values of parts individually and separately from the whole.

8. RESCALED VALUES OF NON-PECUNIARY CHARACTERISTICS

We now demonstrate the practical importance of the decomposition of the ratio of the whole to the sum of its parts into the different categories, each representing different behavior on the part of the respondents. First, we use only data from c^3, which we believe contains the most representative information on these values from the surveys. Within c^3 each individual's component values were rescaled as follows:

$$\tilde{p}_{jm} = v_j \cdot p_{jm}, \tag{6}$$

where $\tilde{p}_{jm}$ is the re-scaled, "true" value of the mth characteristic by the jth respondent. This re-scaling ensures that the re-scaled sum of the "true" values of the part-worths for each respondent equals the stated total value t_j. This property of the rescaling procedure can be shown as follows:

$$\sum_{m=1}^{M} \tilde{p}_{jm} = \sum_{m=1}^{M} v_j \cdot p_{jm} = v_j \cdot \sum_{m=1}^{M} p_{jm} = \frac{t_j}{s_j} \cdot \sum_{m=1}^{M} p_{jm} = t_j \ \forall\, j. \tag{7}$$

This re-scaling was performed for each non-pecuniary characteristic, for each respondent, and in each survey. We report the descriptive statistics of the rescaled values for each non-pecuniary characteristic and each survey in Table 8. We also calculate the share of each characteristic in Table 8. Notice that, as with the v_j's, the distribution of the value of *each* characteristic is positively skewed, indicating that the median value is the best descriptive statistic to use for each characteristic value as well. The corn rootworm survey exhibits the highest degree of skewness for the sum of the parts. The degree of skewness for the sum of the parts in the other surveys is less than half that of the corn rootworm survey.

The standard deviation of the sample distribution is greater than both the mean and the median values for most individual characteristics in each survey. The exceptions are the mean value of total convenience in the national soybean survey and the mean of each characteristic in the NC herbicide-tolerant survey. The standard deviation of the sum of the parts is greater than

Table 8. Stated Values and Re-scaled Values, and Relative Contributions of Parts

Characteristic	Un-scaled Median[a]	Value of the Change in the SMV Re-scaled[b]				Share[c] (%)
		Median	Mean $/acre/year	Std Dev.	Skewness	
Corn Rootworm Survey: c^3 : $J = 367$						
Time savings	1.500	0.588	0.997	1.390	4.047	23.86
Equipment savings	1.000	0.400	0.724	0.969	3.087	17.51
Operator and worker safety	1.000	0.429	0.991	1.623	3.670	17.12
Environmental safety	1.000	0.208	0.787	1.565	4.606	10.88
More consistent stand	2.000	0.800	1.773	2.862	4.111	30.63
Sum of the parts	9.400	3.000	5.272	6.222	3.263	
Total	3.000	3.000				
National Soybean Survey: $c^3 = J = 113$						
Operator and worker safety	3.000	0.913	1.660	2.026	1.367	20.97
Environmental safety	3.000	1.304	1.961	2.201	1.257	24.89
Total convenience	10.000	3.333	4.158	3.690	1.114	54.14
Sum of the parts	17.000	5.000	7.779	6.026	1.266	
Total	5.000	5.000				
North Carolina Herbicide-Tolerant Survey: c^3 : $J = 52$						
Operator and worker safety	6.500	2.361	2.923	2.783	0.884	23.91
Environmental safety	5.000	1.666	2.720	2.660	0.955	20.45
Total convenience	15.000	5.000	7.793	7.818	2.588	55.63
Sum of the parts	28.500	10.000	13.437	10.612	1.608	
Total	10.000	10.000				

[a] Median of stated value over all observations in the sample.
[b] Median and sample moments of re-scaled parts where each respondent's stated value of the part is re-scaled by their individual v_j.
[c] Shares are sample means calculated at every data point.

the median value in all surveys. This implies that the dispersion of the values cannot be ignored in any pricing or R&D decisions using these surveys.

The rescaled median characteristic values range from $5.00 per acre per year for total convenience in the NC herbicide-tolerant survey to $0.21 per acre per year for environmental safety in the corn rootworm survey. Farmers valued the risk reduction achieved, as a result of a more consistent stand of corn in the Yieldgard Rootworm corn relative to conventional corn, more highly ($0.80 per acre per year) than each of the other non-pecuniary characteristic changes from the new technology ($0.21 per acre per year for the additional environmental safety to $0.58 per acre per year for the time savings). The value of this characteristic is over 30 percent of the total value in the corn rootworm survey. Total convenience value takes up over 50 percent of the total value in each of the other surveys. The value of each characteristic is highest in the NC herbicide-tolerant survey with a total value twice as high as the national soybean survey. This is due to the fact that three herbicide-tolerant crops are considered in the North Carolina survey (corn, cotton, and soybeans) and that the values for cotton are higher than those for corn or soybeans. Overall, the median total value of the characteristics ranges from $3.00 per acre per year for the Yieldgard Rootworm technology to $10.00 per acre per year for the herbicide-tolerant crops in North Carolina.

9. CONCLUSION

If we are to present the best information to those who are responsible for making decisions, that is, information most representative of the underlying value, we must consider carefully the potential biases in the data from which we calculate the values. In the case of a good with several dimensions that people value separately, information about important biases is masked if one considers only the overall mean part-whole bias of the sample. We argue here that respondents with $0 < v_j < 1$ give the most representative responses. We must rescale the values of the individual parts to reflect the true total value represented by the value of the whole, but at the same time, retain each part's relative importance in the bundle by using information from this category of respondents. From these rescaled numbers, developed using the methodology outlined in this chapter, innovators can price a new technology product to reflect accurately the additional value placed on it by potential consumers, or government can allocate future project proposals according to their true benefit-cost rankings. Innovators also can use the relative values of the component parts (the shares in Table 8) to help decide in which directions to take future research and development.

Several questions and results from this chapter motivate further work. At least three paths emerge. The first is the empirical finding and arguments made for the idea that there are multiple categories of respondents within the population of respondents to stated valuation surveys. Each category implies a difference in how the individuals respond to valuation questions and the bias that might be introduced as a result. Characteristics of the respondents in the various categories should be investigated further in future research.

The second is the result that the most representative category, c^3, appears to have a distribution of v_j that is positively skewed with a median value of around 0.34–0.38. Further investigation of other survey results should shed light on the robustness of these findings. The positive skewness and median values below 0.40 would imply that most respondents tended to overstate the values of the parts by more than 60 percent compared with the stated total values. We do not know if the magnitude of the rescaling factor is similar in general for goods of this nature or if choice of stated preference method matters. We see some evidence that the similarities do not seem to be affected by whether the SMVs are elicited *ex ante* or *ex post* of the commercialization of the product, by geographic region, or by the degree to which the respondent has adopted similar technologies in the past, but this finding should be studied further.

The third item to be investigated further is the reasonableness and accuracy of the re-scaled parts as estimates of the "true" value of the individual characteristics they represent. Although we have hypothesized and provided some empirical support for the notion that the rescaled values are expected to be close to "the truth," more testing of this hypothesis is needed. The accuracy of these values is critical to pricing further advances in biotechnology with similar characteristics to those already valued. Future studies, possibly comparing stated and revealed preferences for the characteristics or some other method of validity testing, are warranted.

REFERENCES

Alston, J., J. Hyde, M. Marra, and P. Mitchell. 2002. "An Ex Ante Analysis of the Benefits from the Adoption of Corn Rootworm." *Agbioforum* 5(3): 71–84.

Bateman, I., A. Munro, B. Rhodes, C. Starmer, and R. Sugden. 1997. "Does Part-Whole Bias Exist? An Experimental Investigation." *The Economic Journal* 107(441): 322–332.

Boyle, K., W. Devousges, F. Johnson, R. Dunford, and S. Hudson. 1994. "An Investigation of Part-Whole Biases in Contingent-Valuation Studies." *Journal of Environmental Economics and Management* 27(1): 64–83.

Brown, T., P. Champ, R. Bishop, and D. McCollum. 1996. "Which Response Format Reveals the Truth about Donations to a Public Good?" *Land Economics* 72(2): 152–166.

Diamond, P.A. 1996. "Testing the Internal Consistency of Contingent Valuation Surveys." *Journal of Environmental Economics and Management* 30(3): 337–347.

Exxon Valdez Oil Spill Trustee Council. 2005. "Oil Spill Facts: Settlement." Available online at http://www.evostc.state.ak.us/facts/settlement.html (accessed February 20, 2005).

Hoehn, J. 1991. "Valuing the Multidimensional Impacts of Environmental Policy: Theory and Methods." *American Journal of Agricultural Economics* 73(2): 289–299.

Huang, J., and V.K. Smith. 1998. "Monte Carlo Benchmarks for Discrete Response Valuation Methods." *Land Economics* 74(2): 186–202.

Kopp, R.J., and V.K. Smith. 1997. "Constructing Measures of Economic Value." In R.J. Kopp, W. Pommerehne, and N. Schwarz, eds., *Determining the Value of Non-Marketed Goods: Economic, Psychological and Policy Relevant Aspects of Contingent Valuation.* Boston: Kluwer Academic Publishers.

Loomis, J. 1990. "Comparative Reliability of the Dichotomous Choice and Open-ended Contingent Valuation Techniques." *Journal of Environmental Economics and Management* 18(1): 78–85.

Marra, M.C., N.E. Piggott, and G. Carlson. 2004. *The Net Benefits, Including Convenience, of Roundup Ready® Soybeans: Results from a National Survey.* Technical Bulletin No. 2004-3, National Science Foundation Center for Integrated Pest Management, Raleigh, NC.

Marra, M.C., N.E. Piggott, and O. Sydorovych. 2005. "The Impact of Herbicide Tolerant Crops on North Carolina's Farmers." *Tar Heel Economist*, North Carolina Cooperative Extension and Department of Agricultural and Resource Economics, North Carolina State University, Raleigh, NC (March/April).

Mitchell, R., and R. Carson. 1989. *Using Surveys to Value Public Goods: The Contingent Valuation Method.* Resources for the Future, Washington, D.C.

Morrison, G. 2000. "Embedding and Substitution in Willingness to Pay." University of Nottingham Discussion Papers in Economics No. 00/10, School of Economics, University of Nottingham, Nottingham, UK.

Ready, R., J. Buzby, and D. Hu. 1996. "Differences Between Continuous and Discrete Contingent Valuation Estimates." *Land Economics* 72(3): 397–441.

Royal Society of Chemistry. 2001. "Representing Data Distributions with Kernel Density Estimates." AMC Technical Brief No. 4, Analytical Methods Committee, Royal Society of Chemistry, London (January).

U.S. Department of Agriculture 2000. *Agricultural Outlook* (August). Economic Research Service, U.S. Department of Agriculture, Washington, D.C.

_____. 2004. *Acreage* (June). Agricultural Statistics Board, National Agricultural Statistics Service, U.S. Department of Agriculture, Washington, D.C.

APPENDIX

Example of Valuation Questions Used in the Surveys

1. In your opinion, are the herbicides used for Roundup Ready soybeans safer for you and your workers to handle than the herbicides used for non-Roundup Ready soybeans?

Yes   ☐
No   ☐ (go to #3)

2. What value per acre, if any, would you place on this additional safety to you and your workers?

    $____.___ added value per acre

3. In your opinion, are the herbicides used for Roundup Ready soybeans safer to the environment than the herbicides used for non-Roundup Ready soybeans?

    Yes   ☐
    No   ☐ (go to #5)

4. What value per acre, if any, would you place on this added environmental safety?

    $____.___ added value per acre

5. In your opinion, about how much total time do you save per acre in minutes, if any, with the Roundup Ready soybean weed control system compared to your non-Roundup Ready soybean weed control?

    _____ minutes per acre

6. What value per acre, if any, would you place on this time saved per acre?

    $____.___ value per acre

7. In your opinion, are there equipment savings (annual fixed and operating costs separate from the time savings you indicated before) in weed control systems for Roundup Ready soybeans compared to the weed control system for non-Roundup Ready soybeans?

    Yes   ☐
    No   ☐ (go to #9)

8. What value per acre, if any, would you place on these equipment savings?

    $____.___ value per acre

9. What value per acre, if any, would you place on the total additional convenience, including the time and equipment savings plus any other factors you think are important, provided by Roundup Ready soybeans?

    $____.____ value per acre

10. Think again about the added operator, worker, and environmental safety. Now, also consider the convenience provided by the Roundup Ready soybeans compared to non-Roundup Ready soybeans.

    When you think about all of these factors, how much added value per acre, if any, do the Roundup Ready soybeans provide compared to non-Roundup Ready soybeans?

    $____.____ value per acre

Chapter 9

Bt CORN'S REDUCTION OF MYCOTOXINS: REGULATORY DECISIONS AND PUBLIC OPINION

Felicia Wu
Graduate School of Public Health, University of Pittsburgh

Abstract: This chapter analyzes the impact of mycotoxins—toxic and carcinogenic chemicals produced by fungi—in corn, primarily in terms of global trade and health considerations to major corn exporters and importers. It then discusses the state of the evidence for genetically modified Bt corn reducing contamination levels of various mycotoxins, and the implications for human health and markets worldwide. Finally, it speculates on the role that mycotoxin reduction may or may not play in regulatory decisions on commercialization and trade of Bt corn, both today and in the future.

Key words: mycotoxins, Bt corn, international regulation

1. INTRODUCTION

Bt corn and other genetically modified (GM) crops are undergoing intense political scrutiny. Though numerous potential benefits and risks have been discussed, one impact that has been virtually ignored in policy debates on GM crops is mycotoxin reduction, an effect that could improve human and animal health and potentially relieve some food market asymmetries. As adoption of agricultural biotechnology continues to increase on a global scale, policymakers worldwide should consider the economic and health impacts of this secondary benefit of GM pest-protected crops. Mycotoxin reduction has already had significant economic impacts in the United States at current levels of Bt crop planting (Wu, Miller, and Casman 2004). In less developed countries (LDCs), the mycotoxin reduction that Bt crops can provide could have important economic as well as health impacts. Yet for various reasons, the benefit of mycotoxin reduction has thus far been given only tangential consideration in regulatory policies on Bt crops.

This chapter analyzes the impact of mycotoxins in corn, primarily in terms of global trade and health considerations to major corn exporters and importers. It then discusses the state of the evidence for Bt corn reducing contamination levels of various mycotoxins, and the implications for human health and markets worldwide. Finally, it speculates on the role that mycotoxin reduction may or may not play in regulatory decisions on commercialization and trade of Bt corn, both today and in the future.

2. MYCOTOXINS IN CORN

Foodborne *mycotoxins* are secondary metabolites of fungi that accumulate on crops. They are considered unavoidable contaminants in foods in that best-available technologies cannot completely eliminate their presence in pre-harvest or post-harvest crops (Council for Agricultural Science and Technology 2003). As mycotoxins may be toxic or carcinogenic to humans, many nations have established regulatory standards on permissible mycotoxin levels in food. Thus, aside from health risks, mycotoxin contamination can also reduce the price paid for crops or cause wide-scale market rejection.

Losses from mycotoxins in the United States and other industrial nations are typically associated with these economic losses as opposed to illnesses or deaths from the toxins. In less developed countries, however, the economic and health impacts of mycotoxins are far more severe. There, many individuals are not only malnourished but also chronically exposed to high mycotoxin levels in their diet (Miller and Marasas 2002), resulting in deaths from severe toxicoses to various cancers to diseases of malnutrition.

Globalization of food trade has further exacerbated mycotoxin-related losses in two important ways. First, having strict mycotoxin standards imposed by importing nations means that less developed countries are likely to export their best-quality foods while keeping contaminated foods domestically, resulting in higher risk of mycotoxin exposure in those nations (Cardwell et al. 2001). Second, even the best-quality foods produced in these nations may be rejected for export at more strict standards, meaning millions of dollars in losses.

There are several predisposing factors for mycotoxin accumulation in corn. Pre-harvest and post-harvest conditions are both important. In pre-harvest corn, high temperatures, drought stress, and unsuitability of the corn hybrid for the region in which it is planted, insect damage, and other fungal diseases increase mycotoxin levels (Shelby, White, and Burke 1994, Wicklow 1994). High temperature may be the most important weather factor in determining formation and accumulation of fumonisins. Drought stress increases insect herbivory on corn, so it is not really possible to separate these two factors (Miller 2001). In any case, insect damage is well recognized as a collateral factor in fumonisin development. Insects also play a role, both by creating wounds on the corn kernels (leaf feeding may not be important for increasing mycotoxin contamination) and by acting as vectors for certain types of fungal spores (Sinha 1994, Wicklow 1994, Munkvold and Hellmich 1999). In post-harvest corn, storage conditions such as high humidity, pre-harvest presence of mycotoxin-producing fungi, and the presence of stored grain insects contribute to further fungal development and accumulation of mycotoxins in corn (Sinha 1994).

2.1. Background on Fumonisin and Aflatoxin Standards

Two important mycotoxins in corn worldwide are *fumonisins* and *aflatoxins*. Fumonisins are found primarily in corn, and to a limited extent in other commodities such as corn, rice, and sorghum. Aflatoxins are found in a variety of crops including corn, cotton, peanuts (groundnuts), and tree nuts such as pistachios (Robens and Cardwell 2003).

Fumonisins are a recently discovered class of toxins produced by the fungi *Fusarium verticillioides* (formerly *F. moniliforme)*, *Fusarium proliferatum*, and some related species (International Agency for Research on Cancer 2002). Fumonisins were first reported in 1988 in connection with high human esophageal cancer rates in Transkei, South Africa. The following year, interest in these mycotoxins increased dramatically after unusually high horse and swine death rates in the United States (Marasas 1996). Since then, more than 28 types of fumonisins have been isolated and characterized (Rheeder, Marasas, and Vismer 2002). Of these, fumonisin B_1 (FB_1) is the most common in corn worldwide.

While there have been no confirmed cases of acute fumonisin toxicity in humans, epidemiological studies have linked consumption of fumonisin-contaminated grain with elevated human esophageal cancer incidence in various parts of Africa, Central America, and Asia (Marasas et al. 2004) and among the black population in Charleston, South Carolina (Sydenham et al. 1991). Synergistic effects between fumonisin and aflatoxin (discussed below) may also lead to increased risk of liver cancer. Studies of increased rates of neural tube birth defects in Cameron County, Texas, were associated with high corn consumption after a year of high fumonisin in the crop (Hendricks 1999). Because FB_1 reduces the uptake of folate in different cell lines, fumonisin consumption has been implicated in connection with neural tube defects in human babies (Hendricks 1999, Marasas et al. 2004). In addition, elevated levels of fumonisins in animal feed cause diseases such as equine leukoencephalomalacia (ELEM) in horses and porcine pulmonary edema (PPE) in swine (Ross et al. 1992).

To protect consumers from the harmful effects of fumonisins, a number of nations have established regulations for mycotoxins in food and animal feed (van Egmond 2002). In the United States, the U.S. Food and Drug Administration (FDA) has set guidelines to industry for levels of fumonisin acceptable in human food and animal feed. The most stringent of these standards applies to degermed dry-milled corn products for human food, with a recommended total fumonisin maximum level of 2 mg/kg (FDA 2000).

At the moment, very few regulations exist in other nations regarding acceptable fumonisin levels. The fifty-sixth meeting of the Joint FAO/WHO Expert Committee on Food Additives (JECFA) in 2001, however, recommended a provisional maximum tolerable daily intake (PMTDI) of 2 μg/kg bodyweight per day (JECFA 2001). In some parts of the world, such as Latin America and sub-Saharan Africa, corn is a staple in the human diet; thus, meeting the PMTDI for fumonisin would be considerably more difficult in these regions than in the United States or Europe, where corn consumption is much lower (Shephard 2004). At the 2002 Food and Drug Administration/Joint Institute for Food Safety and Applied Nutrition International Workshop on Mycotoxins (July 22–26), a draft for a new EU maximum limit on fumonisins of 0.5 mg/kg had been announced (Food Standards Agency 2003); thus far, however, no such limit has come into legislation.

Aflatoxins are mainly produced by the fungus *Aspergillus flavus*. Aflatoxins are the most potent chemical liver carcinogens known. Moreover, the combination of aflatoxin with hepatitis B and C, prevalent in China and sub-Saharan Africa, is synergistic, raising more than tenfold the risk of liver cancer compared with either exposure alone (Miller and Marasas 2002). Aflatoxins are associated with stunting in children (Gong et al. 2000) and possibly immune system disorders (Turner et al. 2003).

Likewise, aflatoxins can severely damage animal health. Aflatoxin B_1, the most toxic of the aflatoxins, causes a variety of adverse effects in different animal species, especially chickens. In poultry, these include liver damage, impaired productivity and reproductive efficiency, decreased egg production in hens, inferior eggshell quality, inferior carcass quality, and increased susceptibility to disease (Wyatt 1991). In cattle, the primary symptom is reduced weight gain as well as liver and kidney damage. Milk production is also reduced (Keyl 1978). Unfortunately, the loss of income from lower animal production leads to greater poverty, thus reinforcing the conditions conducive to poor human health (Miller and Marasas 2002).

The presence of aflatoxins in foods is restricted in the United States to the minimum levels practically attainable by modern processing techniques. The most stringent of these FDA standards applies again to human food, with a recommended total aflatoxin maximum level of 20 μg/kg. Many other nations have established maximum tolerated levels of aflatoxin in food and feed. Notably, the European Commission has set a total aflatoxin standard of 4 parts per billion (μg/kg) in food and an aflatoxin B_1 standard of 2 μg/kg, considerably more precautionary than any national or international standards currently existing (Lubulwa and Davis 1994).

It is important to note that these maximum tolerated levels vary greatly between countries, requiring harmonization to remove the extreme variability in standards. At the moment, no international standard for aflatoxins exists. Until 1996, JECFA had recommended that dietary aflatoxin be kept to an "irreducible level." Subsequent to this evaluation, there have been a number of attempts to establish standards for aflatoxins in food and feed, but it has been exceedingly difficult to reach consensus on maximum levels that should be included in these standards. Major impediments to consensus are the wide variation in contamination levels worldwide and the relative ability of nations to reduce aflatoxin levels in a cost-effective manner.

2.2. Mycotoxin Exposure Worldwide, and Implications of Regulations

In the majority of growing seasons, the current FDA guidelines for fumonisins are not difficult to meet in the United States. In the last decade, 0.5 percent to 10.5 percent of corn grown in the north central United States had fumonisin B_1 levels of 5 mg/kg or higher. However, only about 3.5 percent of U.S. corn, that devoted to dry milling, masa, popcorn, and corn fed to horses, must meet the lowest recommended fumonisin levels (Munkvold 2001). Otherwise, fumonisin levels rarely are so high that they are rejected for use in other animal feed. Within the animal feed sector, however, the majority of corn is fed to livestock on-farm, without going to market. This means that most of the corn fed to livestock is not inspected for potentially dangerous fumonisin levels.

Throughout the United States, aflatoxins develop on crops primarily when droughts occur, followed by periods of rain before crops are harvested. Crops from anywhere in the United States may be affected, depending on the growth, harvesting, and storage conditions involved; however, aflatoxin contamination is particularly high in warm, dry regions of the Southeast.

Mycotoxins' impacts in the developing world can be far more severe. Low-quality seed, lack of pest control, and poor storage conditions all increase the risk of mycotoxin accumulation in food. Moreover, in the poorest areas of the world, agriculture is largely in the form of subsistence farming rather than commercial operations. Therefore, most of the corn grown is consumed within the farming families and communities without any form of outside inspection or regulatory control. A recent JECFA study (JECFA 2001) shows that in sub-Saharan Africa, over half the diets sampled contained dangerously high levels of fumonisin. Exposure to aflatoxins in West Africa is equally alarming: over 90 percent of corn samples collected from households were contaminated with *A. flavus* (the aflatoxin-producing fungus), and 99 percent of children tested had aflatoxin in their blood (Cardwell et al. 2001). Further complicating the problem is that for the given level of exposure, health effects are more severe in less developed countries because of the prevalence of hepatitis B and C, particularly in China and Africa. A Chinese study associated fumonisin concentrations in corn from particular regions with increases in esophageal cancer rates (Yoshizawa, Yamashita, and Luo 1994), while a more recent report of the International Agency for Research on Cancer (IARC) shows that a causal relationship between fumonisins and esophageal cancer has not been proven (IARC 2002). High aflatoxin concentrations in China have also been linked with liver cancer risk (Zhang et al. 1998).

From the standpoint of compliance with international regulations, the top corn exporting nations worldwide are (in order) the United States, China, and Argentina, together accounting for 89 percent of the total volume of exported corn (USDA 2003a). Hence, these three nations will experience the most serious economic challenges with tightened mycotoxin standards in corn. Table 1 shows the relative sizes of export markets among major corn-exporting nations. By contrast, the European Union is not a major exporter of corn; rather, it is a net importer.

Table 1. Relative Corn Export Market Volumes of Prominent Corn-Exporting Nations, 2002-2003

Nation	Percentage of total world market for corn
United States	53
China	19
Argentina	17
Brazil	4
Hungary	2
South Africa	1
Ukraine	1
All other nations	3

2.3. Integrated Assessment Model and Sensitivity Analysis

Mycotoxin regulations worldwide have largely been based on an analysis of demonstrated health effects to humans and to animals, with additional risk management considerations (e.g., safety factors). Integrated assessment takes the analysis several steps further. It includes available information about health effects, and also considers the questions: What is an acceptable level of risk? What are the economic consequences of the regulation? Among the different stakeholders who are affected by the regulation, who benefits and who loses? Finally, are there any *countervailing* risks and indirect health effects associated with the regulation? All of these questions should be taken into account when considering harmonized mycotoxin regulations, as shown in Figure 1.

To address the question of what the economic consequences are of mycotoxin regulations, an empirical economic model was developed to estimate a nation's total export loss of a particular food crop, given an internationally imposed mycotoxin standard. This is a function of the price of the food crop

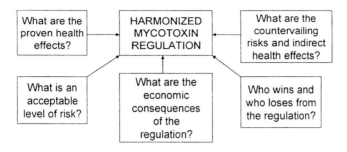

Figure 1. Integrated Assessment to Inform Development of Harmonized Mycotoxin Regulations

per unit volume on the world market, the total volume of that crop exported by a particular nation, and the fraction of that nation's export crop that is rejected as a result of a worldwide mycotoxin standard. The economic model allows a sensitivity analysis on how export losses for food crops in particular nations change as a function of the strictness of the mycotoxin standard. Model equations, parameters, their descriptions, and references for calculating economic impacts for fumonisin are given in Wu (2004).

Next, an assessment was made of which nations would benefit and which would lose as a result of stricter mycotoxin regulations. Logically, those who would experience losses are food exporters, particularly those with higher mycotoxin concentrations in their food crops; while those who would experience the greatest health benefits from stricter standards are food-importing nations. These data were gathered from databases of the Foreign Agricultural Service of the U.S. Department of Agriculture, giving information on food crop imports and exports by nation and year. Potential health effects are considered through epidemiological studies of the effects of moving from one mycotoxin regulation to another.

Countervailing risks and indirect health effects are considered from two angles: the possibility of less developed countries exporting their best crops while keeping the most contaminated food domestically; and the prevalence of predisposing factors among the populations of those countries (such as hepatitis B and C) that make them particularly vulnerable to toxic and carcinogenic effects of mycotoxin consumption. Finally, the question of what is an acceptable level of risk is explored from the integrated findings.

2.4. Results and Discussion: Impact of Mycotoxins on Trade and Health

The top corn exporting nations worldwide are (in order) the United States (53 percent of the total world volume of exported corn), China (19 percent), and Argentina (17 percent) (USDA 2003a). Hence, these three nations will experience the most serious economic challenges with tightened standards for fumonisin and aflatoxin. The European Union is not a major exporter of corn; it is a net importer. Other main importers of corn worldwide are Japan, Korea, Mexico, Egypt, Canada, and Taiwan (USDA 2003b). Africa exports virtually no corn, with the exception of South Africa.

Table 2 shows the export losses to the three major corn-exporting nations—the United States, China, and Argentina—as a function of different international fumonisin regulations in food. If the current FDA guideline of 2 mg/kg fumonisin in food were adopted internationally, the export losses to each of these three nations would range between $30 million to $40 million annually, with a total loss of about $100 million. If a stricter fumonisin regulation of 0.5 mg/kg became a worldwide norm, the export losses would rise to $170 million in the United States, $60 million in China, and $70 million in Argentina; for a total of about $300 million lost export markets annually. Thus, moving from a harmonized fumonisin standard in corn of 2 mg/kg to 0.5 mg/kg would result in an increased worldwide annual market loss of over $200 million through rejected corn. The loss in this case would more than quadruple for the United States, where most of the corn produced has fumonisin levels below the current FDA standard of 2 mg/kg but higher than 0.5 mg/kg. However, percentage-wise, the loss to Argentina is the most severe; compliance with a 0.5 mg/kg standard could result in over 90 percent of their total corn export market being rejected.

Table 3 shows expected corn export losses to those same three nations as a function of international aflatoxin regulations in food. If the current FDA standard of 20 µg/kg aflatoxins in food were adopted worldwide, export losses to these three nations would total about $40 million. If the current EU standard of 4 µg/kg became the worldwide norm, the combined corn export losses in these four regions would reach almost $124 million annually. Moving from a harmonized aflatoxin standard of 20 µg/kg to 4 µg/kg would thus result in a threefold increase in worldwide annual market losses through rejected corn. In fact, at this standard, about a third of the corn being exported worldwide today would be rejected for excessive aflatoxin concentrations. In particular, since China has such heavy aflatoxin contamination levels (Li et al. 2001, Wang et al. 2001), it would lose virtually all of its food corn export market in the latter case. In neither case would Argentina suffer such an enormous loss to its export market, because aflatoxin contamination levels are low in Argentinian corn (Resnik et al. 1996, Etcheverry et al. 1999).

The main beneficiaries of stricter mycotoxin standards should be the food importers; however, the resulting health benefit in these cases described above is negligible. A JECFA study has found that where hepatitis B and C incidence are low, reducing aflatoxin in food from 20 µg/kg to 10 µg/kg would reduce the risk of mortality by 2 in 1 billion annually: undetectable by epidemiological standards (Henry et al. 1999). Thus, nations that would benefit most from more stringent mycotoxin standards are those that are net

Table 2. Expected Export Losses to Top Three Corn-Exporting Nations as a Function of Internationally Imposed Fumonisin Standards in Food

Country	Total annual value of corn exports ($US)	Annual export losses ($US) at fumonisin standard 2 mg/kg; % of total market	Annual export losses ($US) at fumonisin standard 0.5 mg/kg; % of total market
USA	$220 million	$40 million; 19%	$170 million; 77%
China	$80 million	$30 million; 38%	$60 million; 75%
Argentina	$75 million	$30 million; 40%	$70 million; 93%
TOTAL	$375 million	$100 million; 27%	300 million; 80%

Table 3. Expected Export Losses to Top Three Corn-Exporting Nations as a Function of Internationally Imposed Aflatoxin Standards in Food

Country	Total annual value of corn exports ($US)	Annual export losses at aflatoxin standard 20 µg/kg; % of total market	Annual export losses at aflatoxin standard 4 µg/kg; % of total market
USA	220 million	8.8 million; 4%	44 million; 20%
China	80 million	30 million; 38%	72 million; 90%
Argentina	75 million	0.8 million; 1%	7.5 million; 10%
TOTAL	375 million	40 million; 11%	124 million; 33%

importers of corn, and that have high prevalence of hepatitis B and C. As mentioned earlier, the top importers of corn worldwide are Japan, Korea, Mexico, Egypt, Canada, and Taiwan. With the possible exception of Taiwan, all these nations have low hepatitis B and C incidences. Conversely, the nations with high hepatitis prevalence—China and sub-Saharan Africa—may increase the risk to their populations by attempting to export their best-quality food and keeping the most contaminated food to be consumed domestically.

It is unlikely in the near future that even such enormous potential market losses will convince policymakers to relax their standards for food, as these policymakers are accountable to the public citizens they serve, and the public is unlikely to accept a weaker standard to protect their food quality when a stronger standard formerly existed. Therefore, the best solution for less developed countries to the potentially damaging economic losses of food-borne mycotoxins is to find technologies or methods that can reduce mycotoxin contamination effectively. One possible solution lies in agricultural biotechnology, with Bt corn.

3. Bt CORN

Bt corn is one of the most commonly grown genetically modified crops in the world today. It contains a gene from the soil bacterium *Bacillus thuringiensis*, which encodes for formation of a crystal (Cry) protein. This protein is toxic to insects of the order *Lepidoptera*, including the corn pests European corn borer *Ostrinia nubilalis*, Southwestern corn borer *Diatraea grandiosella*, and corn earworm *Helicoverpa zea*, but is harmless to vertebrates and non-lepidopteran insects. (A new variety of Bt corn protects against the corn rootworm *Diabrotica virgifera*.)

Non-GM microbial Bt pest control agents containing Bt Cry proteins were first registered for pesticidal use in the United States in 1961 (International Programme on Chemical Safety 1999). Commercial liquid and dust "natural" insecticides containing Bt are referred to as *microbial Bt sprays*. These Bt sprays are highly regarded by organic farmers and other growers, as they are safe to mammals and birds and safer to non-target insects than conventional insecticides. However, their residual action is limited due to their rapid inactivation by sunlight and removal from leaf surfaces by rain and wind. Also, microbial Bt spray and other sprayed insecticides cannot reach some of the most pest-susceptible interior parts of the corn plant. Bt corn, on the other hand, produces the Cry insect control protein constitutively throughout the plant throughout the growing season, including tissues that are difficult to protect with surface-applied insecticides (Wu, Miller, and Casman 2004).

Bt corn seed was first commercialized in 1996. In 2004, Bt corn was grown on about 27 percent of field corn acres in the United States (USDA 2004). As U.S. regulations do not require segregation of genetically modified grains, Bt corn and traditional grain corn are treated as identical for almost all commercial uses, with the exception of a small number of food companies that will not use genetically modified food, such as Gerber for its baby food. The majority of harvested Bt corn is used for animal feed. A small percentage and specific varieties of corn are designated "food grade" for human consumption. Other uses include non-food items such as paper, adhesives, and pharmaceuticals.

Seven nations other than the United States are planting Bt corn currently. These are: Canada, Germany, Spain, Argentina, Honduras, South Africa, and the Philippines (James 2003). The United States is by far the largest adopter, producing 85 percent of the total global market of Bt corn. Argentina, Canada, and South Africa are also important adopters, comprising 8 percent, 4 percent, and 2 percent respectively of the total global Bt corn market. China has approved field trials, but has not yet allowed commercialization of the crop. The other Bt corn-adopting nations comprise only one percent of the world market. In total, about 25 million acres of Bt corn are planted globally today (James 2003).

3.1. Evidence for Bt Corn Reducing Mycotoxin Contamination

The corn pests European corn borer, Southwestern corn borer, and corn earworm all have been shown to contribute to the occurrence of mycotoxins in corn (Dowd 1998). Even when the larvae do not directly carry the fungi to the corn wounds, spores falling later on the wounded tissue are more likely to infect the plant (Munkvold and Hellmich 1999). Insect-damaged corn is also prone to mycotoxin accumulation in storage (Sinha 1994). Stored grain insects are the problem in this case, creating grain wounds and spreading fungal spores to cause further post-harvest accumulation of mycotoxins. Therefore, to the extent that Bt corn has lower levels of insect damage, it indirectly controls for one of the predisposing factors of mycotoxin accumulation.

Worldwide, field studies have demonstrated that when insect damage from European corn borer or Southwestern corn borer is high, fumonisin concentrations are substantially lower in Bt corn compared with conventional corn. In cases of both a natural European corn borer (ECB) infestation and a manual ECB infestation, field studies in the Corn Belt region of the United States (Munkvold and Hellmich 1999) showed that the amount of Fusarium kernel rot and concentration of fumonisin B_1 were significantly lower in Bt corn events Bt11 and MON810 than in their near-isogenic, non-transgenic counterparts. Perhaps more importantly, in this study, fumonisins in these two events were reduced to safe levels for human consumption according to the FDA standards described earlier. Dowd (2001) also showed that, depending on the control level of pest damage, a 1.8- to 15-fold reduction of fumonisins in Bt corn over conventional corn was achieved. In this study, however, the greatest reductions in fumonisins in Bt corn occurred where European corn borer was the predominant insect pest. Where corn earworm, fall armyworm, western bean cutworm, or other pests were predominant, fumonisin reductions in Bt corn were less dramatic, as Bt corn does not achieve complete control of these pests. Thus, in regions such as the southeastern United States or Texas where really high fumonisin levels occur, fumonisin may not be controlled by Bt corn because of damage by Bt-resistant insects. No studies have yet been published comparing Bt with non-Bt isolines in these regions.

Hammond et al. (2003) found consistently lower levels of fumonisin in Bt hybrids when compared to controls in 288 separate test sites in Argentina, France, Italy, Turkey, and the United States. Fumonisin concentrations in Bt grain were often lower than 4 mg/kg, with a significant proportion of these below 2 mg/kg. Likewise, Bakan et al. (2002) found lower levels of fumonisin contamination in Bt corn than in non-Bt isolines grown in France and Spain.

Compared with the case of fumonisin, insect damage is less strongly correlated with aflatoxin concentrations, as multiple factors predispose corn to accumulation of this mycotoxin. The lepidopteran insects that are controlled by the crystal protein in existing Bt hybrids are not as important in predisposing plants to infection by *A. flavus* as they are for *F. verticillioides* and *F. graminearum* (Miller 1995); *A. flavus* can infect corn not just through kernel wounds caused by insects, but through the silks. This may explain why the effect of Bt corn (which reduces insect damage) on aflatoxin concentration is inconsistent.

Indeed, the experimental record is mixed. Depending on the predominant insect pests in different regions of the United States, Bt corn may or may not have lower levels of aflatoxin than its non-Bt isogenic counterparts. There are a few success stories. Benedict et al. (1998) found that in two locations in Texas, the events of Bt corn that are still registered today consistently had between 2.5 percent to 53 percent lower levels of aflatoxin than its non-Bt isolines. Windham, Williams, and Davis (1999) examined the relationship between insect damage and aflatoxin concentration in different corn hybrids, including a Bt11 hybrid. When corn was manually infested with Southwestern corn borer, which is well controlled by Bt corn, aflatoxin concentration was significantly lower in Bt11 than in conventional corn. However, in the controls (natural insect infestation), both Bt corn and conventional corn had aflatoxin concentrations below the FDA action level. In a follow-on study, Williams et al. (2002) found that the relationship between Bt corn and aflatoxin reduction depends on the *A. flavus* inoculation technique. The non-wounding technique (spraying *A. flavus* inoculum on young ears) and control case resulted in significantly lower aflatoxin levels in Bt corn, while the wounding technique (damaging the kernels) resulted in no difference in aflatoxin levels between Bt and non-Bt corn.

Other studies show no significant effect of Bt corn, or mixed results. Buntin et al. (2001) observed that while Bt11 and MON810 had significantly lower pest damage than non-Bt corn, there was no significant difference in aflatoxin levels between the two groups. Odvody et al. (2000) found significantly lower levels of insect damage in Bt corn in regions of Texas, but inconsistent comparative results on aflatoxin levels in Bt and non-Bt corn. The authors concluded that other factors, such as drought stress and individual hybrid vulnerability, are more important in determining aflatoxin contamination levels than insect damage.

3.2. Economic Impacts of Bt Corn in Mycotoxin Reduction

Three conceptual classes of economic impacts can be identified: market effects, animal health, and human health. High quality corn (i.e., with low levels of mycotoxin contamination) can be sold as human-food–grade corn at the highest market price. Corn contaminated with levels of mycotoxins between the highest permitted levels of food and feed can be sold for animal feed at a lower price, and corn with high levels of mycotoxins is either sold for non-food–non-feed uses at an even lower price or rejected outright. The proportions of the total crop that are rejected at each of these levels depend on the national or international standards for mycotoxins in food and feed. Animal health as a function of mycotoxin contamination is dependent on three factors: the number of animals experiencing mortality or morbidity as a result of mycotoxin ingestion, the cost of treatment for sick animals, and the market value of each animal. Finally, human health impacts can be estimated in a manner similar to animal health impact calculations.

Key parameters to calculate these values, descriptions, assumptions, and references relevant to fumonisins and aflatoxins are summarized in Wu, Miller, and Casman (2004). It was estimated that in the United States, the average total annual loss due to fumonisins in corn is about $40 million ($14 million to $88 million). The annual market loss in the United States from corn rejected either for food or for feed makes up most of this loss: roughly $39 million ($14 million to $86 million). Of this amount, about $38 million of the estimated losses are through corn rejected for food, and slightly less than $1 million of the losses are through corn rejected for feed. The expected loss from corn rejected for feed is relatively low, because the proportion of corn consumed by horses (the animal group most sensitive to fumonisin) is small, because U.S. corn is generally of high enough quality to meet standards for other feed, and because contaminated corn may be blended with clean corn to achieve a safe level for feed.

Assuming that Bt corn contains fumonisins at or below the FDA standard for human consumption 80–95 percent of the time, the savings to U.S. farmers from increased market acceptance is estimated at $8.8 million annually ($2.3 million to $31 million). The total value of animal mortality from fumonisin consumption is relatively small in the United States. This is because in most years, fumonisin levels are sufficiently low for few, if any, animals to be affected in most regions of the United States. We estimate that the annual loss from fatal fumonisin-induced ELEM in horses is $270 thousand ($51 thousand to $2 million). In swine, the annual expected losses from fumonisin-induced PPE are on the order of several tens of thousands of U.S. dollars. These deaths occur on farms that grow their own corn rather than buy commercial feed, which presumably has safe fumonisin levels. The benefit of planting Bt corn from preventing swine and horse mortality is estimated to be $67 thousand annually ($13 thousand to $500 thousand).

Human health benefits from reducing fumonisin in food through Bt corn in the United States are currently impossible to calculate meaningfully, because of weaknesses of the epidemiological literature and the lack of a reliable biomarker for fumonisin exposure. More research is needed in these areas to clear the uncertainties in human health impacts. In any case, it is expected that in the United States, human health losses due to fumonisin consumption would be negligible, because of the diligence of the food production industry and adherence to the FDA guidelines.

It was estimated that in the United States, the total annual loss due to aflatoxin in corn is about $163 million ($73 million to $332 million; here and following, the values in parentheses represent the 95 percent confidence levels). The annual market loss through corn rejected for food is about $31 million ($10 million to $54 million), while the loss through corn rejected for feed and through livestock losses is estimated at $132 million ($14 million to $298 million).

Bt corn would reduce aflatoxin in cases where insect damage from Bt-sensitive insects was the main determinant of aflatoxin development. Given the current level of Bt corn planting in such regions at about 17 percent (USDA 2004), and the assumption that Bt corn is partly effective in reducing aflatoxin only in Texas and the southeastern United States and that 80 percent of the aflatoxin contamination problems occur there, the upper bound of the current benefit is $11 million ($5.0 million to $22 million): at best a 7 percent reduction in total annual costs from aflatoxin in field corn.

Table 4 summarizes our estimates for economic losses due to fumonisin and aflatoxin in corn in the United States, and benefits that Bt corn currently provides in terms of reducing mycotoxin contamination.

Table 4. Estimated Losses Due to Mycotoxins in U.S. Corn and the Benefit from Bt Corn (average and 95 percent confidence intervals; $US millions)

Losses in $US millions	Fumonisin	Aflatoxin
Market loss	39 (14 to 86)	163 (73 to 332)
Animal health loss	0.27 (0.051 to 2)	N/A
Total U.S. losses	40 (14 to 88)	163 (73 to 332)
Benefit from planting Bt corn	8.8 (2.3 to 31)	14 (6.2 to 28)

4. MYCOTOXIN REDUCTION AND POLICIES REGARDING Bt CORN

Given the available evidence that Bt corn planting results in lower mycotoxin contamination, with direct economic benefits and to a more limited extent health benefits, why has this benefit been given minimal, if any, consideration in regulatory decision making regarding Bt corn? There are at least three reasons.

The first is that *any* analysis of benefits of Bt corn and other GM crops is relatively rare in GM crop regulations worldwide. The United States is one of only a few examples of nations in which benefits analysis is part of Bt corn regulation. For the purposes of regulating the technology of GMOs, a Coordinated Framework for Regulation of Biotechnology was created within the Office of Science and Technology Policy (OSTP) in 1986 (51 Federal Register [FR] 23302). However, its outcome was not so much to form new regulations for GMOs as to delegate responsibility for oversight to existing agencies under the framework of existing statutes. Under this framework, Bt corn and other GM crops are subject to the statutes of the Federal Food, Drug and Cosmetic Act (FFDCA), the Federal Plant Pest Act (FPPA), the Federal Plant Quarantine Act (FPQA), and the Federal Insecticide, Fungicide and Rodenticide Act (FIFRA). Hence, it is regulated by three governmental agencies: the U.S. Environmental Protection Agency (EPA), the U.S. Department of Agriculture (USDA), and the Food and Drug Administration (FDA). Table 5 shows the regulatory scheme for coordinating reviews of GM crops. These three agencies have evolved to different levels of involvement in the regulation of Bt corn. GM food or feed is subject to regulation by the FDA, while the U.S. Department of Agriculture is the lead agency for plants grown to produce food or feed crops. The EPA was charged primarily to handle pesticidal microorganisms—the Coordinated Framework of 1986 did not address the regulation of plants that produced their own pesticides, or "plant-incorporated protectants," as they had not yet been submitted for regulatory scrutiny at that point (Carpenter 2001).

Of these statutes, the FIFRA and FFDCA contain explicit language calling for risk-benefit analysis. Indeed, in its 2001 Biopesticides Registration Action Document (EPA 2001), by which Bt corn and cotton were re-registered in the United States, the EPA considers three specific benefits of Bt corn: yield increase, reduction in pesticide usage, and reduction in mycotoxin levels. It is not made explicit how the Agency would have weighed these benefits against any significant risks of Bt corn, of which the EPA found none. The FFDCA is also the statute by which FDA was given authority of assessing GM food risks and benefits. However, in 1992 the FDA determined that GM foods were "substantially equivalent" to non-GM foods, and therefore did not need to be evaluated separately (57 FR 22991). Companies producing GM crops for food undergo *voluntary* food safety analyses

Table 5. Regulatory Scheme for Coordinating Reviews of GM Crops

Product class	Lead agency	Federal statutes
Plants	USDA/APHIS*	FPPA, NEPA, PPA, FPQA
Pesticides	EPA	FIFRA, FFDCA, FPQA
Food and additives	FDA	FFDCA

Source: National Academy of Sciences (2000).

Notes: APHIS = Animal and Plant Health Inspection Service, FPPA = Federal Plant Pest Act, NEPA = National Environmental Policy Act, PPA = Plant Protection Act, FPQA = Federal Plant Quarantine Act, FIFRA = Federal Insecticide, Fungicide and Rodenticide Act, FFDCA = Federal Food, Drug and Cosmetic Act.

with the FDA in the form of flow charts and other instructions on how GM food safety should be evaluated scientifically (57 FR 22993), addressing specifically unexpected or unintended effects, the safety of the host plant, and allergic potential. Finally, the U.S. Department of Agriculture does not do a benefits assessment of GM crops, but a risk assessment of the potential for these GM crops to become plant pests or to introduce diseases in U.S. agriculture (58 FR 17044).

Among other nations that have developed regulations on GM crops, only China, India, and Argentina have policies explicitly incorporating the analysis of potential benefits. China's Implementation Regulation on Agricultural Biological Genetic Engineering (published by its Ministry of Agriculture in 1996) does not assume that GM crops are inherently more dangerous to human health or to the environment than conventional crops; rather, it acknowledges that in some instances GM crops may be more beneficial (Paarlberg 2001). Moreover, this regulation takes into account the economic interests of a given application (Nap et al. 2003). Hence, to the extent that GM crops can provide economic benefits in terms of yield increase and possibly even mycotoxin reduction, they would be evaluated favorably from a regulatory standpoint. China's policies have been deemed "very pragmatic" (Nap et al. 2003).

In India, the Genetic Engineering Approval Committee (GEAC) of the Ministry of Environment and Forests grants permits for commercial production of GM crops. For a crop to be approved for commercial use, not only must there be no proven environmental and agricultural risks, it must also have a proven benefit to farmers economically. In fact, the Indian regulatory system for GM crops is unusual in that the crops that undergo its assessment and management must be shown *not* to have adverse economic impacts on farmers (Pray et al. 2006).

In Argentina, three separate entities of the Secretariat of Agriculture, Livestock, Fisheries and Food regulate GM crops. The National Advisory Commission on Agricultural Biotechnology (CONABIA) evaluates environmental impacts of GM crops, the National Agrifood Health and Quality Service (SENASA) regulates food safety of GM crops, and the National Directorate of Agrifood Markets (DNMA) assesses potential impacts of each GM crop on Argentina's international trade (Nap et al. 2003, Jaffe 2004). The DNMA may specifically consider both the market advantages and disadvantages of each GM crop that is being evaluated for approval (Jaffe 2004). One potential market advantage to evaluate is whether, for example, Bt corn might significantly reduce mycotoxin contamination and hence reduce export market losses.

It is noteworthy that three of the four nations whose regulations incorporate benefits analysis of GM crops—the United States, China, and Argentina—are also the three nations that would stand to benefit most from Bt corn's reduction of mycotoxin contamination. The United States and Argentina also happen to be the two nations with the greatest acreage devoted to Bt corn worldwide (James 2003). While China has not yet commercialized Bt corn, field trials are underway, and it is expected that commercialization should come soon.

The second reason for the lack of policy recognition of Bt corn's mycotoxin-reducing properties is that, even in those nations in which benefits analysis is an explicit part of GM crop regulation, the literature regarding the link between mycotoxins and the impact of Bt corn was fairly sparse until the last several years. In fact, since fumonisin is one of the most newly discovered mycotoxins (1988), human health effects regarding neural tube defects have been hypothesized and studied only within the last six years (Hendricks 1999, Marasas et al. 2004). As described in the previous section, the evidence associating Bt corn with fumonisin reduction in the United States and worldwide is fairly robust; however, the link between Bt corn and aflatoxin reduction is still weak and needs further analysis.

The third reason for the general lack of policy recognition of Bt corn's mycotoxin reduction concerns public opinion, and the attempt of governments to maintain public confidence in their abilities to ensure food safety and quality. If people were aware of the extent to which mycotoxins contaminate their food, even at levels that are generally regarded as harmless to human health, governments and food-related industries might have a much more difficult public perception problem on their hands, one that would outweigh any benefit that might be gained by advertising Bt corn's effect of reducing these mycotoxins.

5. DISCUSSION

If foodborne mycotoxin regulations were based solely on direct health effects, important questions of the economic feasibility of meeting excessively strict standards could be ignored, with potentially disastrous consequences for less developed food-exporting countries.

In the cases presented here of fumonisin and aflatoxin in corn, it is shown that China, Argentina, and the United States might all sustain enormous export market losses if highly precautionary harmonized standards are established. At the same time, such standards would likely reduce the risk of mortality by an amount so small that it would not be detected by epidemiological standards. On the other hand, areas with high incidence of hepatitis B and C—including China—could very well have greater levels of health risk due to precautionary mycotoxin standards.

Until improved agricultural methods of controlling these mycotoxins in crops are available and affordable, such standards will encourage the exportation of their best-quality crops to preserve their export markets. Several control methods, both pre-harvest and post-harvest, are being developed. Until such technologies become widely available and affordable, policymakers should consider the implications of both health and economic outcomes when developing harmonized international standards for mycotoxins.

One potential technology to control mycotoxins is Bt corn. Where Bt corn is planted, it may, depending on the severity of other impacts such as weather conditions, have significantly reduced fumonisin and aflatoxin when pest infestation would otherwise cause high levels of these mycotoxins. In the United States, where roughly a quarter of total field corn acreage is planted with Bt corn, the annual benefits that Bt corn provides in terms of lower fumonisin and aflatoxin contamination are estimated at about $23 million.

Though in the United States the fumonisin and aflatoxin problems are usually not large enough to affect price, this may change as demand for high-quality corn increases in the United States. Bt corn, by its effect on fumonisin and aflatoxin, may be a useful tool, enabling farmers and food producers to meet the increasing demand for high-quality corn. It is likely that animal and human health benefits of Bt corn would be more prominent than market gains in areas such as Latin America and sub-Saharan Africa, where corn is a staple in animal and human diets and mostly exchanged locally.

In spite of this potential benefit, mycotoxin reduction has not played a significant role in the regulation of Bt corn worldwide. Three reasons were discussed in this chapter: the minor or non-existent role that benefits analysis plays in regulatory policies on genetically modified crops; the immaturity of the scientific evidence concerning fumonisin's health effects and the relationship between Bt corn and aflatoxin reduction; and the concern that in making mycotoxin reduction a part of regulatory decision making, governments would reveal to the public that their foods were contaminated with fungal molds.

For these reasons, it is unlikely that Bt corn's reduction of mycotoxin contamination will play a significant role in regulatory decisions on GM crops in the near future. For now, the decision makers that should consider this benefit are those that would experience direct market benefits from Bt corn's mycotoxin reduction: the corn growers themselves. While improvements in public health could also indirectly occur, the public, for reasons described above, is largely unaware of this benefit of Bt corn.

REFERENCES

Bakan, B., D. Melcion, D, Richard-Molard, and B. Cahagnier. 2002. "Fungal Growth and Fusarium Mycotoxin Content in Isogenic Traditional Maize and Genetically Modified Maize Grown in France and Spain." *Journal of Agricultural and Food Chemistry* 50(4): 728–731.

Benedict, J., D. Fromme, J. Cosper, C. Correa, G. Odvody, and R. Parker. 1998. "Efficacy of Bt Corn Events MON810, Bt11 and E176 in Controlling Corn Earworm, Fall Armyworm, Sugarcane Borer and Aflatoxin." Texas A&M University System, College Station, TX (available online at http://lubbock.tamu.edu/ipm/AgWeb/r_and_d/1998/Roy%20Parker/[-] Bt%20Corn/BtCorn.html).

Buntin, G.D., R.D. Lee, D.M. Wilson, and R.M. McPherson. 2001. "Evaluation of Yieldgard Transgenic Resistance for Control of Fall Armyworm and Corn Earworm (Lepidoptera: Noctuidae) on Corn." *Florida Entomologist* 84(1): 37–42.

Cardwell, K.F., A. Desjardins, S.H. Henry, G. Munkvold, and J. Robens. 2001. *Mycotoxins: The Cost of Achieving Food Security and Food Quality* (available online at www.apsnet. org/online/feature/mycotoxin/top.html).

Carpenter, J.E. 2001. "Case Studies in Benefits and Risks of Agricultural Biotechnology: Roundup Ready Soybeans and Bt Field Corn." National Center for Food and Agricultural Policy, Washington, D.C. (available online at www.ncfap.org/pup/biotech/benefitsandrisks. pdf).

Council for Agricultural Science and Technology. 2003. *Mycotoxins: Risks in Plant, Animal, and Human Systems*. Task Force Report No. 139, Ames, Iowa.

Dowd, P.F. 1998. "The Involvement of Arthropods in the Establishment of Mycotoxigenic Fungi Under Field Conditions." In K. Sinha and D. Bhatnagar, eds., *Mycotoxins in Agriculture and Food Safety*. New York: Marcel Dekker.

_____. 2001. "Biotic and Abiotic Factors Limiting Efficacy of Bt Corn in Indirectly Reducing Mycotoxin Levels in Commercial Fields." *Journal of Economic Entomology* 94(5): 1067–1074.

EPA [see U.S. Environmental Protection Agency].

Etcheverry, M., A. Nesci, G. Barros, A. Torres, and S. Chulze. 1999. "Occurrence of Aspergillus Section Flavi and Aflatoxin B1 in Corn Genotypes and Corn Meal in Argentina." *Mycopathologia* 147(1): 37–41.

FDA [see U.S. Food and Drug Administration].

Food Standards Agency (FSA). 2003. "Contaminated Maize Meal Withdrawn from Sale" (available online at www.food.gov.uk/news/newsarchive/maize).

Gong, Y.Y., K. Cardwell, A. Hounsa, S. Egal, P.C. Turner, A.J. Hall, and C.P. Wild. 2000. "Dietary Aflatoxin Exposure and Impaired Growth in Young Children from Benin and Togo: Cross Sectional Study." *British Medical Journal* 325(7354): 20–21.

Hammond, B., K. Campbell, C. Pilcher, A. Robinson, D. Melcion, B. Cahagnier, J. Richard, J. Sequeira, J. Cea, F. Tatli, R. Grogna, A. Pietri, G. Piva, and L. Rice. 2003. "Reduction of Fumonisin Mycotoxins in Bt Corn." *The Toxicologist* 72(S-1): 1217.

Hendricks, K. 1999. "Fumonisins and Neural Tube Defects in South Texas." *Epidemiology* 10(2): 198–200.

Henry, S.H., F.X. Bosch, T.C. Troxell, and P.M. Bolger. 1999. "Reducing Liver Cancer: Global Control of Aflatoxin." *Science* 286(5449): 2453–2454.

IARC [see International Agency for Research on Cancer].

International Agency for Research on Cancer (IARC). 2002. *Some Traditional Herbal Medicines, Some Mycotoxins, Naphthalene and Styrene*. Monograph Volume 82, IARC, Lyon, France (available online at http://monographs.iarc.fr/htdocs/indexes/vol82index.html).

International Programme on Chemical Safety (IPCS). 1999. *Environmental Health Criteria 217: Microbial Pest Control Agent Bacillus thuringiensis.* World Health Organization, Geneva, Switzerland.

Jaffe, G. 2004. "Regulating Transgenic Crops: A Comparative Analysis of Different Regulatory Processes." *Transgenic Research* 13(1): 5–19.

James, C. 2003. *Global Review of Commercialized Transgenic Crops: 2002. Feature: Bt Maize.* International Service for the Acquisition of Agri-Biotech Applications (ISAAA) No. 29, Ithaca, NY.

JECFA [see Joint FAO/WHO Expert Committee on Food Additives].

Joint FAO/WHO Expert Committee on Food Additives (JECFA). 2001. "Safety Evaluation of Certain Mycotoxins in Food." WHO Food Additives Series 47, FAO Food and Nutrition Paper 74, WHO/FAO, Geneva.

Keyl, A.C. 1978. "Aflatoxicosis in Cattle." In T.D. Wyllie and L.G. Morehouse, eds., *Mycotoxic Fungi, Mycotoxins, Mycotoxicoses* (Vol. 2). New York: Marcel Dekker.

Li, F.Q., T. Yoshiazawa, O. Kawamura, X.Y. Luo, and Y.W. Li. 2001. "Aflatoxins and Fumonisins in Corn from the High-Incidence Area for Human Hepatocellular Carcinoma in Guangxi, China." *Journal of Food and Agricultural Chemistry* 49(8): 4122–4126.

Lubulwa, A.S.G., and J.S. Davis. 1994. "Estimating the Social Costs of the Impacts of Fungi and Aflatoxins in Maize and Peanuts." In E. Highley, E.J. Wright, H.J. Banks, and B.R. Champ, eds., *Stored Product Protection: Proceedings of the 6th International Working Conference on Stored-product Protection.* Wallingford, UK: CAB International.

Marasas, W.F.O. 1996. "Fumonisins: History, World-Wide Occurrence and Impact." In L. Jackson, ed., *Fumonisins in Food.* New York: Plenum Press.

Marasas, W.F.O., R.L. Riley, K.A. Hendricks, V.L. Stevens, T.W. Sadler, J. Gelineau-van Waes, S.A. Missmer, J. Cabrera, O. Torres, W.C.A. Gelderblom, J. Allegood, C. Martinez, J. Maddox, J.D. Miller, L. Starr, M.C. Sullards, A.V. Roman, K.A. Voss, E. Wang, and A.H. Merrill, Jr. 2004. "Fumonisins Disrupt Sphingolipid Metabolism, Folate Transport, and Neural Tube Development in Embryo Culture and *in vivo*: A Potential Risk Factor for Human Neural Tube Defects Among Populations Consuming Fumonisin-Contaminated Maize." *Journal of Nutrition* 134(4): 711–716.

Miller, J.D. 1995. "Fungi and Mycotoxins in Grain: Implications for Stored Product Research." *Journal of Stored Product Research* 31(1): 1–16.

_____. 2001. "Factors That Affect the Occurrence of Fumonisins." *Environmental Health Perspective* 109 (Supplement 2): 321–324.

Miller, J.D., and W.F.O. Marasas. 2002. "Ecology of Mycotoxins in Maize and Groundnuts." Supplement to *LEISA* (Low External Input and Sustainable Agriculture) Magazine (March): 23–24.

Munkvold, G.P. 2001. "Potential Impact of FDA Guidelines for Fumonisins in Foods and Feeds." In *Mycotoxins: The Cost of Achieving Food Security and Food Quality.* American Phtopathological Society, St. Paul, MN (available online at www.apsnet.org/online/fea[-]ture/mycotoxin/top.html) (accessed May 2004).

Munkvold, G.P., and R.L. Hellmich. 1999. "Comparison of Fumonisin Concentrations in Kernels of Transgenic Bt Maize Hybrids and Nontransgenic Hybrids." *Plant Disease* 83(2): 130–138.

Nap, J.P., P.L.J. Metz, M. Escaler, and A.J. Conner. 2003. "The Release of Genetically Modified Crops into the Environment, Part I: Overview of Current Status and Regulations." *The Plant Journal* 33(1): 1–18.

National Academy of Sciences. 2000. *Genetically Modified Pest-Protected Plants: Science and Regulation.* Washington, D.C.: National Academy Press.

Odvody, G.N., C.F. Chilcutt, R.R. Parker, and J.H. Benedict. 2000. "Aflatoxin and Insect Response of Near-Isogenic Bt and non-Bt Commercial Corn Hybrids in South Texas." In J.F. Robens, ed., *Proceedings of the 2000 Aflatoxin/Fumonisin Workshop.* Agricultural Research Service, U.S. Department of Agriculture, Beltsville, MD.

Paarlberg, R. 2001. *The Politics of Precaution: Genetically Modified Crops in Developing Countries.* Baltimore, MD: International Food Policy Research Institute.

Pray, C.E., J. Huang, R. Hu, Q. Wang, B. Ramaswami, and P. Bengali. 2006. "Benefits and Costs of Biosafety Regulation in India and China." In R.E. Just, J.M. Alston, and D. Zilberman, eds., *Regulating Agricultural Biotechnology: Economics and Policy.* New York: Springer.

Resnik, S., S. Neira, A. Pacin, E. Martinez, N. Apro, and S. Latreite. 1996. "A Survey of the Natural Occurrence of Aflatoxins and Zearalenone in Argentine Field Maize: 1983–1994." *Food Additives and Contaminants* 13(1): 115–120.

Rheeder, J.P., W.F. Marasas, and H.F. Vismer. 2002. "Production of Fumonisin Analogs by Fusarium Species." *Applied Environmental Microbiology* 68(5): 2101–2105.

Robens, J., and K. Cardwell. 2003. "The Costs of Mycotoxin Management to the USA: Management of Aflatoxins in the United States." *Journal of Toxicology: Toxin Reviews* 22(2–3): 143–156.

Ross, P.F., L.G. Rice, G.D. Osweiler, P.E. Nelson, J.L. Richard, and T.M. Wilson. 1992. "A Review and Update of Animal Toxicoses Associated with Fumonisin-Contaminated Feeds and Production of Fumonisins by Fusarium Isolates." *Mycopathologia* 17(1–2): 109–114.

Shelby, R.A., O.G. White, and E.M. Burke. 1994. "Differential Fumonisins Production in Maize Hybrids." *Plant Disease* 78(6): 582–584.

Shephard, G.S. 2004. "Mycotoxins Worldwide: Current Issues in Africa." In D. Barug, H.P. van Egmond, R. Lopez-Garcia, W.A. van Osenbruggen, and A. Visconti, eds., *Meeting the Mycotoxin Menace.* Wageningen, the Netherlands: Wageningen Academic.

Sinha, A.K. 1994. "The Impact of Insect Pests on Aflatoxin Contamination of Stored Wheat and Maize." In E. Highley, E.J. Wright, H.J. Banks, and B.R. Champ, eds., *Stored Product Protection: Proceedings of the 6th International Working Conference on Stored-product Protection.* Wallingford, UK: CAB International.

Sydenham, E.W., G.S. Shephard, P.G. Thiel, W.F.O. Marasas, and·S. Stockenstrom. 1991. "Fumonisin Contamination of Commercial Corn-Based Human Foodstuffs." *Journal of Agricultural Food Chemistry* 39(11): 2014–2018.

Turner, P.C., S.E. Moore, A.J. Hall, A.M. Prentice, and C.P. Wild. 2003. "Modification of Immune Function Through Exposure to Dietary Aflatoxin in Gambian Children." *Environmental Health Perspectives* 111(2): 217–220.

U.S. Department of Agriculture. 2003a. *World Corn Trade.* Foreign Agricultural Service, Production, Supply and Distribution.

———. 2003b. *World Coarse Grains Situation and Outlook.* Foreign Agricultural Service (available online at http://www.fas.usda.gov/grain/circular/2003/10-03/cgra_txt.htm).

_____. 2004. *Acreage*. USDA National Agricultural Statistics Service, Agricultural Statistics Board, Washington, D.C.

U.S. Food and Drug Administration. 2000. "Background Paper in Support of Fumonisin Levels in Corn and Corn Products Intended for Human Consumption" (available online at http://www.cfsan.fda.gov/~dms/fumonbg3.html).

U.S. Environmental Protection Agency. 2001. *Biopesticides Registration Action Document (BRAD)—Bacillus thuringiensis Plant-Incorporated Protectants*. Available online at www.epa.gov/pesticides/biopesticides/pips/bt_brad.htm.

van Egmond, H.P. 2002. "Worldwide Regulations for Mycotoxins." *Advances in Experimental Medical Biology* 504: 257–269.

Wang, J.S., T. Huang, J. Su, F. Liang, Z. Wei, Y. Liang, H. Luo, S.Y. Kuang, G.S. Qian, G. Sun, X. He, T.W. Kensler, and J.D. Groopman. 2001. "Hepatocellular Carcinoma and Aflatoxin Exposure in Zhuqing Village, Fusui County, People's Republic of China." *Cancer Epidemiological Markers* 10(2): 143–146.

Wicklow, D.T. 1994. "Preharvest Origins of Toxigenic Fungi in Stored Grain." In E. Highley, E.J. Wright, H.J. Banks, and B.R. Champ, eds., *Stored Product Protection: Proceedings of the 6th International Working Conference on Stored-product Protection*. Wallingford, UK: CAB International.

Williams, W.P., G.L. Windham, P.M. Buckley, and C.A. Daves. 2002. "Aflatoxin Accumulation in Conventional and Transgenic Corn Hybrids Infested with Southwestern Corn Borer (Lepidoptera: Crambidae)." *Journal of Agricultural and Urban Entomology* 19(4): 227–236.

Windham, G.L., W.P. Williams, and F.M. Davis. 1999. "Effects of the Southwestern Corn Borer on *Aspergillus flavus* Kernel Infection and Aflatoxin Accumulation in Maize Hybrids." *Plant Disease* 83(6): 535–540.

Wu, F. 2004. "Mycotoxin Risk Assessment for the Purpose of Setting International Regulatory Standards." *Environmental Science and Technology* 38(15): 4049–4055.

Wu, F., J.D. Miller, and E.A. Casman. 2004. "Bt Corn and Mycotoxin Reduction: An Economic Perspective." *Journal of Toxicology, Toxin Reviews* 23(2–3): 397–424.

Wyatt, R.D. 1991. "Poultry." In J.E. Smith and R.S. Henderson, eds., *Mycotoxins and Animal Foods*. Boca Raton, FL: CRC Press.

Yoshizawa, T., A. Yamashita, and Y. Luo. 1994. "Fumonisin Occurrence in Corn from High- and Low-Risk Areas for Human Esophageal Cancer in China." *Applied and Environmental Microbiology* 60(5): 1626–1629.

Zhang, J.Y., X. Wang, S.G. Han, and H. Zhuang. 1998. "A Case-Control Study of Risk Factors for Hepatocellular Carcinoma in Henan, China." *American Journal of Tropical Medicine and Hygiene* 59(6): 947–951.

Chapter 10

CONSUMER ATTITUDES AND MARKET RESISTANCE TO BIOTECH PRODUCTS

Wallace E. Huffman[*] and Matt Rousu[†]
Iowa State University[*] and Susquehanna University[†]

Abstract: Society has had about 11,000 years of experience with domestication of plants and crop improvement, and 10,000 years with domestication of animals and livestock improvement. Breeding practices in plants have progressed in stages from weak forms of selection, to strong forms of selection, and then to hybridization, mutagenesis, and biotechnology with selection. Only the use of modern biotechnology has created consumer concerns. Current genetically modified (GM) products have been developed primarily from so-called input traits, and consumers tend to see little direct benefit and some risks. U.S. consumers are, however, more receptive than European consumers to GM products. Consumers have expressed a strong preference for GM-content labeling when it is costless, but their preferences drop off dramatically when it is costly. Mandatory labeling is not required in the United States and voluntary labeling would require that non-GM products be labeled; they are the superior product currently. However, this would not meet EU labeling and traceability requirements. The empirical evidence is that U.S. consumers discount GM products a little and EU consumers by a larger amount. U.S. consumer demand for GM products has been shown to respond to biotech industry, environmental, and third-party information differently, depending on the amount of prior information that they have. Those who are uninformed are most affected by anti-biotech and pro-tech information.

Key words: consumer acceptance, GM products, GM-labeled products, tolerance levels, information effects, prior beliefs, pro-biotech, anti-biotech, economic experiments, nth-price auction, segregation, input traits, rBST, industry perspective, environmental group perspective, third-party perspective

1. INTRODUCTION

Two decades after the first biotech food product entered the U.S. market and one decade after the first successful transgenic crop products started to come on the market, food made from genetically modified materials remains engulfed in considerable controversy. U.S. dairy farmers have shown a ready willingness to adopt rBGH or rBST, and U.S. cotton, oil seed, and corn producers to adopt Bt cotton, Roundup Ready soybeans, and Bt corn varieties (Fernandez-Cornejo and McBride 2002, James 2003). These biotech traits are what have become known as "input traits," i.e., they can be expected to reduce farmers' cost of production of agricultural products, but they do not have direct benefits for consumers and may be interpreted as having some risks or costs. U.S. and Chinese consumers have been relatively accepting of agricultural biotech products, and Argentinean, Canadian, and Brazilian consumers have been somewhat accepting. In contrast, consumers in the European Union (EU), Norway, Switzerland, and Japan have shown greater opposition to agricultural biotech products (Food and Agriculture Organization 2004).

In our history, other socially useful new consumer goods have sometimes encountered resistance and only after some extended trial period been widely accepted. They include pasteurization of milk a century ago, nuclear power starting fifty years ago, and irradiated meat and poultry over the past decade. In the United States, early opposition to pasteurization of milk was widespread, with opponents saying, among other things, that pasteurization was not needed and that consumers had the "right to drink raw milk" (see Hotchkiss 2001). Full acceptance of pasteurization in the United States was not attained until after 40 years of experience with pasteurization, and long after universal scientific agreement on its benefits. This was for a process that had clear health benefits, and Pirtle (1926) notes that the slow adoption of pasteurization resulted in thousands of deaths that could have been prevented at a very low social cost.

The early prospects for nuclear power were good, but major and persistent resistance developed in Europe and the United States to electricity generated by nuclear power (see Grübler 1996). Although nuclear power is relatively cheap to produce and low in traditional environmental pollutants—e.g., CO_2, nitrous oxides, sulfur oxides—environmental groups like Greenpeace and Friends of the Earth have lobbied and demonstrated against using the technology. Ruttan (2001) indicates that these groups helped increase the public's risk perception of nuclear power in the United States, forcing stringent safety standards to be enacted that contributed to a quadrupling of plant costs in just less than a decade. No new nuclear power plants have been ordered in the United States since 1978, but the dramatic rise in crude oil prices over the past six months from $30 per barrel to $60 and prospects for further increases has ignited general interest in re-examining the economic and political feasibility of nuclear power as a substitute for fossil fuel power.

The image problems of nuclear power have carried over to irradiated foods during the past decade. Although irradiation of meat and poultry essentially eliminates all bacteria harmful to human health, such as *e-coli*, *listeria* and *salmonella*, it has encountered stiff resistance (Nestle 2002; Fox, Hayes, and Shogren 2002; Nayga, Aiew, and Nichols 2005). Failure to irradiate meat and poultry has resulted in a significant number of annual deaths that were preventable at small marginal social cost.

The purpose of this chapter is to provide a critique of the literature on consumer attitudes toward and resistance to genetically engineered or modified food products. This section contains an overview of some key issues. We first present a historical perspective on techniques used for crop and animal improvement. Then we present information on consumer attitudes toward genetically modified (GM) products and consumers' evaluation of GM products. Next we turn to the important issue of effective GM labeling with costly segregation. We then examine the important issue of information effects on the demand for GM food products. The final section contains conclusions.

2. BACKGROUND ON GENETIC IMPROVEMENT

The domestication of wild species of plants and animals began about 11,000 years ago, and it was associated with people moving from a nomadic hunter-gatherer society to a settled agricultural society. These early people saw the potential of some wild species to provide food in larger quantities than wild species, and in the case of animals, they could provide food (e.g., milk and meat), hides and wool, and transportation and power. Domestication involved several steps, but it does include selection against weedy or wild tendencies, and domesticated plants and animals must depend on humans for survival (Wikipedia 2005).

The first plants to be domesticated were wheat and peas in the Middle East. Wild wheat dropped its seeds on the ground when ripe to get positioned to start the next growing season. In contrast, the domesticated wheat retains its seeds on the plant when ripe. It is believed that this change in seed-holding behavior was due to a very early important mutation. About 8,000 years ago early American Indians domesticated the wild plant teosinte to create corn. The native Americans most likely used selective breeding, a crude form of genetic manipulation, in a remarkable way, to create open-pollinated corn varieties with many of the same properties of corn today (Falk et al. 2002). With these open-pollinated varieties, farmers saved their own seed for planting next year's crop.

About 10,000 years ago, sheep, goats, and pigs were domesticated from wild species—sheep and goats in the Middle East and pigs in China. Key attributes of these early domesticated sheep and goats were an ability to survive on poor land with limited grazing potential and an ability to survive outdoors at all times of the years and give birth naturally without human assistance. The prehistoric sheep were smaller than today's sheep and of poorer quality. Domesticated sheep have a pleasant disposition toward humans and a temperament that makes them unlikely to panic and flee, which means they can easily be herded by shepherds, but also makes them vulnerable to predators (Wikipedia 2005). Sheep provided wool and hides for clothing, milk for direct consumption or cheese-making, and meat. Their hides are also useful for clothing. Domesticated goats can survive on an even more varied and meager diet and are smarter than sheep. Domesticated pigs were smaller than wild pigs, had a high reproduction rate, and could consume a large range of organic materials. A large share of the pig could also be consumed as meat by humans.

The cow was domesticated about 8,000 years ago in the Middle East, most likely due to selection from larger-boned wild species, which were better adapted to survive in harsh conditions. Domesticated cows could also survive on relatively low quality feed, but needed better forage than goats or sheep. Milk from cows was useful for drinking and the cream for cheese, meat was a source of food for humans, and hides and leather could be used to make clothing. Also, cattle could be used as a source of power.

Although the domestication of sheep, goats, swine, and cattle occurred 8,000 to 10,000 years ago, little genetic improvement occurred until about 1700. Before 1700, land holdings in Europe (and elsewhere) were under a feudal system, and livestock mingled together as they grazed the "common pastures." There was no way to effectively control mating, and little knowledge of its potential. With enclosure or fencing of the commons, which started about 1700 in Europe, farmers started to consider farm animal improvement, and improved strains began to appear (Huffman and Evenson 1993, pp. 13–15). Early selection focused on adaptation of domesticated farm animals to particular local geoclimatic conditions, e.g., during the eighteenth and nineteenth centuries different strains of sheep are alleged to have inhabited every valley of England and Scotland. In 1760, Robert Bakewell and English farmers were credited with first establishing the patterns of modern animal breeding. He established an early form of inbreeding, or purification of lines. He placed an emphasis on selection for visual traits. This inbreeding led to relatively true breeding strains and became the foundation for "purebred animals."

In 1908, the (single-cross) corn hybrid was discovered. This hybridization involved crossing two inbred corn lines, but the cost of hybrid seed was too expensive because inbred lines are weak seed producers. Donald Jones discovered the double-cross hybrid in 1916–1919, which involved the crossing of two single crosses (Huffman and Evenson 1993, pp. 155–156). Since hybrids cannot reproduce themselves, this created a natural market for annual hybrid seed corn sales to farmers, and the early development of the U.S. private seed industry.

The hybridization of corn, which also provided a method for developing later generation hybrids, did not cause any consumer concerns about acceptance. During the latter part of the twentieth century, plant breeders expanded their tools of genetic manipulation beyond conventional selection and cross-breeding techniques. These new techniques included embryo rescue, chemical and radiation mutagenesis, and somaclonal variation (Falk et al. 2002). These techniques allowed multiple genes to be transferred and required a rigorous selection process to attain desirable and stable traits. The products of these non-conventional breeding techniques, also, have not caused any consumer concerns about acceptance (Falk et al. 2002).

Agricultural biotechnology is the use of living organisms, or parts thereof, to produce food and feed products, frequently involving the transfer of genes across species. Stanley Cohen and Herbert Boyer discovered the technique of recombinant DNA (rDNA) in 1973, and the Cohen-Boyer patent for gene-splicing technology was awarded in 1980. This gene-splicing technology has been a very important technique to the late twentieth century "Gene Revolution."

The first modern successful agricultural biotechnology product was recombinant bovine somatotropin (rBST). BST is a bovine growth hormone that occurs naturally in cows; rBST is BST produced by genetically altered bacteria in the laboratory. Laboratory-produced rBST is then injected into dairy cows to raise the level of rBST in their system, which increases their rate of milk production (Aldrich and Blisard 1998). The first reported use of rBST to increase milk production was in 1982; and in 1984–84, the U.S. Food and Drug Administration (FDA) ruled that milk and meat from rBST-treated cows in experimental herds were safe for human consumption and could be marketed. Thereafter, milk and meat from rBST-treated cows were introduced into the U.S. food supply. In 1993, the FDA approved general use of rBST, and after some delay, this policy went into effect. Some consumers have expressed real concerns about rBST milk, and there has been some long-term consumer resistance due to health concerns and animal rights issues.

One of the important input traits has been the insertion of the Bt gene into cotton and corn. When the Bt gene, *Bacillus thuringiensis*, is inserted into cotton or corn, it causes these plants to express proteins that are produced by the common soil bacteria. The Bt organisms are modified to express a class of proteins that are toxic to certain insects (Lepidoptran of the caterpillar family) and possibly mold and fungal pests, but have been shown to be harmless to humans, mammals, and birds. Furthermore, since this Bt gene is embedded in the plant from the beginning of growth, there is no timing issue or washing off of the Bt protection in contrast to chemical insecticide applications. The Bt gene was introduced successfully into cotton varieties starting in 1996, and has been quite effective in controlling the bud-

and-boll worm complex in cotton in many areas of the United States (Falck-Zepeda, Traxler, and Nelson 2000; Fernandez-Cornejo and McBride 2002; Food and Agriculture Organization 2004). A few years later the Bt gene was introduced into corn to control the European Corn borer and only recently to control the corn root worm.

Biotechnology has been used to insert herbicide tolerance first into soybeans, and then into other field crops—canola, cotton, and corn. The herbicide tolerance or Roundup Ready crop varieties are encoded with a protein, the enzyme mEPSPS, which makes the plant tolerant to glyphosate, the active ingredient in Roundup herbicide. When Roundup is applied to a Roundup Ready crop variety, every plant is killed, except for the Roundup Ready plants. Herbicide tolerance in soybeans and cotton has been extremely popular with U.S. farmers.

A number of other biotech traits have been discovered, but few have come to market or stayed on the market. The biotech tomato having extended shelf life was discovered by scientists at Calgene and brought to market in 1994. Its extended shelf life was a positive consumer attribute, but it was introduced into inferior tomato germplasm and overpriced.[1] Golden rice is another product having a positive consumer attribute of high beta-carotene, which is a source of vitamin A. The golden rice variety was made by moving two genes from daffodil and one from the bacteria *Erwinia uredovora* (Ye et al. 2000, Food and Agriculture Organization 2004). Golden rice seems of greater potential value in very poor countries than to consumers in the United States, where there are other ready sources of vitamin A.

Bt sweet corn has the potential to dramatically reduce insecticide applications, for example from 35 to 3 to 4 applications, similar to that experienced in cotton, but in this case, the ears of corn, which are eaten as a vegetable, have low pesticide residual. Pesticide residual is a so-called negative attribute for consumers. Also, virus resistance has been introduced into potato, yellow and green squash, and papaya, and biotech products have been approved for sale. The virus-resistant potato and papaya carry an input trait valuable to farmers.[2] Biotech squash would dramatically reduce vegetable growers' applications of pesticides.

3. CONSUMERS' ATTITUDES TOWARD GM PRODUCTS

In the United States, consumers were relatively accepting of new GM products over the mid-1990s to 2000, but this support may have weakened a little since then (Hoban 2004). In Europe in the mid-1990s, consumer attitude

---

[1] It failed in 1998 when a sequence of EU food scares set in.

[2] The fast food restaurants and grocery wholesalers, however, would not purchase or handle these products, so they were taken off the market in 2001.

toward genetically modified products was somewhat positive, but their attitude turned more negative in the late 1990s when some food scares surfaced. This negativity has persisted. A recent paper by Chern and Rickertsen (2004) provides insights into consumers' general reaction to GM products, perceived benefits, and perceived risks or ethical concerns. They present results from surveys conducted in 2002 on U.S. and Norwegian consumers.

The Chern and Rickertsen (2004) paper shows that 43 percent of U.S. consumers were extremely or somewhat willing to consume foods produced with GM ingredients. In contrast, only 30.5 percent of Norwegian consumers were in this category (Table 1). Also, 16.4 percent of U.S. consumers were extremely unwilling to consume foods produced with GM ingredients, but a much smaller 4.5 percent of Norwegian consumers were in this category. Hence, we see the generally more willing nature of U.S. than Norwegian consumers to consume foods produced with GM ingredients.

Consumers in these surveys were also asked how willing they would be to purchase GM food having attributes that are valuable to consumers. The new food attributes were the reduction in the amount of pesticide applied to crops or being more nutritious than similar foods that are not GM. For the U.S. consumers, about 70 percent were extremely or somewhat willing to consume GM products with these attributes (Table 1). In contrast, only about 38 percent of Norwegian consumers were extremely or somewhat willing. Norwegian consumers were over three times as likely to be extremely unwilling to consume these products as U.S. consumers.

When asked about the importance of price as a factor when deciding whether or not to buy GM foods, 67.2 percent of U.S. consumers indicated that price was extremely or somewhat important to them. In contrast, price was somewhat or extremely important to only 36 percent of Norwegian consumers (Table 1). One-half of Norwegian consumers indicated that price was extremely unimportant, but only 12.5 percent of U.S. consumers were in this category. Hence, we see that price plays a more positive role in U.S. consumers' decisions on GM foods than for Norwegian consumers.

With so-called input traits, consumers tend to see no direct benefit and some added risks. When asked how willing they would be to purchase GM foods that posed a risk of causing allergic reactions for some people, only one-quarter of U.S. consumers indicated that they were extremely or somewhat willing. The affirmative was a significantly lower 10 percent in Norway. When consumers were asked how risky they thought GM foods are in terms of risk to human health, only 21 to 24 percent of consumers in both countries thought that GM foods were extremely to somewhat safe. However, three times as many Norwegians as Americans believed that GM foods were extremely unsafe (Table 1).

When genes are moved across species, this can raise ethical or religious concerns. When consumers were asked how important ethical or religious

Table 1. Consumer Attitudes Toward GM Foods, Percentage Distribution for Each Question

Question	Country	Willing/Important/Safe			Unwilling/Unimportant/Risky		
		Extremely (1)	Somewhat (2)	Neither (3)	Somewhat (4)	Extremely (5)	Don't Know
1. General reaction							
a. How willing are you to consume foods produced with GM ingredients?	Norway	13.0	17.5	4.0	18.0	4.5	2.0
	USA	4.7	38.3	13.7	23.8	16.4	3.1
1,2 = Willing and 4,5 = Unwilling							
b. How important is it to you that food products are specifically labeled as GM or non-GM?	Norway	94.0	4.5	0.5	0.0	1.0	0.0
	USA	58.6	28.5	4.3	5.9	1.6	1.2
1,2 = Important and 4,5 = Unimportant							
2. Perceived benefits							
a. How willing would you be to consume GM foods if they reduced the amount of pesticides applied to crops?	Norway	17.0	21.5	9.5	11.5	35.5	5.0
	USA	13.7	54.7	9.4	11.3	9.0	2.0
1,2 = Willing and 4,5 = Unwilling							
b. How willing would you be to purchase GM foods if they were more nutritious than similar foods that are not GM?	Norway	17.5	19.5	7.5	10.0	39.0	6.5
	USA	18.0	53.9	5.1	9.4	10.9	2.7
1,2 = Willing and 4,5 = Unwilling							
c. How important is the price factor when you decide whether or not to buy GM foods?	Norway	16.0	20.0	6.0	7.0	50.5	0.5
	USA	29.7	37.5	7.0	12.1	12.5	1.2
1,2 = Important and 4,5 = Unimportant							
3. Perceived risks or ethical concerns							
a. How willing would you be to purchase GM foods if it posed a risk of causing allergic reactions for some people?	Norway	1.5	8.5	2.0	4.0	83.5	0.5
	USA	3.5	21.5	5.9	26.2	41.4	1.6
1,2 = Willing and 4,5 = Unwilling							
b. How risky would you say GM foods are in terms of risk to human health?	Norway	10.5	13.0	8.0	26.0	33.5	9.0
	USA	5.5	15.2	16.0	39.5	9.4	14.5
1,2 = Safe and 4,5 = Risky							
c. How important are ethical or religious concerns when you decide whether or not to consume GM foods?	Norway	21.5	8.0	3.5	2.5	62.5	2.0
	USA	12.5	23.8	15.2	18.0	28.9	1.6
1,2 = Important and 4,5 = Unimportant							

Source: Adapted from Chern and Rickertsen (2004).

concerns are when they decide whether or not to consume GM foods, 36.3 percent of U.S. and 29.5 percent of Norwegian consumers indicated that they were extremely or somewhat important to them. However, a much larger share of Norwegian than American consumers indicated that ethical or religious reasons were extremely unimportant (Table 1). Hence, ethical or religious opposition to GM products may actually be higher in the United States than in Europe.

Food labeling has become an issue with GM products in the market. When consumers were asked how important it was to them that food products are specifically labeled as GM or non-GM, 87.1 percent of U.S. consumers and 99 percent of Norwegian consumers thought that it was extremely or somewhat important (Table 1). These results have been interpreted as preferences for *costless labeling*. An effective food labeling system, however, would use scarce resources and raise the price of food. When consumers were told that an effective GM labeling system would raise the price of food by 5 percent, 44 percent of U.S. consumers and 55 percent of Norwegian consumers continued to prefer labeling. Hence, we see that U.S. and Norwegian consumers' preference for labeling is significantly price-responsive, falling by 44 percentage points for only a 5 percent increase in food prices due to labeling costs.

4. CONSUMERS' EVALUATION OF GM PRODUCTS

With GM products being in the United States and some other markets at least for about a decade, several studies have been conducted that attempt to assess the impact of GM attributes on consumers' willingness to pay, either in an experimental auction or in a state-preference model (contingent valuation) survey. State-preference or contingent-valuation surveys have been criticized because of the existence of hypothetical bias, but they still provide some useful information (Lusk et al. 2001).[3]

First, let us examine the results for studies that have assessed the willingness to pay discount associated with input traits, i.e., GM traits that have no direct benefit to consumers but are expected to reduce the expected cost of production to farmers and thereby lower food prices. If Roundup Ready GM wheat is used for bread in Norway, the average discount is 50 percent, and if it is used for noodles in Japan, the discount is 60 percent (Table 2). If (Bt) GM corn is used for breakfast cereal, it is discounted 38 cents per box in the United States, and 73 cents per box in the United Kingdom. If vegetable

---

[3] Hypothetical bias is due largely to the fact that the consumer is revealing information about hypothetical rather than real transactions, and hence, does not face a budget constraint. This bias leads to over-discounting or over-bidding.

Table 2. Consumer Evaluation of GM Products: Selected Studies

Products	Country	Study Type[a]	Value Attribute	Source
A. Direct consumer benefit				
1. Golden rice (enhanced vitamin A)	U.S.	SP	12¢ / lb (with cheap talk)	Lusk (2003)
2. Corn chips (enhanced shelf life)	U.S. (MS)	SP	1¢ / 14.5 oz.	Lusk et al. (2001)
3. Sweet corn (low pesticide residual)	U.S. (CA)	SP	0	Wolf and Giacalone (2003)
B. Indirect consumer benefit (low price)				
1. GM (Roundup Ready) wheat – bread	Norway	SP	-49.5%	McCluskey, Grimsrud, and Wahl 2004
noodles	Japan	SP	-60%	
2. GM corn – breakfast cereal	U.S.	SP	-38¢ / 24 oz. box	Moon and Balasutramaniau (2004)
	UK	SP	-73¢ / 24 oz. box	Moon and Balasutramaniau (2004)
3. Roundup Ready soybeans – vegetable oil	Norway	SP	NOK -22.13 / liter	Chern and Rickertsen (2004)
	U.S.	SP	$-1.82 / 32 oz.	Chern and Rickertsen (2004)
	U.S. (IA, MN)	EA	$-14¢ / 32 oz.	Huffman et al. (2004a)
4. GM corn (Bt) – chips	U.S. (MS)	SP	-32¢ / 14.5 oz.	Lusk et al. (2001)
	U.S. (IA, MN)	EA	-15¢ / 16 oz.	Huffman et al. (2004a)
5. GM (virus-resistant) potatoes	U.S. (IA, MN)	EA	-13¢ / 5 lbs.	Huffman et al. (2004a)
6. Unlabeled vs. rBST-free milk	U.S.	M	-1¢ / gal [b]	Dhar and Foltz (2005)
7. GM-fed salmon	Norway	SP	NOK -43.42 / kilo	Chern and Rickertsen (2004)
	U.S.	SP	$ -2.75 / lb.	Chern and Rickertsen (2004)
8. GM (rapid growth) salmon	Norway	SP	NOK -53.96 / kilo	Chern and Rickertsen (2004)
	U.S.	SP	$ -4.49 / lb.	Chern and Rickertsen (2004)

[a] Study types are as follows: EA = experimental auction, M = market scanner data, SP = stated preference or contingent valuation.
[b] Discount computed using market shares.

oil is made for Roundup Ready GM beans, the oil is discounted by NOK 22.13 per liter in Norway and by $1.82 per 32 ounces in the United States in one study, and by 14 cents per bottle in another. One reason for the large difference in the U.S. data is that the second study is free from hypothetical bias; consumers had to execute their bids with an exchange of money for oil.

In two U.S. studies of the demand for corn chips, consumers discounted GM chips by 32 cents per bag (14.5 ounces) in one study and by 15 cents per bag (16 ounces) in a second study. The first of these studies used undergraduates and the second used a sample of the adult population. U.S. consumers also discounted GM potatoes by 13 cents per bag (5 pounds). See Table 2.

rBST milk entered the U.S. fresh milk market in large quantities starting in 1994 (Dhar and Foltz 2005). For 1997 to 2002, their data for 12 cities distributed across the United States showed that the average price of unlabeled fresh milk was $2.80 per gallon, but the average price for rBST-free and organic milk was much higher—$4.85 and $4.91 per gallon, respectively. However, the dominant market share went to unlabeled milk, which accounted for 99.1 percent of the fresh milk market. The share for rBST-free milk was 0.5 percent and for organic milk 0.4 percent. Using market share data for unlabeled and rBST-free milk, the average discount for unlabeled (presumably containing some rBST milk) was one cent per gallon.

GM studies have also been undertaken on fish products. Norwegian consumers discounted salmon fed GM feed by NOK 43.42 per kilo, and U.S. consumers discounted GM-fed salmon by $2.75 per pound. The discount for GM salmon, which have genes for especially rapid growth, is even larger. Norwegian consumers discounted GM salmon by NKO 53.96 per kilo, and U.S. consumers discounted it by $4.49 per pound.

There has been a sense that GM products would obtain a favorable consumer image if the GM products delivered enhanced consumer traits. Three studies allow us to assess the value of these traits. First, in a U.S. study of willingness to pay for golden rice, consumers would pay 12 cents per pound (with cheap talk). For corn chips with GM technology used to enhance shelf life, consumers would pay one cent per bag (14.5 ounces), and for Bt sweet corn (low pesticide residual) consumers would not pay any more for it than for regular sweet corn. The latter result suggests that whatever value the low pesticide residual has, it is exactly offset by the perceived consumer risks of GM technology. Hence, the results from the studies of GM food products that have enhanced consumer traits are somewhat pessimistic about the future of GM food products even in the United States.

5. EFFECTIVE GM LABELING WITH COSTLY SEGREGATION

Food products in the United States can be labeled for nutritional claims and for safety. The 1990 Nutrition Labeling and Education Act dramatically changed nutrition labels on packaged foods sold in U.S. supermarkets (Balasubramanian and Cole 2002).[4] This law required packaged foods to display nutrition information prominently in a new label format, namely the Nutrition Facts panel. It also regulates serving size, health claims (that link a nutrient to a specific disease), and descriptor terms (e.g., "low fat") on food packages. The goal of this legislation was to improve consumer welfare by providing nutrition information that will assist consumers in making healthful food choices.

As an indication of the costliness of effective food nutrient labeling, it is estimated that the U.S. food industry spent $2 billion to comply with the 1990 Nutrient Labeling Act (Silverglade 1996). Some attributes are viewed positively by consumers (e.g., enhanced calcium or vitamin A and C, where more is better for the consumer), and these are refereed to as positive attributes. Other food attributes are negative attributes (e.g., salt, fat), where the consumer considers less to be better. Food labels before the Nutrient Labeling Act had a seeming emphasis on negative labeling. Balasubramanian and Cole (2002) suggest that this tendency can be explained by consumers having an asymmetric value function, weighing a dollar of loss more heavily than a dollar of gain, which is Tversky and Kahneman's (1981) prospect theory.

The policies under the Nutrition Labeling and Education Act also tend to emphasize negative rather than positive labeling. First, health claims allowed are ones that associate specific nutrients with reduced risk of specific diseases. Of the 7 health claims approved by the FDA at the onset of the new nutrient labeling act, three linked negative attributes exclusively with deadly diseases—i.e., dietary fat and cancer, sodium and hypertension, and dietary saturated fat and high cholesterol and heart disease—and only one claim featured a positive attribute—calcium and osteoporosis. Later claims have, however, been more balanced. Second, regulations on nutrient-content claims tend to focus more heavily upon negative attributes (calories, sugar, sodium, fat, fatty acid, and cholesterol) than on positive attributes such as fiber and vitamins.

Clearly, with foods made currently from biotech attributes that are input traits, adding a label for GM content would be an example of labeling a negative food attribute. However, GM content has not been proven scientifically to have human health consequences, except for the transport of some known allergens to new locations. Hence, GM food labels today would not

---

[4] Unpackaged foods, for example fresh fruits and vegetables, are not affected.

meet the nutrition labeling law requirement of nutrient intake leading to proven health outcome claims.

Genetically engineered products used for food, however, do have to pass a food safety test. In 1992, the U.S. Department of Health and Human Services issued a regulation that GM food did not have to be labeled if the food product had the same food characteristics as its non-GM counterparts. In January 2001, the FDA issued a "Guidance for Industry" statement reaffirming this policy. In this statement, the FDA stated to the biotech industry that the only GM foods that needed to be labeled are foods that have different characteristics from the non-GM version, e.g., elevated vitamin A levels. In the United States, labeling food for GM content is not required for any other food. However, firms are to notify the FDA at least four month before putting a new GM food product on the market, and the scientific description of the product is posted on the Internet for review during this time. Only minor changes have been made in these guidelines since 2001.

Hence, the U.S. labeling policy for GM can be classified as being voluntary. If a voluntary label is affixed, the FDA has mandated that it cannot use the phrase "genetically modified." The FDA prefers the phrase "genetically engineered" or "made through biotechnology."

In contrast, the European Commission adopted GM food labels in 1997.[5] The Commission requires each member country to enact a law requiring labeling of all new products containing substances derived from GM organisms. Japan, Australia, and many other countries have also passed laws requiring GM labels for major foods. The international environmental lobby has frequently argued that "consumers have the right to know whether their food is GM or not" (Greenpeace International 2001). Effective GM labeling, however, involves real costs, especially the costs of testing for the presence of GM, segregation of GM from non-GM products, variable costs of monitoring for truthfulness of labeling and enforcing the regulations that exist, and risk premiums for being out of contract (Wilson and Dahl 2002).

An effective GM labeling policy includes effective segregation of GM from non-GM commodities. If one or the other of these products could be inexpensively color-coded, segregation might not be very expensive. If, however, identity preservation through the production, marketing, and/or processing chain is required, this system would be substantially more costly (Wilson and Dahl 2002). To the extent that there is a market for non-GM products, buyers would be expected to specify in their purchase contracts some limit on GM content and/or precise prescriptions regarding production/marketing/handling processes. One can envision a marketplace of buyers with differentiated demand according to their aversion to GM content. To

---

[5] In 2003, the European Commission mandated labeling and traceability for food and feed with a 0.9 percent tolerance level.

make this differentiation effective, new costs and risks are incurred. Additional testing involves costs of conducting the tests for which there are several technologies of varying accuracy. The risk is that GM products will be commingled with non-GM products, and so the detection system must test to see that customers' shipments are within the contract limits for GM content. This is a serious economic problem, as agents seek to determine the optimal strategy for testing and other risk-mitigation strategies.

"Tolerances" are an important issue in segregation and identity preservation. GM tolerance refers to the maximum impurity level for GM content that is tolerated in a product that still carries the non-GM label. There are two levels where tolerances apply: one is defined by regulatory agencies such as the FDA, and the other is commercial tolerances. Individual firms can and seem likely to adopt different tolerances, subject to any regulation. Moreover, different countries are likely to have different tolerance levels, and this increases the risks and costs of segregation or identity preservation.

Dual market channels could develop privately without regulated tolerance levels. One option would be that farmers/growers declare GM content at the point of first delivery. This is commonly referred to as "GM Declaration" and has been an important element of the discussion of markets for GM grains (Harl 2001). At the delivery point, a grain elevator could segregate within its own facilities or each elevator could specialize in handling only GM versus non-GM grain. Or, it could be a vertically integrated firm with some delivery points specializing in GM and others in non-GM commodities or different GM commodities.

While private sector handlers routinely segregate and blend grains as a primary function of their business, new risks arise when handling GM and non-GM grains due to the added risk of adventitious commingling. Since GM is the inferior product, growers and handlers of GM products have an incentive to mix GM with non-GM products. Wilson and Dahl (2002) suggest that this risk may be about 4 percent in the United States at the grain elevator level. Farmer-processor contracting in specialty crops, however, could reduce this margin by specializing in the product being delivered, e.g., non-GM or a product with a positive GM trait. Another source of risk is testing, because no test is 100 percent accurate. This risk, however, varies with the technology, tolerance, and variety of products handled, and seems likely to be falling over time as the technology of testing advances.

With current GM products on the market being derived primarily from so-called input traits, and consumers showing an aversion to consuming products made from these traits, non-GM is the superior product. Where voluntary labeling exists, it clearly is not in the interest of the growers and handlers of GM products to signal their "inferior" quality. In fact, it is in their interest to commingle. Hence, growers and handlers of non-GM grains are the ones with the incentive to signal their "superior" quality. Some have charged that this is unfair because the growers of GM products are the ones that changed the marketing environment so that the non-GM producers would need to signal. This signaling is costly, in that testing and tolerance levels need to be set for GM content. Only products destined to be non-GM would need to be tested; producers and handler of GM products would not need to do anything special.

6. AN EXAMINATION OF INFORMATION EFFECTS

In the ongoing GM food controversy, highly conflicting information has been disseminated by interested parties—the agricultural biotech industry (Council for Biotechnology Information 2001) and the environmental groups, e.g., Greenpeace, Friends of the Earth, and Action Aid (Greenpeace International 2001). This diversity of information has undoubtedly contributed to the GM food controversy and may be one factor explaining differences in consumer acceptance across Western developed countries. Also, consistent with consumer education, independent third-party information about agricultural biotechnology may have considerable value if distributed to consumers. This information is anticipated to modify beliefs and impressions left on previous consumers by the biotech industry and environmental groups (Huffman and Tegene 2002).

To test for information effects on consumers' demand for GM foods, we designed and conducted a set of laboratory experiments in 2001. In these lab experiments, we collected survey information and data from experimental auctions where the participants were randomly chosen adults from two major Midwestern cities. The discoveries that resulted from these experiments and associated survey information allowed to us learn a great deal about consumer preferences for GM-labeled food products. Also, we gained insights into how diverse information affects consumer decisions, and the value of independent third-party information to consumers.

7. THE ECONOMIC EXPERIMENTS

The economic experiments are discussed extensively in Rousu et al. (2002) and Huffman et al. (2004a). A brief overview of the design is presented here, but the full design can be found in Huffman et al. (2004a) or obtained from the authors upon request. The participants in these laboratory experiments were adults (Carlson, Kinsey, and Nadav 1998; Katsaras et al. 2001) selected randomly from the population of two major Midwestern cities. Participants came to a central location and bid on 3 foods that were rather dissimilar—russet potatoes, tortilla chips (made from yellow corn), and vegetable oil (made from soybeans)—in two rounds of bidding. In one round participants were bidding on food products that were labeled as genetically modified and in the other round the food products had a plain label.[6] Participants bid on products using the random nth-price auction mechanism, which has been shown to be superior to a 2nd price Vickery auction for eliciting consumers' entire demand curve for new goods (List and Shogren 1999, Shogren et al. 2001).

In Experiment 1, three information types were defined, which were packaged into six information treatments and randomly assigned to experimental units/sessions. The three types were (i) the *industry perspective*—a collection of statements and information on genetic modification provided by a group of leading biotechnology companies, including Monsanto and Syngenta; (ii) the *environmental group perspective*—a collection of statements and information on genetic modification from Greenpeace, a leading environmental group; and (iii) the *independent, third-party perspective*—a statement on genetic modification approved by a third-party group, consisting of a variety of people knowledgeable about GM goods, including scientists, professionals, religious leaders, and academics, who do not have a financial stake in GM foods.

These information treatments were assigned to experimental units (sessions) of 13–16 individuals, and a total of 172 individuals participated in this set of experiments. Once the appropriate information packet was distributed to the participants in a given unit (session), two auction rounds were then conducted. The rounds were differentiated by the food label—either the food had a standard food label or a GM label. In one round, which could be Round 1 or 2 depending on the experimental unit, participants bid on the three food products each with the standard food label. These labels were made as plain as possible to avoid any influence on the bids from the label design. In the other round, participants bid on the same three food products with a GM label, which differed from the standard label by the inclusion of only one extra sentence: "This product is made using genetic modification (GM)."

In Experiment 2, we examined the effects of GM-free food labels. This experiment consisted of 4 experiment units/sessions with a total of 56 participants who now bid on food products with labels that stated "This product is made without genetic engineering." These four experimental units/sessions were combined with the four experimental sessions where both pro- and anti-GM information had been released (sessions 3 and 6). Thus, all participants in this experiment received either (i) pro- and anti-biotechnology information or (ii) pro-biotechnology, anti-biotechnology, and third-party information. By giving all participants both positive and negative information on GM foods, and by giving some participants a third-party perspective, we

---

[6] The sequence was determined randomly.

could examine how newly released information affects consumers' willingness to pay for food products that might be genetically modified and could add a new perspective on the GM food debate. Here, 56 individuals participated in experiment 2, and they were combined with the participants in sessions 3 and 6, who received information treatments of pro- and anti-biotech information.

In Experiment 3, we examined the impact of tolerance levels for GM content. Participants in these sessions bid on three food products that have different tolerance labels. In one trial, all consumers bid on foods with a non-GM label, certified to be completely free of genetically engineered material, and in the other trial consumers bid on foods with a non-GM label, indicating that at most 1 percent or 5 percent of content was genetically modified. Consumers in these treatments did not receive any information on GM food products. This experiment contained 3 experimental units/sessions with a total of 44 participants

8. THE RESULTS

We report and discuss the results from the above experiments.

8.1. Bids for GM- vs. Plain-Labeled Food Products

Using the data from experiment 1, we examined overall bids for GM- vs. plain-labeled food products. As shown in Table 3, on average, consumers discounted the GM-labeled product by about 14 percent relative to its plain-labeled counterpart. The discount that consumers placed on the GM-labeled products did not vary across the three products examined (Huffman, Shogren, Rousu, and Tegene 2003).

To examine this issue more closely, we also examined how many consumers bid less for GM-labeled products than for plain-labeled products. We found that 35–41 percent of participants bid less for any individual product, and that 58 percent of participants bid less for at least one of the three products (Table 4). We also could not reject the hypothesis that demographic and other background characteristics had no effect on consumers' bids for GM-labeled products (Huffman, Shogren, Rousu, and Tegene 2003).

8.2. Does Labeling Structure Matter? Comparing Bids in Mandatory vs. Voluntary Experimental Markets

Using the data from experiment 2, we examine whether consumers behave different in a market with mandatory labeling relative to a market with voluntary labeling (Table 5). We found no evidence that the discount for GM-

Table 3. Mean Bids: Consumers' Willingness to Pay for Food Products That Might Be Genetically Modified, With and Without GM Food Labels (N = 172)

Product	Observations	Mean bid	Std. dev.	Maximum	Number of zero bids
GM Oil	172	0.91	0.84	3.99	39
Oil	172	1.05	0.85	3.79	28
GM Chips	172	0.93	0.86	3.99	37
Chips	172	1.08	0.85	4.99	17
GM Potatoes	172	0.78	0.67	3.00	33
Potatoes	172	0.91	0.67	3.89	14

Source: Huffman et al. (2004a).

Table 4. The Number of Participants Who Bid Less for the GM-Labeled Food Products Than for the Standard-Labeled Food (N = 172)

Product	Number bidding less for the GM-labeled version of the product	Percent bidding less for the GM-labeled version of the product
16-oz. bag of tortilla chips	70	41%
32-oz. bottle of vegetable oil	63	37%
5-lb. bag of potatoes	60	35%
All three items	44	26%
At least one of the three	100	58%

Source: Huffman et al. (2004a).

labeled food products differed in a mandatory labeling market relative to a voluntary labeling market (Huffman et al. 2004b). Because a voluntary labeling regime would be less expensive, our results indicate that the United States has been prudent in fending off calls for mandatory labeling of GM food products.

8.3. Tolerance Thresholds? How Much GM Is Too Much GM?

Using the data from experiment 3, we examine consumers' preferences for small percentages of GM materials in their food products. We found evidence that consumers preferred foods that were 100 percent non-GM relative to food products with small amounts of GM material (1 percent or 5 percent). Consumers bid approximately 10 percent less for the GM-tolerant food products than they did for the certified GM-free products. However, we found that once GM content was present, there was no difference in bids

Table 5. Mean Bids: Markets with Mandatory and Voluntary GM Food Labels

Product	n	Mean Bid	Std. Dev.	Median	Minimum	Maximum
Mean bids for the mandatory GM labeling market						
GM oil	86	0.63	0.65	0.50	0	2.75
Oil	86	0.74	0.75	0.50	0	3.29
GM chips	86	0.61	0.70	0.43	0	3.25
Chips	86	0.69	0.72	0.50	0	2.89
GM potatoes	86	0.59	0.54	0.50	0	2.00
Potatoes	86	0.67	0.54	0.50	0	2.25
Mean bids for the voluntary GM labeling market						
NGM oil	56	0.80	0.80	0.50	0	4.75
Oil	56	0.76	0.68	0.50	0	3.00
NGM chips	56	0.75	0.81	0.50	0	4.00
Chips	56	0.68	0.77	0.50	0	4.00
NGM potatoes	56	0.84	0.75	0.75	0	4.00
Potatoes	56	0.75	0.70	0.68	0	4.00

between foods that tolerated 1 percent vs. 5 percent GM content (Rousu et al. 2004a). Thus, while our findings indicate that a significant percentage of consumers will pay less for GM-labeled food products relative to conventionally labeled food products, it does not appear that the content of GM material matters. Whether or not any GM material is present is more important to consumers.

8.4. The Effects and Value of Information

First, we consider the following issue: *How does information affect consumer decisions on GM-labeled foods?* We were able to make several significant discoveries on how information affects consumer purchasing decisions towards GM foods. We found that biased information, i.e., information from the agricultural biotech industry versus environmental groups, has a major impact on consumer decisions. Consumers who receive only negative (environmental group) information bid far less for GM-labeled food products than for plain-labeled food products (approximately a 35 percent discount). This contrasts with consumers who receive positive (ag biotech industry) information, as they did not bid less for GM-labeled food products. When consumers received information from both interested parties, they placed slightly greater weight on negative information, discounting GM-labeled products by about 20 percent (Rousu et al. 2002, 2004b). This could also be linked to the Tversky and Kahneman's (1981) prospect theory.

A consumer's prior beliefs had a significant impact on his/her bids for GM-labeled food products. Those who had uninformed prior beliefs (as based on a pre-auction survey) were influenced greatly by pro- and anti-biotech information. This occurred even in the presence of verifiable information. Those who had informed prior beliefs placed great weight on independent third-party or verifiable information. In fact, those who considered themselves "informed about GM food products" placed the same discount on GM-labeled food products regardless of the other information they received (Huffman et al 2006).

While verifiable information will be important for consumers to make informed decisions, the source of this information is also important in order for consumers to find it credible. The results are interesting and different based on consumer demographic characteristics. Additional education made a participant more likely to trust verifiable information relative to agribusiness information, government information, more likely to trust verifiable information than other types or sources of information. Those who perceived themselves as informed about GM foods were more likely to trust the government than an independent, third-party source. Those who came from a more formal religious upbringing were more likely to trust independent information than information from a private organization, but were more likely to say they would trust no one than they were to say they would trust an independent third party source of information (Huffman et al. 2004c).

Next, we consider the following question: *What is the value of information on genetic modification?* Objective or verifiable information can have a significant value to society, especially when a market is flooded with conflicting information from interested parties (Huffman and Tegene 2002), like the market for GM foods. We developed a new method to value information on GM foods, and found that verifiable information on GM foods had value in a market where negative information was present. The value of information averaged about 4 cents per person per product that is GM. Based on a rough extrapolation, we found that this information could be worth approximately $2.6 billion annually to U.S. consumers (Rousu et al. 2002).

While verifiable information certainly has value if accurate and objective, what value would information from biased or interested parties have? We investigated this question in Rousu et al. (2004a) and Huffman, Rousu, Shogren, and Tegene (2003). We examined the value of information "through the eyes" of the party that is providing the information. We found that both environmental group information and agribusiness information has value through the eyes of their respective groups in the absence of verifiable information. However, in a market with objective and verifiable information, this biased information no longer had value.

9. SOME RELATED RESULTS

Some recent experiments fill in some remaining knowledge gaps. Most of the willingness-to-pay studies have been for "first-generation" GM foods, i.e., the input traits. This is because of the lack of successful second-generation or GM traits that contained enhanced consumer traits with positive value. Rousu et al. (2005) examined consumer demand for a second-generation GM product, GM cigarettes with lower nicotine levels, a negative consumer trait. They found that in the absence of marketing information, consumers discounted the GM-labeled cigarette by 14 percent relative to the identical cigarettes without a GM label. However, when marketing information was presented, consumers did not discount the GM-labeled cigarettes relative to the plain-labeled version.

Lusk et al. (2004) conducted experiments that provided an interesting extension as well. They used experimental auctions to examine the effects of alternative types of pro-biotech information on the demand for GM cookies. They examined how consumers responded to information discussing the environmental benefits of GM foods, the health benefits of GM foods, and the world benefits of GM foods. They found that consumers in the United States and England increased their bids for GM-labeled products when presented with any of the three positive information treatments, while consumers in France did not increase bids based on the information treatments. These results seem to indicate that if consumers are receptive to pro-biotech information, it does not seem to matter what message is received, only that pro-biotech information is available.

10. CONCLUSIONS

Society has had about 11,000 years of experience with crop improvement, and 10,000 years of experience with livestock improvement. We have seen progress in steps from selection to hybridization, and mutagenic and biotechnology with selection in crop improvement. Only the use of biotechnology in crop improvement has created consumer concerns. Pasteurized milk is an animal product that received considerable consumer resistance about a century ago, and rBST or rBGH milk and GM fish are animal/fish products where there has been consumer resistance. More generally, consumers view input traits as possessing some risks and no real benefits, and hence, they discount them. Only a few biotech food products have come onto the market, and consumers have attributed only small positive value to them.

Consumers have uniformly stated that they prefer that foods show labels for GM content. An effective labeling system, however, is costly. It requires an effective segregation system when both GM and non-GM products are produced and marketed. In one study, the share of participants who wanted GM food labels dropped dramatically when they were informed that the cost of food would rise by 5 percent. And the cost of labels might, however, be significantly more than 5 percent. With voluntary labeling and input traits, the GM product is the inferior type, and the producers of non-GM products have an incentive to reveal their superior type. This, however, has not occurred in the United States, except to the extent that certified organic cannot use GM technology.

Our own research on information effects has shown that biotech industry, environmental group, and third-party information has statistically significant effects on consumers' willingness to pay. We found that prior beliefs about biotechnology were important influences on the way that consumers weighed diverse information. Those participants who had uninformed prior beliefs were most influenced by pro- and anti-biotech information. Those who were informed put great weight on independent third-party information. Hence, differences in consumers' acceptance of GM products across Western developed countries may be due to differences in the amount and intensity of different types of GM information and in consumers' general knowledge about agricultural biotechnology. Also, Western Europe suffers from the ineffectiveness of public institutions to handle food safety issues in a way that gains consumers' trust.

We also found no evidence that the discount for GM-labeled food products differed in a mandatory labeling market relative to a voluntary labeling market. In addition, we found that consumers prefer GM-free to low levels of GM tolerance. Participants discounted food products with 1 and 5 percent impurity by 10 percent relative to ones that were GM-free.

We close by suggesting that consumer acceptance of GM food products might be given a big boost if an innovation would produce a GM product that contains positive attributes that are quite valuable to consumers relative to the non-GM alternatives.

REFERENCES

Aldrich, L., and N. Blisard. 1998. "Consumer Acceptance of Biotechnology: Lessons from the rBST Experience." Agricultural Information Bulletin No. 747-01, Economic Research Service, U.S. Department of Agriculture, Washington, D.C.

Balasubramanian, S.K., and C. Cole. 2002. "Consumers' Search and Use of Nutrition Information: The Challenge and Promise of the Nutrition Labeling and Education Act." Journal of Marketing 66(3): 113–127.

Carlson, A., J. Kinsey, and C. Nadav. 1998. "Who Eats What, When, and From Where?" Working Paper No. 98-05, Retail Food Industry Center, Uniersity of Minnesota, St. Paul.

Chern, W.D., and K. Rickertsen. 2004. "A Comparative Analysis of Consumer Acceptance of GM Foods in Norway and in the USA." In R.E. Evenson and V. Santaniello, eds., Con-

sumer Acceptance of Genetically Modified Foods. Cambridge, MA: CAB International.

Council for Biotechnology Information. 2001. "Frequently Asked Questions." Available online at http://www.whybiotech.com/en/faq/default.asp?MID=10.

Dhar, T., and J.D. Foltz. 2005. "Milk by Any Other Name...Consumer Benefits from Labeled Milk." *American Journal of Agricultural Economics* 87(1): 214–228.

Falck-Zepeda, J.B., G. Traxler, and R.G. Nelson. 2000. "Surplus Distribution from the Introduction of Biotechnology Innovation." *American Journal of Agricultural Economics* 82(2): 360–369.

Falk, M.C., B.M. Chassy, S.K. Harlander, T.J. Hoban, M.N. McGloughlin, and A.R. Akhloghi. 2002. "Food Biotechnology: Benefits and Concerns." *Journal of Nutrition* 132(4): 1384–1390.

Fernandez-Cornejo, J., and W.D. McBride. 2002. "Adoption of Bioengineered Crops." Agricultural Economics Report No. 810, Economic Research Service, U.S. Department of Agriculture, Washington, D.C.

Food and Agriculture Organization. 2004. *The State of Food and Agriculture: Agricultural Biotechnology, 2003–2004.* Rome, Italy: Food and Agriculture Organization.

Fox, J.A., D.J. Hayes, and J.F. Shogren. 2002. "Consumer Preferences for Food Irradiation: How Favorable and Unfavorable Descriptions Affect Preferences for Irradiated Pork in Experimental Auctions." *Journal of Risk and Uncertainty* 24(1): 75–95.

Greenpeace International. 2001. "We Want Natural Food!" Greenpeace International, Amsterdam. Available online at http://www.greenpeace.org/~geneng/.

Grübler, A. 1996. "Time for a Change: On the Patterns of Diffusion." *Daedalus* 125(Summer): 19–42.

Harl, N.E. 2001. "Opportunities and Problems in Agricultural Biotechnology." Paper presented at the Third International Value-Enhanced Grain Conference and Trade Workshop, Portland, OR, July 23.

Hoban, T. 2004. "Societal Perspectives on Agricultural Biotechnology." Invited presentation for the Advisory Committee on Agricultural Biotechnology, U.S. Department of Agriculture, Washington, D.C. (June 2).

Hotchkiss, J.H. 2001. "Lambasting Louis: Lessons from Pasteurization." In A. Eaglesham, S.G. Pueppke, and R.W.F. Hardy, eds. *Genetically Modified Food and the Consumer.* Ithaca, NY: National Agricultural Biotechnology Council.

Huffman, W.E., and R.E. Evenson. 1993. *Science for Agriculture: A Long-Term Perspective.* Ames, IA: Iowa State University Press.

Huffman, W.E., M. Rousu, J.F. Shogren, and A. Tegene. 2003. "The Public Good Value of Information from Agribusinesses on Genetically Modified Foods." *American Journal of Agricultural Economics* 85(5): 1309–1315.

_____. 2004a. "Consumers' Resistance to Genetically Modified Foods in the U.S.: The Role of Information in an Uncertain Environment." *Journal of Agricultural and Food Industrial Organization* 2(2): 1–13.

_____. 2004b. "The Welfare Effects of Implementing Mandatory GM Labelling in the USA." In R.E. Evenson and V. Santaniello, eds., *Consumer Acceptance of Genetically Modified Food.* Cambridge, MA: CAB International.

_____. 2004c. "Who Do Consumers Trust for Information on Genetically Modified Foods: Economics of Trust Formation." *American Journal of Agricultural Economics* 86(5): 1222–1229.

_____. 2006. "The Effects of Prior Beliefs and Learning on Consumers' Acceptance of Genetically Modified Foods." *Journal of Economic Behavior and Organization* 57 (in press).

Huffman, W.E., J.F. Shogren, M. Rousu, and A. Tegene. 2003. "Consumer Willingness to Pay for Genetically Modified Food Labels in a Market with Diverse Information: Evidence from Experimental Auctions." *Journal of Agricultural and Resource Economics* 28(3): 481–502.

Huffman, W.E., and A. Tegene. 2002. "Public Acceptance of and Benefits from Agricultural Biotechnology: A Key Role for Verifiable Information." In V Santaniello, R.E. Evenson, and D Zilberman, eds., *Market Development for Genetically Modified Foods.* New York: CAB International.

James, C. 2003. "Preview: Global Status of Commercialized Transgenic Crops: 2003." ISAAA Briefs No. 30, International Service for the Acquisition of Agri-Biotech, Ithaca, NY.

Katsaras, N., P. Wolfson, J. Kinsey, and B. Senauer. 2001. "Data Mining: A Segmentation Analysis of U.S. Grocery Shoppers." Working Paper No. 01-01, Retail Food Industry Center, University of Minnesota, St. Paul.

List, J.A., and J.F. Shogren. 1999. "Price Information and Bidding Behavior in Repeated Second-Price Auctions." *American Journal of Agricultural Economics* 81(4): 942–949.

Lusk, J.L. 2003. "Effects of Cheap Talk on Consumer Willingness-to-Pay for Golden Rice." *American Journal of Agricultural Economics* 85(4): 840–856.

Lusk, J.L., L.O. House, C. Valli, S.R. Jaeger, M. Moore, B. Morrow, and W.B. Traill. 2004. "Effect of Information about Benefits of Biotechnology on Consumer Acceptance of Genetically Modified Food: Evidence from Experimental Auctions in United States, England, and France." *European Review of Agricultural Economics* 31(2): 179–204.

Lusk, J.L., M. Moore, L. House, and B. Monow. 2001. "Influence of Brand Name, Store Loyalty, and Type of Modification on Consumer Acceptance of Genetically Engineered Corn Chips." *International Food and Agribusiness Management Review* 4(2): 373–383.

McCluskey, K., N. Grimsrud, and T.I. Wahl. 2004. "Comparing Consumer Responses Toward GM Foods in Japan and Norway." In R.E. Evenson and V. Santaniello, eds., *Consumer Acceptance of Genetically Modified Foods.* Cambridge, MA: CAB International.

Moon, W., and S.K. Balasutramaniau. 2004. "Contingent Valuation of Breakfast Cereal Made of Non-Biotech Ingredients." In R.E. Evenson and V. Santaniello, eds., *Consumer Acceptance of Genetically Modified Foods.* Cambridge, MA: CAB International.

Nayga, R.M., W. Aiew, and J. P. Nichols. 2005. "Information Effects on Consumers' Willingness to Purchase Irradiated Food Products." *Review of Agricultural Economics* 27(1): 37–48.

Nestle, M. 2002. *Safe Food: Bacteria, Biotechnology, and Bioterrorism.* Berkeley, CA: University of California Press.

Pirtle, T.R. 1926. *History of the Dairy Industry.* Chicago, IL: Mojonnier Bros. Company.

Rousu, M., W.E. Huffman, J.F. Shogren, and A. Tegene. 2002. "The Value of Verifiable Information in a Controversial Market: Evidence from Lab Auctions of GM Foods." Staff Working Paper No. 344, Department of Economics, Iowa State University, Ames, IA.

_____. 2004a. "Are United States Consumers Tolerant of Genetically Modified Foods?" *Review of Agricultural Economics* 26(1): 19–31.

_____. 2004b "Estimating the Public Value of Conflicting Information: The Case of Genetically Modified Foods." *Land Economics* 80(1): 125–135.

Rousu, M., D. Monchuk, J. Shogren, and K. Kosa. 2005. "Consumer Perceptions of Labels and the Willingness to Pay for 'Second-Generation' Genetically Modified Products." Working paper, Susquehanna University, Selinsgrove, PA.

Ruttan, V.V. 2001. *Technology, Growth, and Development.* Cambridge, MA: Oxford University Press.

Shogren, J.F., M. Margolis, C. Koo, and J.A. List. 2001. "A Random nth-Price Auction." *Journal of Economic Behavior and Organization* 46(4): 409–421.

Silverglade, B. 1996. "The Nutrient Labeling and Education Act: Progress to Date and Challenges for the Future." *Journal of Public Policy and Marketing* 15(1): 148–156.

Tversky, A., and D. Kahneman. 1981. "The Framing of Decisions and the Psychology of Choice." *Science* 211(January 23): 453–458.

Wikipedia. 2005. "Domestication." Wikipedia, available online at http://en.wikipedia.org/wiki/domestication.

Wilson, W.W., and B. Dahl. 2002. "Costs and Risks of Testing and Segregating GM Wheat." Agribusiness & Applied Economics Report No. 501, Agricultural Experiment Station, North Dakota State University.

Wolf, M., and N. Giacalone. 2003. "Purchase Interest in Corn That Was Grown Using Biotechnology." Paper presented at the Seventh ICABR (International Consortium on Agricultural Biotechnology Research) International Conference in Ravello, Italy, June 29–July 3.

Ye, X., S. Al-Babili, A. Kloti, J. Zhang, P. Lucca, P. Beyer, and I. Potrykus. 2000. "Engineering the Provitamin A (Beta-Carotene) Biosynthetic Pathway into (Carotenoid-Free) Rice Endosperm." *Science* 287(January 14): 303–305.

Chapter 11

COMPARISON OF CONSUMER RESPONSES TO GENETICALLY MODIFIED FOODS IN ASIA, NORTH AMERICA, AND EUROPE

Jill J. McCluskey,[*] Kristine M. Grimsrud,[†] and Thomas I. Wahl[*]
Washington State University[*] and University of New Mexico[†]

Abstract: Consumer attitudes toward genetically modified (GM) food products are com-
              plex and differ across cultures. This study uses consumer survey data to compare
              consumer attitudes towards GM food across Canada, China, Japan, Norway, and
              the United States. The comparisons are based on the significance of covariates
              included in country-wise estimations of willingness to pay for GM foods. The
              Canadian respondents were similar to U.S. respondents. Japan and China differ
              more from each other than do Japan and Norway. The Chinese were the most
              favorable toward GM foods. We argue that cultural attitudes including valuing
              tradition and skepticism of science must be considered when marketing GM
              products.

Key words: consumer response, genetically modified food, China, Japan, Norway, Canada,
              United States

1. INTRODUCTION

The introduction of genetically modified (GM) crops to world markets has created a new division between the crop trading countries. The United States and Canada have great economic interests in exporting their transgenic crops; however, lack of public acceptance of genetically modified food products in the European Union (EU) and Japan has already resulted in reduced or curbed demand for GM food products. Many European and Japanese consumers believe GM foods pose a threat to human health, are concerned with

The authors wish to thank without implicating Quan Li, Kynda Curtis, Hiromi Ouchi, and Maria Loureiro for their contributions to some of the specific country studies discussed in this article, and Phil Wandschneider, Ron Mittelhammer, and Kyrre Rickertsen for helpful comments. The authors gratefully acknowledge financial support from the IMPACT Center at Washington State University.

unknown environmental consequences, and are concerned with the ethical dimensions of biotechnology. The Chinese consumer response is mixed but relatively more positive.

Consumer attitudes and behavior toward genetically modified food products are complex and differ across cultures. This study focuses on Canada, China, Japan, Norway, and the United States. We compare consumer attitudes towards GM food across the five countries. We base our comparisons on the significance of covariates included in country-wise estimations of the discount (or in one case the premium) that consumers would need to purchase the GM food products. Since the countries differ widely in their traditional eating habits, we used different food products for each country. In sum, we investigate which factors affect consumer acceptance of GM food for each country and compare the response across countries.

Mandatory labeling of GM foods has obvious implications for trade. The European Union (EU) has imposed mandatory labeling for some foods that contain GM ingredients. In October 1999, the EU gave preliminary approval to a law that requires labels on all foods containing more than one percent GM ingredients. The United States has argued that there is no health-related or scientific reason to reject GM commodities and food products and has challenged EU's mandatory GM labeling as a non-tariff trade barrier. In Japan, authorities have ordered mandatory labeling for 29 categories of food if they contain any GM ingredients. Since June 6, 2001, China has required that all GM products entering China for research, production, or processing have safety certificates from the agricultural ministry to ensure that they are safe for human consumption, animals, and the environment. Since March 20, 2002, labeling has been required in China for listed transgenic biological products.

The Codex committees of the World Trade Organization (WTO) are working on harmonizing international standards and resolving trade disputes associated with food labeling to promote fair trade of foods while protecting consumer health. Since different countries have different attitudes toward GM food products, the Codex framework allows each country to develop its own standards. The challenge of Codex is to set international standards for GM food labeling that both promote fair trade and allow for consumer choice. An important issue in GM labeling policy is scientific versus consumer sovereignty. Although the scientific consensus may be that GM foods are safe for human consumption aside from potential allergens, it may be the case that the majority of the population in a given country prefers to avoid GM foods.

Mandatory labeling would force U.S. producers to segregate crops before being able to claim that food products are "GM-free," which would be difficult and costly. For example, many U.S. grain elevators are not physically equipped to segregate crops. Further, U.S. producers may lose international market share because consumers might reject their GM crops.

2. OUR SURVEYS

Over a three-year period, we collected consumer survey data in Japan, Norway, China, the United States, and Canada. Specifically, in August 2001, we conducted 400 in-person interviews in Japanese at the Seikatsu Club Consumer Cooperative (Seikyou), a grocery store like setting in Matsumoto City, Japan. Matsumoto is a relatively agricultural area where about 13 percent of the population come from farm households compared to 2 percent in all of Japan. Consumer cooperatives usually focus on a strategy of marketing increased food safety. Seikyou has significant market power in the Japanese marketplace. In January 2002, we conducted 400 in-person interviews in Norwegian at the RIMI Liertoppen grocery store in the Oslo region of Norway. This region is the most populated part of Norway and one of the centers of Norwegian economic activity. The RIMI chain of grocery stores has chosen a low-price and limited-selection niche in the market and has in this way gained significant power in the Norwegian marketplace. In August 2002, we performed 599 in-person interviews in Chinese in Beijing, China. The survey was conducted at four separate locations including a supermarket, two outdoor markets, and one shopping area. These locations were chosen to represent a cross section of the Chinese population. In June 2003, we conducted 595 in-person surveys in Seattle, Washington, and 578 in Vancouver, Canada, at six different retail food outlet locations (three in each city), which were chosen to ensure a broad sample of food shoppers.

The surveys solicited respondents' demographic information, their attitudes about the environment and food safety, and their knowledge and perceptions about biotechnology. Summary statistics for these variables are presented in Table 1. Further, respondents were asked if they were willing to pay the same price for a particular GM food as for a corresponding non-GM product. In Japan, we asked about GM noodles and GM tofu; in Norway, we asked about GM bread and GM salmon; and in China, we asked about GM rice and GM soy oil. In the United States and Canada, we asked about products made from GM wheat.

The contingent valuation (CV) method is a standard approach used to elicit consumers' willingness to pay (WTP) through the dichotomous choice, market-type questioning format conducted by direct survey via telephone, mail, or face-to-face (Kanninen 1993). Our surveys included CV questions regarding willingness to pay a premium or accept a discount to purchase genetically modified food products.

Consumers were first asked if they were willing to pay the same price for the GM-product as for the corresponding non-GM products. If the respondent's answer to this question was "no," a follow-up question was asked where the respondent was offered a percentage discount on the GM product relative to the non-GM product. If the respondent's answer to the

Table 1. Summary Statistics

Variable	Description	Distribution of responses				
		Vancouver, Canada	Beijing, China	Nagano, Japan	Oslo, Norway	Seattle, U.S.
Age	Age of respondent	mean = 43.1 s.d. = 15.6	mean = 38.8 s.d. = 13.9	mean = 45.5 s.d. = 13.9	mean = 41.6 s.d. = 12.9	mean = 43.8 s.d. =15.1
Gender	1 if female	63.6%	63.6%	78%	69.3%	43.4%
	0 otherwise	33.1%	36.4%	22%	30.8%	55.3%
Education	1 if at least 4-year college degree	37.3%	10.9%	30.8%	52.2%	60.0%
	0 otherwise	62.7%	89.1%	69.2%	47.8%	40.0%
Children	1 if present	27.9%	61.1%	54%	56.8 %	29.9%
	0 otherwise	67.8%	38.9%	46%	43.3 %	68.9%

first question was "yes," a follow-up question was asked in China, the United States, and Canada only, where the respondent was offered a percentage premium on the GM product relative to the non-GM product. Each level of discount or premium was used for one-fifth of the surveys. The discount was set at one of the following levels: 10, 20, 25, 50, and 75 percent. The premium for the GM rice was set at one of the following levels: 10, 20, 25, 50, and 100 percent. The premium for the other GM food products was set at one of the following levels: 5, 10, 20, 25, and 50 percent. The assignment of survey version (and thus discount or premium) was random to the respondent.

In Japan and Norway, no follow-up question was asked if the customer's answer was "yes" to the initial question and he or she was willing to purchase the GM product at a price equal to the price of the non-GM product. The rationale for no follow-up to a "yes" response is that the genetic modifications associated with the GM products we featured in Japan and Norway are *process attributes*, which reduces production costs, as opposed to *product-enhancing* attributes. An example of a GM product with a product-enhancing attribute is "Golden Rice," which is nutritionally enhanced with Vitamin A. Proponents claim that the GM products with process attributes are identical to non-GM products. Opponents view genetic modification as a negative attribute. Therefore, it would not make economic sense after an initial "yes" to follow up with a question that involves paying a premium for these GM products that have only cost-reducing attributes. In China, we asked about willingness to pay for both Golden Rice and GM soybean oil. Since it could be rational to pay a premium for Golden Rice, we included follow-up questions about paying a premium for both GM products. Surprisingly, some Chinese respondents were also willing to pay a premium for GM soybean oil, which is not an enhanced product. Consequently, we included these questions about paying a premium in the surveys that followed our Chinese study.

3. METHODOLOGY

We use an ordered multinomial choice model (see, e.g., Mittelhammer, Judge, and Miller 2000) with three possible outcomes, each indicating the degree of resistance to the GM-product: (i) the respondent is willing to purchase the GM product at the same price as the non-GM product (positive or indifferent); (ii) the respondent is not willing to purchase the GM product at the same price as the non-GM product, nor at a discount relative to the non-GM product (strong negative attitude); and (iii) the respondent is not willing to purchase the GM product at the same price as the non-GM product, but is willing to purchase the GM product at the random discount offered (lesser negative attitude). The data collected for China, the United States, and Canada split the first category into two. Those are (i) the respondent is willing to purchase the GM product at the same price as the non-GM product, but is not willing to purchase it at a premium (positive or neutral attitudes); and (ii) the respondent is willing to purchase the GM product at the same price as the non-GM product and is also willing to purchase it at a random premium offered relative to the non-GM product (strong positive attitude). The data from China, the United States, and Canada were made comparable to the data from Japan and Norway by merging the additional response categories. That is, for countries where premiums were offered, we merged "yes at no discount" and "yes at a premium" into the "positive or neutral attitude" category. From an economic perspective, this is an acceptable strategy because our research question does not require us to estimate the willingness to pay/accept. Furthermore, we are merging groups of observations where all have a positive or zero willingness to pay (not negative) for the genetically modified food product. Instead, we are interested in comparing the covariates of the regression functions for the different countries.

The survey responses for each country were organized so that we could formulate an ordered multinomial choice model with 3 categories (Agresti 2002). We are left with the following observable discrete outcomes:

$$Y_i = \begin{cases} 3 & \text{If willing to buy GM-food without discount} \\ 2 & \text{If willing to buy GM-food at the offered discount} \\ 1 & \text{If not willing to buy GM-food at discount} \end{cases} \qquad (1)$$

An individual will buy the GM-food product if the discount is sufficiently high, that is, the discount is so high that the price of the GM-food product is less than or equal to the individual's willingness to pay, Y_i^*. The individual's willingness to pay (WTP) is not observable; however, the response to the bid offered gives us an idea of the range of that person's WTP. Based on this we can organize respondents into three groups as follows:

$$Y_i = 3 \text{ if } Y_i^* \geq B_i$$
$$Y_i = 2 \text{ if } B_i \geq Y_i^* \geq B_i - d_i \qquad (2)$$
$$Y_i = 1 \text{ if } Y_i^* < B_i - d_i,$$

where B_i is the initial discount offered for the GM food product ($B_i = 0$ is no price difference between the GM-product and conventional product) and $B_i - d_i$ is the follow-up discount bid offered individual i (d_i represents a discount). The regression function for GM food products for individual i is

$$P_{i1} = F(\alpha - \rho B_i + \lambda' z_i) \qquad \text{if } Y_i = 3$$
$$P_{i2} = F(\alpha - \rho B_i + \lambda' z_i) - F(\alpha - \rho(B_i - d_i) + \lambda' z_i) \qquad \text{if } Y_i = 2 \qquad (3)$$
$$P_{i3} = 1 - F(\alpha - \rho(B_i - d_i) + \lambda' z_{ii}) \qquad \text{if } Y_i = 1,$$

where P_{ij} is the probability of consumer i being in one of the following response categories, $j \in \{1,2,3\}$, and $F(\cdot)$ is a cumulative distribution function. The consumer-choice depends on the bid amount, and it is therefore included explicitly. The vector z_i is a column vector of observable characteristics of the individual, which are the covariates we will consider more closely. The error term ε_i is a random variable accounting for random noise and possibly unobservable characteristics. Unknown parameters to be estimated are α, ρ, and λ. Linearity in z and ε is assumed for all individuals. Furthermore, the distribution of the error term is assumed to follow $\varepsilon \sim F(0, \sigma^2)$, where $F(0, \sigma^2)$ is the standard logistic distribution function.[1] Based on (2), the log-likelihood function can be formulated as

$$L = \sum_{i=1}^{n} \prod_{j=1}^{3} \ln P_{ij}^{I_{(Y_i = j)}(Y_i)}, \qquad (4)$$

where $I_{(Y_i = j)}(Y_i)$ is an indicator function equal to one when $Y_i = j$, that is the jth alternative occurring for the ith consumer. We used the marginal effects (denoted by θ) calculated (see, e.g., Maddala 1983) at the sample average (see Table 3) to compare the consumer attitudes to GM products in the various countries since the estimated coefficients are not directly interpretable in an ordered logit model.

Our comparisons of consumer attitudes were motivated as follows:

Comparison of Canada, China, Japan, and Norway with the United States. The United States is one of the world's largest producer of

---

[1] Mean zero and standard deviation $\sigma = \pi / \sqrt{3}$.

biotechnology. In the United States, consumers have consumed GM products since 1996.

Comparison of Japan with China and Norway. The consumers interviewed in the Japan and Norway studies showed a strong skepticism to GM foods. Our study also indicated that Chinese consumers are very positive about GM food products. The Chinese attitudes may be spurred by the drastic change to a market economy, where Western culture and products are often preferred to consumption patterns associated with the old economic system.

For the above comparisons, we are interested in the average acceptance of GM food as it is explained by the marginal effects of some common variables. First, we consider the responsiveness to price (or the bid offered) for the GM food product. We are also interested in the effect of self-reported knowledge about biotechnology. Specifically, we posed the question, "How knowledgeable are you about GM foods?" The respondents were given the following choices: "very knowledgeable," "somewhat knowledgeable," and "not informed." We merged the categories "somewhat knowledgeable" and "not informed" into one group, which improved the explanatory power of the estimation. We also compare how consumers' choices depend on their formal education. For this variable, we merged college degrees and postgraduate education into one group, and those with less than a college degree into another group. Other demographic variables included are gender and age—higher and lower than average age, and the presence of children less than 18 years of age living in one's household. We included this variable because people with small children are often more concerned about food safety issues.

To test our hypothesis, we proceeded in the following way. For example, for the comparison between the United States and Canada for the kth marginal effect, we tested the following:

$$H_0 : \theta_{CA,k} - \theta_{US,k} = 0 \text{ versus } H_1 : \theta_{CA,k} - \theta_{US,k} \neq 0. \tag{5}$$

Our sample sizes are large enough to justify asymptotic normality for the distribution of the test statistic,

$$Z = \frac{\theta_{k,C} - \theta_{k,US}}{\sqrt{\dfrac{n_C}{T_{C,US}} \operatorname{var}(\theta_{k,C}) + \dfrac{n_{US}}{T_{C,US}} \operatorname{var}(\theta_{k,US})}} \xrightarrow{asy} N(0,1) \text{ under } H_0, \tag{6}$$

where n_C and n_{US} represent the number of observations for the Canadian and the U.S. data sets, respectively, and $\operatorname{var}(\theta_{k,C})$ and $\operatorname{var}(\theta_{k,US})$ represent the estimated variance of marginal effect k for the Canadian and U.S. data sets, respectively. Finally, $T_{C,US} = n_C + n_{US}$, or the total of the sum of observations in both countries.

4. RESULTS

In previous studies, the effects of bid information and attitudinal and demographic information on consumers' willingness to pay for GM food products were estimated. The mean willingness to pay estimates from the original country-specific studies are presented in Figure 1. The estimated coefficients are reported in Table 2 and are discussed briefly below. The corresponding marginal effects of these coefficients are presented in Table 3. More complete discussions are found in the original studies. For all countries, increasing the discount made consumers more willing to purchase the GM food product. The analysis of the Japanese data showed that variables representing self-reported knowledge about biotechnology, formal education, being female, and age all significantly decreased the willingness to purchase GM foods. Furthermore, the Japanese respondents wanted greater than a 50 percent discount to choose the GM product (McCluskey et al. 2003).

From the previous country-specific study of Norway, we see similar negative attitudes toward the GM food product, except that higher levels of formal education had a positive effect on willingness to purchase the GM food product. This may indicate that increasing levels of formal education can improve consumer acceptance. Self-reported knowledge had, on the other hand, a negative effect on consumer acceptance, indicating that the source of information about biotechnology is important for consumer

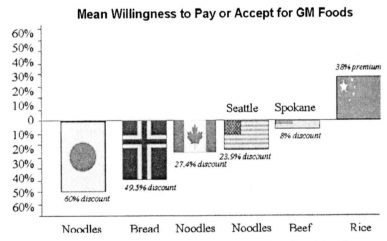

Figure 1. Mean Discounts or Premiums to Purchase GM Food Products

Table 2. Estimated Coefficients on Willingness to Accept a GM Product

Country	Canada	China	Japan	Norway	U.S.
Product	Noodles with GM wheat	Soybean oil with GM soy	Tofu with GM soy	Noodles with GM wheat	Bread with GM wheat
# of observations	498	571	400	399	547
Variable	γ_{CA}	γ_{CH}	γ_{JA}	γ_{NO}	γ_{US}
Intercept	0.2807	1.3283***	-0.0624	0.6308	-.1834
Discount	0.8717***	1.2814***	4.8774***	2.1488***	1.0503***
Knowledge	0.5930***	-0.0161	-0.8669***	-0.3532*	0.6507***
Formal education	-0.7106***	-0.0859	-0.4571	0.1908	0.2575
Female	-0.2735	0.3235*	-0.7906***	-0.5630***	-0.6387***
Age	0.0002	-0.0108*	-2.9094***	-2.5855**	0.0008
Children18	0.2245	0.1065	-0.2138	-0.1714	0.0846

Notes: *** indicates significant at $\alpha = 0.01$, ** indicates significant at $\alpha = 0.05$, and * indicates significant at $\alpha = 0.10$.

Table 3. Estimated Marginal Effects on Willingness to Accept a GM Food Product

Country	Canada	China	Japan	Norway	U.S.
Product	Noodles with GM wheat	Soybean oil with GM soy	Tofu with GM soy	Noodles with GM wheat	Bread with GM wheat
# of observations	498	571	400	399	547
Variable	θ_{CA}	θ_{CH}	θ_{JA}	θ_{NO}	θ_{US}
Intercept	0.06917	0.2283***	-0.00766	0.1486	-.04514
Discount	0.2148***	0.2202***	0.5982***	0.5061***	0.2585***
Knowledge	0.1461***	-0.00276	-0.1063***	-0.0832*	0.1601***
Formal education	-0.1751***	-0.01476	-0.05606	0.04495	0.06337
Female	-0.0674	0.05561*	-0.09696***	-0.1326***	-0.1572***
Age	0.000054	-0.00186	-0.3568***	-0.609**	0.000186
Children18	0.05533	0.01831	-0.02622	-0.04037	0.02082

Notes: *** indicates significant at $\alpha = 0.01$, ** indicates significant at $\alpha = 0.05$, and * indicates significant at $\alpha = 0.10$.

acceptance. Furthermore, women and older people were significantly less accepting of GM foods. The study found that consumers want a just-less-than 50 percent discount on GM bread compared to conventional bread (Grimsrud et al. 2004).

The results for the United States and Canada are very similar. Differing from the findings for Norway, having higher levels of formal education significantly decreases the willingness to accept the GM food product in Canada and has no significant effect in the United States. Interestingly, self-reported knowledge in both countries increases the willingness to purchase the GM food product. In Canada, being female does not significantly decrease the consumer acceptance of GM foods, while in the United States, being female had a significant negative effect. The results indicate that the U.S. respondents on average required a 23.9 percent discount to choose the GM product over the conventional product. Canadian respondents, on average, required a 27.4 percent discount to purchase the GM product. Note that the cities included were Seattle, Washington, and Vancouver, which are urban and more politically liberal. Li, McCluskey, and Wahl (2004) found that the Spokane, Washington, respondents required a mean discount of only 8 percent, with "no discount" included in the 95 percent confidence interval.

Interestingly, the Chinese data present a very different picture. A prevailing positive opinion regarding biotechnology significantly increases consumer confidence in GM foods. In fact, Chinese consumers surveyed were willing to pay a premium for GM foods. Results of a previous country-specific study indicated that Chinese consumers, on average, were willing to pay 38.0 percent more for GM rice over non-GM rice. They were willing to pay a 16.3 percent premium for GM soybean oil over non-GM soybean oil (Li et al. 2002). This is not surprising given that 23 percent of the survey respondents were very positive about the use of biotechnology in foods and 40 percent of the respondents were somewhat positive about the use of bio-technology in foods. It makes sense that consumers in China, who exhibit a low perception of risk associated with GM foods (82 percent felt these products present little or no risk), would be willing to pay a premium for GM products.

Chinese consumer attitudes concerning biotechnology may reflect the Chinese government's traditionally strong support of new technologies, which makes China more competitive on the world market. Thus far, the controversy taking place in Europe and Japan is not evident in China, but new regulations regarding labeling and safety testing will likely lead to increased public awareness. The divergence between the Chinese results and the findings for the other countries in the study may be caused by China's unique history. European countries and Japan gradually developed modern capitalist societies while taking pride in preserving cultural traditions. For the Chinese, a decade of Cultural Revolution from 1966 to 1976 systematically tore down historical and traditional structures in the society. The vacuum remaining was replaced with the doctrines of the Chinese Communist Party. Now, with a highly desired and incredibly rapid transition to capitalism and with much of the old Chinese tradition crushed by the Cultural Revolution, the Chinese are forward-looking. Technological novelties from the rest of the world are often considered much-needed improvements and not reasons for concern.

4.1. Country Comparisons

Results from country comparisons I and II are reported in Table 4. Comparison I involves comparing the United States with all the other countries. Compared to the U.S. sample, the respondents from Japan and Norway were less responsive to the discount offered for the GM food product, while those from Canada and China were not significantly different from those from the United States. Canadian respondents did not differ significantly from U.S. respondents in terms of the effect of self-reported knowledge on the discount needed for GM foods, while self-reported knowledge had less of an influence on the willingness to purchase GM foods for respondents from China, Japan, and Norway than for those from the United States.

From Table 4, the U.S. and Norwegian data have in common a marginal effect of formal education that is positive, indicating that as formal education increases, consumers become more willing to consume GM foods. The U.S. and Norwegian respondents do not differ when comparing the marginal effects of formal education. However, Canadian, Chinese, and Japanese respondents significantly differ from U.S. respondents in terms of the effects on formal education on willingness to pay for GM food products. Respondents in these countries become less willing to accept GM foods as formal education increases than do respondents from the United States.

For the Japanese, Norwegian, and U.S. data, the marginal effect of being female was highly statistically significant (at alpha = 0.01) and negative, which indicates that willingness to purchase GM foods is lower for women. For the Chinese data, women were statistically more willing to purchase GM foods than men (at alpha = 0.05), and for the Canadian data, the gender of the respondent was statistically insignificant. The magnitude of the marginal effect of being female was not statistically different when comparing responses from the United States and Norway. Being female had a significantly more negative effect on the willingness to purchase GM food for the U.S. respondents than for the Chinese, Canadian, and Japan respondents. The largest divergence was between Chinese and U.S. respondents, and the smallest difference was between Japanese female respondents and U.S. female respondents.

In the countries where the marginal effect of age had a significant effect—that is, Japan and Norway—the marginal effect was negative, which indicates that older consumers are less willing to purchase GM foods. For the Japanese and Norwegian data, the marginal effect of age was significantly different from the marginal effect of age for the U.S. data, where the marginal effect of age was not significantly different from zero.

Whether households have children under 18 did not statistically significantly influence the willingness to purchase GM foods in any of the country samples.

Table 4. Country Comparisons

Country comparison Product / Variable	Canada-U.S. Noodles with GM wheat	China-U.S. Soybean oil with GM soy	Japan-U.S. Tofu with GM soy	Norway-U.S. Noodles with GM wheat	China-Japan Noodles with GM wheat	Norway-Japan Soybean oil with GM soy
Discount	1.3675 (0.1715)	-1.1871 (0.2352)	3.9733 (0.0000)	4.5584 (0.0000)	-4.5191 (0.0000)	-0.8971 (0.3697)
Knowledge	-0.2835 (0.7768)	-3.9867 (0.0000)	-7.0069 (0.0000)	-5.1276 (0.0000)	3.6036 (0.0003)	0.6325 (0.5270)
Formal Educ.	-4.6602 (0.0000)	-1.6925 (0.0906)	-2.6854 (0.0072)	-0.3538 (0.7235)	1.1643 (0.2443)	2.4097 (0.0160)
Female	1.9716 (0.0487)	5.606013 (0.0000)	1.6500 (0.0989)	0.5429 (0.5872)	5.6711 (0.0000)	-0.9907 (0.3218)
Age	-0.09327 (0.925688)	-1.4861 (0.1373)	-75.3000 (0.0000)	-6.18 (0.0000)	75.9092 (0.0000)	-2.3500 (0.0188)
Children18	0.729594 (0.465638)	-0.0628 (0.9499)	-1.1391 (0.2547)	-1.2697 (0.2042)	1.3432 (0.1792)	-0.3288 (0.7423)

Notes: The three first columns compare attitudes in Canada, China, Japan, and Norway with those in the United States. The last two columns compare attitudes in China and Norway with those in Japan. The body of the table contains the calculated z-statistics for the pairwise comparisons of marginal effects. The corresponding p-values are given in brackets below each z-value.

5. CONCLUSIONS

This study indicates that compared to the U.S. sample, formal education decreased the willingness to purchase GM foods in Canada, China, and Japan, and had the opposite effect on respondents in Norway. Compared to the U.S. sample, women in the other samples were more accepting of GM foods, except for respondents from the Norwegian sample. Compared to the U.S. sample, the age of the respondent negatively affected the willingness to purchase GM foods in Norway and Japan.

The comparison of China and Norway with Japan produced some interesting results. Japanese consumers seem most restrictive to biotechnology compared to U.S. respondents. Canadian respondents are most similar to U.S. respondents, except that women seem more skeptical in Canada. In all countries but Norway, women seem to be more risk-averse when considering GM food. Age mattered significantly more in Norway and Japan than in other countries.

Interestingly, China and Japan differ more from each other than do Norway and Japan. Norway and Japan both prefer food without GM ingredients. Chinese consumers were more willing to consume food with GM ingredients.

As Curtis, McCluskey and Wahl (2004) discuss, the concerns of food availability and nutritional intake are much greater in lesser-developed countries (LDCs) than in the United States, Europe, and Japan. Increased crop yields and dietary supplements provided by genetically altered foods would be of greater benefit in terms of food availability and nutrition problems for LDCs, while the potential drawbacks are similar. These potential benefits, along with lower perceived risks, have contributed to generally more positive attitudes toward genetically modified foods in developing nations. The stage of economic development, along with cultural attitudes valuing tradition and skepticism of science, must all be considered as firms in the food supply chain consider whether to include GM products.

REFERENCES

Agresti, A. 2002. *Categorical Data Analysis* (2nd edition). New York: John Wiley and Sons.

Curtis, K.R., J.J. McCluskey, and T.I. Wahl. 2004. "Consumer Acceptance of Genetically Modified Food Products in the Developing World." *AgBioForum: The Journal of Agrobiotechnology Management and Economics* 7(1–2): 70–75.

Grimsrud, K.M., J.J. McCluskey, M.L. Loureiro, and T.I. Wahl. 2004. "Consumer Attitudes toward Genetically Modified food in Norway." *Journal of Agricultural Economics* 55(1): 75–90.

Kanninen, B.J. 1993. "Optimal Experimental Design for Double-Bounded Dichotomous Choice Contingent Valuation." *Land Economics* 69(2): 138–146.

Li, Q., K.R. Curtis, J.J. McCluskey, and T.I. Wahl. 2002. "Consumer Attitudes toward Genetically Modified Foods in China." *AgBioForum: The Journal of Agrobiotechnology Management and Economics* 5(4): 145–152.

Li, Q., J.J. McCluskey, and T.I. Wahl. 2004. "Effects of Information on Consumers' Willingness to Pay for GM-Corn-Fed Beef." *Journal of Agricultural & Food Industrial Organization* 2(2): Article 9. Available online at http://www.bepress.com/jafio/vol2/iss2/art9 (accessed May 6, 2005).

Maddala, G.S. 1983. *Limited Dependent and Qualitative Variables in Econometrics.* Cambridge, UK: Cambridge University Press.

McCluskey, J.J., K.M. Grimsrud, H. Ouchi, and T.I. Wahl. 2003. "Consumer Response to Genetically Modified Food Products in Japan." *Agricultural and Resource Economics Review* 32(2): 222–231.

Mittelhammer, R.C., G.G. Judge, and D.J. Miller. 2000. *Econometric Foundations.* New York: Cambridge University Press.

Part II

CONCEPTUAL ISSUES IN REGULATING AGRICULTURAL BIOTECHNOLOGY

Chapter 12

THE ECONOMICS OF BIOTECHNOLOGY REGULATION

David Zilberman
University of California, Berkeley

Abstract: Rational regulations of transgenic products should compare their risks and benefits with the risks and benefits of alternatives. The current regulations ignore the
alternatives. Political-economic considerations govern the establishment of regulatory requirements and tend to lead to overregulation. Optimal testing should
balance the gains and costs of ex ante testing and ex post monitoring. Emphasis
on ex ante testing to control bad products may lead to welfare losses. The ex
ante testing of new transgenic products should occur mostly at the trait level.
Registration requirements at the varietal level, and even the crop level, may be
very costly.

Key words: political economy, risk, overregulation, registration, monitoring

1. INTRODUCTION

Biotechnology utilizes new discoveries in molecular and cell biology to
develop new products and procedures in agriculture, medicine, and industries. It includes many procedures—some (genetically modified organisms,
cloning) are more controversial than others (cell tissue selection). Agricultural biotechnology is in its infancy. Its first commercial products were
introduced in the mid-1990s, it has a limited product line, and it relies on
relatively large ongoing public and private research investments to refine
and expand its basic tools and broaden its applications. The evolution of
biotechnology and its future are likely to be affected by policies and regulations of new products and production processes.

The growing literature on policies affecting biotechnology has emphasized issues of intellectual property, allocation of funding to public research,

The research leading to this paper was partially supported by U.S. Environmental Protection
Agency's STAR Grant No. 829612.

labeling segregation, and identity preservation standards. Here we will address safety regulations that include registration-testing requirements, restriction on use, and monitoring and reassessment of registration in response to mishaps. Before we proceed, we will discuss some of the generic features of the regulatory process affecting transgenic (TG) traits and transgenic varieties (TGVs). The regulatory process affecting TG traits consists of two elements and several agencies. The two main elements are:

- *The registration process.* The initial registration process consists of testing new TG products and defining their use parameters. New products have to pass a battery of tests to evaluate their efficacy and human and environmental safety. Testing new products establishes a foundation with which to determine how they can be used, and for what applications. The costs and impacts of the regulatory process are results of specific features that we will investigate. What types of products should be tested—new TG traits, events, or varieties? At what level should testing occur—global, regional, or national? What should the criteria be for permitting the use of a new product? We will address some of these questions later in this chapter.

- *Monitoring and reevaluation of use parameter.* Adoption of TGVs may result in several types of externalities, and some are related to gene-flow problems. They include outcross of TG materials to wild species or wild crop varieties, and intermingling of genetically modified varieties (GMVs) and non-GMVs. Toxins introduced by TG traits may harm unintended organisms. There is also concern about the buildup of resistance to certain pest-controlling traits. The extent and impacts of these side effects are uncertain. Monitoring and studying side effects is especially valuable during the early stages of the new TGVs. Firms may not have the incentives to conduct this research, and even if they do (for fear of future liabilities), they do not have the incentives to make their findings public. The regulatory process may develop mechanisms to monitor the performance of TGVs and decision criteria to regulate their use in cases where new information on performance or side effects is available.

This chapter aims to answer some basic questions about the regulation of agricultural biotechnology. First, how do registration requirements and safety regulations of TG products square off with economic theory? What is the role of registration requirements? How can these regulations affect the evolution of biotechnologies? The next section will compare the economic decision rules that lead to a socially efficient regulation vs. regulatory criteria that are optimal from the perspective of the regulated industry. That will be followed by an analysis of the political-economic forces that may shape the regulatory process, suggesting that the political process may lead to overregulation of introduction of TGVs. After that, we will introduce economic decision making frameworks to analyze several features of the regulatory process. They include (i) the allocation of regulatory effort throughout the life of TGVs, (ii) the extent to which the registration process will target new traits vs. application of these traits in crops or crop varieties, and (iii) the geographic boundaries of regulating TG products.

2. SOCIAL VS. PRIVATE OPTIMIZATION GIVEN REGULATIONS OF TGVS

It is useful to assess the regulation of TGVs within a stylized welfare economic framework, where first the optimal decision rules are derived and then they are compared to the decision rules that reflect the actual regulatory framework. In analyzing the regulation of TGVs, we examine the three types of choices. They are:

- *The short-term choice of the optimal allocation of land to the TGVs.* The analysis at this stage is a standard externality analysis.
- *The choice of regulatory knowledge.* This is knowledge about the environmental impacts of TGVs and mechanisms that can control these side effects.
- *The long-term decision whether to introduce new TG products.* We will use expected benefit-cost analyses to make this choice. This approach is rather simple. A more advanced treatment may adopt the real-option approach presented in Dixit and Pindyck (1994) and analyze the optimal timing of investment.

These analyses provide a benchmark for policy assessment, and their outcomes are compared to those under existing policies. We will compare heuristically the socially optimal outcomes with those approximating the private choices under the current regulatory environment, namely, the choices when the seed company is a monopoly and subject to safety regulations consisting of registration costs that increase the initial investments, and liability payments for some side effects during the life of the product.

Let A denote the acreage of TGVs. The benefits of the variety, measured in monetary terms as functions of the acreage, is $B(A)$, and the marginal benefits as functions of the acreage is denoted by $MB(A)$. The marginal benefits decline as the area of TGVs increases. The marginal benefit curve can be interpreted as the demand curve for TGV acres. The temporal cost of producing the seeds needed for A acres of TGVs is denoted by $PC(A)$. The marginal cost as a function of the acreage of the TGVs is $MPC(A)$. It is assumed that the marginal cost increases with the area of the TGVs.

The externality costs of A acres of the TGVs are random and denoted by $EC(A,K)$. The externality cost consists of both direct damage to the environment and costs of damage abatement. The variable K denotes the level of regulatory knowledge, i.e., knowledge about the environmental costs and their prevention. Higher levels of knowledge are assumed to reduce environmental costs. The level of knowledge, K, is a result of the initial testing conducted as part of the regulatory process and of learning, and is taken as given as the land allocation decision is made. The marginal externality cost is denoted by $MEC(A,K)$ and is assumed to be nondecreasing in A. It is also reasonable to assume that more knowledge will reduce the marginal externality costs. The net benefit of the TGV acreage per period is $NB(A,K) = B(A) - PC(A) - EC(A,K)$. A^*, the optimal area of the TGVs, is determined by maximizing the expected net benefits, $NB(A,K)$, subject to nonnegativity constraints. The first-order condition at $A^* > 0$ is

$$MB(A^*) = MPC(A^*) + MEC(A^*,K). \tag{1}$$

When $A^* > 0$, the optimal level of the marginal private benefit of the TGVs' $MB(A^*)$, which is also the demand for the acreage of the crop, is equal to the private marginal cost $MPC(A^*)$ of acreage in GMVs plus the marginal externality cost $MEC(A^*,K)$. This condition suggests that more knowledge may lead to increased acreage of the GMVs, as it will reduce the marginal control of environmental costs.

For simplicity, the analysis here of the choice of the TGV acreage is deterministic, even though risk considerations are important in TGV land-use and regulation choices. Introducing risk will not drastically alter the analytical results. The optimality condition then equates the expected value of the marginal benefits of the area of TGVs with the sum of the expected values of the marginal private and externality costs of the acreage of the TGVs.

Thus far, we have analyzed the short-term land allocation decision. To study the long-term choices of the optimal regulatory knowledge and whether or not to introduce a TG product, we need to date the land-use variables. For simplicity, we will consider the long-term choices in the context of a TGV. However, our analysis can be generalized to a more complex and realistic situation where decision makers have to decide whether to invest in the development of a TG trait or the knowledge requirement for regulating a trait (actually, we will later argue that regulatory requirements should target new TG traits rather than TGVs).

Let A_t denote the area of land planted with the TGV seeds at period t, where t assumes values varying from 1 to T (T is the end of the planning horizon). The net benefit of the TGV at period t is a function of K

$NB_t(A_t^*, K) = B(A_t^*) - PC(A_t^*) - EC(A_t^*, K)$. The discounted net operational benefit from introducing a new TGV depends on the knowledge and is

$$DNOB(K) = \sum_{t=o}^{T} \frac{1}{(1+r)^t} NB_t(A_t^*, K).$$

The introduction of a new TGV requires an initial investment I in research and development and infrastructure unrelated to the environmental side effects of the technology. In addition, it requires investment in regulatory knowledge on the externalities of the technology and how to manage them. Let $RC(K)$ denote the registration cost as a function of the initial regulatory knowledge K. The optimal level of the regulatory knowledge, K^*, is determined solving

$$\underset{K^*}{\text{MAX}}\ DNOB(K^*) - I - RC(K^*).$$

At the optimal level of regulatory research, its marginal contribution to discounted net operational benefits is equal to the marginal regulatory cost

$$MDNOB(K^*) = MRC(K^*). \tag{2}$$

Finally, let $Y^S = 1$ denote that it is socially optimal to invest in the genetically modified (GM) seed, and $Y^S = 0$ denote that it is not:

$$Y^S = 1 \text{ if } DNOB(K^*) - I - RC(K^*) > 0. \tag{3}$$

Thus, it is socially optimal to invest in a new GM trait if the discounted net benefit is greater than the sum of the investment and regulatory cost.

3. THE ACTUAL CHOICES VS. THE OPTIMAL CHOICES

The decision rules in equations (1), (2), and (3) do not seem to fully guide the actual regulation of TGVs. The decision whether to invest in TGVs and how much to price the seeds is made by firms that maximize their profits rather than net social benefits. We do not have a formal model of the regulatory process, but clearly it does not apply to the expected net social benefits criterion, presented in (3), to evaluate new technologies. Berwald, Matten, and Widawsky (2006) and Erwin and Welsh (2006) suggest that the regulatory agencies are taking a more lexicographic approach that emphasizes judging new technologies only by their own risk without taking into account the risk-reduction effect of the new technology. We will argue below that these factors result in biased outcomes that frequently lead to underutilization of the technology. We will first investigate the short-term privately optimal choice of the TGV acreage, denoted by A^P. Then we will analyze the private decision whether or not to introduce TGVs. Finally, we will examine the political-economic forces that will determine the initial regulatory knowledge, K^P, and resulting regulatory costs, $RC(K^P)$.

3.1. The TGV Acreage Decision

This decision is frequently determined by seed companies with substantial market power. Here we will consider the case where the seed company is a monopolist, which equates its marginal revenues (MR) that are below the marginal benefits (demand) of the TGV acreage, to its marginal costs. The marginal costs include the marginal private costs of producing and marketing the TGVs and some externality costs, since it is unlikely that the environmental costs of the TGVs are completely ignored. Actually, the regulatory process and the legal system establish liability rules and due care standards that hold seed companies and farmers responsible for some of the gene-flow damage resulting from growing TGVs. The resulting *de facto* private externality costs are denoted by $EC^P(A)$, and the resulting marginal externality cost function is $MEC^P(A)$. We assume that the seed companies carry these externality costs.[1] The area of the TGVs under the private seed company, A^P, is determined according to

$$MR(A^P) = MPC(A^P) + MEC^P(A^P).\qquad(4)$$

Comparison of conditions (1) and (4) suggests that the extent to which there is overuse of TGVs depends on whether the marginal revenue minus the private marginal environmental cost is larger than the marginal benefits minus the true marginal costs. The condition of underutilization of the GM crop becomes

$$A^* > A^P \text{ if } MB(A^*) - MEC(A^*) > MR(A^*) - MEC^P(A^*).\qquad(5)$$

Comparing cases where $A^* > A^P$, the underutilization of the TGVs ($A^* - A^P$) is likely to increase as the gap between $MB - MEC$ and $MR - MEC^P$ increases. That suggests that the underutilization of TGVs will increase as

---

[1] The tendency to go after parties with "big pockets" may shift much of the liability of the side effects of the GMV to the seed company. Aventis paid hundreds of millions of dollars in the case of the Starlink fiasco.

demand for TGV acreage becomes more inelastic (resulting in a larger gap between *MB* and *MR*) and as the private marginal environmental cost increases and the actual marginal environmental cost of the TGV acreage decreases.

The major reason for underutilization of GM crops is that *their true environmental costs may be negative* or very small, and *the regulatory process tends to overestimate these costs*. Since TGVs in most cases replace traditional crop varieties, the environmental costs of the GM crops should be computed, taking into account the alternative. When a farmer considers the benefits of a TGV, she considers the yield increase and cost reduction relative to a traditional variety. Ameden, Qaim, and Zilberman (2005) show that pesticide cost savings have been a major cause of the adoption of GMVs. Computation of the net social benefits of TGVs should add the direct environmental costs of the TGV and subtract the environmental costs of the pesticides saved by the TGV. Moreover, if the TGV increases yield significantly and demand is not infinitely elastic, the transition to TGVs is likely to lead to a reduction of cultivated acreage and the associated environmental costs. This environmental cost savings reduces the net environmental cost of the TGV, which may be negative. Thus, taking into consideration the net of the TGVs, including the environmental costs of the technology replaced by the TGVs, may actually increase the social gain from the adoption of the TGVs.

3.2. The Product Introduction Decisions

The introduction of TG traits and varieties is generally initiated by private companies, which have to meet the requirements of the registration process, including paying the regulatory costs $RC(K^P)$ (where K^P denotes the knowledge requirement when the product is introduced by a private company). Again we will address the choices in the context of a TVG, but the results can be generalized to decisions and regulations of TG traits. The firm's net discounted operational profits from the introduction of the TGV are denoted by

$$DOP = \sum_{t=0}^{T} \frac{1}{(1+r)^t} \left[R(A_t) - PC(A_t) - EC^P(A_t) \right],$$

where $R(A_t)$ is the revenue in period t, and $EC^P(A_t)$ is the externality cost (the liability and due care cost) paid by the private seed company. Let $Y^P = 1$ if it is optimal for the seed company to introduce a new GM trait, and $Y^P = 0$ otherwise. A firm will introduce a new GM trait if the net discounted profit is positive, namely,

$$Y^P = 1 \text{ if } DOP - I - RC^P(K^P) > 0. \tag{6}$$

New GM traits will be introduced when the net operational discounted profits are greater than the initial investment plus the regulatory costs.

Our analysis suggests two plausible situations of underutilization of the GM trait:

Underutilization of the trait when it is introduced (if $Y^S = Y^P = 1$, $A^* > A^P$). Equation (5) suggests that, when the introduction of a new GM trait is optimal both socially and privately, under plausible conditions the GM acreage under the private solution will be suboptimal.

Not introducing the technology when it is socially optimal ($Y^S = 1$, yet $Y^P = 0$). Comparison of equation (3) with (6) yields the conditions for these outcomes,

$$Y^S = 1, \text{ yet } Y^P = 0 \text{ if } DOP - RC^P(K^P) < I, \text{ but } DNOB(K^*) - RC(K^*) > I. \tag{7}$$

Socially beneficial GM traits may not be introduced for several reasons:

- The firms consider the revenue rather than all the benefits from adopting the GM crop. The difference between $B(A_t)$ and $R(A_t)$ is the consumer surplus, which may be quite substantial (especially when demand for the GM area is inelastic). This difference will contribute to the difference between $DNOB(K^*)$ and DOP.
- Underadoption occurs because of differences between MB and MR and environmental cost considerations which occur when (5) holds and $A^* > A^P$. The lower acreage contributes to the difference between $DNOB(K^*)$ and DOP.
- The TG trait replaces a technology with high environmental costs (pesticides), and the firm does not consider the social environmental cost saving. This factor will contribute to underadoption if $EC^P(A_t) > EC(A_t)$, and thus it will add to the positive difference between $DNOB(K^*)$ and DOP.
- The private knowledge requirement and the resulting registration costs are higher than the socially optimal levels of these variables $[RC^P(K^P) > RC(K^*)]$. Higher registration costs will contribute to making the private returns to the investment in the GM trait lower than the social returns.

Our analysis suggests that the pursuit of profits and monopoly power of the seed companies will contribute to the under-introduction of these technologies. The impacts of the regulatory process are complementary when it imposes extremely high registration and externality costs on the GM trait. A regulatory process [resulting in $RC(K^P) < RC(K^*)$ and $EC^P(A_t) < EC(A_t)$] that is too lax may encounter the negative impacts of the behavioral rules of the seed company from introduction of GMVs, and may even lead to over-introduction. Thus, a better understanding of the determination of the registration requirement will allow assessment of their effect on the introduction of GM traits. The determination of the fate of GM crops is a *political issue,* and large segments of the population may not subscribe to the economic criteria we present here. Thus, it is useful to assess the outcome of the regulatory process within a political-economic framework.

4. THE SUBVERSION OF THE REGISTRATION PROCESS

The research findings of environmental science, public health, and statistics establish much of the intellectual foundation for the design of the registration requirement and other regulations. However, the actual determination of the details of the regulatory process and requirements is the result of political processes. Political-economic theories were introduced to explain regulatory outcomes.

The first is the Capture Theory (Posner 1974). It states that the regulators may be controlled by the regulated industry. In the case of agricultural biotechnology, this theory seems to suggest that Monsanto and other major companies capture the regulatory process. Assuming that it is true, it is not clear whether capture by the major biotechnology will lead to light or heavy regulation. On the one hand, Monsanto and the other multinational biotechnology firms will benefit from lighter regulation, since it will reduce their cost and accelerate the introduction of products to the market. On the other hand, light regulation may lead to substantial objections from environmentalists and lead to obstacles in introducing products. Furthermore, one of the relative advantages of the major biotechnology companies is their financial resources and their capacity to carry out demanding regulatory tests. Most of the new innovations in agricultural biotechnology were originated by small start-up companies, and with light regulatory requirements they may be tempted to develop and commercialize their innovations themselves. With heavier requirements, the small companies and startups may be less competitive in product development and be inclined to either sell the rights to their products or to be taken over by Monsanto and other multinationals. Thus, the multinationals may not object to heavy regulations, as they will serve as barriers to entry and reduce competition.

The Capture Theory is incomplete. First, in democracies the ruling party changes over time. In the case of the United States, corporations are likely to have more influence under Republican administrations, while the environmentalists may have more impact on the regulatory process when the Democrats are in power. Second, industries are one of many groups that are interested in the regulation of biotechnology, including farmers, food retailers, etc.

Becker (1983) and Zusman (1976) introduced more generalized approaches to political economy. They argue that regulations are the result of interactions between various groups with varying resources and interests. These interactions may be modeled through market-like processes or through games. However, the final outcomes reflect the relative power and resolution of different parties. For example, Zusman suggested that the regulatory outcomes could be derived from solving optimization problems where the objective function is a weighted sum of the welfare of all the affected parties. In our case we assume that the decision variable determined by the regulatory process is K, the regulatory knowledge required for product registration. Higher K increases the cost of registration. The value of K is constrained to be between 0 (light) and $\overline{K}$ (heavy) registration requirements. Thus, the optimal K will be determined solving

$$\max_{0 \le K \le \overline{K}} \alpha_1 RW(K) + \alpha_2 CS(K) + \alpha_3 FS(K) + \alpha_4 FR(K) + \qquad (8)$$
$$\alpha_5 GMM(K) + \alpha_6 SU(K) + \alpha_7 EB(K) + \alpha_8 CIN(K),$$

where the α_i denotes the weight of each group.

$RW(K)$ is the welfare of the regulator as a function of the level of regulatory knowledge requirement. The professionals at regulatory agencies have power and knowledge, and they use their perspective to establish policy parameters. When the regulators of safety have training in public health, they will have the tendency to pursue a very strict regulatory agenda that emphasizes reducing environmental and health risks. In certain countries the regulators may benefit from strict regulations as they give them more power, and they can utilize them for self-interest. Thus, $RW(K)$ reflects both professional satisfaction from higher standards and personal benefits from the power, and sometimes the income, that higher standards bestow on the regulators.

$CS(K)$ represents the consumer perspective on the impact of more rigorous registration process on their well-being (consumer surplus). If consumers do not perceive biotechnology products to improve their welfare but, on the contrary, view them as a source of risk, they will take a strong position for heavy regulation.

$FS(K)$ is farmers' surplus as a function of the severity of regulation. Based on the experience of the current generation of biotechnology products, which saves costs and increases yields, we expect farmers to support light regulation and FS to decline with K.

$FR(K)$ is the food retailers' profit as a function of K. Food retailers want minimal exposure to both food safety problems and negative publicity. Thus, they are likely to be supportive of strict regulation if they perceive it as providing "industrial quiet" and as enhancing the profit.

$GMM(K)$ is the economic welfare of the major biotechnology firm as a function of the severity of regulation. As we mentioned before, it is not clear *a priori* whether major biotechnology companies will favor strict or light regulation because of the conflict between the desire to reduce compliance cost and the gain from barrier to entry that strict regulations provide.

$SU(K)$ is economic welfare of start-up biotechnology firms as a function of the strictness of knowledge requirement for registration. For the most part, small biotechnology companies will favor less-restrictive regulation.

$EB(K)$ is the environmental benefit as a function of severity of registration requirement. Under current conditions, environmental groups tend to push for the strictest regulation.

$CIN(K)$ is the economic welfare of chemical input manufacturers as a function of the severity of knowledge requirement for registration of TG products. Chemical manufacturers with a weak position in biotechnology are likely to be the great losers from a less strict registration requirement that will accelerate the spread of agricultural biotechnology. While these companies may strive to reduce the regulatory burden in the chemical area, they would not oppose stricter standards for competing biotechnology products.

The political process may result in a balance of power that will favor strict regulation. A coalition of environmentalists, producers of products competing with biotechnology (chemicals, bio-control), risk-averse food marketers, large multinationals that would like to establish barriers to entry, and concerned consumers may collaborate with sympathetic regulators and dominate the interest of farmers and start-up companies to introduce strict regulation. However, regulation may change with circumstance. Good safety records of the TG products, combined with the increase in the price of food and concerns about food supplies, may lead to reduced restrictions. Various groups may change their perception of the cost of regulations and may use their influence to modify the regulatory requirements. For example, the major biotechnology companies may give less weight to concerns about entry and more to the regulation costs, and may exert pressure for regulatory relief. Similarly, once second-generation TG products with direct consumer benefits are introduced, consumer groups may support a lighter regulatory load.

Despite the likelihood that the regulatory process will result in excessive regulations, it is quite likely that it is serving an important purpose and is generating significant social benefits. The social optimum is one benchmark for assessing the outcomes of the regulatory process, but it presents an idealized and improbable situation. Another benchmark to consider is the introduction of the GM traits without addressing their side effects. The registration process provides incentives to consider and addresses externality problems, which leads to the generation of information that would not have been generated otherwise. It may prevent the introduction of technologies that are profitable but may have negative overall social side effects. Furthermore, the value of an institution is measured by comparisons to alternatives. Without the current regulatory framework, political objections might have led to even less introduction of GM crops. As the paper by Gaskell et al. (1999) suggests, having a trusted regulatory system is crucial in overcoming objections to the new technology. Public concern and trust in the regulatory process are based on its capacity to handle environmental and health side effects, which thus should be the focus of the regulatory process.

A better understanding of the safety regulations of biotechnology requires knowledge and even quantification of the perspectives and power of the various groups that affect these regulations. It may be especially valuable to know the disciplinary perspectives guiding scientists who affect the regulatory decisions and their implications. The works of Lichtenberg (2006), Ervin and Welsh (2006), and James (2003) are important early steps in this direction. Economic decision making criteria are not likely to be used to establish the regulatory framework, but economic analysis could be used to assess the impact of existing regulations and the details of the regulatory process in order to identify potential gains from reform.

5. THE ECONOMICS OF REGULATORY FEATURES OF TG PRODUCTS

The regulatory process of TG products has many features, each affecting its overall effectiveness. Here we will provide a perspective on determining several of these features.

5.1. Allocation of Regulatory Efforts During the Life of a Product

A new TG trait may have negative environmental or health side effects that will dominate its benefits, and one of the challenges of the regulatory process is to determine the screening and monitoring efforts for controlling the side effects of new TG products. The screening efforts (that are part of the registration requirements) can be viewed as ex ante regulation, and monitoring and restriction of product use are ex post regulation. Here we introduce a simple model to determine optimal levels of these regulations.

Let B denote the economic benefits of a new TG product once introduced. The new product may have the externality and health side effects, and in this case the externality costs are denoted by E. We assume that $E > B$ and let q be the probability of having negative side effects. The costs of product development are denoted by D. Without regulation, a product should be introduced if $B - qE - D > 0$. But we assume that the ex ante screening can identify TG products with negative side effects prior to use. Let C denote the costs of testing, and let $h(C)$ denote the probability of detecting a bad product as a function of the cost of screening. We assume that the detection probability increases with the costs, but the marginal detection probability declines with costs $h'(C) > 0$, $h''(C) \leq 0$.

When new TG products are screened, once it is discovered that a product is generating externalities, it is not used. The distribution of net returns of a new product thus becomes

$$
\begin{array}{lll}
-D & \text{with probability} & h(C)q \\
B - E - D & \text{with probability} & [1 - h(c)]q \\
B - D & \text{with probability} & 1 - q.
\end{array}
$$

The probability that a bad product is discovered is $h(C)q$ and, by not introducing it, there is a gain of $E - B$ dollars. When the bad product is not discovered, the net loss is $E + D - B$. The gain if there are no externalities is $B - D$.

Based on this distribution of net benefits, the optimal expenditure of ex ante testing is determined solving

$$
\underset{C}{MAX}\{(1 - q)B - [1 - h(C)]q(E - B) - D - C\}.
$$

If the problem has an internal solution, the optimality condition is

$$
h'(C)q(E - B) = 1. \tag{9}
$$

$$
h'(C) = \frac{1}{q(E - B)}
$$

The condition states that the optimal value of marginal gain from testing [which is the product of the marginal probability of detecting $h'(C)$, the probability of externality, q, and the loss per product $(E - B)$] is equal to the marginal cost of (1). The second presentation of condition (9) is depicted in Figure 1. At the optimal level of testing expenditures, the marginal probability of detection is equal to the reciprocal of the expected cost of a bad product $[q(E - B)]$. This condition and Figure 1 suggest that the ex ante testing increases as the gap between the externality cost and the benefits of a TG crop widens and as the probability of the externality increases.

The regulatory process controls damage by not registering products that failed the ex ante test. However, the damage of bad products not detected by

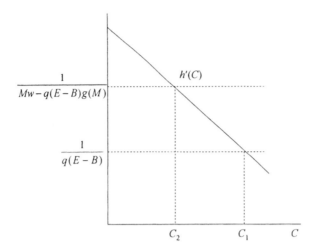

Figure 1. Regulatory Effort With and Without Ex post Monitoring

the ex ante testing can be reduced if they are banned once they are discovered by a monitoring process.

Let M denote the amount of monitoring and w the price of a unit of monitoring. High monitoring costs allow early discovery of a bad product and thus early banning of its use. The fraction of the benefits and costs of a bad product that is materialized before it is banned as a result of monitoring is denoted by the function $g(M)$, where $g' < 0$ and $g'' \geq 0$. Higher monitoring accelerates the discovery and banning of a bad product and thus reduces $g(M)$. With monitoring, the distribution of the costs and benefits becomes

$$
\begin{array}{lll}
-D & \text{with probability} & h(C)q \\
g(M)(B-E)-D-Mw & \text{with probability} & [1-h(c)]q \\
B-D-Mw & \text{with probability} & 1-q.
\end{array}
$$

Monitoring costs apply whenever the new product is introduced, but affects the net benefits only when the product is bad. With this distribution, an optimal policy requires determination of both ex ante testing costs and the monitoring costs, and the optimization problem becomes

$$
\underset{C,M}{MAX}\{(1-q)(B-Mw)-[1-h(C)]q[(E-B)g(M)+Mw]-D-C\}.
$$

The first-order optimality conditions determining the optimal testing cost become

$$h'(C)[q(E-B)g(M)+Mw]=1, \tag{10}$$

$$h'(C) = \frac{1}{q(E-B)g(M)+Mw}$$

and the optimal monitoring is determined where

$$-(E-B)g'(M)=w. \tag{11}$$

$$-g'(M) = \frac{w}{(E-B)}$$

Condition (10) is similar to condition (9), and it states that the optimal ex ante testing will be at a level where the marginal benefits from increased expenditure on testing are equal to its cost. Comparison of condition (10) with (9) suggests that when monitoring is introduced, the expected cost because of the bad product is lower since its use will be banned once the damage is discovered through monitoring. This cost reduction is not free, and it is paid by the monitoring cost. However, altogether we expect that monitoring will reduce the cost of bad products, and the results are reflected in Figure 1, where C_2 (corresponding to the case of monitoring) is smaller than C_1. Namely, the introduction of monitoring will reduce the optimal level of ex ante testing.

Condition (11) determines the optimal level of ex post monitoring, where the marginal reduction in the net cost of bad products because of monitoring is equal to the cost of monitoring. It is easy to show that lower costs of monitoring will increase optimal monitoring, reduce the damage of bad products, and [from condition (10)] reduce the ex ante optimal testing. Our analysis thus suggests that less ex ante testing is more optimal in societies with ongoing assessment of environmental impacts of new technologies and effective mechanisms to ban bad products than in societies with minimal environmental monitoring or constraints on banning chemicals. Policies aiming to use only ex ante testing to control bad products may backfire, as they will be too expensive and lead to welfare losses.

5.2. What Categories of New TGVs Should Require Registration?

The regulatory costs depend on the extent to which new products are subject to registration requirements. It is useful to distinguish among new TG traits (that may apply to several related crops), an event (a new trait applied to a specific crop), and a new TGV. Regulation at a more detailed level may be justified if the cost of the extra testing is smaller than the extra benefit that results from the testing. For example, registration requirements at the varietal level rather than the event level would have been justified if pre-introductory testing of every variety of Roundup Ready soybeans in the United States led to the prevention of negative side effects that were not discovered by the generic screening of Roundup Ready soybeans, and if the externalities were more costly than the extra testing.

The arguments introduced in Ameden, Qaim, and Zilberman (2005) suggest that registration requirements at the varietal level may be very costly and even have negative environmental side effects. When every TGV has to be registered, and registration is costly, the technology seller is likely to introduce the TG trait to small numbers of varieties. Farmers who grow varieties that have not been modified have a choice between using their own unmodified variety and adopting one of the varieties that has been modified. They have to trade the gains from the TG modification with the losses associated with the varietal switch. For example, suppose a local variety is producing y_1 tons of output per acre, but with a probability α it loses a fraction of the output d due to pest damage. Suppose that a trait that fully controls the damage is inserted in a generic variety that produces $y_0 < y_1$ at our location. Suppose that the price of output is p and that the extra price paid for the TGV is v. A risk-neutral farmer will adopt this variety if

$$p[y_0 - y_1(1 - \alpha d)] - v .$$

Farmers will adopt the technology with sufficiently large probability and magnitude of damage to compensate for the permanent yield loss and the extra trait costs. The varietal switch may result in years of minimal gain from pest protection. Thus, in our case there is a probability $1 - \alpha$ that the yield of the TGV will be lower than that of the local variety (it will then be $y_0 < y_1$). There are indeed observations that adoption of TGVs in, say, India, where the number of Bt cotton varieties is low, may lead to significant variations in the yield effect of adoption resulting from variations in pest damage and the impact of "yield drag" due to varietal transition (Ameden, Qaim, and Zilberman 2005).

In addition, Qaim, Yarkin, and Zilberman (2005) show that when only a small subset of the varieties is genetically modified, the overall rate of adoption is likely to be smaller, and then the welfare gain from the new technologies will be much smaller than if the genetic trait had been inserted in a large number of varieties. Furthermore, when registration requirements at the varietal level significantly reduce the number of TGVs, that may also lead to the loss of agricultural biodiversity. Qaim, Yarkin, and Zilberman argue that under plausible conditions, some local varieties may not be used and be replaced by generic TG varieties. Currently, there are 1,000 varieties of Roundup Ready varieties and close to a 1,000 varieties of Bt corn in the United States, where registration for every variety is not required.

Another drawback of requiring registration of TGVs is that there will be a significant time lag between the introduction of new varieties by selective breeding and the time that they will be available in the TG form, which may again lead to a significant loss in yield and excessive expense of chemicals and other inputs.

Thus, unless there are very significant environmental gains from investigating the impact of inserting a trait in individual varieties (which is quite unlikely), the registration requirement at the varietal level seems to be very inefficient.

The same logic suggests that there is significant potential for gain from imposing strong registration requirements at the trait level rather than the event level. This is likely to be especially pertinent in the case of specialty crops that are closely related (e.g., watermelon, cantaloupe, etc.). The potential markets for TGVs of specialty crop seeds are likely to be small. Imposing high registration requirements at each event may make it unprofitable to invest in the development of such TGVs. Even though the overall welfare effect of their introduction may be significant, the environmental risk for each event may be very small.

Thus, the analysis here suggests that understanding the factors affecting the environmental side effects of new TG traits in a way that will reduce the overall burden of registration requirements is quite important both in enhancing the introduction and adoption of new TGVs and in improving social and sometimes environmental gains of these new technologies.

5.3. Geography and Regulation

Political borders rather than ecological and agricultural considerations generally determine regulatory regimes. Repetitive registration requirements may be imposed for introducing traits in regions with similar conditions that belong to different countries. One advantage that large countries like the United States, Canada, and China have is that registration costs of new TG traits apply to a large territory. On the other hand, small and relatively poor countries in Asia, Africa, and Latin America may insist on imposing costly registration requirements as a result of testing in nearby countries, and other regions may be applicable. As we argued earlier, political-economic considerations may provide policymakers with the incentive to impose extra registration that will enhance their power and well-being, but will not benefit the countries that these policymakers are supposed to serve. Thus, one of the challenges of designing regulatory regimes is to establish regional organizations that may encompass several countries that share similar agroecological conditions to provide economies of scale in registering new TGVs.

6. CONCLUSIONS

The introduction of TGVs has been a source of controversy and will require a regulatory process that will serve both to address possible environmental side effects and to provide peace of mind to the public. We have to recognize that the design of the regulatory process is the outcome of a political process and does not necessarily follow economic principles. We have argued that political-economic reasons may cause excessive registration requirements and underadoption of TGVs and, as a result, underinvestment in research and development.

Economics can play a major role in assessing the outcome of the current regulatory process and provides insight for its improvement and reform. One major flaw of the current regulations is that each new trait is evaluated separately, without taking into account the alternatives. Even if TGVs pose some risks, rational regulations that determine their fate should compare this risk as well as the associate benefits with the risks and benefits of alternatives. Economists are challenged to educate the public and regulators of the importance of evaluating new TGVs within the context of existing new alternatives. Economics can also contribute to providing the guidelines for allocating effort between ex ante and ex post regulation of new TGVs. It can provide evidence that may lead to reducing excessive layers of regulations.

The research on the economics of regulating TGVs is in its infancy. While we emphasize normative aspects and the role of political-economic considerations, it is also important to understand consumers' perception and preferences regarding the regulatory process. Although it is desirable that regulation will be based on the best available science, an important objective of the regulatory process is to address consumers' concerns, and thus it has to be designed with these concerns in mind.

REFERENCES

Ameden, H., M. Qaim, and D. Zilberman. 2005. "Adoption of Biotechnology in Developing Countries." In J. Cooper, L. Lipper, and D. Zilberman, eds., *Agricultural Biodiversity and Biotechnology in Economic Development*. Norwell, MA: Springer Publishers.

Becker, G.S. 1983. "A Theory of Competition among Pressure Groups for Political Influence." *Quarterly Journal of Economics* 98(3): 371–400.

Berwald, D., S. Matten, and D. Widawsky. 2006. "Economic Analysis and Regulating Pesticide Biotechnology at the U.S. Environmental Protection Agency." In R.E. Just, J.M. Alston, and D. Zilberman, eds., *Regulating Agricultural Biotechnology: Economics and Policy*. New York: Springer.

Dixit, A., and R. Pindyck. 1994. *Investment Under Uncertainty*. Princeton, NJ: Princeton University Press.

Ervin, D.E,, and R. Welsh. 2006. "Environmental Effects of Genetically Modified Crops: Differentiated Risk Assessment and Management." In R.E. Just, J.M. Alston, and D. Zilberman, eds., *Regulating Agricultural Biotechnology: Economics and Policy*. New York: Springer.

Gaskell, G., M.W. Bauer, J. Durant, and N. Allum. 1999. "World Apart? The Reception of Genetically Modified Foods in Europe and the U.S." *Science* 285: 384–387.

James, C. 2003. *Global Review of Commercialized Transgenic Crops*. International Service for the Acquisition of Agri-Biotech Applications, Ithaca, NY.

Lichtenberg, E. 2006. "Regulation of Technology in the Context of Risk Generation." In R.E. Just, J.M. Alston, and D. Zilberman, eds., *Regulating Agricultural Biotechnology: Economics and Policy*. New York: Springer.

Posner, R.A. 1974. "Theories of Economic Regulation." *Rand Journal of Economics* 5(2): 335–358.

Qaim, M., C. Yarkin, and D. Zilberman. 2005. "Impact of Biotechnology on Crop Genetic Diversity." In J. Cooper, L. Lipper, and D. Zilberman, eds., *Agricultural Biodiversity and Biotechnology in Economic Development*. Norwell, MA: Springer Publishers.

Zusman, P. 1976. "The Incorporation and Measurement of Social Power in Economic Models." *International Economic Review* 17(2): 447–462.

Chapter 13

LABELING REGULATIONS AND SEGREGATION OF FIRST- AND SECOND-GENERATION GM PRODUCTS: INNOVATION INCENTIVES AND WELFARE EFFECTS

GianCarlo Moschini and Harvey Lapan
Iowa State University

Abstract: We review some of the most significant issues and results on the economic effects of genetically modified (GM) product innovation, with emphasis on the question of GM labeling and the need for costly segregation and identity preservation activities. The analysis is organized around an explicit model that can accommodate the features of both first-generation and second-generation GM products. The model accounts for the proprietary nature of GM innovations and for the critical role of consumer preferences vis-à-vis GM products, as well as for the impacts of segregation and identity preservation and the effects of a mandatory GM labeling regulation. We also investigate briefly a novel question in this setting, the choice of "research direction" when both cost-reducing and quality-enhancing GM innovations are feasible.

Key words: identity preservation, labeling, market failure, product differentiation, welfare

1. INTRODUCTION

The first nine years of genetically modified (GM) crops, since their introduction in 1996, have been a mixed success. Adoption has been fast and extensive by any standard, reaching a worldwide area of 200 million acres in 2004 (James 2005). But large-scale adoption has been confined to a handful of countries,[1] and, perhaps most important, the advent of GM crops has met

---

[1]The United States, Argentina, Brazil, Canada, and China accounted for about 96 percent of total GM crop cultivation in 2004 (James 2005).

The support of the U.S. Department of Agriculture, through a National Research Initiative grant, is gratefully acknowledged.

with considerable public opposition and a flurry of new restrictive regulations. In the European Union (EU), in particular, the initial laissez-faire attitude, which allowed several GM products to be approved, was reversed in 1998 with the introduction of a *de facto* moratorium on new GM products. Only in 2004 did progress appear with the unveiling of a new and extensive framework for GM approvals and marketing. Ostensibly meant to foster food safety, protect the environment, and ensure consumers' "right to know," the new (and already controversial) system is centered on the notions of mandatory GM labeling and traceability (European Union 2004). Meanwhile, the strain that the EU moratorium and GM regulations can have on trade has become apparent (Lapan and Moschini 2001, Sheldon 2002) and the prospects for its resolution are rather uncertain. A central question, it seems, concerns the economic effects of the GM product innovation, including both intended and unintended effects.

Assessing the economic implications of the introduction of GM products continues to be a challenging endeavor. It has become clear over time that a critical element of this new technology concerns consumers' acceptance. A portion of consumers clearly has a negative perception of food produced from GM products, at least based on what one can conclude from consumer surveys (e.g., Gaskell, Allum, and Stares 2003) and experimental results (e.g., Huffman et al. 2003; Noussair, Robin, and Ruffieux 2004). Furthermore, the first generation of GM crops, characterized by agronomic traits such as herbicide resistance and pest resistance, offered no direct benefit to consumers. Hence, from a consumer perspective, GM innovation has produced what Lapan and Moschini (2004) call "weakly inferior" substitutes. The fact that GM food is not a perfect substitute for conventional food per se simply implies a smaller potential market for the new GM products. But the introduction of first-generation GM crops means that, to deliver traditional GM-free food, additional costs must be incurred (relative to the pre-innovation situation). That is, costly (and hitherto unnecessary) segregation and identity preservation activities are required. Essentially, therefore, the GM innovation process has also introduced a new market failure, a type of externality on the production of traditional food products (Lapan and Moschini 2001).

Consumer acceptance is likely to be different for GM products that offer output traits of direct interest to end users, such as improved nutritional content (e.g., increased vitamin content, as in the widely publicized "golden rice"). This defines so-called second-generation GM products (Pew Initiative on Food and Biotechnology 2001). But whereas the attribute of the innovation may be of interest, per se, to consumers, the fact remains that the GM nature of the innovation is likely to continue to play a role in consumer acceptance. Hence, a sound economic assessment of the effects of GM product innovation needs to address directly the question of consumer preferences and how these interact with the nature of the market failure discussed in the foregoing.

In this chapter we propose to review some of the most significant issues and studies that have dealt with the economic effects of GM product innovation. We will pay particular attention to the question of GM labeling and its relation to the need for costly segregation and identity preservation activities. To organize some of the main findings to date, we develop an explicit, simple model that can accommodate the features of both so-called first-generation and second-generation GM products. This model explicitly accounts for the effects of consumer preferences vis-à-vis GM products, as well as for the distinct impacts of segregation and identity preservation, and the effects of an EU-style GM labeling regulation. We also investigate briefly a novel question in this setting, specifically, the choice of "research direction" when both cost-reducing and quality-enhancing GM innovations are feasible.

2. THE ECONOMICS OF LABELING

Much has been written on the scope, merit, and effects of food labeling regulations.[2] An important distinction, for our purposes, is between "voluntary labeling" and "mandatory labeling." Voluntary labeling strategies naturally arise as firms compete in the marketplace and try to differentiate their products from those of competitors. The underlying assumption is that firms' products are in fact differentiated (in some dimension) and that consumers may value a product's specific attributes that labeling emphasizes. Here it is assumed that firms have some information that may be useful to consumers, that such attributes are not easily observable by consumers prior to the purchase, and that a label can credibly disclose the information about the "quality" of the good that consumers desire. Thus one is dealing with "experience goods" or "credence goods," rather than "search goods."[3] An issue that arises in this setting is whether firms disclose truthful information and whether they disclose all of the information. For positive attributes it is obviously in the strategic interest of firms to disclose the information, but more generally Grossman (1981) shows that, when consumers make rational inferences and assume that undisclosed attributes are of the worst possible quality, there is a powerful market incentive for full disclosure of information. The credibility of voluntary labeling can be enhanced by third-party services (producer associations, consumer groups, governments) that may supply standards, testing services, and certification.

---

[2] Golan, Kuchler, and Mitchell (2001) provide a useful introduction and a review of the literature.

[3] See Tirole (1988, chapter 2) for definitions and an introduction.

Mandatory labeling is typically harder to justify on economic grounds, for a number of reasons. The presumption again is that there is asymmetric information: firms know something that consumers do not, and the latter would benefit from disclosure. But to advocate mandatory disclosure, one has to postulate that firms would not reveal the information without government intervention. Thus one must assume that various forms of "screening" or "signaling" that are feasible in the marketplace do not yield a desirable outcome in this setting. Often there may be much better policy tools, depending on the specific situation (bans, production standards, etc.).

2.1. Labeling of GM Products: Segregation and Identity Preservation

Some general issues concerning biotech labeling are discussed by Teisl and Caswell (2003) and Golan, Kuchler, and Mitchell (2001). One recurrent hypothesis, in discussions of GM food labeling, is that the good in question is a pure "credence good," whereby the true attribute of interest to the consumer cannot be observed after consumption. Many studies uncritically presume that non-GM goods in this case are credence goods, and that is taken as sufficient evidence of a market failure to warrant government intervention. But, arguably, GM products are not really prototypical "credence goods." It is in fact possible to uncover the nature of the product by "testing." Testing could be done by organizations, rather than by individuals, and need not be systematic: not every unit needs to be tested insofar as the outcome of testing can implicate a "brand."[4]

The presumption of "asymmetric" information (between firms and consumers) may oversimplify the issue as well in the context of GM labeling. Unlike Akerlof's (1970) classic problem, here the relevant information of interest to the consumer (i.e., the non-GM nature of the superior product) needs to be "produced" (by costly segregation, identity preservation, and systematic testing). Thus, in this context it is critical to distinguish between the "information" that needs to be created to supply consumers with a meaningful choice and the actual information disclosed by a label.

Segregation and identity preservation systems are sometimes held to mean different things, the latter entailing a higher degree of traceability for instance (e.g., Smyth and Phillips 2002). Here, however, there is no point in separating these concepts, and thus we will think of a segregation and identity preservation (SIP) system as the set of production, handling, processing, and

---

[4] The StarLink case is a good example. Traces of an unapproved (for human consumption) GM maize were found in taco shells sold in U.S. grocery stores by tests carried out by an independent lab on behalf of a coalition of consumer and environmental organizations, prompting Kraft Foods to recall 2.5 million boxes of Taco Bell brand taco shells. See Taylor and Tick (2003) for more details and a complete chronology of the StarLink case, and for a discussion of related regulatory issues.

distribution practices that maintain the purity of the good under consideration. To ensure the non-GM nature of the product, various costly activities
need to be undertaken at various stages of the vertical production chain,
from "farm to fork." Such activities may involve the need for seed of an appropriate degree of purity, isolation measures at the growing stage to prevent
cross-pollination, clean and/or dedicated equipment for planting and harvesting, clean and/or dedicated storage and transportation facilities, segregated
handling and processing facilities, and so on. In addition, of course, record
keeping and multiple testing at various stages may be necessary (Bullock
and Desquilbet 2002, Sundstrom et al. 2002).

The nature of such SIP activities has direct implications for the working
of a GM labeling system. In some sense it is true that, because of the binary
nature of the information (a product either is or is not GM), both positive
and negative labels, when present, should convey the same information to
consumers. But one cannot ignore the SIP costs that are necessary for the
label "non-GM," or the absence of a label "GM," to be meaningful or credible in this setting. In particular, it is clear that simply requiring that GM products identify themselves as such by an EU-type mandatory labeling requirement does not diminish the costly segregation activities that are required by
the suppliers of the (unlabeled) non-GM product.

Golan, Kuchler, and Mitchell (2001) conclude that the potential of GM
labeling for the purpose of addressing problems of missing or asymmetric
information is limited. The question of the appropriate type of GM labeling
was also discussed by Runge and Jackson (2000) in the context of a choice
between "positive labeling" (e.g., this product contains GM organisms) and
"negative labeling" (e.g., this product does not contain GM organisms). Crespi
and Marette (2003) contrast some of the implications of voluntary and mandatory labeling regimes but neglect to consider explicitly SIP costs. As emphasized by Lapan and Moschini (2001), it is critical to understand the in-
centive-compatibility requirements of alternative labeling systems. The first
generation of GM products essentially confers no attribute that is directly
desirable from the consumers' point of view. Hence a positive labeling for
first-generation GM products would need to be mandatory, whereas a negative labeling system in this setting could be voluntary. But either labeling
system, to be credible, must impose SIP costs on the non-GM good. An
explicit two-country trade model with costly SIP and GM labeling is developed by Lapan and Moschini (2001, 2004), where mandatory GM labeling is
taken as adding costs to GM producers without detracting from SIP costs
incurred by non-GM suppliers. Fulton and Giannakas (2004) analyze labeling and no-labeling regimes, with IP costs impacting the marketing margin
for the non-GM product (but with no differentiation between voluntary and
mandatory labeling systems).

2.2. The New GM Labeling and Traceability Rules in the European Union

Whereas many countries are introducing GM labeling requirements (Carter and Gruère 2003), the sweeping nature of EU rules deserves special attention. Since April 2004, GM food and feed in the EU have been regulated under Regulation (EC) No. 1829/2003 "on genetically modified food and feed." This framework provides for a single EU procedure for the authorization of all food and feed derived from GM products and of GM products themselves. Furthermore, since April 2004, products consisting of or containing GM organisms and food products obtained from GM organisms have been also subject to traceability and labeling requirements, as established in Regulation (EC) 1830/2003 "concerning the traceability and labeling of genetically modified organisms and the traceability of food and feed products produced from genetically modified organisms and amending Directive 2001/18/EC" (European Union 2004).

The EU "mandatory labeling" of GM products specifically requires that all pre-packaged products consisting of or containing (authorized) GM material and food products produced from GM products must carry a label stating that "This product contains genetically modified organisms" or "This product contains genetically modified [name of organism(s)]." In the case of non-prepackaged products (such as food offered by restaurants), these words must appear with the display of the product. GM foods must be labeled regardless of whether or not DNA or proteins derived from genetic modification are contained in the final product, and thus the GM labeling requirement also pertains to highly refined products (e.g., vegetable oil). The same labeling rules apply to animal feed, including any compound feed that contains GM products (e.g., soybeans or maize) or that is derived from GM products (e.g., corn gluten feed).

The mandate of "traceability" states that all persons who place a GM product on the market or receive a GM product placed on the market within the EU must be able to identify their supplier and the companies to which the products have been supplied. Operators handling GM product must transmit in writing to those receiving the product information to the effect that the product in question is of GM origin, and the unique identifier(s) assigned to those GM products. Operators must hold the information for a period of five years from each transaction and be able to identify the operator by whom and to whom the products have been made available. The regulation covers all GM products that have received EU authorization for their placing on the market, including previously authorized GM product transacted in bulk quantities (e.g., soybean and maize).

Exemption from the requirement of GM labeling and traceability includes products obtained from animals fed with genetically modified feed (e.g., meat, milk, or eggs). Conventional products are also not subject to traceability and labeling. Conventional products that are accidentally contaminated by GM products must carry the GM label only if the (authorized) GMO content exceeds 0.9 percent, provided the presence of this material is adventitious or technically unavoidable. In this case, operators must be able to demonstrate that they have taken adequate measures to avoid the presence of GM material.[5]

To summarize, it seems clear that the new EU rules impose substantial costs on the suppliers of GM products. In order to fulfill the traceability requirements, for instance, each operator must have in place an information system capable of documenting for public authorities, on demand, each transaction that took place for the last five years. For example, a company selling GM seed would have to inform buyers that the seed is genetically modified and provide more information so the specific GMO can be precisely identified. The company is also obliged to keep a register of business operators who have bought the seed. The farmer would have to inform any purchaser of the harvest that the product is GM and keep a register of operators to whom he has made the harvest available. Downstream handlers and processors also need to undertake similar steps as they carry out market transactions that involve GM products. As noted by Buckwell, Brookes, and Bradley (1999), a critical element determining the cost of SIP activities is the purity threshold level that is sought; the 0.9 percent level prescribed by EU rules appears to be very strict.

The mandatory nature of the EU system is in sharp contrast with the regulatory approach pursued in the United States, where at most a voluntary GM labeling system can be envisioned (U.S. Food and Drug Administration 2001). What is unclear is whether, by mandating explicit disclosure of everything GM that goes through the system, the EU rules may in fact decrease somewhat the implementation of an SIP system for non-GM products. Many of the real costs of such an SIP system would seem to be unaffected. Thus, to a first approximation at least, we will construe the new rules as (i) increasing the cost of supplying GM products, and (ii) leaving unchanged the SIP costs of supplying non-GM products. Finally we will note that at this point it is unclear what sort of monitoring system will be in place to enforce the new EU system. That this may be a challenging task is apparent, for example, when one notes that the mandatory disclosure for highly refined products (e.g., vegetable oils) appears to be an open invitation to cheat.

---

[5] The presence of GM products that are not yet approved in the EU, but which have received a favorable scientific assessment, is tolerated up to the stricter threshold of 0.5 percent (marketing of products with more than 0.5 percent of such material is prohibited).

3. THE MODEL

The simple model we develop here captures the main economic elements of interest and, in fact, can accommodate the features of both first-generation and second-generation GM innovations. The pre-innovation situation is characterized by a conventional product that is supplied competitively. We simplify the analysis substantially here by considering a constant-returns-to-scale industry. Whereas this assumption may be consistent with an individual (small) agricultural industry, it clearly cannot apply to the agricultural sector as a whole (because of the inelastic supply of land, for example), so a generalization of the setup to an increasing-cost industry (as in Lapan and Moschini 2004, for example) is desirable. But with our simplifying assumption, the pre-innovation conventional product is assumed to be produced with a constant unit production cost equal to c. This conventional product also has a given quality, and without loss of generality we normalize that quality to equal unity.

In this framework, a GM innovation can work in two directions: it can increase efficiency by reducing the unit production cost c and/or it can increase the quality level of the product. The first type of efficiency-enhancing innovations characterizes the so-called first-generation GM products, which embody agronomic traits such as herbicide-resistant soybeans and cotton, and Bt maize and cotton (for example).[6] An increase in the quality of the product, on the other hand, is what so-called second-generation GM products are attempting to do (Pew Initiative on Food and Biotechnology 2001). For such products the attribute contributed by GM innovation is of direct interest to the user of the product, such as improved nutritional content (e.g., increased vitamin content, as in the widely publicized "golden rice").

Specifically, a given GM innovation (labeled by the subscript i) is modeled as decreasing unit cost from c to $c - a_i$ and increasing quality from 1 to $1 + b_i$. Of course, the polar cases of a pure first-generation GM product ($b_i = 0$) and a pure second-generation GM product ($a_i = 0$) are readily encompassed by this framework. The pre-innovation conventional product and the potential new GM product are illustrated in Figure 1.

3.1. Preferences

Consumers are assumed to have heterogeneous preferences with respect to the new product. Following Mussa and Rosen (1978), we represent individual preferences through a simplified vertical product differentiation model (see also Tirole 1988, chapter 7). Specifically, the individual consumer with

---

[6] Some of these innovations may be better thought of as increasing expected yields (e.g., Bt maize). But given our constant unit cost assumption, by duality an increase in yield is fully equivalent to a decrease in unit cost.

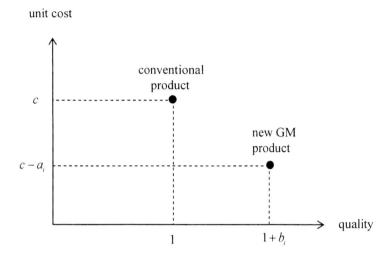

Figure 1. A GM Innovation Combining Efficiency Enhancement and Quality Improvement

preference parameter $\theta \in [0,1]$ obtains the following utility levels by consuming the two possible products:

$$V_0 = u - p_0 \text{ if the consumer buys one unit of conventional product,} \quad (1)$$

$$V_i = (1 + b_i)u - \theta\delta - p_i \text{ if the consumer buys one unit of GM product,} \quad (2)$$

where $u > 0$ is the utility of a unit of conventional product, $b_i \geq 0$ is the quality augmentation parameter of the GM product discussed earlier, and $\delta > 0$ is the maximum disutility of a unit of GM product (for the consumers with $\theta = 1$). The prices of non-GM and GM products are denoted p_0 and p_i, respectively. Consumer differences vis-à-vis GM product acceptance is captured by postulating that the parameter θ, in the population of consumers, is distributed on $[0,1]$ with a distribution function $F(\theta)$. A direct interpretation of equation (2) is that consumers all place the same value on the quality enhancement but have different disutility from the GM attribute. In any event, what formulation (2) maintains is that, *ceteris paribus*, consumers dislike the GM nature of the product irrespective of whether the GM innovation is of the first or second generation type.[7]

---

[7] Note that this formulation can capture the opposition to GM products that arises because of perceived shortcomings that are directly borne by the individual (such as, for example, the

3.2. Segregation and Identity Preservation

In the post-innovation situation, after a GM product has been introduced, SIP costs are necessary for the conventional non-GM product to be sold as such to consumers. Some SIP activities are also necessary for the quality-improved product to retain enough purity of its valuable character (Bender et al. 1999). We model such costs by a constant unit segregation cost undertaken by the non-GM product (s_i) and by the GM-product (σ_i). Of course, when the GM product is a first-generation product, with $b_i = 0$, no segregation cost is required for the GM product (i.e., in that case $\sigma_i = 0$).

Furthermore, we wish to account for the additional costs of a system of mandatory GM labeling and GM traceability such as the one implemented by the European Union. This additional burden is represented by the unit cost t_i that must be incurred by the GM product (as in Lapan and Moschini 2004). Because the suppliers of a pure first-generation product (with $b_i = 0$) have no incentive to do any segregation at all, for such a producer the EU regulations can be seen as adding a new additional cost. For the suppliers of a second-generation good, on the other hand, some of the activities required by the EU rules are likely to be undertaken voluntarily as part of the effort to capture the additional value of the innovation.[8]

3.3. Innovating Firm

Assuming that the GM product is fully protected by appropriate intellectual property rights (a patent, for example), the innovator has a temporary monopoly that allows it to profit from the innovation (Moschini and Lapan 1997). Because we have assumed a constant-returns-to-scale agricultural industry, it is not necessary that we explicitly model the innovation adoption by farmers. Instead, we can think of the innovator as producing the GM product directly and selling it to final users.

The demand for the innovation can be derived from the preference structure that was postulated. To this end, we postulate that the individual preference parameter θ, in the population of consumers, is distributed on [0,1] with a distribution function $F(\theta)$. The distribution function $F(\theta)$ is assumed to be strictly increasing and twice differentiable. We also assume that u is large enough to have a "covered market" outcome such that all consumers buy one

---

risk of an adverse health effect). But in this setting what consumers may also care about could be a "public good"—e.g., the environmental implications of GM products. Arguably, such concerns would not be reflected in the willingness to pay displayed by private consumption decisions.

[8] Thus we may want to think of such costs as being related to the quality of the innovation, i.e., $t_i = t_i(b_i) \geq 0$, with $t_i'(b_i) \leq 0$, but in this chapter we will not pursue this hypothesis further.

unit of the good (either conventional or GM product). From our consumer preference specification, an individual with preference parameter θ will buy the GM product if and only if $V_0 \leq V_i$, which requires that

$$u - p_0 \leq u(1 + b_i) - \theta\delta - p_i \quad \Leftrightarrow \quad \theta \leq \frac{ub_i - p_i + p_0}{\delta} \equiv \hat{\theta}_i. \tag{3}$$

At given prices p_i and p_0, the quantity of GM product sold on the market is $Q_i = NF(\hat{\theta}_i)$, where N is the market size (e.g., the number of consumers). Without loss of generality, we can normalize the market size and put $N = 1$. Given that, the profit of the GM innovator is

$$\pi_i \equiv F(\hat{\theta}_i) \cdot \left(p_i - (c - a_i) - \sigma_i - t_i \right). \tag{4}$$

Noting that from (3) we can write $\hat{\theta}_i \delta \equiv ub_i - p_i + p_0$, and putting $p_0 = c + s_i$ as dictated by the assumed competitive conditions for the farm sector, the innovator's profit maximization problem can be stated as

$$\max_{\hat{\theta}_i} \quad \pi_i = F(\hat{\theta}_i)\left(r_i - \delta\hat{\theta}_i \right), \tag{5}$$

where we have used the definition $r_i \equiv ub_i + a_i + s_i - \sigma_i - t_i$.

4. RESULTS

The simple model outlined in the foregoing permits us to derive some important conclusions. The optimality condition for the program in (5), for an interior solution $0 < \hat{\theta}_i^* < 1$, is[9]

$$f(\hat{\theta}_i^*) \cdot \left(r_i - \delta\hat{\theta}_i^* \right) - \delta F(\hat{\theta}_i^*) = 0, \tag{6}$$

where $f(\theta) \equiv F'(\theta)$ denotes the density function of the distribution of consumer types. The second-order sufficient condition for an interior solution is

$$-\delta \cdot \left[1 + \frac{d}{d\theta}\left(\frac{F(\hat{\theta}_i^*)}{f(\hat{\theta}_i^*)} \right) \right] < 0. \tag{7}$$

---

[9] An interior solution is guaranteed if $0 < r_i < (1 + 1/f(1))\delta$.

Thus, a sufficient condition that guarantees that (7) holds is

$$\frac{d}{d\theta}\left(\frac{F(\theta)}{f(\theta)}\right) \geq 0. \tag{8}$$

This condition that the ratio $F(\theta)/f(\theta)$ is nondecreasing, sometimes referred to as the "monotone hazard rate property," is often invoked in the mechanism design literature (e.g., Laffont and Tirole 1993, chapter 1) and it is satisfied if the distribution function $F(\theta)$ is log-concave, a property enjoyed by most commonly used distributions (such as the uniform, the exponential, and the normal). The condition in (8) will be assumed to hold from this point onward.

Given the optimality condition in (6), the maximized profit of the innovator is

$$\pi_i^* \equiv F(\hat{\theta}_i^*) \cdot \left(r_i - \delta\hat{\theta}_i^*\right) \ = \frac{\delta \cdot \left[F(\hat{\theta}_i^*)\right]^2}{f(\hat{\theta}_i^*)}. \tag{9}$$

From the optimality condition in equation (6), and given that the condition in (8) holds, it follows that

$$\frac{\partial\hat{\theta}_i^*}{\partial r_i} \geq 0 \quad \text{and} \quad \frac{\partial\hat{\theta}_i^*}{\partial\delta} \leq 0. \tag{10}$$

Recalling the definition $r_i \equiv ub_i + a_i + s_i - \sigma_i - t_i$, the comparative statics results in (10) immediately establish the behavior of the adoption rate with respect to the parameters of the problem. Furthermore, from (9), sign $(\partial\pi_i^*/\partial r_i) =$ sign $(\partial\hat{\theta}_i^*/\partial r)$ if the monotone hazard rate condition in (8) holds. Hence, we can conclude the following:

Result 1. Adoption of the GM product, and the profit of the innovator, are (i) an increasing function of the quality improvement b_i and of the efficiency gain a_i; (ii) a decreasing function of the consumer GM disutility parameter δ; (iii) inversely related to the (GM product) segregation cost σ_i and directly related to the (conventional product) segregation cost s_i; and (iv) inversely related to the "regulation cost" t_i.

This result clearly summarizes some of the main features of GM product innovation. Both quality improvements and efficiency gains can further GM product adoption and provide profit incentives for innovators. Segregation costs—unnecessary in the pre-innovation situation but critically necessary in the post-GM innovation case—play a significant and subtle role. In particular, segregation costs that have to be borne by GM producers discourage adoption, but the extent of the segregation costs that have to be incurred by the producers of the conventional product to supply non-GM product has a positive impact on GM crop adoption and innovators' profit.

4.1. Welfare

For a GM innovation characterized by the cost-decreasing parameter a_i and the quality-increasing parameter b_i, if W_0 represents the level of welfare prior to the innovation and W_i represents the welfare after the innovation, for a given adoption level $\hat{\theta}$ we have

$$W_0 = u - c \qquad (11)$$

$$W_i = (u - c - s_i)\left(1 - F(\hat{\theta})\right) + \int_0^{\hat{\theta}} (u(1 + b_i) - c + a_i - \sigma_i - t_i - \theta\delta)dF(\theta). \quad (12)$$

Evaluating the latter at the innovator's profit-maximizing solution $\hat{\theta} = \hat{\theta}_i^*$ we obtain

$$W_i^* = (u - c - s_i) + H(\hat{\theta}_i^*) + \pi_i^*, \qquad (13)$$

where π_i^* is given by equation (9) and $H(\hat{\theta}_i^*) \equiv \int_0^{\hat{\theta}_i^*} \delta(\hat{\theta}_i^* - \theta)dF(\theta)$. From (11) to (13) we therefore obtain

$$W_i^* - W_0 = -s_i + H(\hat{\theta}_i^*) + \pi_i^*. \qquad (14)$$

Hence, we have the following:

Result 2. For negligible segregation cost for the non-GM product (i.e., $s_i \rightarrow 0$) the GM innovation is welfare-increasing. For a given disutility parameter $\delta > 0$, however, the need for segregation costs may entail that the GM innovation decreases welfare.

The first part of Result 2 follows from the observation that $H(\hat{\theta}_i^*) > 0$ and $\pi_i^* > 0$ whenever $\hat{\theta}_i^* > 0$. To see the second part of Result 2 it suffices to observe the behavior of the welfare function for low equilibrium adoption rates: because $H(\hat{\theta}_i^*) \rightarrow 0$ and $\pi_i^* \rightarrow 0$ whenever $\hat{\theta}_i^* \rightarrow 0$, then $(W_i^* - W_0) \rightarrow -s_i < 0$

as $\hat{\theta}_i^* \to 0$. But it should be clear that welfare can be decreased by the innovation even for large adoption rates. Indeed, note that if s_i is large enough we
obtain the corner solution with $\hat{\theta}_i^* = 1$. Specifically, complete adoption obtains
when $r_i/\delta \geq 1 + 1/f(1)$. And with complete adoption the equilibrium welfare is

$$W_i^* = (u - c - s_i) + \pi_i^* + \int_0^1 \delta(1 - \theta)dF(\theta). \tag{15}$$

Given that in this corner solution case we have $\pi_i^* = r - \delta$, and recalling that
$r_i \equiv ub_i + a_i + s_i - \sigma_i - t_i$, the condition

$$\delta(1 + 1/f(1)) \leq (ub_i + a_i + s_i - \sigma_i - t_i) < s_i + \delta\int_0^1 \theta dF(\theta) \tag{16}$$

would ensure both a corner solution with complete adoption and the decreased
welfare result $W_i^* < W_0$.

Thus, Result 2 displays the conclusion that an efficiency-enhancing (or
quality-enhancing) innovation may turn out to be welfare-decreasing
because it brings about a novel market failure, a type of externality (i.e., the
need for hitherto unnecessary SIP activities for the pre-existing non-GM
product to be available as such to consumers).[10]

Moving on to analyze the impact of an EU-style mandatory labeling
regime, note that, because $\partial\hat{\theta}_i^*/\partial t_i < 0$ and $\partial\pi_i^*/\partial t_i < 0$, from (13) it is clear
that $\partial W_i^*/\partial t_i < 0$. Hence, we have the following:

Result 3. Taking for given that GM products are introduced, regulation that
increases the cost of GM product marketing but does not affect the SIP costs
for the non-GM product, such as the EU labeling and traceability requirements, reduces welfare.

Results 1 and 2 summarize conclusions that, in one form or another, have
appeared in various studies that have attempted an assessment of the economic implications of the introduction of GM products. Earlier studies documented sizeable efficiency gains from new GM crops (e.g., Moschini, Lapan,
and Sobolevsky 2000; Falck-Zepeda, Traxler, and Nelson 2000) but ignored
the critical element of this new technology discussed in the introduction:
consumer preferences and the inferior-substitute nature of first-generation
GM products. Once the "unintended" economic effects of GM crop innova-

---

[10] The "market failure" is not an externality in the usual sense because—to a first approximation at least—it is the presence of GM products, not the extent of their cultivation, which is
the problem (and thus it is essentially a nonconvexity of the aggregate production set).

tions are accounted for, the efficiency and welfare implications of first-generation GM products are ambiguous at best (e.g., Fulton and Giannakas 2004; Furtan, Gray, and Holzman 2003; Lapan and Moschini 2004; Sobolevsky, Moschini, and Lapan 2005). A version of Result 3 can be found in Lapan and Moschini (2004). None of these studies concerned the potential impact of second-generation GM products. The model that we have outlined provides a useful starting point in that direction.

4.2. Choice of Research Direction

Many a commentator has lamented the fact that input-trait GM products, which offered no direct benefit to the consumer, were the first and most visible output of the biotechnology industry. The (somewhat plausible but untested) presumption here is that, had output-trait GM product been marketed first, consumer acceptance would have been different. Whether or not that is true, there are technological reasons as to why the biotechnology industry went the way it did: input traits based on a single-gene transformation (such as Roundup Ready soybeans or Bt maize) are easier than the multiple-gene transformations often associated with quality improvements. But the question remains as to whether there are other explanations, and in particular whether the inherent market failure of GM innovation (the concomitant creation of the need for costly SIP) also had an effect.

To try to address this question in the context of our simple model, consider an innovator facing two possible innovations, a purely efficiency-enhancing innovation and a purely quality-enhancing innovation. Thus, $i = 1$ denotes a "first-generation" GM product, whereby $a_1 \equiv a > 0$ and $b_1 = 0$. Similarly, $i = 2$ denotes a "second-generation" GM product, whereby $a_2 = 0$ and $b_2 \equiv b > 0$. Figure 2 illustrates. Given this choice of research direction, we want to know what factors determine the choice of the innovators and whether the private choice of the innovating firm is consistent with the direction that maximizes social welfare.

Ex post, from the innovating firm's point of view, the first-generation (i.e., cost-reducing) innovation is more attractive if $\pi_1^* > \pi_2^*$. From equation (9) we have

$$\pi_1^* > \pi_2^* \quad \Leftrightarrow \quad \frac{\left[F(\hat{\theta}_1^*)\right]^2}{f(\hat{\theta}_1^*)} > \frac{\left[F(\hat{\theta}_2^*)\right]^2}{f(\hat{\theta}_2^*)}, \tag{17}$$

and thus, given that the monotone hazard rate condition in equation (8) holds, we can conclude that the innovation providing the highest profit to the innovator is the one that would attain the highest adoption rate (in the

unit cost

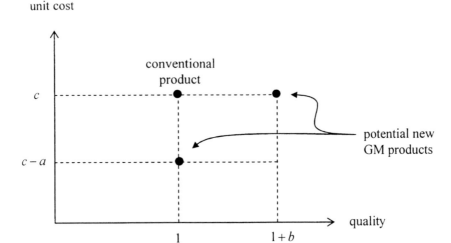

Figure 2. First-Generation and Second-Generation GM Products

monopoly pricing equilibrium); that is, $\pi_1^* > \pi_2^* \Leftrightarrow \hat{\theta}_1^* > \hat{\theta}_2^*$. Hence, given the choice between the two innovations, the condition for $\pi_1^* > \pi_2^*$ reduces to

$$(18)$$

$$a + s_1 - \sigma_1 - t_1 > ub + s_2 - \sigma_2 - t_2 \Leftrightarrow a - ub > (s_2 - s_1) - (\sigma_2 - \sigma_1) + (t_1 - t_2).$$

Without externality effects, i.e., with $s_i = \sigma_i = t_i = 0$, the choice of research direction depends only on the magnitude of cost reduction relative to the quality enhancement, so that $\pi_1^* > \pi_2^*$ iff $a > ub$. With external effects, however, it is clear that the choice of the innovator is affected by the presence of segregation costs.

To consider the welfare-maximizing research direction, conditional on the innovation being provided by an innovator-monopolist, the condition is

$$W_1^* > W_2^* \quad \Leftrightarrow \quad \left(\pi_1^* - \pi_2^*\right) + \left(H(\hat{\theta}_1^*) - H(\hat{\theta}_2^*)\right) > \left(s_1 - s_2\right). \qquad (19)$$

From equation (19) we can conclude the following:

Result 4. With $s_1 = s_2$ the social ordering of the research directions is the same as the private ordering (based on the innovator's profit functions). With $s_1 \neq s_2$, however, that need not be the case. Specifically, with $s_1 > s_2$ the rule for the privately chosen research direction is tilted in favor of efficiency-enhancing innovations.

To show Result 4, recall the definition

$$H(\hat{\theta}_i^*) \equiv \int_0^{\hat{\theta}_i^*} \delta\left(\hat{\theta}_i^* - \theta\right) dF(\theta),$$

so that it follows that sign $(H(\hat{\theta}_1^*) - H(\hat{\theta}_2^*)) = $ sign $\hat{\theta}_1^* - \hat{\theta}_2^*$. Also, because we have shown that $\pi_1^* > \pi_2^* \Leftrightarrow \hat{\theta}_1^* > \hat{\theta}_2^*$, then sign $(\hat{\theta}_1^* - \hat{\theta}_2^*) = $ sign $(\pi_1^* - \pi_2^*)$. Hence, with $s_1 = s_2$ the social ordering of the research directions is the same as the private ordering (based on the innovator's profit functions). The case $s_1 > s_2$ is of interest under the presumption that when the GM product carries out its own SIP (the incentive for which exists for quality-improving innovations), it is less costly for the non-GM product to achieve a given SIP level.

5. CONCLUSION

GM product innovations clearly increase the efficiency of production and also have the potential to offer new and/or quality-improved products to the consumer. But because some consumers are apparently opposed to the GM technology of the new products, a portion of the market has a preference for the pre-existing conventional products. Regulations aimed at ensuring the consumers' "right to know" about the GM nature of the food consumed, so as to preserve their ability to choose, require some form of GM labeling. In particular, for example, the new 2004 EU regulations have introduced mandatory labeling and traceability of all GM food and food ingredients and of GM feed. To fulfill such requirements requires costly steps to be undertaken by the suppliers of GM product. Somewhat paradoxically, however, it does not seem that the new regulations make it easier to supply non-GM products to consumers, because costly segregation and identity preservation (SIP) activities are still required.

In this chapter we have reviewed some of the significant issues concerning the effects of GM product innovation, with an emphasis on the issues of GM regulations that focus on labeling. We have developed a simple model that allows a characterization of the main features of both first- and second-generation GM product innovation, and we have used that model to offer an interpretative review of some of the existing studies that have dealt with the assessment of the economic impacts of GM product innovation. We can conclude that introduction of GM products entails the real possibility of a welfare-decreasing innovation because of the externality-like effects that it has on the agricultural and food system's ability to deliver non-GM products.

But because the costly SIP activities need to be undertaken by the suppliers of the superior (non-GM) products, it is also apparent that EU-style mandatory labeling of GM products cannot help (taking for given that GM products are introduced), and indeed it is itself a wasteful regulation in our model. We have also shown that the existence and nature of SIP costs may have a role in the choice of research directions by GM innovating firms.

REFERENCES

Akerlof, G.A. 1970. "The Market for Lemons: Quality Uncertainty and the Market Mechanism." *Quarterly Journal of Economics* 84(3): 488–500.

Bender, K., L. Hill, B. Wenzel, and R. Hornbaker. 1999. "Alternative Market Channels for Specialty Corn and Soybeans." Report No. AE-4726, Department of Agricultural Economics, University of Illinois.

Buckwell, A., G. Brookes, and D. Bradley. 1999. "Economics of IP for Genetically Modified Crops." Report for Food Biotechnology Communication Initiative (FBCI), London.

Bullock, D., and M. Desquilbet. 2002. "The Economics of Non-GMO Segregation and Identity Preservation." *Food Policy* 27(1): 81–99.

Carter, C., and G.P. Gruère. 2003. "International Approaches to the Labeling of Genetically Modified Foods." *Choices* (2nd quarter): 1–4.

Crespi, J.M., and S. Marette. 2003. " 'Does Contain' vs. 'Does Not Contain': Does It Matter Which GMO Label Is Used?" *European Journal of Law and Economics* 16(3): 327–344.

European Union. 2004. "Question and Answers on the Regulation of GMOs in the EU." MEMO/04/85, European Union, Brussels (April 15).

Falck-Zepeda, J.B., G. Traxler, and R.G. Nelson. 2000. "Surplus Distribution from the Introduction of a Biotechnology Innovation." *American Journal of Agricultural Economics* 82(2): 360–369.

Fulton, M., and K. Giannakas. 2004. "Inserting GM Products into the Food Chain: The Market and Welfare Effects of Different Labeling and Regulatory Regimes," *American Journal of Agricultural Economics* 86(1): 42–60.

Furtan, W.H., R.S. Gray, and J.J. Holzman. 2003. "The Optimal Time to License a Biotech 'Lemon'." *Contemporary Economic Policy* 21(4): 433–444.

Gaskell. G., N. Allum, and S. Stares. 2003. "Europeans and Biotechnology in 2002." *Eurobarometer* 58.0. Report to the EC Directorate General for Research (March).

Golan, E., F. Kuchler, and L. Mitchell. 2001. "Economics of Food Labeling." *Journal of Consumer Policy* 24(2): 117–184.

Grossman, S. 1981. "The Informational Role of Warranties and Private Disclosure about Product Quality." *Journal of Law and Economics* 24(3): 461–483.

Huffman, W.E., J.F. Shogren, M. Rousu, and A. Tegene. 2003. "Consumer Willingness to Pay for Genetically Modified Food Labels in a Market with Diverse Information: Evidence from Experimental Auctions." *Journal of Agricultural and Resource Economics* 28(3): 481–502.

James, C. 2005. "Preview: Global Status of Commercialized Biotech/GM Crops: 2004." 2005. ISAAA Brief No. 32-2004, International Service for the Acquisition of Agri-Biotech

Applications, Ithaca, NY.

Laffont, J.-J., and J. Tirole. 1993. *A Theory of Incentives in Procurement and Regulation.* Cambridge, MA: MIT Press.

Lapan, H., and G. Moschini. 2001. "GMO Labeling and Trade: Consumer Protection or (Just) Protectionism?" Paper presented at the International Agricultural Trade Research Consortium (IATRC) conference entitled "Globalization, Biotechnology and Trade," Tucson, AZ (December).

_____. 2004. "Innovation and Trade with Endogenous Market Failure: The Case of Genetically Modified Products." *American Journal of Agricultural Economics* 86(3): 634–648.

Moschini, G., and H. Lapan. 1997. "Intellectual Property Rights and the Welfare Effects of Agricultural R&D." *American Journal of Agricultural Economics* 79(4): 1229–1242.

Moschini, G., H. Lapan, and A. Sobolevsky. 2000. "Roundup Ready Soybeans and Welfare Effects in the Soybean Complex." *Agribusiness* 16(1): 33–55.

Mussa, M., and S. Rosen. 1978. "Monopoly and Product Quality." *Journal of Economics Theory* 18(2): 301–317.

Noussair, C., S. Robin, and B. Ruffieux. 2004. "Do Consumers Really Refuse to Buy Genetically Modified Food?" *Economic Journal* 114(492): 102–120.

Pew Initiative on Food and Biotechnology. 2001. *Harvest on the Horizon: Future Uses of Agricultural Biotechnology.* Pew Initiative on Food and Biotechnology, Washington, D.C. Available online at http://pewagbiotech.org/research/harvest/harvest.pdf (accessed May 26, 2005).

Runge, C.F., and L.A. Jackson. 2000. "Labelling, Trade and Genetically Modified Organisms: A Proposed Solution." *Journal of World Trade* 34(1): 111–122.

Sheldon, I. 2002. "Regulation of Biotechnology: Will We Ever 'Freely' Trade GMOs?" *European Review of Agricultural Economics* 29(1): 155–176.

Smyth, S., and P.W.B. Phillips. 2002. "Product Differentiation Alternatives: Identity Preservation, Segregation and Traceability." *AgBioForum* 5(2): 30–42.

Sobolevsky, A., G. Moschini, and H. Lapan. 2005. "Genetically Modified Crops and Product Differentiation: Trade and Welfare Effects in the Soybean Complex." *American Journal of Agricultural Economics* 87(3): 621–644.

Sundstrom, F.J., J. Williams, A. Van Deynze, and K.J. Bradford. 2002. "Identity Preservation of Agricultural Commodities." Publication No. 8077, Agricultural Biotechnology in California Series, Davis, CA.

Taylor, M.R., and J.S. Tick. 2003. "Post-Market Oversight of Biotech Foods: Is the System Prepared?" Report commissioned by the Pew Initiative on Food and Biotechnology, Washington, D.C., and prepared by Resources for the Future, Washington, D.C.

Teisl, M.F., and J.A. Caswell. 2003. "Information Policy and Genetically Modified Food: Weighing the Benefits and Costs." Working Paper No. 2003-1, Department of Resource Economics, University of Massachusetts at Amherst.

Tirole, J. 1988. *The Theory of Industrial Organization.* Cambridge, MA: The MIT Press.

U.S. Food and Drug Administration. 2001. "Voluntary Labeling Indicating Whether Foods Have or Have Not Been Developed Using Bioengineering." *Draft Guidance for Industry* (January).

Chapter 14

REGULATION OF TECHNOLOGY IN THE CONTEXT OF RISK GENERATION

Erik Lichtenberg
University of Maryland

Abstract: This chapter uses a generic model of the risk-generating process and an approach to accommodating uncertainty about risk to draw some inferences about regulating transgenic crops. The analysis suggests that regulation should concentrate on reducing uncertainty about environmental impacts of transgenic crops. Thus, it may be cost effective for the United States to de-emphasize restrictions on planting but expand post-commercialization monitoring. The analysis also indicates that there are tractable ways of incorporating firms' and consumers' reactions to regulation into models of the risk-generation process. The main impediment to doing so is a lack of empirical models, which suggests a need for empirical work aimed at producing results that can be extrapolated to biotechnology regulation. The model can be extended easily to accommodate multiple, heterogeneous sites.

Key words: biotechnology, transgenic crops, uncertainty, precautionary principle, environmental risk, environmental regulation, food safety

1. INTRODUCTION

The specific environmental and human health problems that have raised red flags about agricultural biotechnology are in some ways novel in environmental regulation: gene flow, insect resistance, and the potential introduction of allergens into the food supply have not been traditional concerns of environmental policy. Even so, these problems have some fundamental commonalities with other technologies thought to pose environmental risk, most notably (i) a high degree of uncertainty about the magnitude and extent of the risks they pose and (ii) a distinct aversion to that uncertainty on the part of regulators and the public at large. This chapter derives some lessons for agricultural biotechnology regulation from an approach designed for analyzing policies for addressing environmental risks in the presence of uncertainty utilizing the kinds of quantitative risk assessment procedures typically used in regulatory contexts.

The causes of uncertainty about the risks posed by transgenic crops (and other agricultural biotechnology products) are common to most environmental risks: limits on scientific understanding, dependence on a complex of natural conditions that can be highly variable, and susceptibility to random events. Take gene flow, for instance: while it is known in principle that genes from crops can migrate into other plant species and that escaped domesticated organisms can cross-breed with other wild organisms, there is a great deal of uncertainty about the conditions under which these phenomena actually occur and about the extent to which they occur under various conditions. Gene flow from conventionally bred crops has rarely, if ever, been studied extensively, for one thing. For another, like most ecological phenomena, they are heavily influenced both by variability in natural conditions (for example, the prevalence of various weed species in and around fields, which depends on climate, soils, location, and other factors) and by stochastic factors (e.g., weather events), which makes it difficult to generalize about them with much confidence (see, for example, National Research Council 2000, 2002).

The sheer novelty of agricultural biotechnology products is also often cited as a critical source of policy-relevant uncertainty. For example, the two recent National Research Council reports on the regulation of transgenic crops cited above concluded that there is no difference in the types and magnitudes of environmental and human health risks posed by crops developed using genetic engineering and those derived from conventional breeding practices. Nevertheless, both reports advocated regulating transgenic crops differently than conventionally bred crops (which are largely unregulated) on the grounds that transgenic crops are less familiar and thus deserve closer scrutiny.

These National Research Council reports reflected a common policy response to uncertainty about technological risks: an aversion to uncertainty and a corresponding desire for policies that are precautionary in nature. Like many others, both reports cite aversion to uncertainty in endorsing a precautionary approach to regulation of transgenic crops. Regulatory agencies may favor such an approach from bureaucratic caution. A precautionary approach also helps maintain favorable public perception of how well an agency is doing its job. In particular, agencies may design policies that limit the occurrence of adverse outcomes large enough to cause scandals and thus undermine public confidence (consider for example the impact of revelations about the deaths of patients using Vioxx and Celebrex on public confidence in the U.S. Food and Drug Administration's regulation of pharmaceuticals). Regulators may also be responding to demand from the general public, which tends to favor a precautionary approach from a desire for assurance of safety. Studies of the public's attitudes about pesticide residues on foods, for instance, suggest that a large majority in the United States simply want assurance that their food has been determined to be safe by a trusted entity and have little or no willingness to pay for reductions in residues below levels that have been determined to be safe by a trusted entity (for a brief survey, see Roberts, Buzby, and Lichtenberg 2002).

Much of the environmental legislation in force in the United States (and elsewhere) similarly mandates a specific form of precautionary approach that focuses on prevention of adverse environmental and human health effects and stipulates that the degree of protection provided be attained with an adequate margin of safety. Put another way, most of this environmental regulation involves choosing maximum acceptable levels of adverse outcomes; because the realizations of those outcomes are uncertain, the best that regulators can do is limit the frequency with which those maximum acceptable levels are exceeded. In what follows, I derive implications for biotechnology regulation from a formal model corresponding to this precautionary approach and utilizing a method of quantitative risk analysis commonly used in regulatory applications that was introduced into the literature by Lichtenberg and Zilberman (1988) and elaborated further by Lichtenberg, Zilberman, and Bogen (1989), Zivin and Zilberman (2002), and Lichtenberg and Penn (2003).

2. A MODEL OF RISK GENERATION AND RISK REGULATION

The factors that generate environmental risks can be divided into three general categories. The first category is the rate at which a potentially damaging substance or organism is introduced into the environment, denoted I. The second category is the rate at which the substance or organism disseminates through the environment, denoted F; this category is often referred to as environmental fate and transport. The third category is the susceptibility of humans or other organisms to the substance, denoted S. In the case of gene flow, for instance, the rate of introduction is determined by the land allocated to growing transgenic crops; environmental fate is determined by the movement of pollen and/or vectors of genetic material like viruses; and susceptibility is determined by the rate at which genetic material from the crop is incorporated into the genomes of non-crop plant species present in or around fields of transgenic crops. In the case of allergenicity (or other human health effects), the rate of introduction is determined by the volume of transgenic crops grown or imported; environmental fate is determined by the volume of transgenic crops entering the food supply (either intentionally or from cross-contamination in storage, transport, and processing facilities); and susceptibility is determined by human sensitivity to transgenetic material in foods made from transgenic crops.

The incremental risk created by potentially damaging substances is often modeled as the product of these three factors, $R = IFS$. This multiplicative model of risk generation is a reasonable approximation when the incremental risk is relatively small and when there is significant background risk, as is often the case with environmental risks (see for example Van Ryzin 1980, Crouch and Wilson 1981, Crump and Howe 1984). Each of these three factors is subject to uncertainty, so that each is a random variable. The resulting environment risk, R, is thus random as well.

We can categorize policies in the same way that we categorize risk-generating factors. Some policies involve restrictions on the introduction of potentially damaging material into the environment, e.g., restrictions on planting or importing transgenic crops (including, of course, complete bans). Other policies attempt to limit the dissemination of the potentially damaging material throughout the environment. Measures that might limit gene flow include requiring that production occur only in confined facilities or requiring setbacks for fields of transgenic crops. Measures that might limit dissemination of allergens include segregation of storage, transport, and processing facilities. Policies that limit susceptibility include limiting plantings to areas where transgenic plants have few wild relatives and labeling requirements that allow people with high sensitivity to allergens to take protective measures. Following Lichtenberg and Zilberman (1988), let x_i, x_f, and x_s denote the social cost of implementing policies limiting the introduction, dissemination, and susceptibility of potentially damaging substances, respectively.

Most environmental legislation in the United States specifies a precautionary approach in the face of uncertainty about environmental risks derived from sanitary engineering by requiring that regulators act to provide sufficient protection of public health (or the environment) with an adequate margin of safety. Lichtenberg and Zilberman (1988) formalize this approach as an application of Kataoka's (1963) safety-fixed model. The sufficient degree of protection is the nominal regulatory standard M. The margin of safety is the probability that the standard is achieved, α. The goal of regulation can thus be expressed as a constraint ensuring that the probability of violations of the nominal standard does not exceed the margin of safety,

$$Pr\{R \geq M\} = Pr\{I(x_i)F(x_f)S(x_s) \geq M\} \leq 1 - \alpha. \qquad (1)$$

A logarithmic transformation of this constraint can be used to simplify the analysis without any loss of generality. Letting lower case letters represent natural logarithms, we can write the safety-fixed constraint (1) as

$$Pr\{r(x_i,x_f,x_s) = i(x_i) + f(x_f) + s(x_s) \geq m\} \leq 1 - \alpha. \qquad (2)$$

If the three risk-generating factors I, F, and S are lognormal, this constraint can be written

$$r(\alpha) \equiv \mu_i(x_i) + \mu_f(x_f) + \mu_s(x_s) + Z(\alpha)[\sigma_i^2(x_i) + \sigma_f^2(x_f) + \sigma_s^2(x_s) \qquad (3)$$

$$+ \rho_{if}\sigma_i(x_i)\sigma_f(x_f) + \rho_{is}\sigma_i(x_i)\sigma_s(x_s) + \rho_{fs}\sigma_f(x_f)\sigma_s(x_s)]^{1/2} \leq m,$$

where $Z(\alpha)$ is the standardized value of a normal random variable exceeded with probability $1 - \alpha$ and μ_k and σ_k represent the expected value and standard deviation of risk-generating factor k. Similar expressions of the level of risk exceeded with probability $1 - \alpha$ as a weighted sum of the mean and standard deviation of risk can also be derived for distributions other than the normal.

This approach to adjusting for uncertainty comes from classical statistics rather than the Bayesian approaches typically used in economics. The weight on the standard deviation of risk, $Z(\alpha)$, is the upper limit of a one-tailed α percent confidence interval for a standard normal random variable. The safety-fixed approach involves choosing policies to limit the probability of Type I error, in this case, a false determination of risk as below the maximum acceptable level m when it is in fact greater. In keeping with apparent regulatory and public preferences, it focuses on unacceptably bad outcomes. (In a sense, it is an extreme form of downside risk aversion.)

This approach produces risk estimates that are adjusted for uncertainty in a statistically meaningful way. Regulatory agencies are often criticized for adjusting for uncertainty using arbitrary safety factors applied to individual risk-generating factors. For example, the Food Quality Protection Act amended the Federal Insecticide, Fungicide, and Rodenticide Act to require the use of specific arbitrary safety factors for exposure and toxicological responses in analyzing dietary risks from exposure to pesticides (including transgenic plants) for children and other groups thought to have an especially high degree of susceptibility. The meaning of the resulting risk "estimate" is not clear: it is not really an estimate of risk, and the adjustments for uncertainty have no rigorous statistical meaning. As a result, quantitative assessments of different risk are not comparable to each other, making it difficult to set regulatory priorities on an objective basis. The arbitrary nature of these uncertainty adjustments also renders the results of quantitative risk assessments non-comparable to estimates of risk derived from epidemiological or ecological data as well, making verification difficult. A widely advocated alternative is to use probabilistic risk assessments along the lines of the simple one discussed above, which allow the use of safety factors applied to overall risk that have a clear statistical meaning, making it possible to conduct reality checks of risk assessments against real-world data as well as making them comparable across risks.

3. A NOTE ON RISKS OF TRANSGENIC CROPS

The risk-generation model presented above does not require the contribution of every risk-generating factor to risk on average to be positive, i.e., $\mu_k > 0$ $\forall$ k. In some cases, for instance, a new technology might reduce risk on average while increasing uncertainty about risk. If the increase in uncertainty about risk is sufficiently great, introducing that technology will heighten the uncertainty-adjusted level of risk attained with an agreed-on margin of safety, in which case it would be considered risk-increasing in the eyes of policymakers (and the public).

The latter is quite possibly an accurate characterization of transgenic crops like those expressing Bt toxins and some herbicide-tolerant varieties. Bt crops express a toxin that acts only on certain species of insects and is delivered only to organisms actually feeding on the plant. As a result, these crops have markedly less extensive spillover effects on non-target organisms than the chemical pesticides they replace. Similarly, a selling point for herbicide-tolerant varieties like Monsanto's is that they allow farmers to use herbicides that are shorter lived and less mobile in the environment (and thus pose less environmental hazard) than the alternatives.

More generally, varieties produced by genetic engineering methods are in some respects less risky than those developed via conventional breeding methods. Genetic engineering involves introducing a small amount of carefully selected genetic material into a crop. Its unintended effects on the crop itself derive from interactions of that genetic material (including promoters that increase expression of the products it codes for) with pre-existing genes. By contrast, conventional breeding involves moving around large chunks of genetic material whose exact functions are not completely known. As a result, conventional breeding may have unintended effects that turn out to be severe. Well-documented examples include a number of conventionally bred potato varieties that turned out to be toxic to humans, and corn varieties that turned out to be highly susceptible to disease (National Research Council 2000).

It remains generally accepted that transgenic crops may increase certain types of environmental risk relative to that posed by current varieties and production methods. But, as noted above, the strongest recurring argument for subjecting transgenic crops to extra regulatory scrutiny has focused on uncertainty about those risks rather than on expected risk.

4. COST-EFFECTIVE RISK REGULATION

Consider the choice of regulatory instruments x_i, x_f, and x_s to minimize the social cost $(x_i + x_f + x_s)$ of meeting the safety-fixed constraint (3). The necessary condition for any regulatory instrument k in use $(x_k > 0)$ can be written as

$$\frac{\partial r(\alpha)}{\partial x_k} = \frac{\partial \mu_k}{\partial x_k} + \frac{Z(\alpha)\Sigma}{x_k}\tau_k\eta_k = \frac{1}{\lambda}, \ k = i, f, s, \tag{4}$$

where Σ denotes the standard deviation of the uncertainty adjusted risk level $r(\alpha)$, $\tau_k = [\sigma_k^2(x_k) + \sum_{l \neq k}\rho_{kl}\sigma_k(x_k)\sigma_l(x_l)]/\Sigma$ is risk-generating factor k's share of total uncertainty about risk, $\eta_k = [x_k/\sigma_k][\partial\sigma_k/\partial x_k]$ is the elasticity of uncertainty about risk-generating factor k with respect to policy, and λ, the shadow price of the constraint (3), is the marginal cost of attaining the nominal standard m with a margin of safety α.

The marginal effect of stricter regulation of any risk-generating component k (modeled as an increase in the social cost of x_k) has two components: an impact on risk on average, $\partial\mu_k/\partial x_k$, and an impact on uncertainty about risk, $Z(\alpha)\Sigma\tau_k\eta_k/x_k$, which depends on the risk-generating factor's share of total uncertainty, τ_k, the tractability of uncertainty about the risk-generating factor, η_k, the overall level of uncertainty about risk, Σ, and the adjustment for aversion to uncertainty, $Z(\alpha)$. Since $Z(\alpha)$ is increasing in α, a greater margin of safety α means a greater weight placed on reductions in uncertainty, so that increases in the margin of safety can be said to correspond to greater aversion to uncertainty.

Equation (4) indicates that if it is cost-effective to use more than one form of regulation, all regulations in effect will have equal marginal effects on uncertainty adjusted risk. The resulting policy will be a portfolio of interventions, some of which have a comparative advantage in reducing risk on average while others have a comparative advantage in reducing uncertainty about risk. A regulatory instrument's impact on uncertainty about risk is increasing in the overall level of uncertainty about risk Σ, in the share of that total uncertainty of the risk-generating factor affected by that policy τ_k, and by the amenability of uncertainty about risk to policy-induced reduction η_k. Thus, policies with a comparative advantage in reducing uncertainty about risk are likely to play a more prominent role in cases where there is a great deal of uncertainty about risk, when they affect risk-generating factors that contribute a large share of that uncertainty, or when they can effect relatively large reductions in uncertainty.

Note that the model does not require that all policies in use induce first- or second-order stochastic dominant shifts in the probability distribution of overall risk r. In particular, the model does not rule out the use of policies that permit increases in the mean contribution of a given risk-generating factor ($\partial\mu_k/\partial x_k > 0$) as long as they result in sufficiently large decreases in uncertainty about that factor. It is readily seen from equation (4) that policies of this kind are more likely to be used ($x_k > 0$) when uncertainty matters more in the decision process, specifically, when the margin of safety α is high (so that $Z(\alpha)$ is large), when uncertainty about risk Σ is large, and when the relevant risk-generating factor accounts for a large share of total uncertainty about risk overall.

A symmetric argument obviously holds for policies that increase uncertainty about risk while simultaneously effecting large reductions in the mean of a risk-generating factor.

Lichtenberg and Zilberman (1988) derive additional implications for a simplified case in which susceptibility is not amenable to policy intervention and thus represents a source of uncontrollable background risk. One is that a more precautionary posture (higher margin of safety α) makes it cost-effective to increase the stringency of policies that have a comparative advantage in reducing uncertainty, possibly to the extent of decreasing the stringency of policies that have a comparative advantage in reducing risk on average. A second is that greater background uncertainty makes it cost-effective to increase the stringency of policies that have a comparative advantage in reducing risk on average, possibly to the extent of decreasing the stringency of policies that have a comparative advantage in reducing uncertainty about risk. A third is that greater background mean risk makes it cost-effective to increase the stringency of all regulatory instruments in use, a result that provides an alternative rationale for the commonly observed phenomenon that people express a desire for more stringent regulation of risks perceived to be less controllable (see for example Slovic 1986).

5. PRE-COMMERCIALIZATION TESTING VERSUS POST-COMMERCIALIZATION MONITORING

The current U.S. regulatory system relies on a combination of pre-commercialization screening and, if warranted to keep risk acceptably low, limits on commercialization. It utilizes a patchwork of legislation passed before the advent of transgenic crops, with oversight divided among three agencies [the Animal and Plant Hazard Inspection Service (APHIS) of the U.S. Department of Agriculture, the U.S. Environmental Protection Agency (EPA), and the U.S. Food and Drug Administration (FDA)] in what is known as the Coordinated Framework (for a more extensive description, see National Research Council 2000).

APHIS requires that field trials of new transgenic crops conform to specific guidelines to determine risks from gene flow, weediness of escaped transgenic crops, and other potential hazards for agriculture and the landscape generally. If the field trials indicate that such risks are acceptably low, APHIS approves the product for unregulated commercial release; if the field trials indicate some cause for concern, APHIS may place restrictions on commercial release.

FDA evaluates transgenic crops for potential allergenicity by comparing transgene-created substances with known allergens and by testing their capacity to survive stomach acid. Substances with no similarities to known allergens or that cannot survive treatment with stomach acid are judged to pose no threat of allergic reaction and are allowed without restrictions in food crops. Substances that raise some concerns may not be approved as food additives, hence forbidden in food-use crops; an example is the Starlink variety of Bt corn, which expressed a Bt toxin that did not decompose completely in stomach acid and hence could not be cleared as a non-allergen.

EPA has jurisdiction over transgenic crops sold with claims of pesticidal activity. EPA relies on information derived from field trials conducted under norms set by APHIS, on allergenicity and other food-use testing by FDA, and on information derived from other forms of testing to determine whether transgenic crops of this kind can be registered as pesticides and hence sold commercially. It has also placed restrictions on commercial use in some cases; one example is the requirement of refuge set-asides for resistance management in Bt crops.

As the preceding discussion indicates, the current U.S. regulatory system uses only measures that (i) are strictly preventive and (ii) rely primarily on information about risk on average. The Lichtenberg-Zilberman results suggest that this regulatory system might be well served by adding policies aimed at reducing uncertainty about risk. One such policy is post-commercialization monitoring, which reduces uncertainty about risk by providing information on potential adverse effects like gene flow and allergic reactions under actually existing conditions rather than in controlled field trials or laboratory tests that lack natural variability and randomness, but which may increase these risks on average. As noted above, a more precautionary stance (as measured by a higher margin of safety α) suggests the cost-effectiveness of greater reliance on uncertainty-reducing policies like post-commercialization monitoring. In fact, one of the principal recommendations of the most recent National Research Council (2002) study on transgenic plants was that the United States should require post-commercialization monitoring in order to assess the adequacy of pre-commercialization testing, i.e., to reduce uncertainty about risk. That recommendation was made on strictly ecological grounds. The Lichtenberg-Zilberman results provide an economic rationale for this recommendation; they also indicate that policies like this are especially needed to respond to the public's concern with uncertainty about the ecological risks generated by transgenic plants.

The sites in which transgenic crops are introduced commercially may differ in their susceptibility to risk. Some sites may contain larger populations of weedy relatives of the crop, for instance, putting them at greater risk of adverse effects from gene flow. At other sites it may be more difficult to segregate transgenic crops from non-transgenic ones, creating a greater background risk of adverse spillovers. Still other sites may have larger populations of non-target organisms susceptible to adverse effects from exposure to transgenic plants (e.g., monarch butterflies). The Lichtenberg-Zilberman results indicate that it is cost-effective to impose more stringent regulatory restrictions on commercialization of transgenic crops at sites with greater background susceptibility (i.e., mean risk). Policies that would enhance the stringency of regulation include more rigorous pre-commercialization testing, stricter limits on planting, and greater setbacks. The Lichtenberg-Zilberman results also indicate the desirability of increasing the use of policies that reduce uncertainty about risk at such sites. Thus, enhanced post-commercialization monitoring is just as warranted at more susceptible sites as stricter pre-commercialization screening and greater limitations on planting.

The current U.S. system, which emphasizes pre-commercialization testing and restrictions on commercial release of transgenic crops, can be cost-effective in some circumstances. The most notable are situations in which there is a great deal of uncontrollable background uncertainty about risk, e.g., because of variability in natural conditions that cannot be observed or because risk is heavily influenced by random factors like specific types of weather events. Even under these circumstances, though, it is cost-effective to reduce the use of policies with a comparative advantage in reducing uncertainty about risk (like post-commercialization monitoring) only in cases of extreme specialization where those policies provide little or no reduction in risk on average.

The marginal effects of post-commercialization monitoring may also differ across transgenic crops. Post-commercialization monitoring will be most desirable for transgenic crops for which monitoring can effect large marginal reductions in uncertainty about risk. As noted above, the magnitude of reductions in uncertainty about risk depends on factors such as the overall magnitude of uncertainty about risk (Σ), the relative contribution of a given risk-generating factor to that overall uncertainty (τ), and the tractability of that uncertainty with respect to policy (η). Post-commercialization monitoring is thus likely to be most desirable for transgenic crops whose ecological effects are highly uncertain and for which environmental fate accounts for a large share of uncertainty. It is also likely to be most desirable in situations where potentially confounding factors like natural variability are more observable or where risks are less sensitive to randomness like particular kinds of weather events, since the tractability of uncertainty is greater in such circumstances.

6. WHEN ARE BANS JUSTIFIED?

The European Union (EU) has invoked the precautionary principle to justify banning raising and importing transgenic crops, arguing that uncertainty about potential adverse effects on human and animal health from consuming these crops renders them unacceptably risky. The EU argues further that it is justified in taking such restrictive measures with respect to transgenic crops than other countries because its citizens have a greater aversion to uncertainty.

In the Lichtenberg-Zilberman model, a ban on imports and plantings is a policy portfolio that specializes exclusively in regulations limiting the introduction of potentially hazardous substances into the environment, in other words, a high level of $x_i > 0$ and $x_e = x_s = 0$. The EU position may be consistent with the qualitative results of the Lichtenberg-Zilberman analysis, as can be seen by considering a baseline in which imports and plantings of transgenic crops are allowed. Removal of these options may increase risk on average from the use of synthetic insecticides and herbicides that persist longer and are more mobile in the environment but can reduce uncertainty about environmental risk. As noted above, a more precautionary posture (higher margin of safety α) makes it cost-effective to increase the stringency of policies that have a comparative advantage in reducing uncertainty, possibly even to the extent of allowing increases in risk on average.

The necessary conditions characterizing the solution to the cost minimization problem (4) indicate that whether a corner solution such as a ban on imports and plantings holds depends as much on the cost of policies limiting the introduction of transgenic crops as on aversion to uncertainty. Import bans may be quite costly if domestically produced substitute crops are substantially more expensive than imported transgenic crops. The EU may perceive the costs of non-transgenic crops to be low enough to make it cost-effective to rely on banning imports and production of transgenic crops as the sole form of regulation. Other countries are unlikely to find the economic and political costs of such a policy equally low. For example, banning the production of transgenic crops could make countries like China or India unacceptably dependent on imports of food and feed, while countries like the United States, Argentina, and Brazil might find losses of export revenue too costly to make a ban on plantings acceptable. As a result, countries with an aversion to uncertainty equal to that of the EU may well choose more balanced policy regimes.

7. INCORPORATING BEHAVIORAL RESPONSES INTO RISK ASSESSMENTS

One thing that standard quantitative risk assessments do not typically do is incorporate human behavior. Rates of introduction, dissemination, and susceptibility are usually assumed to be invariant with respect to regulatory choices. In contrast, standard economic reasoning suggests that firms and consumers respond actively to policy choices. Economists believe that firms act to minimize the total costs of compliance, which consist of expected penalties for non-compliance as well as the costs of actions taken to meet regulatory requirements [see Polinsky and Shavell (2000) for a survey of law enforcement generally and Cohen (1999) for a survey of applications to environmental regulations]. Consumers may respond to stricter safety regulation by taking fewer personal precautions. Consumers also respond to safety information; using safety information as a regulatory instrument requires knowledge of how different kinds of consumers react (see, for example, Viscusi and Magat 1987).

The risk-generation model outlined here provides a structure that allows incorporation of behavior into quantitative risk assessments. Consider for example the issue of regulated firms' compliance with restrictions on field tests and commercialization of transgenic crops. Compliance is known to be imperfect. To take one example, as part of a recent initiative to improve public information, APHIS let it be known that Monsanto and its partners violated federal regulations on planting transgenic crops 44 times between 1990 and 2001 (Melcer 2003).

To incorporate compliance choices on the part of firms into the risk-generation process, suppose that both the introduction of potentially hazardous substances from transgenic crops into the environment $i(.)$ and their dissemination through the environment $f(.)$ depend on precautions taken by the regulated firm, y, as well as regulatory measures (setback and other safety requirements and post-commercialization monitoring x_i and x_f), so the means and variances of introduction and dissemination are $\mu_k(y,x_k)$ and $\sigma_k^2(y,x_k)$, $k = i, f$. Suppose further that firms choose the level of compliance to minimize total compliance costs $C(y) + L(y,x_i,x_f)$, where $C(y)$ denotes direct compliance costs and $L(y,x_i,x_f)$ denotes expected penalties from non-compliance, which may depend on the degree of non-compliance (not modeled here), the firm's good faith efforts y, and the probability of detection due to regulatory monitoring x_i and x_f.

This cost minimization problem implicitly defines the firm's compliance effort $y(x_i,x_f)$ as a function of regulatory policy, so the mean and variance of introduction and dissemination are $\mu_k(y(x_i,x_f),x_k)$ and $\sigma_k^2(y(x_i,x_f),x_k)$, $k = i, f$. The resulting cost-efficient mix of regulatory instruments clearly depends on the sensitivity of the means and variances of introduction and dissemination to firms' precautionary efforts and to firms' reactions to those instruments $\partial y(x_i,x_f)/\partial x_k$ in addition to the direct effects of those regulatory instruments on the means and variances of these risk-generating factors.

Modifying the risk-generation model to incorporate the behavior of firms and consumers is straightforward conceptually. The principal difficulty in carrying out such modifications for policy applications is a lack of sound empirical studies. Studies of regulatory enforcement have concentrated on the effects of regulatory agency monitoring on compliance with occupational safety and health regulations, maritime transportation safety regulations, and pollution control regulations [see Cohen (1999) for a survey and Helland (1998), Stafford (2002), and Earnhart (2004) for more recent studies]. How biotechnology firms and farmers planting transgenic crops respond to various forms of regulation has not been studied to date, however, so there is little solid information to use in modeling behavioral responses on the production side.

Labeling transgenic products has received a considerable amount of attention both in policy debates and among economists. Such a policy can be viewed as a means of providing consumers with information allowing them to take self-protective measures in the face of whatever risks transgenic products may pose. In terms of the risk-generation model used here, labeling could be considered as a policy affecting susceptibility indirectly via its impact on consumer choices, denoted a, so that the mean and variance of susceptibility could be written as $\mu_s(a(x_s))$ and $\sigma_s^2(a(x_s))$. The social cost of labeling x_s would depend in the amount and type of information contained in labels.

As in the case of firms' compliance, lack of solid empirical information constitutes the central difficulty in modifying the risk-generation model to incorporate consumers' responses to alternative forms of labeling. Conceptual discussions of labeling among economists typically assume that consumers have well-defined preferences for attributes such as product safety or the environmental quality impacts of alternative methods of producing goods and services. But it is well known that risk perceptions are highly biased: the public systematically overestimates the frequency of low-probability, high-visibility events, of events that are perceived to be less controllable, and of events that are highly unfamiliar—all of which apply to risks from transgenic crops (Slovic 1986). Moreover, a sound understanding of those risks requires more sophisticated ecological and biomedical knowledge than most of the public possesses. Such a context makes it difficult to design labels that would actually help consumers make more informed choices. Thus, incorporating consumer behavior into risk assessment models requires careful use of information from the empirical literatures on risk perception and the psychology of choice.

8. INDIVIDUAL VERSUS AGGREGATE RISK

The preceding discussion considers a single site or set of homogeneous sites. It is equally valid when a transgenic crop is grown at multiple heterogeneous sites, provided that regulators act to ensure that the risk of adverse environmental consequences is kept sufficiently low with an adequate margin of safety at every site. Regulators in the United States often take such an approach: for example, EPA makes its pesticide regulation decisions on the basis of whether a given set of pesticide uses poses an unacceptable risk to a random individual, often one belonging to a highly susceptible group such as children. An alternative approach is to regulate on the basis of aggregate risk across ecosystems or human populations, as envisaged in standard cost-benefit analysis.

The model is easily extended to encompass multiple heterogeneous sites. Following Zivin and Zilberman (2002) and Lichtenberg and Penn (2003), assume for convenience that sites differ according to susceptibility s, so that the average susceptibility at each site is fixed at s while the means and variances of introduction and dissemination vary with s. To accommodate the possibility of uncertainty about susceptibility at each site, let the variance of susceptibility be fixed at σ_s^2. To simplify the exposition, assume that susceptibility is uncorrelated with other risk-generating factors. Let $g(s)$ denote the number of sites or organisms (e.g., people) of susceptibility s. The uncertainty-adjusted population risk in this case is

$$r(\alpha) = \int_s \left[\mu_i(x_i,s) + \mu_f(x_f,s) + s \right] g(s)ds \; + \tag{5}$$

$$Z(\alpha)\sqrt{\int_s \left[\sigma_i^2(x_i,s) + \sigma_f^2(x_f,s) + \rho_{if}\sigma_i(x_i,s)\sigma_f(x_f,s) + \sigma_s^2(s) \right] g(s)ds} \; .$$

The necessary conditions for cost minimization are

$$\left[\frac{\partial \mu_k(s)}{\partial x_k(s)} + \frac{Z(\alpha)\Sigma}{x_k(s)} \tau_k(s)\eta_k(s) \right] g(s) = \frac{1}{\lambda}, \; k = i,\, f,\, s. \tag{6}$$

Cost-efficient regulation in this case will be differentiated by site, i.e., the portfolio of policies addressing introduction and environmental fate will vary according to site susceptibility.

Adding heterogeneity to the model expands the range of regulatory choices in two ways. First, if the first-order condition for cost minimization (6) is monotonic in susceptibility s, there may exist cutoff sites below which there is no need for regulation; Zivin and Zilberman (2002) and Lichtenberg and Penn (2003), for example, analyze cases of water contamination where no (additional) treatment is necessary for some sites. Second, adding firms' behavioral responses to the model raises the possibility that regulators may need to address adverse selection problems where site susceptibility is private (hidden) information.

9. CONCLUDING REMARKS

This chapter presents a generic model of the risk-generating process and an approach to accommodating uncertainty about risk, which are used to draw some inferences about regulating transgenic crops specifically and about assessing the impacts of environmental risk regulation more generally. Most notably, the analysis suggests that if uncertainty is the principal cause of concern regarding transgenic crops because of their novelty and because of aversion to uncertainty, regulation should concentrate on reducing that uncertainty. Extensive post-commercialization monitoring is one way of reducing uncertainty about risks posed by transgenic crops generally as well as those posed by individual varieties. Thus, it may be cost effective for the United States to de-emphasize restrictions on planting somewhat but expand post-commercialization monitoring substantially. The analysis also indicates that there are tractable ways of incorporating firms' and consumers' reactions to regulatory options into models of the risk-generation process. The main impediment to doing so is a lack of empirical models, which suggests a need for empirical work aimed at producing results that can be extrapolated to biotechnology regulation.

There are obviously many other ways of modeling the risk-generating process and of incorporating preferences regarding uncertainty into decision models. The specific model of the risk-generating process used here is quite simple and the preferences regarding uncertainty differ in important ways from those typically used in cost-benefit or risk-benefit analyses. While the specific inferences drawn are of interest in their own right, this exercise has a more important general lesson, namely the importance of modeling the ways in which policies affect the risk-generating process—both directly in the physical and biological sense and indirectly via impacts of human behavior. Economics is often given a very restricted role in assessments of environmental risk regulation, notably putting a price tag on an extremely narrow set of policy options selected largely on physical or biological grounds. In other words, economists are typically asked to provide accounting services rather than decision analysis. As a profession, we have tended to adapt to this role by focusing on issues like estimating willingness to pay for risk reduction or regulatory costs and ignoring what production economics can teach us about managing environmental risks under uncertainty.

An alternative to the current approach of conducting risk assessments and risk-benefit analyses in isolation from each other is to bring economic analysis to bear on the risk-generating process, i.e., to study the production of environmental risk. Studies of this kind can demonstrate prospects for a richer, more flexible approach to achieving adequate protection of human health and the environment with sufficient provision for uncertainty. Additionally, they are likely to produce more information about a richer set of tradeoffs than tends to enter policy debates at present, which should result in better informed debates and a more transparent policy-making process.

REFERENCES

Cohen, M.A. 1999. "Monitoring and Enforcement of Environmental Policy." In H. Folmer and T. Tietenberg, eds., *International Yearbook of Environmental And Resource Economics: 1999/2000*. Northampton, MA: Edward Elgar.

Crouch, E., and R. Wilson. 1981. "Regulation of Carcinogens." *Risk Analysis* 1(1): 47–57.

Crump, K., and R. Howe. 1984. "The Multi-Stage Model with Time-Dependent Dose Patterns: Application to Carcinogenic Risk Assessment." *Risk Analysis* 4(3): 163–176.

Earnhart, D. 2004. "Regulatory Factors Shaping Performance at Publicly-Owned Treatment Plants." *Journal of Environmental Economics and Management* 48(1): 655–681.

Helland, E. 1998. "The Enforcement of Pollution Control Laws: Inspections, Violations, and Self-Reporting." *Review of Economics and Statistics* 80(1): 141–153.

Kataoka, S. 1963. "A Stochastic Programming Model." *Econometrica* 31(1–2): 181–196.

Lichtenberg, E., and T.M Penn. 2003. "Prevention versus Treatment Under Precautionary Regulation: A Case Study of Groundwater Contamination Under Uncertainty." *American Journal of Agricultural Economics* 85(1): 44–58.

Lichtenberg, E., and D. Zilberman. 1988. "Efficient Regulation of Environmental Health Risks." *Quarterly Journal of Economics* 103(1): 167–178.

Lichtenberg, E., D. Zilberman, and K.T. Bogen. 1989. "Regulating Environmental Health Risks Under Uncertainty: Groundwater Contamination in California." *Journal of Environmental Economics and Management* 17(1): 22–34.

Melcer, R. 2003. "Monsanto Broke U.S. Planting Rules 44 Times over 12 Years." *St. Louis Post-Dispatch*, October 18, p. 3 of Business Section.

National Research Council. 2000. *Genetically Modified Pest-Protected Plants*. Washington, D.C.: National Academy Press.

_____. 2002. *Environmental Effects of Transgenic Plants*. Washington, D.C.: National Academy Press.

Polinsky, A.M., and S. Shavell. 2000. "The Economic Theory of Public Enforcement of Law." *Journal of Economic Literature* 38(1): 45–76.

Roberts, T., J. Buzby, and E. Lichtenberg. 2002. "Economic Consequences of Foodborne Hazards." In R. Schmidt and G. Rodrick, eds., *Food Safety Handbook*. New York: McGraw-Hill.

Slovic, P. 1986. "Informing and Educating the Public about Risk." *Risk Analysis* 6(4): 403–415.

Stafford, S.L. 2002. "The Effect of Punishment on Firm Compliance with Hazardous Waste Regulations." *Journal of Environmental Economics and Management* 44(2): 290–308.

Van Ryzin, J. 1980. "Quantitative Risk Assessment." *Journal of Occupational Medicine* 22(5): 321–326.

Viscusi, W.K., and W.A. Magat. 1987. *Learning About Risk*. Cambridge: Harvard University Press.

Zivin, J.G., and D. Zilberman. 2002. "Optimal Health Regulations with Heterogeneous Populations: Treatment versus 'Tagging'." *Journal of Environmental Economics and Management* 43(3): 455–476.

Chapter 15

ENVIRONMENTAL EFFECTS OF GENETICALLY MODIFIED CROPS: DIFFERENTIATED RISK ASSESSMENT AND MANAGEMENT

David E. Ervin[*] and Rick Welsh[†]
Portland State University[*] and Clarkson University[†]

Abstract: The environmental risks and benefits of genetically modified crops have vary-ing degrees of certainty. The U.S. regulatory system evaluates a suite of hazards for the crops primarily by minimizing type I error. However, genetically modi-fied crops vary widely in their potential for environmental harm. We develop a differentiated risk assessment process using three models that shift from pri-mary emphasis on controlling type I error for crops similar to conventionally bred varieties to stringent control of type II error for the most novel genetically modified crops. Parallel risk management approaches with economic consid-erations are discussed. Crop biotechnology development implications are drawn.

Key words: environmental risk, genetically modified, precautionary, type I error, type II error

1. INTRODUCTION

The growth in transgenic crop plantings is the most rapid technology revo-lution in the recent history of U.S. agriculture. Beginning from zero in 1996, USDA data show that farmers intend to plant approximately 80 percent of soybean acreage, 70 percent of cotton, and 38 percent of corn to transgenic varieties (USDA 2003). Barring a serious environmental or human health problem linked to the crops, these plantings likely will grow and spread across ecosystems throughout the United States over the next decade.

Senior authorship is shared equally. This chapter was originally published in J. Wesseler, ed., *Environmental Costs and Benefits of Transgenic Crops* (Wageningen UR Frontis Series, Vol. 7, Springer, Dordrecht). The authors are grateful to the publisher for permission to include the paper in this volume as well. We express our appreciation to Elizabeth Minor for her excel-lent editorial assistance.

Given the rapid pace of adoption, perhaps it should not be surprising that knowledge of the environmental risks and benefits of the crops is immature. Independent appraisals have concluded that the science on the environmental risks of transgenic crops is small and incomplete (Ervin et al. 2001, Wolfenbarger and Phifer 2000). Estimates of the benefits also are crude, mostly aggregate changes in pesticide use. A root cause of the science deficiencies is inadequate monitoring of environmental effects at field or ecosystem scales (Ervin et al. 2001, NRC 2002).

The central question addressed in this chapter is how to make sound regulatory decisions about releasing transgenic crops under such information deficiencies. We suggest the development of risk assessment and management approaches that are tailored to the nature of the ecological risks posed by the genetically modified (GM) plant. Two reasons underpin the need for such a differentiated approach. First, the organisms inserted into transgenic crops vary and expose the environment to quite different hazards. The distance between the engineered organism and the source of the genetic variation may be a useful measure for assessing the novelty of the introduced genetic changes and risks (Nielsen 2003). A second related reason is the varying amount of information about the environmental risks and benefits of transgenic crops. For example, field studies have documented growing resistance to highly used pesticides. In contrast, the risks of gene flow and deleterious effects on non-target organisms mostly have not been evaluated at large field scales. Reduced pesticide use and toxicity have been estimated for some transgenic crops in some areas, albeit not in relation to ecological conditions. But the effects of herbicide-resistant crops on yields, soil erosion, carbon loss, and supplemental water use have not been measured or estimated.

We begin with an interpretative summary of the latest evidence on the environmental risks and benefits of transgenic crops in the United States. After a brief review of the current U.S. regulatory process and its limitations, we develop the framework for a differentiated risk assessment approach. We close with a discussion of the implications of more effective regulation on private and public R&D for GM crops.

2. ENVIRONMENTAL RISKS[1]

Transgenic crops do not present new categories of environmental risk compared to conventional methods of crop improvement. "However, with the

---

[1] Risk is used here to convey the combination of the probability of occurrence of some environmental hazard and the harm associated with that hazard (NRC 2002, p. 54). This section updates a similar review in Ervin et al. (2003). For more complete discussions of the risks, readers may consult Ervin et al. (2001), Wolfenbarger and Phifer (2000), NRC (2002), and Barrett et al. (2001).

long-term trend toward increased capacity to introduce complex novel traits into the plants, the associated potential hazards, and risks, while not different in kind, may nonetheless be novel" (NRC 2002, p. 63). The nature of the risks varies depending on the characteristics of the crop, the ecological system in which it is grown, the way it is managed, and the private and public rules governing its use. Three categories of hazard emerge from the interaction of these factors.[2] Table 1 shows often-mentioned environmental concerns for herbicide-tolerant, virus-resistant, and insect-resistant crops.

Table 1. Selected Transgenic Traits and Environmental Concerns

Genotype	Environmental Concerns
Herbicide tolerance (HT)	• Increased weediness of wild relatives of crops through gene flow
	• Development of HT weed populations through avoidance and selection
	• Development of HT "volunteer" crop populations
	• Negative impact on animal populations through reduction of food supplies
Insect resistance (IR)	• Increased weediness of wild relatives of crops through gene flow
	• Development of IR populations
	• Toxicity to non-target and beneficial insect and soil microorganism populations
Virus resistance (VR)	• Increased weediness of wild relatives of crops through gene flow
	• Disease promotion among plant neighbors of VR crops through plant alteration
	• Development of more virulent and difficult to control viruses through virus alteration

2.1. Resistance Evolution

Current commercial transgenic crops emphasize effective pest control via the increased use of certain pesticides, such as Bt. Crops bred to resist herbicides, viruses, and insects have the potential to dramatically change agricultural practices. The lack of long-term studies poses a serious obstacle to performing an adequate assessment of the potential environmental effects. Nevertheless, some studies have assessed the impacts.

---

[2] The National Research Council (NRC) (2002) defines four categories of hazards: (i) resistance evolution, (ii) movement of genes, (iii) whole plants, and (iv) non-target effects. Categories (ii) and (iii) are combined here.

Herbicide-tolerant crops. The primary environmental concern from herbicide-tolerant (HT) crops is the development of weed populations that are resistant to particular herbicides. This resistance can occur from the flow of herbicide-resistant transgenes to wild relatives or to other crops, or from the development of feral populations of herbicide resistant crops (NRC 2002). Also, if farmers rely on only one or a few herbicides, weed populations can develop that can tolerate or "avoid" certain herbicides, which enables them to out-compete weeds that do not manifest such tolerance. Weed scientists find the latter development likely. In fact, Owen (1997) reports that in Iowa, "common waterhemp (*Amaranthus rudis*) populations demonstrated delayed germination and have 'avoided' planned glyphosate applications. Velvetleaf (*Abutilon theophrasti*) demonstrates greater tolerance to glyphosate and farmers are reporting problems controlling this weed." And VanGessel (2001) reports that horseweed (*Conyza canadensis*) has been found to be resistant to glyphosate through experiments conducted in a farmer's field in Kent County, Delaware.

Virus-resistant crops. There is a relative dearth of research on the ecological risks associated with these crops. However, scientists have voiced several environmental concerns related to virus-resistant (VR) transgenic crops. First, these bioengineered varieties may promote disease in neighboring plants by altering such plants so that they become hosts for particular viruses, when such plants were not previously susceptible to infection by the viruses of concern. Second, VR transgenic crops may alter the methods through which viruses are transmitted (Rissler and Mellon 1996, Royal Society of United Kingdom 1998). These changes could result in the development of stronger viruses (Hails 2000, Rissler and Mellon 1996, Royal Society of United Kingdom 1998). Scientists are also concerned that the genome in VR crops may recombine with the plant virus genome (which is comprised of RNA in most/all plant viruses) during viral replication (Rissler and Mellon 1996, Royal Society of United Kingdom 1998). Researchers believe that such recombination could lead to genetically unique viruses that may be difficult to control (Greene and Allison 1994). Third, the flow of VR transgenes may enhance the weediness of wild relatives of VR crops (see section 2.2). A National Research Council (NRC) assessment (NRC 2000) found that the USDA's assumption that transgenic resistance to viruses engineered in cultivated squash will not result in enhanced weediness of wild squash through gene flow needs verification through longer-term studies. The NRC study also concluded that the USDA's assessment of the potential for virus-protective transgenes in cultivated squash to affect wild populations of squash "is not well supported by scientific studies," especially for transgenic squash engineered to be resistant to several viruses instead of three or fewer (NRC 2000, p. 124). In a new report, the NRC (2002, p. 134) argued that the evidence collected to date is "scientifically inadequate" to support the conclusion of USDA's Animal and Plant Health Inspection Service (APHIS) that gene flow from VR squash would not result in increased weediness of free living *Cucubita pepo.*

Insect-resistant crops. The innate ability of insect populations to rapidly adapt to pest protection mechanisms poses a serious threat to the long-term efficacy of insect-resistant (IR) biotechnologies. Such adaptations can have environmental impacts. For example, adaptation by insect populations to a more environmentally benign pest control technique, such as Bt, could result in the use of higher toxicity pesticides (NRC 2000). The Canadian Expert Panel on the Future of Food Biotechnology (Barrett et al. 2001) finds that it is important to account for insect movement when devising resistance management plans. Regional or interregional scale plans, rather than local, are needed if the insect of concern is highly mobile (Gould et al. 2002, Hails 2000). Field outbreaks of resistance to Bt crops have not yet been documented (Morin et al. 2003). The body of science to inform resistance management is limited to laboratory studies of specific insect pests (Ervin et al. 2001, Morin et al. 2003). Such studies show the potential for resistance to develop. Indeed, the NRC (2002, p. 76) finds that the evolution of "insect resistance to Bt crops is considered inevitable." Similarly, Tabashnik and colleagues write that eventually insects will develop resistance to IR crops, and therefore, "any particular transgenic crop is not a permanent solution to pest problems" (Tabashnik et al. 2001, p. 1). A recent laboratory study found that *resistant* populations of diamondback moth larvae may be able to use a toxin derived from Bt "as a supplementary food protein, and that this may account for the observed faster development rate of Bt resistant insects in the presence of the Bt toxin" (Sayyed, Cerda, and Wright 2003).

One of the few field studies by Tabashnik et al. (2000, 2001) found that in 1997 approximately 3.2 percent of pink bollworm larvae collected from Arizona Bt cotton fields exhibited resistance. This level was far above what was expected, raising fears that rapid resistance development would occur. However, data collected in 1998 and 1999 showed no increase in resistant populations of pink bollworm. Tabashnik et al. (2001) conclude that there might be high fitness costs for insects to develop resistance to Bt. In addition, Carrière et al. (2003, p. 1523) found that widespread and sustained use of Bt cotton can suppress regional pink bollworm populations and thus that "Bt cotton could reduce the need for pink bollworm control, thereby facilitating deployment of larger refuges and reducing the risk of resistance."

The potential for insect resistance implies that integrating transgenic crops into a multiple-tactic pest management regime may prove to be a more effective long-term strategy. The exact path of the emergence of insect resistance is yet to be characterized. However, progress has been made on identifying resistant alleles in pests of certain crops, such as the pink boll-worm in cotton (Morin et al. 2003). Therefore research is needed to better define the parameters of resistance development, as well as to design crops that minimize the opportunities for resistance to develop in the first place. This latter point on fostering precautionary technology development is dis-cussed in the concluding section.

2.2. Transfer of Genes—Gene Flow

There is little doubt in the scientific community that genes will move from crops into the wild (Hails 2000, NRC 2000, Snow and Palma 1997). The relevant research questions are whether transgenes will thrive in the wild, and how they might convey a fitness advantage to wild plants that makes them more difficult to control in areas (Hails 2000; Keeler, Turner, and Bolick 1996; NRC 2000; Barrett et al. 2001; Snow and Palma 1997).

Generally, crops with wild relatives in close proximity to the areas where the crops are grown pose higher risk for gene flow to wild relatives. U.S. examples include sunflower (*Helianthus annuus L*) and oilseed rape (*Brassica napus*) (Hails 2000; Keeler, Turner, and Bolick 1996; Snow and Palma 1997). Gene transfer could become a problem if the transferred genes do not have deleterious effects on the crop-wild hybrids, but instead confer an ecological advantage (Hails 2000, Barrett et al. 2001, Snow and Palma 1997). Gene flow from classically bred crops to wild plants has been docu-mented. Ellstrand (2001) finds that classically bred crop-to-wild gene flow has enhanced the "weediness" of weeds for seven of the world's thirteen most important crops [e.g., Johnson grass (*Sorghum halepense*) from culti-vated sorghum (*Sorghum bicolor*)].

Snow and Palma (1997) argue that widespread cultivation of transgenic crops could exacerbate the problem of gene flow from cultivated to wild crops, enhancing the fitness of sexually compatible wild relatives. Tradi-tional breeding typically results in the inclusion of deleterious alleles (i.e., alternative forms of a gene at a given site on the chromosome) linked to the desired beneficial genes. The inclusion of the deleterious genes decreases the likelihood that crop-to-wild outcrossing will result in enhanced weediness of the wild plants. In contrast, biotechnological methods enable solitary genes to be selected without including neutral or deleterious genes (Snow and Palma 1997; NRC 2000, p. 85).

Scientists generally expect that herbicide-resistant transgenes will not result in increased weediness of wild relatives, as such genes tend to impose a cost, or are neutral, to wild relatives. Nonetheless, in situations where her-bicides are typically used to control weedy plants, herbicide resistance could confer a competitive advantage to unwanted volunteer crops (Keeler, Turner, and Bolick 1996). Indeed, the flow of herbicide-resistant transgenes has already become a problem regarding within-crop gene flow. Hall, Huffman, and Topinka (2000) reported the presence in a Canadian farmer's field of volunteer oilseed rape resistant to three herbicides: glyphosate, imidazolinone, and glufosinate. The "triple-resistant" oilseed rape developed from gene flow among three oilseed rape varieties designed to resist each of the herbicides, which were planted in close proximity to each other. The Canadian expert panel finds that "herbicide-resistant volunteer canola plants are beginning to develop into a major weed problem in some parts of the Prairie Provinces of Canada" (Barrett et al. 2001, p. 122). They expressed special concern about the potential for "stacked" resistance to multiple herbicides, which could force farmers to employ older herbicides that are often more environmentally harmful than newer classes (Barrett et al. 2001).

In general, though, ecologists tend to be more concerned about potential fitness advantages of insect- and virus-resistant transgenes (Hails 2000). For example, recent research has shown that the Bt gene for lepidopteran resistance can increase seed production in wild sunflowers (Snow 2002, Snow et al. 2003). Another study showed that crossing Bt oilseed rape with a wild relative (*Brassica rapa*) did not enhance the weediness of the resulting plant relative to unmodified *Brassica rapa* (Adam 2003). To understand the contrasting findings, more studies along with monitoring and testing are needed to detect potential ecological problems. However, reactions by Pioneer Hi-Bred International and Dow AgroSciences to the Snow et al. (2003) study may impede such research. The firms blocked a follow-up study by denying access to the materials they controlled that were needed to conduct further investigation (Dalton 2002). The firms denied three requests to continue studying Bt sunflower using the scientists' research funding. Therefore, Snow and her colleagues are legally barred from continuing their investigations (personal e-mail communication, April 25, 2003).

2.3. Impacts on Non-Target Animals and Plants

While crops bred to resist pests may suffer less damage and lead farmers to use less insecticide, there is concern that the toxins these plants produce may harm non-target organisms, including animals and plants that are not pests (Royal Society of United Kingdom 1998, NRC 2002). Laboratory research confirms that transgenic insecticidal crops can have negative impacts on potentially beneficial non-target organisms, including lacewings (Hilbeck et al. 1998a, Hilbeck et al. 1998b), ladybird beetles (Birch et al. 1997), monarch butterfly larvae (Losey, Raynor, and Carter 1999), and soil biota (Watrud and Seidler 1998).

Tabashnik (1994) asserts that reductions in pest populations due to transgenic crops may negatively affect available numbers of desirable natural predators. Similarly, the NRC finds that "Herbicide tolerant crops might cause indirect reductions on beneficial species that rely on food resources associated with the weeds killed by the herbicides" (NRC 2002, p. 70). The potential negative effect on some birds has been modeled (Watkinson et al. 2000). In contrast, research sponsored by the European Commission on the safety of genetically modified organisms (GMOs) found no negative impacts on honeybees from transformed oilseed rape plants. Also, GNA (*Galanthus nivalis agglutinin*) lectin accumulation in aphids did not result in acute toxicity to ladybird beetles or prevent *Eulophus pennicornis* from successfully parasitizing tomato moth larvae (Kessler and Economidis 2001, Pham-Delegue et al. 2000). A field study in Wisconsin found that populations of predators and parasites were higher in Bt potato fields than in conventional potato fields where conventional insecticides were used. Non-chemical or less-intensive chemical treatments were not evaluated (Hoy et al. 1998). This finding points to the need to evaluate the impacts of transgenic crops relative to conventional chemically intensive practices and alternative systems (Dale, Clarke, and Fontes 2002, NRC 2002).

Generalizations may well be inappropriate as to the impact on non-target organisms, with each crop and region requiring specific research. For example, in a widely publicized laboratory study, Losey, Raynor, and Carter (1999) found a 44 percent mortality rate in monarch butterfly larvae fed on milkweed leaves dusted with Bt corn pollen. No mortality occurred in monarchs fed on leaves with non-Bt corn pollen. These laboratory findings on the toxicity of Bt corn to monarch butterflies generated significant controversy and prompted responses as to the applicability of the findings to field settings (Beringer 1999), follow-up research supporting the original findings (Hansen-Jesse and Obrycki 2000), and a risk assessment study finding that monarchs are not at risk from Bt corn since "overall exposure of monarch larvae to Bt pollen is low" (Sears et al. 2001). In turn, the NRC panel asserted that "In the upper Midwest, herbicide tolerant soybeans might cause indirect reductions of monarch populations because their milkweed host plants are killed by the herbicides" (NRC 2002, p. 71). Though many consider the debate over monarchs and Bt corn closed, there are still questions being raised about the effects of long-term and low-level exposure to Bt in corn pollen on monarch larvae survival and fitness (Stanley-Horn et al. 2001).

Watkinson et al. (2000) modeled the potential impacts on skylarks (*Alauda arvensis*) from a reduction in seeds of a weed of sugar beet (primary food supply of skylarks) from the introduction of HT sugar beet. They found that the weed populations could be almost completely eradicated depending on the conditions surrounding adoption of such transgenic sugar beets, such as the management practices. Severe reductions in weed populations could significantly affect the skylark's use of fields as a food source. Conversely, some bird populations may increase if farmers replace broad-spectrum synthetic herbicides, which have cut into the birds' food supply, with transgenic crops (NRC 2000, p. 80).

Insects and other animals are not the only organisms potentially affected by transgenic crops. The Canadian expert panel found that the cultivation of transgenic crops could impact the diversity and abundance of soil microflora, even if the impacts are "minor relative to the natural variability" (Barrett et al. 2001, p. 111). They observed that transgenic manipulation aimed at modifying biogeochemical cycles should receive more scrutiny. The NRC (2002) largely concurs by arguing that no effects on soil organisms have been found to date, though it has been discovered that Bt toxin "leaks out" of corn roots and can persist in the soil for months (Saxena and Stotzky 2000; Dale, Clarke, and Fontes 2002).

2.4. Risk Summary

In general, the environmental hazards associated with transgenic crops are potential risks. However, research results provide emerging parameters with which to evaluate the relative magnitude of the potential risk. For example, good evidence is emerging that the combination of natural promiscuity regarding gene flow among crop varieties and engineered herbicide resistance is a serious concern. Likewise, it is becoming clearer that HT crops will probably not create "superweeds" through crop-wild flow of genes that enable plants to tolerate particular herbicides. Rather, weed problems will be enhanced by the selection of resistant weed populations through increased use of herbicides tied to particular transgenic crops, such as glyphosate-resistant soybeans. Research efforts should concentrate on the latter potential risks.

Also of concern is the enhanced weediness of wild relatives of crops from the flow of genes enabling plants to resist insects and viruses. However, the research to evaluate the extent of these risks is incomplete. More study is needed to assess the potential for the widespread adoption by farmers of IR sunflower and VR squash to promote the development of wild plants with improved fitness relative to other wild plants. The improved fitness of particular plants in wild populations could alter plant and animal ecosystems. The controversy over the potential for Bt corn to harm monarch butterfly populations also illustrates the need to move beyond laboratory studies to comprehensive field scale when assessing the potential negative impact on susceptible but beneficial populations—that is, the need for studies that account for the temporal and spatial interaction between the introduced technology and the organism of interest.

That field outbreaks of resistance to Bt crops have not yet been documented despite widespread adoption of such crops deserves more investigation. Potential questions include:

- Have the refugia plans prevented the development of such resistance?
- If so, can such plans be developed for herbicide-resistant technologies to delay the development of weed populations resistant to herbicides linked with HT crop varieties?

3. ENVIRONMENTAL BENEFITS

3.1. Reduced Pesticide Use and Toxicity

Data to assess the effects of transgenic crops on pesticide use should capture the full range of climate, pest, and economic conditions. The data should also be linked to environmental conditions to estimate changes in acute and chronic toxicity on ecological systems (Antle and Capalbo 1998). The impacts of changes in pesticide use for transgenic crops on the environment can be determined only by comparing the fate, transport, and toxicity of the full array of compounds available to farmers, and how they are applied. The following estimates for three major U.S. transgenic crops do not yet measure up to these standards.

Results from farm surveys generally indicate that farmers who plant Bt cotton apply fewer insecticides than those who plant conventional cotton (Carlson, Marra, and Hubbell 1998; Hubbell, Marra, and Carlson 2000; USDA 1999a, 1999b; Fernandez-Cornejo and McBride 2002). USDA analysts recently estimated that Bt cotton plantings in 1997–98 reduced insecticide use by approximately 250,000 pounds of active ingredients (a.i.) (Fernandez-Cornejo and McBride 2002). Other studies have found that the reductions vary by area and year depending on pest pressures and other factors. For example, Carlson, Marra, and Hubbell (1998) report that the average number of insecticide applications by farmers who adopted Bt cotton in 1996 was 3.29 on their traditional acres in the upper South, but only 2.58 in the lower South, a difference that likely reflects different insect conditions in the two regions. Farmers who plant Bt cotton likely use more conventional insecticides in the first place and can save more money than farmers who apply lower levels. The long-term effects of Bt cotton on insecticide use may require analyses of 10 years or more to cover the cycles of pest, climate, and economic variations.

The latest national analysis estimated that planting of HT soybeans in 1997–98 increased overall pesticide use by approximately 2.5 million pounds a.i. (Fernandez-Cornejo and McBride 2002). The increase was largely in the form of glyphosate, which is anywhere from 3 to 16 times less toxic than the herbicides it has replaced and 1.6 to 1.9 times less likely to persist in the environment (Heimlich et al. 2000). Whether and how long these shifts in herbicide composition on soybeans will persist depends on how fast the resistance problems discussed above unfold. If glyphosate becomes ineffective, farmers will use other herbicides to control weeds that develop resistance to it, and the environmental implications depend on the substitute compound, location, and other conditioning factors.

Over 1997–98, U.S. farmers who planted HT corn were estimated to decrease their herbicide use by approximately 5 million pounds a.i. (Fernandez-Cornejo and McBride 2002).

The overall effect for the three crops was a net reduction of approximately 3 million pounds of pesticide a.i. in 1997–98. This is characterized as "a small but statistically significant effect" (Fernandez-Cornejo and McBride 2002, p. 36). Since nearly 80 percent of the reduction in pesticide treatments is attributed to switching to glyphosate, the overall toxicity quotient decreased by more than the simple volume reduction. Since 1997–98, the adoption of transgenic crops has spread and the decrease in pesticide volume and toxicity likely has increased as well, but likely still remains a small percentage of overall pesticide use.

3.2. Reduced Tillage, Erosion, Carbon Loss, and Water Savings

Several other potential environmental benefits of transgenic crops have been hypothesized. For example, manufacturers and advocates of transgenic crops have asserted that HT varieties will increase use of conservation tillage. However, the USDA analysis could not support this hypothesis, concluding that farmers already using conservation tillage were more likely to adopt HT crops (Fernandez-Cornejo and McBride 2002). Evidence on the other potential environmental benefits has been nil.

4. THE CURRENT ENVIRONMENTAL REGULATORY PROCESS

The U.S. Environmental Protection Agency (EPA) and USDA's APHIS share responsibility for assessing the environmental risks of transgenic crops before their release for testing and commercialization. EPA evaluates plant-incorporated protectants, such as Bt in cotton and corn, and regulates them in the same way it regulates conventional chemical pesticides. The APHIS evaluation, depending upon the particular plant line, covers a broad range of potential environmental effects: (i) the potential for creating plant pest risk, (ii) disease and pest susceptibilities, (iii) the expression of gene products, new enzymes, or changes to plant metabolism, (iv) weediness, and impact on sexually compatible plants, (v) agricultural or cultivation practices, (vi) effects on nontarget organisms, including humans, (vii) effects on other agricultural products, and (viii) the potential for gene transfer to other types of organisms (McCammon 2001).

Each agency conducts a risk analysis of the biophysical effects of the crop. We discuss the more comprehensive APHIS process here. The risk analysis includes three stages—hazard identification (as covered in section 2), risk assessment, and risk management. The risk assessment stage is the focus of this section. One or more of several techniques may be used to perform the assessment: (i) epidemiological analysis, (ii) theoretical models, (iii) experimental studies, (iv) expert judgments, and (v) expert regulatory judgments. APHIS generally uses expert regulatory judgment, a less rigorous technique than (i), (ii), or (iii) according to the NRC (2002, p. 60).

APHIS uses a two-part model in which a transgenic plant is divided into (i) the unmodified crop and (ii) the transgene and its product (NRC 2002, p. 90). The theoretical reasoning behind this choice is that the transgene is a small genetic change that is likely to have only a small phenotypic effect. Invoking that principle accepts the simple linear model of "precise" single gene modifications that do not significantly alter other plant processes. Experts have reservations about this rationale, in that "unanticipated changes can be induced by expression of a novel gene, and their phenotypic consequences need to be assessed empirically across time and environments" (Barrett et al. 2001, p. 185). The NRC panel also noted that the assumption that single gene changes have small ecological effects is not always true (NRC 2002, p. 91).

The APHIS risk assessment process evaluates the risk of the unmodified crop separate from the risks of the transgene and its products. The test for biophysical risk basically attempts to control the type I error, α, of rejecting the null hypothesis that the transgene has *no* effect on the environment when in fact the null is true.[3] Minimizing the frequency of type I error (a "false positive") by requiring a high level of certainty, most often 95 percent, is the dominant approach used for hypothesis testing in science. It is also considered to be a conservative approach in detecting an ecological effect because it is difficult to construct studies with sufficient power to reject the null hypothesis under $\alpha = .05$. This is especially true for studies dealing with environmental issues where natural variation, both in space and time, is typically great (Buhl-Mortensen 1996, NRC 2000). The effect of this high variation is to spread out the distribution of the random variable for testing the null hy-

---

[3] We assume for ease of exposition that the environmental hazards can be described by reliable probability distributions to illustrate the differences in the tests. For some hazards, probability distributions cannot be estimated, either objectively or subjectively, i.e., true uncertainty exists. For those cases, nonparametric analytical techniques must be used. For still other hazards, the outcome space is undefined, i.e., surprises may occur, and game theory techniques can be used (Bishop and Scott 1999).

pothesis and increase the range of values for which the sample test statistic is interpreted as not rejecting the null hypothesis.

The NRC panel (2002), noting several limitations of the two-part model, recommended more evaluation of "fault-tree" and "event-tree" risk analyses that systematically search for potential ecological risks. These and other techniques place more emphasis on understanding and controlling type II error, β, the error of failing to reject the null hypothesis when it is in fact false, i.e., a "false negative" when in fact the transgene actually causes an ecological hazard. Lemons, Shrader-Frechette, and Cranor (1997) argue that scientists and decision makers should be more willing to minimize type II errors (and accept higher risk of committing type I errors) because of the pervasiveness of scientific uncertainty in complex ecological processes. They explain that the normal argument to minimize type I error when adding new scientific knowledge does not apply equally to environmental regulatory decisions. Type I error with a 95 percent confidence level is appropriate for evaluating new scientific results in order to prevent or minimize the inclusion of "speculative knowledge to our body of [scientific] knowledge" (Lemons, Shrader-Frechette, and Cranor 1997, p. 228). However, knowledge generation to support environmental regulatory decisions is different in-kind as it is based on finding whether negative environmental or health outcomes are likely, or unlikely, to occur, and not on generating new scientific results per se (Lemons, Shrader-Frechette, and Cranor 1997, pp. 224–230). Buhl-Mortensen (1996) also notes that since ecosystem models, with a few exceptions, have low predictive power, it is prudent to control for type II error when evaluating the potential ecological impacts of industrial processes or technologies. In a similar vein, Jasanoff (2000, p. 279) concludes that the current U.S. system has "biased the assessment exercise away from large, holistic questions," instead focusing on relatively precise genetic manipulations. Part of the reason for the bias may be the lack of postmarket monitoring and testing of biotechnology crops to provide adequate data for examining the larger ecological questions (NRC 2002, Taylor and Tick 2003).

Figure 1 illustrates the differences in risk assessment tests under control of type I versus type II errors. Assume for purposes of illustration that the horizontal axis measures the difference in seed production by wild relatives containing a transgene compared to seed production in wild plants without the transgene. The hypothetical distributions measure the probability density of various values. The distributions centered over $\mu = 0$ and $\mu = SD_a$ are the null and alternative population distributions. If a value of SD* is observed from the sample test, the APHIS risk assessment would fail to reject the null hypothesis of no significant gene flow, because SD* lies to the left of the critical test statistic for a one-tailed test, assuming $\alpha = .05$ (and $\beta = .20$). However, under a criterion to minimize type II error by setting β at .10 (i.e., the power of the test = .90), the alternative hypothesis of a positive effect on

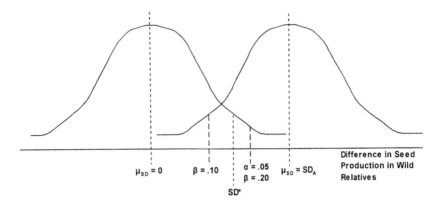

Figure 1. Differences in Statistical Tests to Control Type-I and Type-II Errors

seed production in wild relatives would fail to be rejected. Note also that if the natural processes and the assumed distribution become more variable or flatter, the range under which the alternative hypothesis is not rejected would expand. That is, the probability of a "false negative" increases, ceteris paribus.

The NRC analysis repeatedly emphasized that "for purposes of decision support, risks must be assessed according to the organism, trait and environment" (NRC 2002, p. 63). The review of potential environmental risks also suggests that information deficiencies could be added to this list. The recognition that GM crops vary in their potential for environmental risk invites consideration of a risk assessment process that uses different methods and different standards of proof for different types of genetic modification.

5. A DIFFERENTIATED RISK ASSESSMENT PROCESS

A differentiated risk assessment process captures the novelty of the ecological hazard(s) from the GM crop and the quality of information about the potential hazards and their occurrences. Different models might cover a range, from controlling type I error for GM crops using high quality information that shows little potential ecological risk to a very high standard of avoiding type II errors for crops judged to pose serious potential ecological disruptions with little scientific evidence to assess the nature of the risk(s).

To differentiate the risk assessment, a robust method for characterizing the nature of the ecological risks of GM crops is needed. We adopt an approach suggested by Nielsen (2003), who argues for conceptual diversification in discussing and regulating genetically engineered organisms. She claims that the current process-based categorization is imprecise and does not adequately convey the sources, extent, and novelty of the GMO. For example, GM crops with simple nucleotide changes are unlikely to generate serious ecological concerns beyond those of their traditional counterparts. In contrast, species-foreign genes, synthetic genes, and some other changes in GM crops deviate substantially from what classical selective-based breeding has achieved. The latter organisms have genetic compositions that do not reflect evolutionary processes occurring under natural conditions. She cites Bt corn, derived by genetic engineering of several unrelated DNA segments, as an example of an organism that cannot be replicated by natural processes within the same time scale. Thus, the "genetic distance" between the engineered organism and the source of the new genetic variation could serve as a functional criterion for determining the type of risk assessment conducted. This approach dovetails with earlier arguments by Snow and Palma (1997) and the NRC (2000, p. 85) that the differences between classical breeding approaches and transgenic methods can justify differences in risk assessment.

It is important to note that we are not asserting that classical breeding inherently produces safer products and that deviations from this approach produce more dangerous products. Rather, our argument is subtler. We believe that our long experience and familiarity with classical breeding techniques makes it reasonable to assume that as ecological or other problems potentially develop, we are more likely to recognize such problems and take corrective measures. It follows that our relatively brief experience with transgenic and other recently developed techniques and the scant science base makes it more likely that if a biosafety problem develops with these new crops, we may not recognize the problem as quickly because our ability to discern potential problems is primarily based on our experience with traditional breeding techniques. Our rationale and approach are consistent with the finding of the NRC stated at the beginning of section 2 that "the associated potential hazards, and risks [of transgenic crops], while not different in kind, may nonetheless be novel" (NRC 2002, p. 63).

Nielsen (2003) proposes five categories that vary from low to high genetic distance: (i) intragenic (within genome), (ii) famigenic (species in the same family), (iii) linegenic (species in the same lineage), (iv) transgenic (unrelated species), and (v) xenogenic (laboratory-designed genes). We combine categories (i) and (ii) and categories (iii) and (iv) along with category (v) to develop three risk models based on increasing genetic distance.[4]

---

[4] Because transgenic crops vary considerably in traits and biological makeup, a disaggregated taxonomy of transgenic crops would improve the analytical power of the differentiated assessment.

1. *Intragenic and famigenic.* These two categories of genetic modification respectively include those from directed mutations or recombinations including those arising in classical, selection-based breeding, and from the taxonomic family, including those arising from applying cellular techniques in classical breeding. The risk assessment process for these cases could reasonably presume no substantial ecological risks from releasing the crops beyond those from conventional breeding. Thus, it would include a straight-forward review of evidence submitted by the applying entity and application of the "probability rule" criterion for all relevant effects reviewed by APHIS (Mooney and Klein 1999). A test of the null hypothesis of no significant effect would be conducted to control type I error, i.e., failing to accept the null when it is in fact true, at the standard .05 level of significance.

2. *Linegenic and transgenic.* These two categories include organisms that contain genetic variability *beyond that* possible with conventional breeding. Linegenic includes species in the same lineage and the recombination of genetic material beyond what can be achieved by classical breeding methods. Transgenic covers those plants that contain DNA from unrelated organisms, and includes most of the GM plants commercialized today. For these plants, the test shifts the framing hypothesis to one that assumes a significant environmental effect because of the increased novelty of the crop and less information. The standard of proof would be set at a specified power of test, for example .90 or $\beta = .10$. The decision of setting the standard of proof moves beyond science into the realm of public input and political decisions because the standard reflects society's general preference for avoiding such risks, i.e., the degree of precaution (Van den Belt 2005).

3. *Xenogenic.* This category includes laboratory-designed genes for which no naturally evolved genetic counterpart can be found or expected, e.g., synthetic genes and novel combinations of protein domains. This class is the furthest of the three from natural genetic variability, and therefore poses the greatest potential for ecological hazard and risk. For the hazards that can be characterized with objective or subjective probability distributions, the bar for approval to release would be highest for such plant organisms. Since the genetic distance from classically bred crops is greatest for this category, a higher standard of proof would be applied to control type II error than for category 2—for example, $\beta = .05$.

To implement this risk assessment framework for GM crops, a group of experts with sufficient breadth across ecological and other relevant sciences would be assembled. The composition and independence of the groups is critical if reliable risk assessments are to be completed. To counter criticisms that the expert panels used by USDA inappropriately favor releases, both government and university scientists would be involved and each would face sanctions if their contributions were subsequently determined to be biased, e.g., a bonding mechanism for liability. Due to the scant knowledge that exists for many GM crops, especially new transgenic varieties, the groups would at first conduct case-by-case assessments. However, over time with the accumulation of more systematic knowledge on the potential environmental risks due to the search for type II errors, the assessments likely would shift to broader categories and become more routine and cost-efficient over time.

6. DIFFERENTIATED RISK MANAGEMENT

The final stage of risk analysis is the management decision taken, including commercial release and regulatory measures that may accompany the releases, such as refugia requirements and post-commercialization monitoring and testing. The differentiation of the risk management process parallels that taken for risk assessment in that increasing degrees of precaution are imposed on more novel organisms. However, risk management decisions would involve weighing environmental and economic considerations for the organisms that do not pose potential catastrophic and irreversible hazards. The answers to six questions (see Table 2) summarize the differences for the APHIS-like process, a risk-benefit evaluation, and the precautionary approach for the three risk management categories.

There is a substantial body of science on the potential environmental effects of intragenic and famigenic GMOs or products from similar techniques. Therefore, the evaluation would be the least precautionary by controlling type I error using the standard 95 percent confidence level. If no significant effects are detected, the organism would be approved for release. However, if evidence is found to support the hypothesis of a significant ecological risk, the crop, though not rejected automatically for release, would be passed through a risk-benefit test. The estimated value of the ecological damages would be compared to the potential net benefits of releasing the crop, including production, human health, and any positive environmental effects, such as pesticide toxicity reductions. Note that the production benefits must incorporate the relevant social value of added production due to lowering the supply curve, which for the United States and EU countries may be negative if excess supplies are creating deadweight losses. The decision to release is made by a comparison of the estimated ecological

Table 2. Comparison of APHIS, Risk-Benefit, and Precautionary Risk Management

Questions	APHIS-like Approach	Risk-Benefit Approach	Precautionary Approach
What is the framing hypothesis?	No significant environmental hazard (the null hypothesis)	Significant environmental effect (alternative hypothesis)	Significant environmental effect (alternative hypothesis)
What rule is used to test the hypothesis?	Probability of rejecting null hypothesis when it is true (type I error) is less than critical value, e.g., $\alpha = 05$	Power of test to correctly reject null hypothesis (1- type II error) is high, e.g., .90	Power of test to correctly reject null hypothesis, (1- type II error) is very high, e.g., .95
What party is responsible for the burden of proof?	U.S. government	Shared between the U.S. government and entity introducing crop	Entity introducing the transgenic crop (as certified by independent party)
What costs are considered?	Lost production, environmental, and health benefits from not releasing the crop	Lost production, environmental, and health benefits from not releasing the crop	Potential ecological risks from releasing the organism outweigh economic considerations
What is the general rule for making release decisions?	Permit release if test to minimize type I error at standard $\alpha = .05$ level indicates no significant ecological risk, or if net benefits exceed the ecological risks/costs	Permit release if test to minimize type II error at high power level does not indicate significant ecological risk, or if net benefits exceed the ecological risks/costs	Permit release if test to minimize II error at very high power level is passed, but avoid irreversible risks until information is available to assure adequate ecological safety
Will compensation be provided to negatively affected parties?	Collect some of the net benefits to compensate "losers" or remediate damages	Collect some of the net benefits to compensate "losers" or remediate damages	Not applicable

risks/costs against the potential social benefits. Nonmarket valuation methods would be applied to those ecological effects for which reliable monetary values could be estimated. It is doubtful that all effects could be reliably monetized. Thus, expert scientific and policy judgments would be necessary to compare order-of-magnitude effects and implement the decision rule.

If the estimated benefits outweigh the potential costs, then release would be permitted. To turn the cost-benefit decision rule into a real rather than potential Pareto improvement, a portion of the net benefits would be used to compensate for associated losses, such as from contamination from genetic drift. The burden of proof lies with the government using information provided by the applying entity for this least precautionary category. Because novel risks are unlikely and good quality information about the ecological risks is likely available for these familiar crops, this risk assessment based on minimizing type I error will result in more commercialization decisions than the following models. However, APHIS currently uses this probability rule to approve the field testing of approximately 99 percent of most transgenic crops, which would fall into the next model in our differentiated approach.

For linegenic and transgenic organisms, more stringent tests for ecological risk would be applied. Because of our relative lack of experience with these crops, they conceivably could introduce serious ecological risks. An independent scientific panel would first screen the crops for potentially serious irreversible impacts, and any such organisms would move to the precautionary risk assessment process (model 3) with higher standards for release. For the remaining crops in this category, the framing hypothesis that the crop causes significant ecological risks would be tested to control type II error by specifying a minimum power of the test, e.g., 90 percent ($\beta = .10$).[5] Adequate ecological risk information is a prime requirement to frame and test the alternative hypotheses. However, this task presents a conundrum. Small-scale field trials before commercialization can detect order-of-magnitude differences in ecological effects, but low-probability and low-magnitude effects likely will escape detection (NRC 2002). Evidence collected from large-scale field trials would be required. Thus, the test may have to be conducted in progressive stages of field experiments followed by limited releases to gather sufficient data to assess all potential ecological impacts. This process would address a weakness in current ecological monitoring of GM crops (NRC 2000, p. 19). The USDA and the entity proposing release would share responsibility in gathering the ecological risk data under scientifically certified protocols. The entity requesting permission to release could conduct the tests, if the experimental design and measurement were inde-

---

[5] Knudsen and Scandizzo (2001) offer a similar approach by reversing the null hypothesis from one of no significant effect to one of presuming a significant effect.

pendently certified. Alternatively, the tests could be conducted by an independent certifying body.

If the test indicates that the crop does not cause significant ecological risk, its release would be permitted. Further monitoring of crops that pose unknown long-term effects, such as cases of uncertain resistance development, would be conducted. Just as for intragenic and famigenic crops, if evidence is found to support the hypothesis of a significant ecological hazard, the estimated production, environmental, and health benefits of the linegenic or transgenic crop would be compared to its estimated ecological damages before making a decision to release. Expert and diverse scientific panels would be used to evaluate the ecological effects because of less knowledge about the impacts of these organisms than with the first category. In cases where the science and evidence are not robust, the regulating authority may choose to permit release but require periodic review with new monitoring data to improve the analysis of risks and may later renew or revoke commercialization. As for the first model, some of the net benefits would be used to compensate for associated losses, such as transgenic contamination of organic fields from genetic drift.

The final model applies to transgenic crops judged to hold the potential to cause serious irreversible ecological effects and to xenogenic crops. As for linegenic and transgenic crops, the framing hypothesis is for significant environmental effects and where type II error is controlled. However, the standard for approving release of these crops is extremely high. For example, the required power level could be increased over model 2 to 95 percent ($\beta = .05$). The entity applying for release must prove beyond scientific doubt that the organism is safe. Expert scientific panels with representation from all relevant ecological sciences would be used to implement the model and make decisions concerning release. The potential social benefits would not be considered until minimum levels of safety were assured for all ecological hazards.

It is important to note that a well-designed environmental regulatory process does more than minimize the potential for unwanted environmental hazards from new technologies. If implemented properly, environmental regulations can provide incentives and disincentives to beneficially influence the research and technology development process. Under the differentiated risk assessment framework, imposing higher regulatory costs on organisms that pose higher ecological risks stimulates research and development of GM varieties with traits that provide production benefits with acceptable environmental risk and perhaps ecological benefit. To realize this outcome, a new set of bioengineered traits would be developed and inserted into important agronomic crops: traits that are less likely, for example, to result in resistant insect populations or to harm non-target organisms. Increased and targeted involvement of the public sector agricultural research and regulatory branches also is necessary to achieve these types of outcomes (Ervin et al. 2003).

7. CONCLUSIONS

There is a substantial need for increased public research funding on the environmental effects of transgenic crops and for research of a different character. It is natural to ask why more public research is needed when private research on transgenic crops has increased so dramatically. Under current U.S. biosafety regulatory policies, private industries have scant incentive to invest in the research to understand the environmental impacts of transgenic crops, especially the ecosystem effects beyond the farm boundary. Most environmental risks stem from missing markets; there are few or no market incentives for reducing the environmental risks of transgenic crops. Thus, private research to control the full range of negatively affected environmental services will not be triggered by current market and regulatory signals (Batie and Ervin 2001).

Evidence in support of this argument is provided by the recent decision by Pioneer Hi-Bred and Dow AgroEvo to deny access to the proprietary materials required by independent scientists to conduct biosafety analysis of Bt sunflower (Dalton 2002)—a decision made even more problematic by the fact that it was made after the firms had initially cooperated with Snow and her colleagues. Permission to access the material was withdrawn only after the scientists' preliminary findings indicated potential biosafety risks from Bt sunflower (Snow 2002, Snow et al. 2003).

Likewise, it is unrealistic to assume that most private firms will develop transgenic crops that provide ecological benefits and minimize potential risks in line with social preferences. The development of such crops would suffer from the same missing markets dilemma since the environmental benefits would not merely accrue to the farmer that purchased the transgenic seed. Rather, other farmers, the general public, and even future generations would enjoy the benefits from such crops. For example, a vehicle for addressing many of the identified potential risks from insect pest resistant crops is to develop crops that are pest damage tolerant rather than toxic to the pest, as are Bt crops (Hubbell and Welsh 1998, Pedigo 2002). The difference between tolerance of damage and resistance to pests is fundamental. Tolerance does not rely on toxicity to kill pests and therefore does not negatively impact non-target organisms or promote resistance development (Welsh et al. 2002). Pedigo (2002) finds that certain crops display tolerance to pest damage. This characteristic has been used commercially with great success for decades with no public controversy. For example, cucumbers with stable tolerance to cucumber mosaic virus have dominated the industry since the 1960s. Genetic modification could be used to amplify these types of properties in several other important crops (Pedigo 2002). The publicness of the environmental benefits potentially derived from such crops dampens private sector enthusiasm to develop and commercialize them.[6]

However, if regulatory policies effectively controlled type II error for transgenic crops, the private sector would receive signals and incentives to more fully assess environmental risks and develop crops that cause less risk while providing production, health, and other potential market benefits. If, for example, governments assigned liability for the deleterious environmental effects to the biotechnology company, perhaps through the posting of a significant bond upon commercialization, more private R&D resources would likely be devoted to controlling adverse effects either through risk assessment research or through developing technologies such as damage-tolerant crops. In essence, this approach forces firms to take into account the shadow price of environmental risks when making decisions about attempting to commercialize a transgenic technology or investing in the development of crops with particular sets of characteristics or traits.

REFERENCES

Adam, D. 2003. "Transgenic Crop Trial's Gene Flow Turns Weeds into Wimps." *Nature* 421(6922): 462.

Antle, J.M., and S.M. Capalbo. 1998. "Quantifying Agriculture-Environment Tradeoffs to Assess Environmental Impacts of Domestic and Trade Policies." In J.M. Antle, J.N. Lekakis, and G.P. Zanias, eds., *Agriculture, Trade and the Environment: The Impact of Liberalization on Sustainable Development.* Cheltenham, UK: Edward Elgar.

Barrett, S., J. Beare-Rogers, C. Brunk, T. Caulfield, B. Ellis, M. Fortin, A. Ham Pong, J. Hutchings, J. Kennelly, J. McNeil, L. Ritter, K. Wittenberg, R. Wyndham, and R. Yada. 2001. *Elements of Precaution: Recommendations for Regulation of Food Biotechnology in Canada.* Ottawa: Royal Society of Canada.

Batie, S., and D. Ervin. 2001. "Transgenic Crops and the Environment: Missing Markets and Public Roles." *Environment and Development Economics* 6(4): 435–457.

Beringer, J.E. 1999. "Cautionary Tale on Safety of GM Crops." *Nature* 399(6735): 405.

---

[6] Some new products follow this approach, but they are clearly the exception rather than the rule. The biopesticide *Messenger* is an interesting example of this approach to technology development. *Messenger* is a biopesticide that acts as a non-hormone growth regulator for a wide variety of plants. The active ingredient is the naturally occurring Harpin bacterial protein. Applying *Messenger* topically, as a spray, essentially signals plants to activate their natural plant defenses against a variety of diseases and boosts plant development and growth. The primary mode of action is not toxicity to the invading pathogen, but rather entrapment or the creation of a physical barrier to the movement of the pathogen through localized cell death (see EDEN Bioscience Corporation 2002a, 2002b; Wei et al. 1992).

Birch, A.N.E., I.E. Geoghegan, M.E.N. Majerus, C. Hackett, and J. Allen. 1997. "Interactions Between Plant Resistance Genes, Pest Aphid Populations and Beneficial Aphid Predators." In *Scottish Crop Research Institute Annual Report 1996/97*. Scottish Crop Research Institute, Dundee, Scotland.

Bishop, R.C., and A. Scott. 1999. "The Safe Minimum Standard of Conservation and Environmental Economics." *Aestimum* 37: 11–40.

Buhl-Mortensen, L. 1996. "Type-II Statistical Errors in Environmental Science and the Precautionary Principle." *Marine Pollution Bulletin* 32(7): 528–531.

Carlson, G.A., M.C. Marra, and B. Hubbell. 1998. "Yield, Insecticide Use, and Profit Changes from Adoption of Bt Cotton in the Southeast." *Beltwide Cotton Conference Proceedings* (2): 973–974. National Cotton Council, Memphis, TN.

Carrière, Y., C. Ellers-Kirk, M. Sisterson, L. Antilla, M. Whitlow, T.J. Dennehy, and B.E. Tabashnik. 2003. "Long-Term Regional Suppression of Pink Bollworm by *Bacillus thuringiensis* Cotton." *Proceedings of the National Academy of Sciences of the United States of America* 100(4): 1519–1523.

Dale, P.J., B. Clarke, E.M.G. Fontes. 2002. "Potential for Environmental Impact of Transgenic Crops." *Nature Biotechnology* 20(6): 576–574.

Dalton, R. 2002. "Superweed Study Falters as Seed Firms Deny Access to Transgene." *Nature* 419(6908): 655.

EDEN Bioscience Corporation. 2002a. "What Is Harpin?" EDEN Bioscience Corporation, Bothell, WA. Available online at http://www.edenbio.com/tk/tkmain_whitepaper.html (accessed June 26, 2003).

———. 2002b. "What Is Messenger?" EDEN Bioscience Corporation, Bothell, WA. Available online at http://www.edenbio.com/tk/tkmain_whitepaper.html (accessed June 26, 2003).

Ellstrand, N. 2001. "When Transgenes Wander, Should We Worry?" *Plant Physiology* 125(4): 1543–1545.

Ervin, D., R. Welsh, S. Batie, and C.L. Carpentier. 2001. "Public Research for Environmental Regulation of Transgenic Crops." Unpublished paper, Environmental Sciences and Resources Program, Portland State University, Portland, OR.

———. 2003. "Towards an Ecological Systems Approach in Public Research for Environmental Regulation of Transgenic Crops." *Agriculture, Ecosystems and the Environment* 99(1/3): 1–14.

Fernandez-Cornejo, J., and W. McBride. 2002. "Adoption of Bioengineered Crops." Economic Research Service, U.S. Department of Agriculture, Washington, D.C.

Greene, A.E., and R.F. Allison. 1994. "Recombination Between Viral RNA and Transgenic Plant Transcripts." *Science* 263(5152): 1423–1425.

Gould, F., N. Blair, M. Reid, T.L. Rennie, J. Lopez, and S. Micinski. 2002. "Bacillus Thuringiensis-Toxin Resistance Management: Stable Isotope Assessment of Alternate Host Use by *Helicoverpa zea*." *Proceedings of the National Academy of Sciences of the United States of America* 99(26): 16581–16586.

Hails, R.S. 2000. "Genetically Modified Plants: The Debate Continues." *Trends in Ecology and Evolution* 15(1): 14–18.

Hall, L.M., J. Huffman, and K. Topinka. 2000. "Pollen Flow Between Herbicide Tolerant Canola (Brassica napus) Is the Cause of Multiple Resistant Canola Volunteers." *Weed*

Science Society of America Abstracts 40. Weed Science Society of America, Lawrence, KS.

Hansen-Jesse, L.C., and J.J. Obrycki. 2000. "Field Deposition of Bt Transgenic Corn Pollen: Lethal Effects on the Monarch Butterfly." *Oecologia* 125(2): 241–248.

Heimlich, R.E., J. Fernandez-Cornejo, W. McBride, C. Klotz-Ingram, S. Jans, and N. Brooks. 2000. "Adoption of Genetically Engineered Seed in U.S. Agriculture: Implications for Pesticide Use." In C. Fairbairn, G. Scoles and A. McHughen, eds., *The Biosafety of Genetically Modified Organisms*. University Extension Press, University of Saskatchewan, Saskatoon.

Hilbeck, A., M. Baumgartner, P.M. Fried, and F. Bigler. 1998a. "Effects of Transgenic Bacillus Thuringiensis Corn-fed Prey on Mortality and Development of Immature Chrysoperla Carnea (Neuroptera: Chrysopidae)." *Environmental Entomology* 27(2): 1–8.

Hilbeck, A., W.J. Moar, M. Pusztai-Carey, A. Filippini, and F. Bigler. 1998b. "Toxicity of *Bacullus thuringiensis* Cry1AB Toxin to the Predator *Chrysoperla Carnea* (Neuroptera Chrysopidae)." *Environmental Entomology* 27(5): 1255–1263.

Hoy, C.W., J. Feldman, F. Gould, G.G. Kennedy, G. Reed, and J.A. Wyman. 1998. "Naturally Occurring Biological Controls in Genetically Engineered Crops." In P. Barbosa, ed., *Conservation Biological Control*. New York: Academic Press.

Hubbell, B.J., M.C. Marra, and G.A. Carlson. 2000. "Estimating the Demand for a New Technology: Bt Cotton and Insecticide Policies." *American Journal of Agricultural Economics* 82(1): 118–132.

Hubbell, B.J., and R. Welsh. 1998. "Transgenic Crops: Engineering a More Sustainable Agriculture?" *Agriculture and Human Values* 15(1): 43–56.

Jasanoff, S. 2000. "Commentary: Between Risk and Precaution: Reassessing the Future of GM Crops." *Journal of Risk Research* 3(3): 227–282.

Keeler, K.H., C.E. Turner, and M.R. Bolick. 1996. "Movement of Crop Transgenes into Wild Plants." In S.O. Duke, ed., *Herbicide Resistant Crops: Agricultural, Environmental, Economic, Regulatory, and Technical Aspects*. Boca Raton, FL: CRC/Lewis Publishers.

Kessler, C., and L. Economidis. 2001. *EC-Sponsored Research on Safety of Genetically Modified Organisms: A Review of Results*. European Commission, Brussels. Available online at http://europa.eu.int/comm/research/quality-of-life/gmo/index.html (accessed May 14, 2003).

Knudsen, O., and P. Scandizzo. 2001. "Evaluating Risks of Biotechnology: The Precautionary Principle and the Social Standard." In *International Consortium on Agricultural Biotechnology Research*. Proceedings of the International Consortium on Agricultural Biotechnology Research (ICABR), Ravello, Italy, June 15–18.

Lemons, J., K. Shrader-Frechette, and C. Cranor. 1997. "The Precautionary Principle: Scientific Uncertainty and Type I and Type II Errors." *Foundations of Science* 2(2): 207–236.

Losey, J.E., L.S. Raynor, and M.E. Carter. 1999. "Transgenic Pollen Harms Monarch Larvae." *Nature* 399(6733): 214.

McCammon, S. 2001. "APHIS' Review of Biotechnology Products." Animal and Plant Health Inspection Service, U.S. Department of Agriculture, Washington, D.C. Available online at http://www.usda.gov/gipsa/millennium/mccammon.htm (accessed May 14, 2003).

Mooney, S., and K. Klein. 1999. "Environmental Concerns and Risks of Genetically Modified Crops: Economic Contributions to the Debate." *Canadian Journal of Agricultural Economics* 47(4): 437–444.

Morin, S., R.W. Biggs, M.S. Sisterson, L. Shriver, C. Ellers-Kirk, D. Higginson, D. Holley, L.J. Gahan, D.G. Heckel, Y. Carrière, T.J. Dennehy, J.K. Brown, and B.E. Tabashnik. 2003. "Three Cadherin Alleles Associated with Resistance to *Bacillus Thuringiensis* in Pink Bollworm." *Proceedings of the National Academy of Sciences of the United States of America* 100 (9): 5004–5009.

National Research Council. 2000. *Ecological Monitoring of Genetically Modified Crops: A Workshop.* Washington, D.C.: National Academy Press.

_____. 2002. *Environmental Effects of Transgenic Plants: The Scope and Adequacy of Regulation.* Washington, D.C.: National Academy Press.

Nielsen, K. 2003. "Transgenic Organisms: Time for Conceptual Diversification?" *Nature Biotechnology* 21(3): 227–228.

NRC [see National Research Council].

Owen, M. 1997. "North American Developments in Herbicide Tolerant Crops." *The British Crop Protection Conference.* Brighton, England. Available online at http://www.weeds.iastate.edu/weednews/Brighton.htm (accessed April 19, 2003).

Pedigo, L. 2002. *Entomology and Pest Management* (4th ed.). New York: Prentice Hall.

Pham-Delegue, M.H., L.J. Wadhams, A.M.R. Gatehouse, A. Toppan, and J.-L. Deneubourg. 2000. "Environmental Impact of Transgenic Plants on Beneficial Insects." In C. Kessler and L. Economidis, eds., *EC-Sponsored Research on Safety of Genetically Modified Organisms: A Review of Results.* European Commission, Brussels. Available online at http://europa.eu.int/comm/research/quality-of-life/gmo/01-plants/01-07-project.html (accessed May 14, 2003).

Rissler, J., and M. Mellon. 1996. *The Ecological Risks of Engineered Crops.* Cambridge, MA: MIT Press.

Royal Society of United Kingdom. 1998. *Genetically Modified Plants for Food Use.* London: Carlton House Terrace.

Saxena, D., and G. Stotzky. 2000. "Insecticidal Toxin is Released from Roots of Transgenic *Bt* Corn *In Vitro* and *In Situ.*" *FEMS Microbiology Ecology* 33(1): 35–39.

Sayyed, A.H., H. Cerda, and D.J. Wright. 2003. "Could Bt Transgenic Crops Have Nutritionally Favorable Effects on Resistant Insects?" *Ecology Letters* 6(3): 167–169.

Sears, M.K., R.L. Hellmich, D.E. Stanley-Horn, K.S. Oberhauser, J.M. Pleasants, H.R. Mattila, B.D. Siegfried, and G.P. Dively. 2001. "Impact of Bt Corn Pollen on Monarch Butterfly Populations: A Risk Assessment." *Proceedings of the National Academy of Sciences of the United States of America* 98 (21): 11937–11942.

Snow, A. 2002. "Transgenic Crops: Why Gene Flow Matters." *Nature Biotechnology* 20(6): 542.

Snow, A., and P.M. Palma. 1997. "Commercialization of Transgenic Plants: Potential Ecological Risks." *BioScience* 47(2): 86–96.

Snow, A., D. Pilson, L.H. Rieseberg, M. Paulsen, N. Pleskac, M.R. Reagon, D.E. Wolf, and S.M. Selbo. 2003. "A Bt Transgene Reduces Herbivory and Enhances Fecundity in Wild Sunflowers." *Ecological Applications* 13(2): 279–286.

Stanley-Horn, D.E., G.P. Dively, R.L. Hellmich, H.R. Mattila, M.K. Sears, R. Rose, L.-C.H. Jesse, J.E. Losey, J.J. Obrycki, and L. Lewis. 2001. "Assessing the Impact of Cry1Ab-expressing Corn Pollen on Monarch Butterfly Larvae in Field Studies." *Proceedings of the National Academy of Sciences of the United States of America* 98(21): 11931–11936.

Tabashnik, B.E. 1994. "Evolution of Resistance to Bacillus Thuringiensis." *Annual Review of Entomology* 39: 47–79.

Tabashnik, B.E., T.J. Dennehy, Y. Carrière, Y. Liu, S.K. Meyer, A. Patin, M.A. Sims, and C. Ellers-Kirk. 2001. "Resistance Management: Slowing Pest Adaptation to Transgenic Crops." *Acta Agriculturae Scandinavica* B 53(Suppl. 1): 57–59.

Tabashnik, B.E., A.L. Patin, T.J. Dennehy, Y. Liu, Y. Carrière, M.A. Sims, and L. Antilla. 2000. "Frequency of Resistance to *Bacillus Thuringienses* in Field Populations of Pink Bollworm." *Proceedings of the National Academy of Sciences of the United States of America* 97(24): 12980–12984.

Taylor, M., and J. Tick. 2003. *Postmarket Oversight of Biotech Foods: Is the System Prepared?* Pew Initiative on Food and Biotechnology, Washington, D.C.

U.S. Department of Agriculture. 1999a. "Genetically Engineered Crops for Pest Management." Economic Research Service, U.S. Department of Agriculture, Washington, D.C. Available online at http://www.econ.ag.gov/whatsnew/issues/biotech (accessed March 4, 2000).

_____. 1999b. "Impacts of Adopting Genetically Engineered Crops in the U.S.: Preliminary Results." Economic Research Service, U.S. Department of Agriculture, Washington, D.C. Available online at http://www.ers.usda.gov/emphases/harmony/issues/genengcrops/genengcrops.htm (accessed March 4, 2000).

_____. 2003. "Prospective Plantings." National Agricultural Statistics Service, Agricultural Statistics Board, U.S. Department of Agriculture, Washington, D.C. Available online at http://usda.mannlib.cornell.edu/reports/nassr/field/pcp-bbp/pspl0303.txt (accessed April 3, 2003).

Van den Belt, H. 2005. "Biotechnology, the U.S.-EU Dispute and the Precautionary Principle." In Wesseler, J., ed., *Environmental Costs and Benefits of Transgenic Crops* (Wageningen UR Frontis Series, Vol. 7). Dordrecht: Springer.

VanGessel, M.J. 2001. "Glyphosphate Resistant Horseweed from Delaware." *Weed Science* 49(6): 703–705.

Watkinson, A.R., R.P. Freckleton, R.A. Robinson, and W.J. Sutherland. 2000. "Predictions of Biodiversity Response to Genetically Modified Herbicide-Tolerant Crops." *Science* 289(5484): 1554–1557.

Watrud, L.S., and R.J. Seidler. 1998. "Nontarget Ecological Effects of Plant, Microbial, and Chemical Introductions to Terrestrial Systems." In P.M. Huang, ed., *Soil Chemistry and Ecosystem Health*. Soil Science Society of America, Madison, WI.

Wei, Z.M., R.J. Laby, C.H. Zumoff, D.W. Bauer, S.Y. He, A. Collmer, and S.V. Beer. 1992. "Harpin Elicitor of the Hypersensitive Response Produced by the Plant Pathogen Erwinia Amylovora." *Science* 257(5066): 85–88.

Welsh, R., B. Hubbell, D. Ervin, and M. Jahn. 2002. "GM Crops and the Pesticide Paradigm." *Nature Biotechnology* 20(6): 548–549.

Wolfenbarger, L., and P. Phifer. 2000. "The Ecological Risks and Benefits of Genetically Engineered Plants." *Science* 290(5499): 2088–2093.

Chapter 16

IRREVERSIBILITY, UNCERTAINTY, AND THE ADOPTION OF TRANSGENIC CROPS: EXPERIENCES FROM APPLICATIONS TO HT SUGAR BEETS, HT CORN, AND Bt CORN

Sara Scatasta,[*] Justus Wesseler,[*] and Matty Demont[†]
Wageningen University, the Netherlands,[*] and Catholic University, Leuven, Belgium[†]

Abstract: This study applies a real option approach to quantify, *ex-ante*, the maximum incremental social tolerable irreversible costs (MISTIC) that would justify immediate adoption of HT and Bt corn in the European Union (EU). The results are compared with previous ones for HT sugar beets. In total, according to our analysis, the EU gives up about €309 million on average per year due to the quasi moratorium on transgenic crops for Bt corn, HT corn, and HT sugar beets alone. On the other hand, the MISTIC per household and year for Bt corn, HT corn, and HT sugar beets is €0.27, €0.46, and €1.10, respectively, or €1.83 for all three crops. The low MISTIC provides a strong economic argument for prohibiting the immediate introduction of the three transgenic crops. The validity of the argument will largely depend on consumer attitudes towards transgenic crops.

Key words: Bt corn, HT corn, HT sugar beets, real option approach, irreversible social costs

1. INTRODUCTION

Regulating the introduction of new technologies and genetically modified crops in particular requires a good understanding of the social benefits and costs of the technology (Ervin and Welsh 2005). An immediate release of a transgenic crop is expected to provide both instant and future benefits, through the positive effects on yields, product quality, production costs,

Financial support from the VIB–Flanders Interuniversity Institute for Biotechnology and from the European Union under the EUWAB and ECOGEN projects are gratefully acknowledged.

and/or other characteristics of the crop.[1] However, an immediate release may also have negative but highly uncertain effects, notably potential environmental or health risks (Kendall et al. 1997). Thus, decision makers are faced with a dilemma: they can release the crop(s) and receive (part of) the benefits mentioned above, but at a risk of irreversibly harming human beings and the environment. Alternatively, they may delay/reject the release, thereby avoiding human and environmental risks, waiting for new information but also missing out on potential benefits. The decision maker has to weigh the expected benefits from an immediate release against its potential risks, and the option of delaying the decision until a future time.

Traditional *ex-ante* assessment of the costs and benefits of a new product of agriculture biotechnology does not take into consideration that the adoption of a new technology might be associated with higher risks and uncertainty with respect to both its costs and its benefits. Some of these costs and benefits might be irreversible in nature. Irreversible costs and benefits imply that, once the decision is taken, it is not possible to go back to the economic equilibrium that existed before such a decision was made. Examples of irreversible costs associated with the adoption of genetically modified organisms (GMOs) are losses in biodiversity and development of pest resistance. Examples of irreversible benefits are gains in human health due to reduced poisonings from pesticide use and gains in biodiversity from reduced pesticide use.[2] In this context the option to delay the release of a GMO until more information on its risks becomes available may become of value to society. The value of the possibility of delaying the decision to release transgenic crops into the environment can be explicitly taken into consideration by analysts via a *real option approach* (Demont, Wesseler, and Tollens 2004, Morel et al. 2003, Wesseler 2003).

The *real option* decision criteria for releasing GMOs immediately requires that reversible private net-benefits from GM crops, such as net-benefits accruing to farmers, be greater than irreversible social net-costs by a factor, the hurdle rate, that depends on the uncertainty associated with the adoption of a new technology.

Following Dixit and Pindyck (1994), hurdle rates associated with GM crops can be quantified by assuming that additional private net-benefits from transgenic crops follow a geometric Brownian process. The hurdle rate then becomes a well-specified function whose parameters can be inferred from time-series data on farmer gross margins and secondary literature, by assuming that GM crops constitute a normal technological change.

---

[1] Throughout the chapter we will use the terms transgenic crops, GMOs (genetically modified organisms), and GM crops (genetically modified crops) interchangeably, following the FAO (Food and Agriculture Organization) glossary (Zaid et al. 2001).

[2] For a detailed discussion of irreversible benefits and costs and their impact on the decision to release transgenic crops consult, Demont, Wesseler, and Tollens (2004, 2005).

As hurdle rates are always greater than one, the *real option* decision criteria for releasing transgenic crops immediately differs from the traditional decision criteria as it requires that reversible social net-benefits be greater than irreversible social net-costs. The traditional decision criteria for releasing GM crops immediately requires, instead, that reversible private net-benefits be at least equal to irreversible social net-costs.

Demont, Wesseler, and Tollens (2004) compute hurdle rates for herbicide-tolerant (HT) sugar beets and reassess whether the 1998 moratorium of the European Union (EU) on HT sugar beets is justified from a cost-benefit perspective. The authors conclude that such a moratorium would be justified if transgenic sugar beets caused annual irreversible social costs above €121 euros per hectare planted.

The objective of this study is to carry out a similar assessment for pest resistant (Bt) corn and herbicide-tolerant (HT) corn to identify, *ex-ante,* potential social welfare impacts of adoption of Bt and HT corn in the 15 member states of the EU (EU-15) before the enlargement in the year 2004. In section 2 we present some background information on Bt and HT corn for Europe. In section 3 we describe the theoretical model. In section 4 we quantify the maximum incremental social tolerable irreversible costs. In section 5 we discuss the results in light of the current regulations on co-existence in the EU. Section 6 summarizes our findings and conclusion.

2. BACKGROUND

The EU-15 produces about 3 percent of the world's corn. The corn production is concentrated in France (40 percent), immediately followed by Italy (30 percent). The EU-15 also is a net importer of corn for human consumption, 6.4 percent of its consumption being imported (mainly from Argentina [4 percent] and Hungary [2 percent]), while only 0.4 percent of domestic production is exported outside the EU-15 (FAOSTAT 2004).

Corn is grown in the EU-15 mainly for animal feed (80 percent). Corn for human consumption (20 percent) is used to produce corn oil, starch, and sweeteners, which are common ingredients in many processed foods such as breakfast cereals and dairy goods; only a small amount is used for direct consumption (see Essential Biosafety 2004, EUROSTAT 2004).

2.1. Corn Resistant to Pests (Bt corn)

Bt corn in the EU is currently grown in Spain with an adoption rate of about 17.5 percent (0.1 million hectares), and a small amount in Germany and France (James 2004). Bt corn has been genetically engineered to contain a gene of the soil bacterium *Bacillus thuringiensis* (Bt). This bacterium produces a crystal-like (Cry) protein that is toxic to the European corn borer (ECB-*Ostrinia nubilalis*), a pest that can cause damage to corn plants by penetrating the stalk and excavating large tunnels into the plant.

Conventional ECB pest control strategies are difficult to manage because a correct timing of insecticide applications is crucial to their effectiveness. Insecticides are effective only when the ECB is in its larval status but has not yet penetrated the stalk or is migrating to neighboring plants (Jansens et al. 1997). The insertion of the Bt gene into the corn plant potentially improves a farmer's ability to manage the ECB and other serious insect pests (Pilcher et al. 2002). Hence, Bt corn is expected to benefit farmers through reduced harvest losses due to ECB infestation. Bt corn is also expected to benefit the environment through reduced insecticide use.[3] In addition, due to the protection of Bt varieties against physical insect damage, it has been widely reported that Bt varieties are associated with a lower incidence of secondary Fusarium contamination (Munkvold, Hellmich, and Rice 1999, Munkvold, Hellmich, and Showers 1997, Dowd 2000, Wu, Miller, and Casman 2004). On the other hand, due to higher costs for Bt seeds it is not undisputed that the associated yield improvements will also translate into increased farmer income. The development of ECB resistance to Bt due to the commercialization of Bt corn, furthermore, could reduce the benefits of the technology (Hurley 2005, Laxminarayan and Simpson 2005) and be a problem for organic farmers who currently use this bacterium, incorporated into sprays, as a natural crop protection tool (Demont and Tollens 2004).

2.2. Herbicide-Tolerant Corn

Effective weed control is more crucial to economic corn production than control of ECB. The influence of weed cover on yields is often the most important and significant, independent of other factors involved. To control weeds, conventionally a tank mix of soil-active and leaf-active herbicides in pre- to early post-emergence of the crop is used.

The post-emergence herbicides glyphosate and glufosinate-ammonium provide a broader spectrum of weed control than current herbicide programs, while at the same time reducing the number of active ingredients. Glyphosate was first introduced as a herbicide in 1971. The gene that confers tolerance to glyphosate was discovered in a naturally occurring soil bacterium. Glufosinate-ammonium was first reported as a herbicide in 1981. The gene that confers tolerance to glufosinate is also derived from a naturally occurring soil bacterium (Dewar, May, and Pidgeon 2000).

By inserting these herbicide-tolerant (HT) genes into a plant's genome, two commercial transgenic HT systems resulted: the Roundup Ready® sys-

---

[3] This is of lesser importance as many farmers do not control for the ECB in any case, due to the above-mentioned management problems.

tem, providing tolerance to glyphosate, and the Liberty Link® system, providing tolerance to glufosinate-ammonium. The combination of transgenic seed with a post-emergence herbicide offers farmers broad-spectrum weed control, flexibility in the timing of applications, and a reduction in the need for complex compositions of spray solutions. Glyphosate and glufosinate-ammonium have a low toxicity and are metabolized fast, without leaving soil residues, and therefore have better environmental and toxicological profiles than most of the herbicides they replace.

At the social level, the impact of HT corn systems on biodiversity and human health has been questioned, while at the same time positive impacts on biodiversity can be expected due to reduced use of persistent herbicides and the possibility of reduced tillage systems (Heard et al. 2003). Furthermore, a net reduction in pesticide use on HT corn (and Bt corn as well) can have a positive impact on farmers' health (Antle and Pingali 1994, Waibel and Fleischer 1998). It is important to note that the effects of pesticide use on farmers' health and biodiversity are irreversible. If the introduced transgenic crop results in a lower pesticide application, it provides additional irreversible benefits.

The observations for corn are supported by the Environmental Impact Quotient (EIQ) for Bt and HT corn.[4] A recent study by Nillesen, Scatasta, and Wesseler (2005) compared field rate EIQs for Bt corn, HT corn, and conventional practices using the data from ECOGEN field trials in Narbonne, France, and from the study by Gianessi, Sankula, and Reigner (2003). Results show that the EIQ for pesticides used is 43.36 for Bt corn and 47.26 for conventional crop management. The EIQ for glufosinate-tolerant corn is 25.43. The EIQ indicates social benefits from reduced pesticide use in Bt and HT corn. Hence, the introduction of Bt and HT corn varieties can provide important private and social irreversible benefits due to changes in pesticide use. Including the irreversible benefits and irreversible costs in a theoretically consistent decision making framework will be done in the following section.

---

[4] The EIQ is a simple measure to attach a numerical value to environmental impacts of pesticides, which enables one to compare different pesticides regarding their toxicity or harmfulness (Kovach et al. 1992). The EIQ incorporates the impacts on farm worker (application and harvest worker), consumer, and ecology (non-target organisms such as fish, birds, honeybees, and other beneficial insects).

3. THE REAL OPTION APPROACH

3.1. Defining the Maximum Incremental Social Tolerable Irreversible Costs (MISTIC)

The decision to release transgenic crops in the EU involves both uncertainty and irreversibility. The social planner, in this case the EU, has to decide whether to release a transgenic crop immediately or to postpone the release until further information about benefits and costs of the new crop is available. The objective of the planner is assumed to be to maximize the welfare of EU citizens. Impacts of the decision on the rest of the world (ROW) are not considered. Also, benefits for the R&D sector, whether European or ROW, are not considered. Only downstream costs and benefits are of importance for the decision maker.

Given this, the decision maker needs to compare the benefits of an immediate release with those from a postponed decision. Only if the benefits of an immediate release, the *value of the release*, outweigh those of the *option to release*, should the option to release be exercised.

In the case of GMOs it is important to differentiate between the incremental reversible and irreversible benefits and costs at the private as well as the public level. In general, information about potential benefits and costs from GMOs, even though they are uncertain, is available from field trials and experiences in other countries. Also available is information about incremental irreversible benefits, such as health and environmental benefits from changes in pesticide and fuel use. Less available is information about the irreversible costs, such as impacts on biodiversity and ecosystems in general. Under these circumstances, the *real option* decision criteria can be stated as the maximum incremental social tolerable irreversible costs (I^*) being no greater than the sum of the incremental irreversible social benefits (R) and reversible social net-benefits (W) from GM crops weighted by the hurdle rate $\beta/(\beta-1)$ (Wesseler 2003):

$$I^* \leq \frac{W}{\beta/(\beta-1)} + R \tag{1}$$

with

$$\beta = \frac{1}{2} - \frac{r-\delta}{\sigma^2} + \sqrt{\left[\frac{r-\delta}{\sigma^2} - \frac{1}{2}\right]^2 + \frac{2r}{\sigma^2}} > 1,$$

where r is the riskless rate of return, δ the convenience yield defined as $\delta = \mu - \alpha$, with μ as the risk adjusted rate of return and α the drift rate of a geometric Brownian motion,[5] σ the variance rate of a geometric Brownian motion, and β the positive root of the solution for a Fokker-Planck equation.

Since $[\beta/(\beta-1)] > 1$, the *real option* decision criteria is more restrictive than the *traditional* decision criteria $I^* \leq W + R$.

The use in practice of the *real option* decision criteria specified in equation (1) requires quantification of the following factors:

- social incremental reversible benefits from GM crops (W)
- hurdle rate, $\beta/(\beta-1)$
- social incremental irreversible benefits, R

In the following sections the quantification of these three factors for Bt corn and HT corn will be discussed. For the quantification of these three factors for HT sugar beets we refer to the paper by Demont, Wesseler, and Tollens (2004). To simplify the use of terminology, from now on we will call the maximum incremental social tolerable irreversible costs "MISTIC," the social incremental irreversible benefits "SIIB," the social incremental reversible benefits "SIRB," and the private incremental reversible benefits "PIRB."

3.2. Defining the Social Incremental Reversible Net Benefits (SIRB)

To make our results comparable with those for HT sugar beets by Demont, Wesseler, and Tollens (2004), the value of SIRB for country i over time will be given, in our case, by the present value of those benefits in 1995 terms such that

$$\text{SIRB}_i = \text{SIRB}_{i,95} = \int_0^\infty \text{SIRB}_i(t)e^{-\mu t}dt , \qquad (2)$$

where $\mu = 10.5$ is the capital asset pricing model (CAPM) risk adjusted rate of return, as in Demont, Wesseler, and Tollens (2004).

SIRB$_i$ at time t, SIRB$_i(t)$, will be given by the maximum amount of social incremental reversible net-benefits obtainable at time t at complete adoption, SIRB$_{i,\text{MAX}}(t)$, times the adoption rate at time t, $\rho(t)$, such that

$$\text{SIRB}_i(t) = \text{SIRB}_{i,\text{MAX}}(t)\rho(t). \qquad (3)$$

As we were not able to identify, on the basis of the available literature, non-private reversible net-benefits of transgenic corn, in this study the maximum

---

[5] Lognormality of the Brownian motion is not a problem, assuming technology adopters can temporarily suspend planting HT sugar beets and plant non-HT sugar beets instead, without bearing any additional costs. This follows from Dixit and Pindyck (1994 , pp. 187–189).

amount of SIRB$_i$ obtainable at time t at complete adoption, SIRB$_{i, \text{MAX}}(t)$, and the present value of SIRB$_i$ over time entail only private reversible net-benefits, that is SIRB$_i$ = PIRB$_i$.

To quantify the maximum amount of SIRB$_i$ obtainable at time t at complete adoption, SIRB$_{i, \text{MAX}}(t)$, we consider only the most relevant corn market, the market for green corn, corn used for fodder such as silage and corn-cob-mix. Based on EUROSTAT data there is no trade in the market for green corn, thus we choose a closed economy framework composed of the 15 EU member states.

Our model is also framed to recognize the presence of the price support system for corn provided, through a regime of levies and export subsidies, by the European Common Agricultural Policy (CAP). This price support system implies that the price paid by corn buyers is lower than that received by corn sellers. We allow our model to take this difference into consideration (Katranidis and Velentzas 2000).

We consider two sets of market agents: buyers and sellers. We limit the analysis to two types of technologies, transgenic and conventional, without taking organic production into consideration.

As suggested in Katranidis and Velentzas (2000) and Moschini and Lapan (1997), the corn supply function is best represented in constant elasticity log-linear form. The demand for corn is modeled as a function of own price and the quantity produced of animal feed products (e.g., meat and milk). This function is also represented in constant elasticity log-linear form. Potential losses in consumer welfare due to the introduction of GM-feeds may arise if a section of consumers demand animal products derived from non-GM feeds. The EU does require mandatory labeling for GM feeds but not for animal products derived from those. We assume that consumers with revealed preferences for non-GM animal products are the same consumers that buy organic animal products. Therefore, we do not expect a shift in demand with the introduction of GM corn, and consider a pooled market for GM and non-GM corn.

Figure 1 presents a graphical representation of the economic model used to derive private incremental reversible benefits of transgenic corn in the EU-15. The details of the model are laid out in the appendix. In Figure 1 demand for corn, D, intersects the supply curve before the introduction of GM corn, S, at point b. Producers receive a direct subsidy per hectare corn, illustrated by the parallel upward shift of the supply curve to S'. The difference between S and S' at the horizontal axis indicates the amount of direct payment converted to a payment per unit of quantity produced before the introduction of GM corn. The introduction of GM corn reduces the marginal production cost and results in a pivotal downward shift of the supply function S to S_g. The new quantity produced is the amount at equilibrium point d. The subsidy is indicated by the parallel shift from S_g to

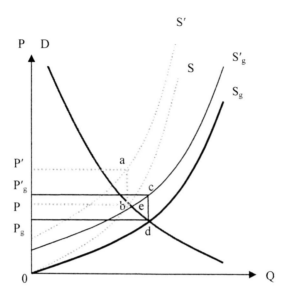

Figure 1. Partial Equilibrium Displacement Model for Transgenic Corn in the EU-15

S'_g. The subsidy paid per unit of corn decreases as the yield of GM corn increases. The consumer surplus[6] of the introduction of GM corn is the area $PbdP_g$, the producer surplus is the difference between $0P'_g cd$ and $0P'ab$, and the welfare gain the sum of the two.

3.3. Defining the Social Incremental Irreversible Benefits (SIIB)

The SIIB per hectare, r_i, of transgenic sugar beets is approximated by

$$r_i = \omega \Delta A_i + \psi \Delta n_i Dc \,, \tag{4}$$

where ΔA_i is the change in volume of pesticide active ingredients per unit of land in country i due to the switch from conventional crop protection to

---

[6] The consumer surplus is the surplus that will be distributed over the whole supply chain. As there is almost no trade in green corn, the producer and consumer surplus will remain within one farm during the early years of planting transgenic corn. Under the current CAP, producer and consumer surplus from lower production costs will be kept within the agricultural sector. Over time the benefits will be distributed along the supply chain, as the agriculture policy sector will also react by changing policies such as reducing subsidies, as observed in the EU. Whether final consumers will benefit due to lower prices will depend on the competitiveness of the downstream food processing and retailing sector. Therefore, the interpretation of the consumer surplus without considering a change in agriculture policies is as a surplus that remains within the agriculture sector.

transgenic corn, ω the average external social cost of pesticide use per unit of active ingredient, Δn_i the change in the number of applications per hectare, D the average diesel use per application and per unit of land, c the average CO_2 emission coefficient per unit of diesel, and ψ the average external social cost per unit CO_2 emission. We assume that the per hectare $SIIB_i$ function is proportional to the adoption function for the new crop:

$$\text{SIIB}_i(t) = r_i \frac{\rho_{\max,i}}{1 + \exp(-a_{\rho,i} - b_{\rho,i}t)}. \tag{5}$$

The 1995 present value of the $SIIB_i$ can be written as

$$\text{SIIB}_{95,i} = \int_0^\infty R_i(t)e^{-\mu_i t}dt. \tag{6}$$

4. DATA

The data used for this analysis are from EUROSTAT New Cronos database (ECD) and the FAOSTAT-Agriculture database (FAD) (EUROSTAT 2004, FAOSTAT 2004). From ECD we obtained data on produced quantities and input and output prices (with 1995 base) for green corn. Output prices received by corn sellers include subsidies to agricultural producers. From FAD we obtained produced quantities of milk and meat.

Data for estimating the Bt and HT corn adoption curve was obtained from ISAAA. Data for estimating the proportionate vertical supply shift, K, in the supply function for Bt corn was obtained from field trials from the EU-funded ECOGEN project in Narbonne, France. Information on the vertical supply shift for HT corn was obtained from Gianessi, Sankula, and Reigner (2003).

Where no estimates were available from secondary literature, demand and supply elasticities were taken from the European Simulation Model (ESIM) derived from behavioral equations. Suggested green corn demand elasticities in ESIM range from -0.7 to -1.0, so we use the average -0.85 as our base case. Suggested elasticities of land allocation to green corn are 0.77, so we approximated supply elasticities to this value in our base case outlined in Table 1 (see Banse, Grethe, and Nolte 2005).

From field trials (16 plots) carried out for the ECOGEN project in Narbonne, France, we obtained the following information about the percentage shift in the supply function (see appendix equation A9):

$$K = \frac{mc_c/y_c - mc_{Bt}/y_{Bt}}{mc_c/y_c} = \frac{729/11.37 - 727/12.4}{729/11.37} = 0.0852. \tag{7a}$$

Table 1. Base Demand and Supply Elasticities for Green Corn

Country	Source	Demand elasticity	Supply elasticity
Austria	ESIM[a]	-0.85	0.77
Belgium/Luxemburg			
Denmark	ESIM[a]	-0.85	0.77
France	ESIM[a]	-0.85	0.77
Germany	ESIM[a]	-0.85	0.77
Greece	(Katranidis and Velentzas 2000)	-0.92	0.6
Italy	ESIM[a]	-0.85	0.77
The Netherlands	ESIM[a]	-0.85	0.77
Spain	supply elasticity (Lekakis and Pantzios 1999); demand elasticity: ESIM[a]	-0.85	2.5
Portugal			
United Kingdom	ESIM[a]	-0.85	0.77

[a] European Stimulation Model.

Variable operational costs for conventional technology are calculated as the average over the 8 plots managed with the conventional technology using conventional seeds. The average value of variable costs over three plots (one plot was destroyed by protestors) managed with the Bt technology was used as the indicator for the costs of the Bt technology.

With respect to HT corn we assume a percentage shift in the supply function of 0.12 (Gianessi, Sankula, and Reigner 2003) such that:

$$[K]_{Ht} = \frac{mc_c / y_c - mc_{Ht} / y_{Ht}}{mc_c / y_c} = 0.12 . \tag{7b}$$

Estimates of the adoption curves for Bt and HT corn were obtained assuming an adoption rate ceiling of 30 percent for Bt corn and an adoption rate ceiling of 40 percent for HT corn but with half the speed of the U.S. adoption:[7]

$$\ln\left(\frac{\theta(t)}{0.3 - \theta(t)}\right)_{Bt} = 2.41 - 0.335t \tag{8a}$$

---

[7] The original estimated speed of adoption was 0.67 for Bt corn and 0.18 for HT corn.

$$\ln\left(\frac{\theta(t)}{0.3-\theta(t)}\right)_{HT} = 2.15 - 0.09t \,. \tag{8b}$$

Given the information in equations (7), (8a), and (8b), we computed the SIRB of Bt and HT corn as the sum of the changes for the EU-15 in producer and consumer surplus assuming no change in the buyers' demand for corn. We also assumed that Bt corn would be adopted only in France, Italy, Portugal, Spain, and Greece, where, based on expert opinions, the ECB infestation levels would justify adoption of Bt corn. We considered that HT corn would be adopted in each country currently producing green corn. Due to lack of data we could analyze only the EU-15 countries, except for Finland, Ireland, and Sweden. As those three countries plant only a small amount of corn, leaving them out does not have a great impact on the results for the EU-15.

SIRB is presented in annuities, $SIRB_a$, in million euros for the EU-15:

$$\left[SIRB_a = \mu SIRB_{95} = \mu \int_0^\infty \Delta PS(t) + \Delta CS(t) e^{-\mu t} dt \right]_{Bt} \tag{9a}$$

for Bt corn, and

$$\left[SIRB_a = \mu SIRB_{95} = \mu \int_0^\infty \Delta PS(t) + \Delta CS(t) e^{-\mu t} dt \right]_{HT} \tag{9b}$$

for HT corn.

For estimating the drift rate, α, and the variance rate, σ, of the new technology, we compute the maximum likelihood estimator assuming continuous growth (Campbell, Lo, and MacKinlay 1997). We use time-series data on annual gross margin differentials in corn production from 1973 to 1995 as a proxy for estimating the drift and variance rate of future SIIB. The data are extracted from the EU/SPEL dataset (EUROSTAT 1999) for all EU-15 member states and deflated and converted into real terms using the GDP deflators published by the World Bank (2002). The country-specific hurdle rate is calculated using the estimated drift and variance rate per country and choosing a risk-free rate of return, r, of 4.5 percent, and a risk-adjusted rate of return, μ, of 10.5 percent for all countries. The results for individual countries differ, as the estimated drift rates, $\hat{\alpha}_i$, and variance rates, $\hat{\sigma}_i$, vary between EU member states depending on the time series for the gross margin per member state. Finally, data on areas planted to corn, numbers of corn holdings, and currency rates are extracted from the ECD, while household data are reported by the European Environment Agency (2001). The estimated and chosen parameter values are used to calculate β_i which is then used for the calculation of the MISTIC in equation (1). Country-specific hurdle rates can be obtained so far only for France, Germany, Greece, Italy, and Spain. We used the average EU hurdle rates for those countries for which hurdle rates could not be obtained.

We computed the total SIIB based on information from the ECOGEN field trials in Narbonne, France, and from the data published by Gianessi, Sankula, and Reigner (2003) for HT corn, using the same approach as in Demont, Wesseler, and Tollens (2004).

5. RESULTS AND DISCUSSION

5.1. MISTIC and SIRB for the EU-15

The results per country and for the EU-15 are presented in Tables 2–4 for Bt corn, HT corn, and HT sugar beets, respectively. The results for HT sugar beets are adopted from Demont, Wesseler, and Tollens (2004).

For Bt corn the change in producer surplus results in a loss of €4.65 million per year. The loss in producer surplus is driven by the somewhat inelastic demand function, which drives also a gain in consumer surplus equal to € 51.66 million. The total SIRB is about €47.02 million per year. For HT corn the change in producer surplus also results in a loss; the amount is about €12.95 million per year. The loss in producer surplus is again driven by the somewhat inelastic demand function, which drives also a gain in consumer surplus equal to €92.61 million. The SIRB for corn is much lower than that reported for HT sugar beets, with about €169 million per year. In total, according to our analysis, the EU gives up about €295 million on average per year in 1995 values due to the quasi moratorium on transgenic crops for Bt corn, HT corn, and HT sugar beets alone.

The results also show the marginal importance of the adoption of Bt corn under Narbonne conditions in terms of SIIB (€0.03 per hectare). Slightly more important is the SIIB for HT corn, equal to about €2.50 per hectare. These observations are supported by the Environmental Impact Quotient (EIQ) for Bt corn and HT corn discussed in section 2. The SIIB for HT sugar beets is about €1.59 per hectare on average in the EU. The results for sugar beets differ by country, as national data on input uses for assessing the incremental irreversible benefits were used in that study.

The average hurdle rate for corn, at 1.17, is slightly higher than for sugar beets, at 1.04. The hurdle rate for corn is below 2 for all countries considered except for Germany, with a hurdle rate of 4.27. The average MISTIC is highest for HT sugar beets (€121/ha), followed by HT corn (€113/ha) and Bt corn (€110/ha). The results differ substantially between EU countries for corn. The MISTIC for HT corn for the Netherlands is about €174/ha, whereas it is only €51/ha for Germany. Interestingly, the MISTIC per household is €0.27 for Bt corn, €0.46 for HT corn, and €1.10 for HT sugar

Table 2. Hurdle Rates, Annual Social Incremental Reversible Net Benefits (SIRB$_a$), Social Incremental Irreversible Benefits (SIIB$_a$), and Maximum Incremental Social Tolerable Irreversible Costs (MISTIC$_a$) per Hectare of Bt Corn, per Household and per Corn-Growing Farmer

Member State	SIRB$_a$ (€/ha)	SIIB$_a$ (€/ha)	Hurdle rate	MISTIC$_a$ (€/ha)	MISTIC$_a$ (million €)	MISTIC$_a$ (€/household)	MISTIC$_a$ (€/farmer)
Austria							
Belgium/Lux.							
Denmark							
Finland							
France	111.05	0.03	1.70	65.36	18.78	0.82	137
Germany							
Greece	0.15	0.03	1.46	0.14	0.03	0.01	
Ireland							
Italy	35.70	0.03	1.82	19.64	5.74	0.26	183
Netherlands							
Portugal	18.40	0.03	1.92	9.61	2.81	0.78	48
Spain	15.80	0.03	1.92	8.26	2.41	0.16	64
Sweden							
UK							
EU[a]	128.52	0.03	1.17	109.87	40.21	0.27	195

[a] The hurdle rate is estimated based on the average gross margin for the whole EU.

Table 3. Hurdle Rates, Annual Social Incremental Reversible Net Benefits (SIRB$_a$), Social Incremental Irreversible Benefits (SIIB$_a$), and Maximum Incremental Social Tolerable Irreversible Costs (MISTIC$_a$) per Hectare of HT Corn, per Household and per Corn-Growing Farmer

Member State	SIRB$_a$ (€/ha)	SIIB$_a$ (€/ha)	Hurdle rate	MISTIC$_a$ (€/ha)	MISTIC$_a$ (million €)	MISTIC$_a$ (€/household)	MISTIC$_a$ (€/farmer)
Austria	159	2.50	1.17	138	1.56	0.47	49
Belgium/Lux.	171	2.50	1.17	148	2.85	0.67	103
Denmark	186	2.50	1.17	162	0.73	0.32	159
Finland							
France	111	2.50	1.70	71	13.19	0.58	96
Germany	152	2.50	4.27	51	8.21	0.22	60
Greece	85	2.50	1.46	61	0.03	0.01	3
Ireland							
Italy	219	2.50	1.82	123	4.10	0.18	131
Netherlands	200	2.50	1.17	174	4.63	0.72	145
Portugal	253	2.50	1.92	134	2.05	0.57	35
Spain	253	2.50	1.92	134	2.41	0.16	64
Sweden							
UK	129	2.50	1.17	113	1.47	0.07	200
EU[a]	129	2.50	1.17	113	69.24	0.46	152

[a] The hurdle rate is estimated based on the average gross margin for the whole EU.

Table 4. Hurdle Rates, Annual Social Incremental Reversible Net Benefits (SIRB$_a$), Social Incremental Irreversible Benefits (SIIB$_a$), and Maximum Incremental Social Tolerable Irreversible Costs (MISTIC$_a$) per Hectare of HT Sugar Beets, per Household and per Sugar-Beet–Growing Farmer

Member State	SIRB$_a$ (€/ha)	SIIB$_a$ (€/ha)	Hurdle rate	MISTIC$_a$ (€/ha)	MISTIC$_a$ (million €)	MISTIC$_a$ (€/household)	MISTIC$_a$ (€/farmer)
Austria	251	3.36	2.88	91	1.84	0.56	156
Belgium/Lux.	168	2.09	1.26	135	5.85	1.38	379
Denmark	178	2.06	1.73	105	2.86	1.25	363
Finland	251	0.74	3.69	69	0.98	0.46	249
France	179	1.05	1.25	145	24.96	1.09	737
Germany	179	1.57	1.36	134	27.85	0.75	527
Greece	264	7.97[b]	3.12	93	1.77	0.49	84
Ireland	116	-0.96[b]	2.29	50	0.69	0.61	164
Italy	330	2.32	1.82	183	22.68	1.02	361
Netherlands	121	0.83	1.31	94	4.63	0.72	241
Portugal	354	-0.65[b]	1.67[c]	212	0.62	0.17	769
Spain	252	0.53	2.10	121	7.26	0.48	260
Sweden	150	0.18	3.01	50	1.23	0.31	233
UK	127	1.78	1.76	74	5.14	0.24	461
EU[a]	199	1.59	1.04	192	163.36	1.10	587

Source: Demont, Wesseler, and Tollens (2004).

[a] The hurdle rate is estimated based on the average gross margin for the whole EU.

[b] The extreme estimates for Greece, Ireland and Portugal are probably due to data inconsistencies in the Eurostat (2000) dataset. These countries only cover 4 per cent of total EU area allocated to sugar beets, such that the EU average is almost not affected.

[c] For Portugal, no data on margins has been found. The EU area-weighted average has been used as a proxy for its hurdle rate.

beets, or €1.83 for all three crops together. The MISTIC per transgenic crop planting farm-household is €195 for Bt corn, €152 for HT corn, and €587 for HT sugar beets, at the EU level. If households in the EU on average expect the incremental social irreversible costs per hectare and year and crop to be higher than the values calculated, then from a purely economic point of view, the crops should not be released. As the numbers are extremely low, it can be expected that a survey among households will indicate a higher willingness-to-pay to avoid the introduction of those crops. This indicates a general problem with introducing new technologies. The average benefits per household are often very small, resulting in low MISTIC. This can explain the reluctance of a large segment of society to adopt new technologies. The question then arises whether there is an indication that the irreversible costs are indeed beyond this level. A useful approach is to compare the maximum incremental social tolerable irreversible costs with the social incremental irreversible benefits, as many of the irreversible costs of planting transgenic crops are similar to the irreversible benefits, such as impacts of pesticides on health and the environment. The comparison indicates that the MISTIC at the EU 15 level is 3,681 times higher than SIIB for Bt corn, 120 times higher for HT sugar beets, and 45 times higher for HT corn. The number for HT corn indicates, for example, that social incremental irreversible costs can be up to 45 times higher than social incremental irreversible benefits and still justify a release of HT corn.

The total MISTIC per year is €40.21 million for Bt corn, €69.24 million for HT corn, and €102.63 million for HT sugar beets, or €212.08 million per year for all three crops together. This is almost 10 percent of the total expenditure in 1995 (€2,259 million) of EU institutions for environmental protection.

5.2. Implications for Regulating the Release of GMOs in the European Union

As GMOs will be released only if considered safe for human consumption, the major regulatory issues concerning their release are issues related to environmental impacts of GM crops (e.g., impacts on biodiversity) and issues of co-existence with non-GM crops (e.g., pollen flow). Space does not allow us to discuss those issues in detail. The interested reader is referred to Beckmann and Wesseler (forthcoming), Soregaroli and Wesseler (2005), and Beckmann, Soregaroli, and Wesseler (2005) for a discussion on the implication of co-existence rules and regulations in the EU on the adoption of GMOs. Here we merely report the information our analysis can provide for the discussion.

In Europe there are three type of rules and regulations evolving for governing the co-existence of non-GM and GM crops. Spain has almost no rules and regulations, and farmers can grow Bt corn without having to comply with additional planting requirements that differ from non-GM corn. In Denmark, farmers have to register areas allocated to GM crops, keep a minimum distance to neighboring non-GM crops, and pay a certain amount into a trust fund that will be used to compensate for any damages. In Germany, GM farmers need to register all areas allocated to GM crops in a publicly available database, have to keep a minimum distance to neighboring non-GM crops, and will be liable for any damages to non-GM farmers under a system of joint liability. Other EU member states are still developing their own co-existence laws that follow either the Danish or the German model.

The incremental benefits per hectare provide a first indicator about the maximum costs farmers are willing to bear for complying with regulations. The average SIRB per hectare, at about €129 euros, is almost the same for Bt and HT corn, and is about €199 for HT sugar beets. Adopting farmers will not be willing to pay more than this amount to comply with co-existence rules and regulations. As those amounts are relatively small on a per-hectare basis to cover additional co-existence costs—for example, as under the German regulations—adoption will become economically attractive only if farmers can realize economies of scale with respect to co-existence costs. In general, looking at the average SIRB per farmer for the EU-15, the chances for immediate adoption of Bt corn and HT corn are almost the same assuming constant co-existence costs for both. The highest potential, of course, can be expected for stacked varieties, where farmers can benefit from both pest resistance and herbicide tolerance. As the co-existence regulations for GM sugar beets are expected to be less restrictive than for GM corn, the potential for immediate adoption of HT sugar beets will be higher.[8] The SIRB for the different countries indicates that on average there will be sufficient gain from HT corn and HT sugar beets to pay into a fund to compensate for potential damages, a policy that, as mentioned above, has been introduced in Denmark.

6. CONCLUSION

In this study we estimated the MISTIC, the maximum incremental social tolerable irreversible costs, associated with the immediate adoption of Bt corn and HT corn in the EU-15, using a real option approach and data from field trials carried out in 2004 in Narbonne, France. The MISTIC is an

---

[8] Note that the sugar industry is very reluctant to accept HT sugar beets, as currently many customers reject sugar derived from HT sugar beets.

amount that would cover irreversible benefits from Bt corn and irreversible private net-benefits weighted by an estimated hurdle rate.

SIRB accruing to producers of Bt corn and HT corn is found to be about €129 per hectare and year. The estimated SIRB for Bt corn and HT corn is lower than for HT sugar beets. The highest economic potential according to the MISTIC per farm for Bt corn is found to exist for France and Italy, for HT corn, the UK, Denmark, and the Netherlands, and for sugar beets, France, Germany, and the UK.

The private incremental reversible benefits on a per-hectare basis can hardly cover *ex-ante* regulatory and *ex-post* liability costs of planting transgenic corn and sugar beets in Germany and other EU member states that adopt the same type of rules and regulations. Economies of scale can lower the regulatory costs per hectare, set incentives for regional agglomeration of transgenic crops, and induce adoption. The case reported by Achilles (2005) supports this argument.

The low MISTIC per household for all three crops over all countries included in the analysis provides a strong economic argument for prohibiting the immediate introduction of the three transgenic crops. The validity of the argument will largely depend on consumer attitudes towards transgenic crops. However, consumer attitudes may change over time, e.g., if scientific evidence shows higher environmental benefits or lower irreversible costs of transgenic crops.

REFERENCES

Achilles, D. 2005. "GMO Situation in Germany 2005." GAIN Report No. GM5011. U.S. Department of Agriculture, Washington, D.C.

Antle, J., and P. Pingali. 1994. "Pesticides, Productivity and Farmers' Health: A Philippine Case Study." *American Journal of Agricultural Economics* 76(2): 418–430.

Banse, M., H. Grethe, and S. Nolte. 2005. *European Simulation Model (ESIM) in GAMS: User Handbook.* European Commission, DG AGRI, Brussels.

Beckmann, V., C. Soregaroli, and J. Wesseler. 2005. "Coase and Co-existence Under Irreversibility and Uncertainty." Unpublished paper, Wageningen University.

Beckmann, V., and J. Wesseler. Forthcoming. "Spatial Dimension of Externalities and the Coase Theorem: Implications for Coexistence of Transgenic Crops." In W. Heijman, ed., *Regional Externalities.* Berlin: Springer.

Campbell, J.Y., A.W. Lo, and A.C. MacKinlay. 1997. *The Econometrics of Financial Markets.* Princeton, NJ: Princeton University Press.

Demont, M., and E. Tollens. 2004. "First Impact of Biotechnology in the EU: Bt Corn Adoption in Spain." *Annals of Applied Biology* 145(3): 197–207.

Demont, M., J. Wesseler, and E. Tollens. 2004. "Biodiversity Versus Transgenic Sugar Beet: The One Euro Question." *European Review of Agricultural Economics* 31(1): 1–18.

_____. 2005. "Reversible and Irreversible Costs and Benefits of Transgenic Crops." In J. Wesseler, ed., *Environmental Costs and Benefits of Transgenic Crops*. Dordrecht: Springer.

Dewar, A.M., M.J. May, and J. Pidgeon. 2000. "GM Sugar Beet—The Present Situation." *British Sugar Beet Review* 68(2): 22–27.

Dixit, A., and R.S. Pindyck. 1994. *Investment Under Uncertainty*. Princeton, NJ: Princeton University Press.

Dowd, P.F. 2000. "Indirect Reduction of Ear Molds and Associated Mycotoxins in *Bacillus thuringiensis* Corn Under Controlled and Open Field Conditions: Utility and Limitations." *Journal of Economic Entomology* 93(6): 1669–1679.

Ervin, D., and R. Welsh. 2005. "Environmental Effects of Genetically Modified Crops: Differentiated Risk Assessment and Management." In J. Wesseler, ed., *Environmental Costs and Benefits of Transgenic Crops*. Dordrecht: Springer.

Essential Biosafety. 2004. Crop Database: DBT418. Available at http://www.essentialbiosafety. info/dbase.php?action=ShowProd&data=DBT418 (accessed February 21, 2005).

European Environment Agency (EEA). 2001. *Household Number and Size*. European Environment Agency, Copenhagen.

EUROSTAT. 1999. *SPEL/EU Data for Agriculture on CR-ROM: 1973–1998 Data*. Office for Official Publications of the European Communities, Luxembourg.

_____. 2000. *Plant Protection in the EU—Consumption of Plant Protection Products in the European Union*. Office of Official Publications of the European Communities, Luxembourg.

_____. 2004. Eurostat New Cronos database (theme 5). Available at http://europa.eu.int/new[-] cronos (updated October 2004).

FAOSTAT. 2004. "Faostat-Agriculture." Available at http://faostat.fao.org/faostat/col[-] lections?subset=agriculture (updated February 2004).

Gianessi, L., S. Sankula, and N. Reigner. 2003. "Plant Biotechnology: Potential Impact for Improving Pest Management in European Agriculture." The National Center for Food and Agricultural Policy (NCFAP), Washington, D.C.

Heard, M.S., C. Hawes, G.T. Champion, S.J. Clark, L.G. Firbank, A.J. Haughton, A.M. Parish, J.N. Perry, P. Rothery, R.J. Scott, M.P. Skellern, G.R. Squire, and M.O. Hill. 2003. "Weeds in Fields with Contrasting Conventional and Genetically Modified Herbicide-Tolerant Crops—I. Effects on Abundance and Diversity." *Philosophical Transactions of the Royal Society London* B358: 1819–1832.

Hurley, T. 2005. "*Bacillus thuringiensis* Resistance Management: Experiences from the USA." In J. Wesseler, ed., *Environmental Costs and Benefits of Transgenic Crops*. Dordrecht: Springer.

James, C. 2004. "Global Status of Commercialized Biotech/GM Crops 2004." Brief No. 32-2004, International Service for the Acquisition of Agri-Biotech Applications (ISAAA), Ithaca, NY.

Jansens, S., A. Van Vliet, C. Dickburt, L. Buysse, C. Piens, B. Saey, A. De Wulf, V. Gosselé, A. Paez, E. Göbel, and M. Peferoen. 1997. "Transgenic Corn Expressing a Cry9C Insecticidal Protein from Bacillus thuringiensis Protected from European Corn Borer Damage." *Crop Science* 37(6): 1616–1624.

Katranidis, S., and K. Velentzas. 2000. "The Markets of Cotton Seeds and Corn in Greece: Welfare Implications of the Common Agricultural Policy." *Agricultural Economic Review* 1(2): 80–95.

Kendall, H.W., R. Beachy, T. Eisner, F. Gould, R. Herdt, P.H. Raven, J.S. Schell, and M.S. Swaminathan. 1997. *Bioengineering of Crops. Environmental and Socially Sustainable Development Studies and Monograph Series 23*. The World Bank, Washington, D.C.

Kovach, J., C. Petzoldt, J. Degni, and J. Tette. 1992. "A Method to Measure the Environmental Impact of Pesticides." New York Food and Life Sciences Bulletin No. 139, New York State IPM Program, Cornell University, Ithaca, NY.

Laxminarayan, R., and D. Simpson. 2005. "Biological Limits on Agricultural Intensification: An Example from Resistance Management." In J. Wesseler, ed., *Environmental Costs and Benefits of Transgenic Crops*. Dordrecht: Springer.

Lekakis, J.N., and C. Pantzios. 1999. "Agricultural Liberalization and the Environment in Southern Europe: The Role of the Supply Side." *Applied Economics Letters* 6(7): 453–458.

Morel, B., S. Farrow, F. Wu, and E. Casman. 2003. "Pesticide Resistance, the Precautionary Principle, and the Regulation of *Bt* Corn: Real Option and Rational Option Approaches to Decision-Making." In R. Laxminarayan, ed., *Battling Resistance to Antibiotics and Pesticides*. Washington, D.C.: Resources for the Future.

Moschini, G., and H. Lapan. 1997. "Intellectual Property Rights and the Welfare Effects on Agricultural R&D." *American Journal of Agricultural Economics* 79(4): 1229–1242.

Moschini, G., H. Lapan, and A. Sobolevsky. 2000. "Roundup Ready Soybeans and Welfare Effects in the Soybean Complex." *Agribusiness* 16(1): 33–55.

Munkvold, G.P., R.L. Hellmich, and L.G. Rice. 1999. "Comparison of Fumonisin Concentrations in Kernels of Transgenic Bt Maize Hybrids and Nontransgenic Hybrids." *Plant Disease* 83(2): 130–138.

Munkvold, G.P., R.L. Hellmich, and W.B. Showers. 1997. "Reduced Fusarium Ear Rot and Symptomless Infection in Kernels of Maize Genetically Engineered for European Corn Borer Resistance." *Phytopathology* 87(10): 1071–1077.

Nillesen, E., S. Scatasta, and J. Wesseler. 2005. "Assessing the Environmental Impact of Bt Corn and HT Corn vs. Conventional Corn Production with Regard to Pesticide Use." Unpublished, Wageningen University.

Pilcher, C.D., M.E. Rice, R.A. Higgins, K.L. Steffey, R.L. Hellmich, J. Witkowski, D. Calvin, K.R. Ostlie, and M. Gray. 2002. "Biotechnology and the European Corn Borer: Measuring Historical Farmer Perceptions and Adoption of Transgenic Bt Corn as a Pest Management Strategy." *Journal of Economic Entomology* 95(5): 878–892.

Soregaroli, C., and J. Wesseler. 2005. "The Farmer's Value of Transgenic Crops under Ex-Ante Regulation and Ex-Post Liability." In J. Wesseler, ed., *Environmental Costs and Benefits of Transgenic Crops*. Dordrecht: Springer.

Waibel, H., and G. Fleischer. 1998. *Kosten und Nutzen des chemischen Pflanzenschutzes in der deutschen Landwirtschaft aus gesamtwirtschaftlicher Sicht*. Kiel: Wissenschaftsverlag Vauk.

Wesseler, J. 2003. "Resistance Economics of Transgenic Crops: A Real Option Approach." In R. Laxminarayan, ed., *Battling Resistance to Antibiotics: An Economic Approach*. Washington, D.C.: Resources for the Future.

———— (ed.). 2005. *Environmental Costs and Benefits of Transgenic Crops*. Dordrecht: Springer.

World Bank. 2002. *World Development Indicators 2002 on CD-ROM*. The World Bank, Washington, D.C.

Wu, F., J.D. Miller, and E.A. Casman. 2004. "The Economic Impact of Bt Corn Resulting from Mycotoxin Reduction." *Journal of Toxicology–Toxin Reviews* 23(2&3): 397–424.

Zaid, A., H.G. Hughes, E. Porceddu, and F. Nicholas. 2001. "Glossary of Biotechnology for Food and Agriculture." FAO Research and Technology Paper No. 9, Food and Agriculture Organization (FAO), Rome.

APPENDIX

Specification of the partial equilibrium model for green corn

Country j's supply of green corn (f), $Q^s_{f,t}$, is given below:

$$Q^s_{f,t} = A^s_{f,t} \left[P^s_{f,t} \right]^{\varepsilon_{P,f}},$$ (A1)

where the subscript j is dropped for ease of notation, $P^s_{f,t}$ is the producer (or output) price received by corn sellers at time t, and $A^s_{f,t}$ is a technology-specific constant term for the associated product and function.

The aggregate demand for green corn, $Q^d_{f,t}$, is modeled as a constant elasticity function of green corn price,

$$Q^d_{f,t} = A^d_{f,t} \left[P^d_{f,t} \right]^{\eta_{P,f}},$$ (A2)

where $P^d_{f,t}$ is the buyers' (or input) price paid for corn at time t, and $A^d_{f,t}$ is a parameter capturing qualitative aspects of the product such as the type of technology used in its production process.

The market clears with the following requirements:

$$Q^d_{f,t} = Q^s_{f,t}$$ (A3)

$$P^d_{f,t} \left[1 + \tau_{f,t} \right] = P^s_{f,t},$$ (A4)

where

$$\tau_{f,t} = \frac{\left[P^s_{f,t} - P^d_{f,t} \right]}{P^d_{f,t}}$$

represents the proportional CAP price support coefficient identifying the relative difference between the output and the input price of corn due to the CAP corn price support regime.

Based on EUROSTAT data on the value of production calculated at the seller's price and the value of production calculated at the buyer's price, we observe that the variation in support received by corn sellers per unit of the product does not vary with the quantity produced. The price support system, therefore, reduces marginal production costs for corn sellers, causing a parallel downward shift in the supply function.

At any time period, the equilibrium price, $P_{f,t}^*$ and quantities, $Q_{f,t}^*$, are given by

$$\begin{cases} P_{f,t}^{s*} = P_{f,t}^{d*}\left[1+\tau_{f,t}\right] \\[2mm] P_{f,t}^{d*} = \left[\dfrac{A_{f,t}^d}{A_{f,t}^s}\right]^{\frac{1}{\varepsilon_{Pf}-\eta_{Pf}}} \left[\dfrac{1}{1+\tau_{f,t}}\right]^{\varepsilon_{Pf}} \\[2mm] Q_{f,t}^* = \left[A_{f,t}^d\right]\left[\dfrac{A_{f,t}^d}{A_{f,t}^s}\right]^{\frac{\eta_{Pf}}{\varepsilon_{Pf}-\eta_{Pf}}} \left[\dfrac{1}{1+\tau_{f,t}}\right]^{\eta_{Pf}\varepsilon_{Pf}} . \end{cases} \tag{A5}$$

Producer surplus, $PS_{f,t}$, at the equilibrium conditions in (A5), is given by

$$PS_{f,t} = P_{f,t}^{d*}\left[1+\tau_{f,t}\right]Q_{f,t}^* - \tag{A6}$$

$$\int_0^{Q_{f,t}^*}\left[\frac{Q_{f,t}^s}{A_{f,t}^s}\right]^{\frac{1}{\varepsilon_{Pf}}}\frac{1}{\left[1+\tau_{f,t}\right]}dQ_{f,t}^s = P_{f,t}^{s*}Q_{f,t}^* - P_{f,t}^{d*}Q_{f,t}^*\frac{\varepsilon_{Pf}}{\varepsilon_{Pf}+1} .$$

Consumer surplus, $CS_{f,t}$, at the equilibrium conditions in (9) is given by

$$CS_{f,t} = \int_{P_{f,t}^{d*}}^{\infty} A_{f,t}^d\left[P_{f,t}^d\right]^{\eta_{Pf}} dP_{f,t}^d . \tag{A7}$$

Following Moschini, Lapan, and Sobolevsky (2000), we assume that the adoption of a technological innovation, such as transgenic corn, causes a pivotal shift in the inverse supply function by changing the value of the technology-specific constant term, α. The proportional vertical shift in the inverse supply function, f and t, will be given by

$$\frac{\left[\dfrac{1}{A_0^s}\right]^{1/\varepsilon_{Pf}} - \left[\dfrac{1}{A_1^s}\right]^{1/\varepsilon_{Pf}}}{\left[\dfrac{1}{A_0^s}\right]^{1/\varepsilon_{Pf}}} = \theta(t)K, \tag{A8}$$

where subscripts f and t are dropped to simplify notation; $\theta(t)$ is the transgenic corn adoption rate over time, t; A_0^s is the direct supply function constant coefficient with conventional technology; and A_1^s is the direct supply function constant coefficient with transgenic technology and

$$K = \frac{[mc_c/y_c] - [mc_g/y_g]}{[mc_c/y_c]}, \tag{A9}$$

where mc_c are variable operating costs (euros per hectare) associated with the conventional technology, mc_g are variable operational costs (euros per hectare) associated with the transgenic technology, y_c is production (in metric tons) under conventional technology, and y_g is production (in metric tons) under the Bt technology.

Given equations (A5) to (A7) we can compute changes in the equilibrium price and quantities due to adoption of transgenic corn as a function of the vertical shift in the inverse supply function and the CAP price support coefficient:

$$\begin{cases} \Delta P^{s*} = \Delta P^{d*} * \left[1 + \tau\right] \\[2mm] \Delta P^{d*} = P_1^{d*} - P_0^{d*} = \left[\left[1 - \theta(t)K\right]^{\frac{\varepsilon_{Pf}}{\varepsilon_{Pf} - \eta_{Pf}}} - 1\right]P_0^* \\[2mm] \Delta Q^* = Q_1^* - Q_0^* = \left[\left[1 - \theta(t)K\right]^{\frac{\varepsilon_{Pf}\eta_{Pf}}{\varepsilon_{Pf} - \eta_{Pf}}} - 1\right]Q_0^*. \end{cases} \tag{A10}$$

In equation (A10) the subscripts f and t are dropped again to simplify notation. It should be noted that according to experts' opinions, green corn is consumed mainly in the dairy sector. In this sector, the milk quota regime, established by the EU, makes the demand for green corn perfectly inelastic, $\eta_{Pf} = 0$, and the change in the equilibrium conditions in (A10) becomes

$$\begin{cases} \Delta P^{s*} = \Delta P^{d*} * \left[1 + \tau\right] \\ \Delta P^{d*} = P_1^{d*} - P_0^{d*} = -\theta(t)KP_0^* \\ \Delta Q^* = Q_1^* - Q_0^* = 0. \end{cases} \quad \text{(A10a)}$$

As we lack data on corn use in the dairy sector, we allow input demand to be not perfectly inelastic, and refer to the changes in the equilibrium conditions in (A10) for our welfare analysis. The change in producer surplus, in particular, will be given by

$$\Delta PS = PS_1 - PS_0 = \left[\tau + 1 - \frac{\varepsilon_{Pf}}{\varepsilon_{Pf} + 1}\right]\left[\left[1 - \theta(t)K\right]^{\frac{\varepsilon_{Pf}\left[1 + \eta_{Pf}\right]}{\varepsilon_{Pf} - \eta_{Pf}}} - 1\right]Q_0^* P_0^{d*}. \quad \text{(A11)}$$

The total change in producer surplus can be decomposed into two parts: the change in producer surplus accruing from the government due to the price support system, ΔPS^{gov}, and the change in producer surplus accruing from the market,

$$\Delta PS^{gov} = PS_1^{gov} - PS_0^{gov} = \quad \text{(A11a)}$$

$$\tau\left[P_1^{d*}Q_1^* - P_0^{d*}Q_0^*\right] = \tau\left[\left[1 - \theta(t)K\right]^{\frac{\varepsilon_{Pf}\left[1 + \eta_{Pf}\right]}{\varepsilon_{Pf} - \eta_{Pf}}} - 1\right]Q_0^* P_0^{d*}$$

and

$$\Delta PS^{mkt} = PS_1^{mkt} - PS_0^{mkt} \quad \text{(A11b)}$$

$$= \Delta PS - \Delta PS^{gov} = \left[1 - \frac{\varepsilon_{Pf}}{\varepsilon_{Pf} + 1}\right]\left[\left[1 - \theta(t)K\right]^{\frac{\varepsilon_{Pf}\left[1 + \eta_{Pf}\right]}{\varepsilon_{Pf} - \eta_{Pf}}} - 1\right]Q_0^* P_0^{d*}.$$

Assuming that the introduction of transgenic corn does not cause shifts in the demand function, the consumer surplus changes as follows:

$$\Delta CS = CS_1 - CS_0 = \left[P_0^{d*} - P_1^{d*}\right]Q_0^* + \int_{Q_0^*}^{Q_1^*}\left[\frac{Q^d}{A^d}\right]^{\frac{1}{\eta_{Pf}}} dQ^d \quad \text{(A12)}$$

$$= \left[\left[1 - \theta(t)K\right]^{\frac{\varepsilon_{Pf}\left[1 + \eta_{Pf}\right]}{\varepsilon_{Pf} - \eta_{Pf}}} - 1\right]P_0^{d*}Q_0^* \frac{\eta_{Pf}}{\eta_{Pf} + 1} - \left[\left[1 - \theta(t)K\right]^{\frac{\varepsilon_{Pf}}{\varepsilon_{Pf} - \eta_{Pf}}} - 1\right]P_0^{d*}Q_0^*.$$

The transgenic corn adoption curve is assumed to follow a logistic pattern over time, such that

$$\theta(t) = \frac{\theta_{MAX}(t)}{\exp(-a - bt)} \; .$$

(A13)

Equation (A13) can be transformed into

$$\ln\left(\frac{\theta(t)}{\theta_{MAX}(t) - \theta(t)}\right) = -a - bt \; .$$

(A14)

The coefficients in equation (A14) can be estimated with ordinary least squares (OLS) using data from the adoption rates in the United States. Following Demont, Wesseler, and Tollens (2004), the speed of adoption b will be assumed to be half of the speed of adoption in the United States to obtain conservative estimates of the social reversible benefits.

Chapter 17

ANTICOMPETITIVE IMPACTS OF LAWS THAT REGULATE COMMERCIAL USE OF AGRICULTURAL BIOTECHNOLOGIES IN THE UNITED STATES

Richard E. Just
University of Maryland

Abstract: This chapter examines economic inefficiencies that have occurred under the Federal Insecticide, Fungicide and Rodenticide Act (FIFRA) because it is now used to regulate some biotechnologies in the United States. This statute has regulated pesticides for over half a century. Before a new pesticide can be sold it must be registered with the U.S. Environmental Protection Agency, which requires the registrant to bear the cost of many regulatory tests to ensure environmental safety. The pesticide can then be sold monopolistically either under a patent or under a 10-year exclusive use provision of FIFRA. Thereafter, generic firms can register and sell by offering to share the costs of prior regulatory tests (or by duplicating tests, which is typically prohibitive). However, FIFRA provides no standard for sharing these test costs, so many cases ultimately require litigation. This chapter shows that the share of regulatory test costs imposed on generic firms has major implications for post-patent competition because generic entrants typically capture a small share of the market and sell under competitive rather than monopolistic prices. Consumers and farmers are the main beneficiary of generic competition but their benefits are often delayed or never realized when the share of test costs borne by generic entrants is too large.

Key words: generic entry, regulatory cost sharing, compensation, competitive price effects

1. INTRODUCTION

As discussed in Chapter 1, the major laws regulating biotechnologies in the United States include the Plant Protection Act (PPA), the Federal Food, Drug, and Cosmetic Act (FFDCA), the Federal Insecticide, Fungicide, and Rodenticide Act (FIFRA), and the Toxic Substances Control Act (TSCA). Because the Coordinated Framework for Regulation of Biotechnology placed regulation of new biotechnologies under these laws, which have long been used to regulate other substances such as pesticides, analysis of how those regulations have performed previously can give insights into how well they will work for regulation of biotechnologies. This chapter discusses the effects of regulations imposed under these laws on economic and distributional efficiency. In some cases, the regulations appear to operate relatively efficiently, while in others they appear to facilitate anticompetitive behavior and associated inefficiency. These effects are explored through simple theoretical constructs and anecdotal empirical calculations. In-depth empirical analysis is precluded because most critical data are proprietary. The empirical calculations focus on chemical pesticide regulation. The interesting issues relate to market efficiency beyond patent expiration. Although agricultural biotechnology product markets are yet in infant stages, the longer-run implications of regulation will likely be similar for successful biotechnologies. Results suggest that the potential for private gains motivates vigorous lobbying efforts that may have as much impact on the way regulations operate as the regulations have on industry structure.

2. PERTINENT PROVISIONS OF SELECTED REGULATIONS

Because most regulations that apply to agricultural biotechnologies after commercialization are administered by the U.S. Environmental Protection Agency (EPA), the discussion in this chapter focuses on the associated regulatory effects of FIFRA and TSCA. Before a pesticidal substance can be marketed and used, whether developed through biotechnology or not, FIFRA requires the EPA to evaluate the substance thoroughly to ensure safety for human health and the environment. The expense of tests to develop the regulatory data required by the EPA are borne by product registrants. For some substances, original entrants claim that the required tests can cost tens of millions of dollars. Thus, the cost of ensuring product safety can be substantial. A critical issue is: Who should bear these regulatory costs? Certainly, once a product is found to be safe, society is not well served by requiring each new market entrant to repeat the same costly tests. But if later generic entrants benefit from selling the product, should they also share in the one-time cost of ensuring public safety?

2.1. FIFRA Provisions and Amendments

The original FIFRA was enacted in 1947 and required pesticides to be registered with the U.S. Department of Agriculture. Many new pesticides were developed under these provisions. Regulatory data costs were minor, and no provisions were made for sharing regulatory test costs. With the 1972 amendments, registration was shifted to the newly created EPA and regulatory data requirements increased. While the 1972 and 1978 FIFRA amendments clearly had an intent of promoting post-patent competition and avoiding duplication of tests, the magnitude of regulatory data costs, potential sharing issues, and implications for competitive efficiency after patent expiration were not well anticipated,.

For example, in passing the 1972 Amendment, the U.S. Senate recognized that "unnecessary duplicative testing would represent a wasteful, time-consuming, and costly process resulting in a substantial misallocation of resources," and specifically noted that one of the purposes for the amendments was to "prevent unnecessary repetitive testing by subsequent applicants" (S. Rep. No. 92-838, Part II, 1972, pp. 72–73). But while the Congressional intent to avoid duplicative testing was clear, many loopholes remained that led to a great deal of litigation between original and generic entrants about how regulatory test costs should be shared. The associated delay imposed continuing high costs and risk on generic entry, which discouraged competition. Incentives for duplicative testing continued.

Some original entrants contended that the compensation provisions of FIFRA required the generic entrant to compensate the original entrant for the value of lost monopoly profits in addition to the cost of regulatory tests. In some cases, this was characterized as compensation for the privilege of early market entry (avoiding the regulatory delay that would be required with duplicative testing). Since the benefits of competition flow primarily to consumers through lower prices rather than to generic entrants (as shown below), this interpretation can require more compensation than generic entrants' aggregate future profit potential. Such an interpretation would thus allow FIFRA to be used as an artificial means to permanently extend monopoly conditions.

When enacting the 1975 Amendment, the members of the House Agricultural Committee specifically recognized this attempt to base compensation on market value rather than regulatory cost, noting that

> considerable attention was focused on the anti-competitive aspects [which] ... could seriously and substantially lessen effective competition in the pesticide field by preventing or delaying the entrance of qualified manufacturers because of their inability to purchase data from the first applicant or to bear the expense of duplicating the research data [H.R. Rep. No. 94-497, 1975, p. 62].

A further view was that

> Reasonable compensation should be an equitable sharing of the direct costs of producing governmentally required test data. It should not be based upon a "value" basis. No profit should accrue to the original applicant ... [H.R. Rep. No. 94-497, 1975, p. 65].

In the wake of the 1972 amendments, original registrants contended that the regulatory test data contained trade secrets, which should prevent use by generic entrants. In response, when enacting the 1978 FIFRA amendments, the Senate noted that "the Department of Justice has indicated that the provisions in present law are 'needlessly anti-competitive'" and expressed the intent to "eliminate the prohibition in present law against the use of data submitted by a person to the Environmental Protection Agency in order to obtain a registration which contains 'trade secret' information" (S. Rep. No. 95-334, pp. 2–3).

Some original registrants made further attempts to impose a high entry fee on generic entrants by claiming compensation based on the costs of developing a pesticide as well as the cost of developing the regulatory data. But the Senate also made clear with the 1978 Amendment that this was not the intent of FIFRA.

> The amendments in S. 1678 ... keys the amount of payment by subsequent registrants to the costs of developing data necessary for government approval, not the costs of developing the pesticide [S. Rep. No. 95-334, pp. 30–31].

2.2. The Regulatory Cost Sharing Experience Under FIFRA

In spite of this legislative history, litigation between original and generic entrants regarding regulatory cost sharing continues under the contention that FIFRA does not provide an explicit standard regarding how regulatory test costs are to be shared. In 1984 and 1985 the U.S. Supreme Court had occasion to address FIFRA's data sharing scheme. In *Ruckelshaus v. Monsanto* 467 U.S. 986 (1984), the Court stated that

> the public purpose behind the data-consideration provision is clear from the legislative history. Congress believed that the provisions would eliminate costly duplication of research and streamline the registration process, making new end-use products available to consumers more quickly. Allowing the applicant for registration, upon payment of compensation, to use data already accumulated by others, rather than forcing them to go through the time-consuming process of repeating the research, would eliminate a significant barrier to entry into the pesticide market, thereby allowing greater competition among producers of end-use products [pp. 1014–1015].

In *Thomas v. Union Carbide* 473 U.S. 568 (1985), the Court further stated:

> The 1972 Act established data-sharing provisions intended to streamline pesticide registration procedures, increase competition, and avoid unnecessary duplication of data-generation costs ... Although FIFRA's language

does not impose an explicit standard, the legislative history of the 1972 and 1978 amendments is far from silent ... [pp. 571, 593].

With this legislative and judicial history, the amount of continuing litigation regarding regulatory cost sharing, and the wide range of compensation awards in such litigation has been both surprising and disturbing. As patents of a number of registered pesticides began to expire in the late 1970s and 1980s, generic entrants, encouraged by the 1972 and 1978 FIFRA amendments, began to compete. Under these amendments, generic entrants could cite the data submitted by others and make a written offer to pay compensation, and necessary EPA registrations were granted relatively quickly (usually in about a year rather than the six years or so required to generate regulatory data). However, generic entrants were subject to considerable risk because the amount of compensation was usually subject to subsequent determination, including, as required by FIFRA, binding arbitration if the parties could not agree.

In practice, original entrants' best interests are not served by agreeing on a compensation amount before generic entrance because delaying agreement imposes risk on the generic entrant, which can serve as a deterrent to entry. Furthermore, potential generic entrants can be intimidated or discouraged from entering by inflating claimed compensation with questionable tests, using questionable accounting practices, and liberal estimation of test costs by questionable methods in lieu of invoices. More importantly, because an explicit standard for cost sharing was missing in FIFRA, a generic entrant expecting to be a 5 percent player in the market and pay a corresponding 5 percent share of the regulatory cost has been often asked to pay a 50 percent equal share by the original entrant—or even more.

In cases reaching arbitration, an equal sharing of test costs is typically claimed even though (i) the original entrant has been able to rely on the test data to sell in the market for many years before generic entrance, (ii) generic entrants typically never capture an equal market share, (iii) the test data (including their potential for scientific discovery, product promotion, and liability defense) are kept confidential from the generic entrant, (iv) the generic entrant must develop duplicative tests to register in states such as California and Arizona that require their own regulatory data if the original registrant refuses authorization, and (v) duplication of some studies is required to compete in other countries that require comparable regulatory data.

Decision precedent for FIFRA awards has not seemed to converge on some of the most important issues. For awards that have been made public, Table 1 illustrates the wide variation in awards, compensation shares, and the basis for them. Some restrict the basis for compensation to actual test costs, while others add royalties, early entry fees, or other considerations. Some allocate a market share, others allocate an equal share, while yet others

Table 1. Awards in FIFRA Compensation Cases with Public Information

Arbitration	Award Date	Claimed Data Cost	Claimed Compensation	Cost Share Claimed	Cost Share Awarded	Awarded Amount	Percent of Claim
Ciba-Geigy v. Farmland	1980	$2,636,024	$8,110,000	100.0%	9.5%	$240,682	3.0%
Union Carbide v. Thompson-Hayward	1983	$689,000	$1,317,500	50.0%	33.3%	$51,760	3.9%
Stauffer v. PPG	1983	$2,920,000	$1,465,000 + royalty	50.5%	50.0%	$1,465,000 + 25% of 10-yr. profit	100.0%
American Cyanamid v. Aceto	1987	$3,283,000	$1,971,500	50.0%	35.0%	$1,149,050	58.3%
Griffin & Drexel v. DuPont	1988						
Griffin		$15,700,000	$7,000,000	25.0%	18.3%	$495,178[a]	7.1%
Drexel		$15,700,000	$5,000,000	25.0%	2.83/10.0%	$125,986[a]	2.5%
Ciba-Geigy v. Drexel	1994	$25,075,056	$6,673,560				32.0%
Atrazine		$14,688,486	$3,672,122[b]	25.0%	5.5%	$2,137,348	32.0%
Simazine		$10,386,570	$3,462,190[b]	33.3%	12.8%	$1,329,481	38.4%
Enviro-Chem v. Lilly	1999	$612,000	$306,000 + royalty	50.0%	33.3%	$18,398	<6.0%
Phosphine TF v. Bernardo & Midland	1998						
Bernardo		$915,015	$228,754 + interest + risk	25.0%	20.0%	$171,759	<75.1%
Midland		$915,015	$228,754 + interest + risk	25.0%	20.0%	$171,759	<75.1%
Microgen v. Lonza	2000	$2,250,456	$10,500,000	50.0%	37.5/40.0%	$147,000	1.4%
Avecia v. Amareva	2002	$5,829,842	$4,372,000	50.0%	33.3%	$2,217,571	50.7%

[a] Additionally, the same market share (18.29 percent for Griffin, 2.83% for Drexel) was awarded on 28 other studies not included in the original claim and a future correction in market share was included.

[b] Adjusted downward by 6.5% to reflect amounts already received as compensation.

determine an arbitrary share. Some award inflation while others award interest. The fact that original entrants are able to rely on the tests for many more years in the market has received little or no consideration.

Under these conditions, FIFRA continues to provide an opportunity to artificially discourage generic competition. After an exceptional award in 1983, royalties have come to be widely regarded as inappropriate. However, few other legal precedents have been established. Furthermore, new ways of using FIFRA to prolong monopoly markets or discourage generic entry have been developed.

Common practices that discourage generic entry are to inflate claimed costs with high inflation rates, interest charges, opportunity costs at the stock market rate of return, management fees (whether actual or estimated), and risk premiums. Beginning with the case of *Dow Elanco and the Trifluralin Development Consortium v. Albaugh* (awarded June 1, 1998), original entrants have often claimed an additional risk premium beyond actual test costs because tests are allegedly undertaken with some risk of failure, adverse outcomes, or non-acceptance by the EPA. While some awards have rejected a risk premium, others have awarded anywhere from 5 percent to 60 percent risk premium markups of test costs. Whether the risk premium is a "cost" as defined by FIFRA has been debated. However, the basis for a risk premium is particularly dubious because many of the tests are performed over many years during which the original entrant is already selling in the market, and the original entrant would have had to cease production during those years if it had not commenced the testing. Risk claims ignore the fact that all test costs are not put at risk simultaneously. Rather, tests are spread over many years, with an ability to curtail future testing if early tests fail. In some cases the more risky tests can be undertaken first so that only a small share of test costs are at risk. Furthermore, because annual test costs are typically far less than the annual profit earned from (monopoly) sales, the test costs are effectively a required expense of continuing in the business (prior to generic entry) rather than a risk against profits in future years.[1] Additionally, most claims involve no precise calculations of risk or a monetizing of the risk using appropriate premium rates. For example, in many cases the

---

[1] Under FIFRA, two major cases of compensation are differentiated: Section 3(c)(1)(F) considers regulatory data generated for purposes of original registration while Section 3(c)(2)(B) considers regulatory data generated in response to later EPA data call-ins. While tests associated with original registration are often scattered over a decade or more after conditional registration and sales have commenced, many recent cases have involved "call-in" data in response to the re-registration of all pesticides (to bring them up to current registration standards) initiated in 1975 and completed under the 1988 Amendment, which imposed a 9-year schedule for completion. More recently, the Food Quality Protection Act of 1996 has tightened standards that eliminated some substances (organophosphates, carbamates, and certain carcinogens) and initiated another round of review of all existing registrations to be completed in 10 years.

justification is based on value-related arguments (e.g., the risk of lost profits if the EPA registration is canceled).

2.3. The Regulatory Cost Sharing Experience Under TSCA

Enactment occurred much later for TSCA than for FIFRA. TSCA covers chemical substances and, since 1988, genetically modified substances, specifically excluding those regulated by FIFRA and the FFDCA. It also requires registration and regulatory testing of substances to ensure public safety, comparable to FIFRA. Perhaps because TSCA was developed later after more government regulatory experience, it contains provisions that avoid many of the regulatory test cost sharing problems encountered under FIFRA.

Under TSCA, Section 4(c), any person required to conduct tests and submit data may apply to the Administrator for an exemption. If the chemical substance is equivalent to one for which data has been submitted, the Administrator is instructed to exempt the applicant from conducting tests and submitting data. If the exemption is granted and the parties do not agree on compensation, then the Administrator is instructed to order equitable reimbursement. This much is similar to FIFRA. But the important additional instruction in Section 4(c)(3)(A) is that, "In promulgating rules for the determination of fair and equitable reimbursement ... the Administrator shall ... consider all relevant factors, including the effect on the competitive position of the person required to provide reimbursement in relation to the person to be reimbursed and *the share of the market* for such substance or mixture of the person required to provide reimbursement in relation to the share of such market of the persons to be reimbursed" (emphasis added).

The ensuing rules and regulations adopted by the EPA set out a specific formula for sharing:

§791.40 (b) "In general, each person's share of the test cost shall be in proportion to its share of the total production volume of the test chemical"

§791.48 (a) "Production volume will be measured over a period that begins one calendar year before publication of the final test rule in the Federal Register and continues up to the latest data available upon resolution of the dispute"

§791.48 (b) "For the purpose of determining fair reimbursement shares, production volume shall include amounts of the test chemical imported in bulk form and mixtures, and the total domestic production of the chemical including that produced as a byproduct" (40 CFR Ch. 1, 7-1-92 edition).

This simple establishment of a test cost sharing standard under TSCA has brought about a completely different outcome with respect to economic efficiency and encouragement of post-patent competition. This approach proved to avoid the expense of litigation, the delay in entry, and the risk of failure in the event of generic entry that has been artificially imposed under FIFRA. Based on well over a decade of experience, the EPA stated:

> EPA has extensive experience under TSCA section 4 with cost-sharing for testing. EPA has found that persons conducting testing under section 4 have chosen in each instance to date to work out their own arrangements for cost-sharing or reimbursement without any need for EPA involvement. EPA issued regulations in 40 CFR part 791 for data reimbursement. In spite of the significant number of test rules issued under TSCA, no one has invoked any of the formal procedures for data reimbursement under the regulations [*Federal Register*, Vol. 55, No. 152, August 1990, p. 32221].

3. THE ROLE OF PATENT POLICY

Any economic analysis of regulatory cost sharing between original and generic entrants must consider the role of patent policy. Typically, each new substance regulated by FIFRA comes to market initially under patent protection. Patents provide an important incentive to original registrants for product innovation. They allow recovery of product development costs through monopoly pricing or licensing until patent expiration.[2] However, a regulatory system that requires commercial sellers to have product registrations can have the same effect as a patent. Any time after patent expiration that a registration is withheld or delayed for the first generic entrant, the effect is the same as extending the period of patent protection.

Upon patent expiration, other firms in unregulated markets can copy successful products and competition begins. With competition, consumers typically gain more than the monopoly profit lost by the innovator because prices are not only lower but the benefits of greater quantities of consumption can be enjoyed at lower prices. Generic firms also gain but, as shown below, their gains are usually small by comparison. Social welfare improves through competition if the product is still useful to society. Achieving optimal social benefits requires finding a proper balance between incentives for and benefits of new product innovation, which are greater with a longer patent period, and the benefits of competition following patent expiration, which are greater with a shorter patent period. Choice of the length of the patent period represents a tradeoff between these two benefits. This choice is made by Congress in the form of patent law.

---

[2] Sometimes, due to time required for generating data and EPA review, part of the patent period is past by the time a new product can be introduced. For this purpose, the 1978 Amendment of FIFRA instituted a 10-year exclusive use period under which a new product can be sold under monopoly conditions even if the patent has expired.

While development costs are high for many substances regulated under FIFRA, patents for agricultural pesticides have also been found to be highly effective as a means of recovering development costs. For example, in an examination of 130 industries, Levin et al. (1984) found patents to be more effective for pesticides than any other industry except pharmaceuticals. Their study confirms that

> to have the incentive to undertake research and development, a firm must be able to appropriate returns sufficient to make the investment worthwhile. On the other hand, if consumers are to benefit substantially from innovation, sooner or later competitors must be able to imitate, respond to, or build upon an innovator's initiative to assure erosion of the innovator's monopoly and wide diffusion of the innovation on favorable terms [p. 13].

Their study examines a comprehensive set of approaches used by firms to appropriate the returns from innovations. In some industries, patents do not provide great protection because similar products can be developed and legal enforcement is difficult. In such industries, means of appropriation other than patents are used, such as trade secrets, quick movement on the learning curve, and development of superior sales or service efforts. Levin et al. (1984) found the pesticide industry to rate patents consistently higher than all industries other than pharmaceuticals and inorganic chemicals as a means of preventing duplication. The effectiveness of patents for pesticides explains three general facts about pesticide markets investigated further below: (i) why profit margins for pesticides tend to be high prior to generic competition, (ii) why original pesticide registrants face a substantial incentive to achieve a patent extension or some arrangement that acts like a patent extension, and (iii) why aggregate social well-being is best served by achieving competition as soon as possible after pesticide patent expiration.

3.1. Profitability Under Patent Protection

Most successful substances under FIFRA regulation are developed and sold by a single firm under monopoly conditions until patent expiration.[3] While profits under monopoly pricing are believed to be high due to inelastic demand, profit margins of firms are proprietary and, if released, can affect a firm's ability to compete. Thus, understandably, public profit margin data are not readily available. However, some clear indications of large profit margins are available. One of the clearest indicators is a comparison of domestic prices with offshore (or world) prices. In many cases, the same domi-

---

[3] In a few cases, sales by other firms take place under licensing and royalty payments, but the welfare implications are similar because the product developer is able to impose a monopoly profit margin in the form of a royalty.

nant firm sells the same product both domestically and abroad. Assuming both prices are above cost, the difference represents a lower bound on domestic profit margins. A comparison of several domestic and foreign prices have been given in testimony before Congress, as shown in Table 2. These prices show that world prices have been 27.5–61.8 percent below U.S. prices. Thus, to the extent firms are competing in both domestic and international markets, and given that registration compliance costs have been running no more than about 3.5 percent of sales, domestic profit margins on these pesticides were at least 24.0–58.3 percent, respectively. Such price comparisons are typical of pesticides that lack patent protection abroad and have not yet had domestic generic competition.

Table 2. Comparison of U.S. and Foreign Prices

Product	U.S. Price	Foreign Price	Apparent U.S. Gross Profit Margin
Malathion	$1.60/lb.	$0.89/lb.	> 44.4%
Methyl Parathion	$1.55/lb.	$0.99/lb.	> 36.1%
Carbaryl	$2.55/lb.	$1.85/lb.	> 27.5%
Treflan	$26.00/gal.	$16.00/gal.	> 38.5%
Paraquat	$34.00/gal.	$13.00/gal.	> 61.8%
Roundup	$68.00/gal.	$43.00/gal.	> 36.8%

Note that midpoints of price ranges are given here to facilitate calculation of margins.

While Table 2 suggests some lower bounds on profit margins, monopoly conditions support profit margins as high as 60–80 percent for successful substances, and generate hundreds of millions of dollars in profit during the patent period. For example, a Florida public court document filed on behalf of ISK Biotech in Antitrust Civil Investigation 92-404 reveals that ISK Biotech's gross profit margin on Bravo (chlorothalonil) in the United States in 1991, 1992, and the first half of 1993 varied between 62.1 percent and 63.2 percent. High profit margins are due in large part to sales under patent protection. For such products, the costs of generating FIFRA regulatory data for the EPA ultimately become relatively small. However, for a generic entrant who gains access to the market for a much shorter period of time after patent expiration, often in a declining stage of the market and under more competitive pricing conditions, the regulatory costs are relatively large. Senate discussions recognized this problem when enacting the 1978 Amendment to FIFRA:

While the cost impact of FIFRA on the larger basic manufacturers can be shown to be relatively small, this is not true when extended to the smaller manufacturers and the formulators. The current law in essence treats every firm in exactly the same way and every registration as an autonomous entity. This has resulted in duplication of costs and significant reduction in competition. This was not the intent of Congress ... The Congress tried to avoid unnecessary duplication of testing, and to prevent any unique competitive advantages as a result of government regulation. ... [S. Rep. No. 95-334, pp. 30–31].

3.2. The Incentive for Patent Extension

With these market conditions, original registrants are understandably interested in postponing competition by generic entrants. One means of postponing competition is through preventing registration of generic products through control of required regulatory data. But postponement not only deprives generic firms from competing. It also deprives farmers and consumers of the benefits of competition. This problem was also recognized by Congress when enacting the 1978 Amendment:

It is perfectly understandable that a developer firm would desire to lengthen the period of exclusive control of production of the chemical beyond the 17 years the patent law allows. This may be especially true when it is remembered that a portion of the 17-year period may have been lost to preregistration testing.

One can equally easily understand the lack of enthusiasm over this kind of behavior on the part of the developer's potential competitors, those who desire to enter the technical-grade market. From their standpoint, the developer has obtained his due reward under the patent law; now he wants to start all over in reliance on a different kind of protection, that which can be obtained by denying access to the test data required for FIFRA registration. If a prospective competitor can be required to perform duplicate tests as a condition of market entry, in most cases the potential profits will not justify [sic] the expense of this duplicative testing and the developer will retain control over production and price levels [S. Rep. No. 95-334, pp. 37–38].

Of course, patent protection for new products in the United States has now been extended to 20 years from the date of application (or 17 years from the date of granting the patent, whichever is later), but the point remains valid.

Responding to complaints that registering a new pesticide may require 5 to 7 years to develop necessary regulatory data, the 98th Congress considered but did not act on a proposal to provide a special patent extension for pesticides. In spite of regulatory delay, industry data suggest that even the 17-year patent period in effect until recently acted as a substantial incentive to innovate new pesticide products. The pesticide industry allocates a healthy

15–16 percent of revenues to research and development expenditures (Aspelin, Grube, and Torla 1992, p. 13). This level of research and development has been sufficient to bring 10 to 15 new products to the point of FIFRA registration in most years since 1980. The number of newly registered pesticides averaged about 14 per year for 1967–1979 and 13 per year for 1980–1993 (Aspelin 1994, p. 22) but then increased to about 30 per year for 1994–1997 as a result of emphasis on reduced-risk pesticides (Aspelin and Grube 1999, p. 27). This increase is evidence of high profitability in pesticide markets for successful products under patent. More importantly, the specific Congressional consideration and rejection of an increase in the patent period for pesticides suggests that Congress does not intend to move the balance between incentives for new product innovation and consumer benefits of post-patent competition further toward incentives for product innovation. Thus, artificial use of agency rules and regulations employed under FIFRA that enable original registrants to achieve the same effect as extending patent life deserve careful scrutiny.

3.3. The Adverse Interaction of Patent Policy with Regulation Under FIFRA

The holder of a patent typically has a significant post-patent advantage in competition because investments to establish the brand name often outlive the patent. The original registrant usually has well over a decade to advertise a product by its brand name (rather than its generic name) and to develop the brand name's reputation and customer recognition. This makes competition for post-patent competitors difficult, especially initially. For example, when the brand name Treflan came off patent, few users knew it by its generic equivalent, trifluralin. New brand names developed by generic firms for trifluralin required time to gain buyer recognition and trust. Faced with this hurdle, generic competitors typically must sell their products for less than those of original market registrants to be competitive. These price disadvantages for generic pesticide firms are usually 5–10 percent even when generic firms obtain modest market shares.

For these reasons, post-patent generic competitors are rarely successful in capturing more than a small market share. For example, a single generic firm may capture only 2–5 percent in the early years of competition and rarely exceeds an ultimate share of 20–30 percent unless the generic producer has a distinct cost advantage.[4] Nevertheless, such a small share of

---

[4] For example, the FIFRA data compensation award in the arbitration of *DuPont v. Griffin and Drexel* specifies maximum linuron market shares prior to 1988 for the two generic entrants as 2.83 percent for Drexel and 18.29 percent for Griffin, with the rest held by DuPont. See *E.I. DuPont de Nemours and Company v. Griffin Corporation and Drexel Chemical Corporation*, Docket No. 16 171 0080 86M, December 22, 1988, pp. 37–38. Similarly, the

competition is typically sufficient to bring substantial benefits to farmers and consumers through lower prices (as shown below). With generic competition, prices typically fall dramatically by as much as 15–50 percent.

The problem of facing long-established brand name loyalty for generic pesticide competitors is no different than in other markets. However, with pesticide markets, these realities have special implications regarding the ability to spread additional regulatory costs over the relatively shorter time and smaller market share potential of a generic competitor compared to the original registrant. In contrast, the original registrant can spread regulatory costs over the entire time since tests were required and over a 100 percent market share prior to generic entry.

Aside from regulatory costs, generic entrants normally have low overhead and can compete profitably relative to their production and marketing costs with a very small market share. For example, many generic competitors contract for toll manufacturing by general purpose chemical manufacturers or import from manufacturers abroad who do not require large profit margins. In these cases, generic entrants have no plant overhead so their per-unit profit margin is almost independent of market volume (even though average costs may be somewhat higher due to transportation and import duties). But when a large regulatory fee independent of market volume is incurred, their profit margin can become heavily dependent on market volume. Thus, the lingering brand loyalty in post-patent markets for regulated substances can become a great deterrent to generic entry unless the regulatory cost is divided among firms on the basis of market volume rather than equally shared as per capita aggregates among all eventual competitors. This is particularly true for markets that are in or threatened by serious decline, either due to regulation (EPA cancellation) or innovation of alternative products. This adverse impact is due to the interaction of patent policy with the magnitude of aggregate regulatory cost requirements and the lack of a volume-based regulatory cost sharing standard in FIFRA (such as the one in TSCA).

3.4. Effects of Per Capita Test Cost Sharing on Speed of Generic Entry

An additional competition-discouraging aspect of FIFRA's lack of a regulatory cost sharing standard relates to the dynamics of generic entry. Because equal (per capita) sharing of aggregate regulatory test costs is typically claimed by original entrants, and equal sharing has been granted in some

---

FIFRA data compensation award in the arbitration of *Ciba-Geigy v. Drexel* identifies Drexel's average market share as 12.8 percent for 1984 to 1992 in the simazine market, where it was Ciba-Geigy's only competitor, and as 5.5 percent for 1989 to 1992 in the atrazine market, which had two other generic competitors. See *Ciba-Geigy Corporation v. Drexel Chemical Company*, Docket No. 16 171 00321 92G, August 15, 1994, p. 20.

cases, generic firms face an incentive to delay entry. Under a per capita claim, the first generic entrant faces a potential obligation to pay half the compensable test costs, but claims against the second, third, and fourth generic entrants drop to one-third, one-fourth, and one-fifth of the test costs, respectively. Thus, the threat of an equal or per capita sharing outcome imposes a large incentive for potential generic entrants to delay entry until several other generic firms have already entered, simply due to the lack of an explicit sharing standard under FIFRA.[5] But when all generic firms face this incentive, generic entry can sometimes be delayed many years after patent expiration. For this reason, the lack of a standard for test cost sharing that promotes generic entry has the effect of extending monopoly pricing in some markets regulated by FIFRA, and delaying the associated consumer benefits, for many years.

3.5. The Adverse Effect of the Petitioning Process on Generic Competition

Another common approach used to prolong the process of generic registration and extend monopoly pricing is for original entrants to petition the EPA to deny individual generic registrations because of generic product characteristics such as purity. The EPA must give serious consideration to such claims whether well-founded or not, but the process of registration is thereby slowed. And clearly the original entrant has an incentive to slow the registration process for generic entry that is directly proportional to the incentive to extend a monopoly (see estimates below).

Similarly, the original entrant has an incentive to circulate adverse rumors about the quality of a generic product to instill uncertainty among potential buyers. Being able to claim that a petition to EPA will likely prevent a generic entrant from obtaining a registration in a given crop season gives the original entrant a powerful lever to pull. By careful timing, such a strategy can scare distributors away from stocking a generic product at the beginning of what is typically a short season of sales each year, effectively causing a full year's delay in generic marketing.

While these effects are temporary, such tactics are effective given the customer confidence that an original entrant can accumulate during the patent period. The adverse impact of delaying generic entry and extending monopoly profits is due to the implicit opportunity given to original entrants to slow the process of generic registration through the petitioning process—a

---

[5] Awards can include clauses to adjust compensation if compensation is received from future generic entrants, but future entrance can be facilitated by selling an additional technical registration held by the original entrant, or future compensation can be implicit in the form of supply agreements or business agreements on other substances, which potentially makes such clauses ineffectual.

process provided as an approach to ensuring product safety, but open to manipulation for strategic reasons.

3.6. Should a Regulatory Cost Sharing Standard Be Added to FIFRA?

These issues raise the possibility that adding an explicit regulatory cost sharing standard to FIFRA consistent with relative marketing opportunities, and eliminating potential strategic use of FIFRA provisions, could greatly reduce required litigation and greatly promote competitive efficiency for post-patent substances regulated by FIFRA. But which standard is the right standard? In addition to a cost sharing such as under TSCA, clear statements about applicability of inflation, interest, and risk premiums are needed. Another major component of litigation has involved determination of which tests should be included and how their costs should be established when invoices have not been preserved.

One possibility is to include only the tests listed (by identification numbers) by the EPA as supporting a registration. These lists have been clarified by the re-registration of pesticides under the 1988 Amendment. Without an explicit standard under FIFRA, original entrant claims for compensation typically include many other tests. Perhaps only invoiced test costs or tests for which cost records have been kept should be reimbursable as a means of requiring cost verification. The importance of keeping records has been emphasized in several awards beginning with the 1988 case of *Griffin and Drexel v. DuPont*, which set the precedent for applying discounts when records were not kept and, instead, costs were estimated ex post. Thus, the importance of keeping records is understood even though an original entrant may find imposing the risk of estimated test costs on generic entrants a useful deterrent to competition.

Congressional hearings have been held to consider such changes.[6] However, the interests of generic firms differ greatly from the interests of large firms that tend to be original registrants. When the industry cannot present a united view, and given the relatively better developed lobbying efforts of large firms, such changes in FIFRA that would reduce possibilities for extending monopoly marketing were not and are not likely to be adopted. Two points appear applicable. First, unanticipated provisions (or lack thereof) in policy for regulated substances can have distinct effects on the organizational structure of regulated markets. Second, once entrenched in prior policy, provisions that tend to promote monopoly pricing for large firms are unlikely to be changed by Congress due to the organized lobbying to which they lead.

---

[6] See Testimony on Generic Pesticide Registration, Committee on Small Business, Subcommittee on Energy and Agriculture, U.S. House of Representatives, June 24, 1987.

To some extent, an illusion that small market players do not matter much seems to pervade policy revision. The remainder of this chapter examines the behavior of and market response to small generic market players and shows that small generic entrants typically have major impacts on pricing and consumer benefits in markets for regulated substances if regulations are not used artificially to prevent them from competing.

4. UNDERSTANDING THE MAGNITUDE OF PRICE REDUCTIONS WITH MINOR GENERIC COMPETITION

Because of large profit margins, post-patent competition in pesticide markets typically brings substantial price reductions—as long as generic entry is not blocked. Examples of substantial price declines following generic entry in pesticide markets are numerous. Several such declines have been cited in testimony before Congress, as shown in Table 3.[7] These price reductions of 10–40 percent occurred at a time when the General Pesticide Index was rising and the prices of many other important pesticides such as Furdan, Lasso, and Lorsban were increasing substantially (by an average of 9.9 percent).[8]

Table 3. Price Reductions Following Generic Entry

Product	Pre-Generic Price	Post-Generic Price	Price Reduction
Atrazine	$2.63/lb.	$2.00/lb.	> 24.0%
Diuron	$3.25/lb.	$2.40/lb.	> 26.2%
Simazine	$15.50/gal.	$9.50/gal.	> 38.7%
Phostoxin	$30.20/kg.	$24.28/kg.	> 19.6%
Treflan	$28.50/gal.	$21.43/gal.	> 24.8%
Sutan	$19.89/gal.	$17.90/gal.	> 10.0%

Note that midpoints of price ranges are given here to facilitate calculation of price reductions.

4.1. Minor Generic Competition Often Causes Large Price Reductions

"When generic pesticide producers enter the market following the expiration of a patent, they generally capture only 10–20 percent of the existing mar-

---

[7] See Federal Insecticide, Fungicide, and Rodenticide Act, Hearings before the Committee on Agriculture, Nutrition, and Forestry, U.S. Senate, 100th Congress, First Session, on S. 1516, Part II, April 29, May 20, and July 30, 1987, pp. 187–188.

[8] An account of these price changes is summarized in Federal Insecticide, Fungicide, and Rodenticide Act, Hearings before the Committee on Agriculture, Nutrition, and Forestry, U.S. Senate, 100th Congress, First Session, on S. 1516, Part II, April 29, May 20, and July 30, 1987, pp. 188–189.

ket."[9] Nevertheless, obtaining a small market share is usually sufficient to cause large price declines. A typical situation is where the generic competitor holds only a small share but the large original registrant cannot raise the price too much above the generic firm without losing substantial market share. As a result, prices find an equilibrium nearer competitive levels than monopolistic levels. Also, as a result, farmers enjoy lower prices and most of the savings are passed along to consumers in the form of lower food prices because of competitive conditions in farm product markets.

4.2. A Dominant-Firm Price Leadership Explanation

In this chapter, the standard dominant-firm price-leadership oligopsony model is used to explain theoretically the dramatic price effects that have been observed for regulated substances. While a wide range of economic theories could be used for this purpose, many add significant complications without adding substantial intuition, and those that produce significantly different results tend to employ inapplicable assumptions. For example, a Nash bargaining model might be used to represent the effects of generic entrance of a small number of firms. But the assumptions of a cooperative game required by Nash bargaining do not appear to apply.[10] As another example, Baumol, Panzar, and Willig's theory of contestable markets implies that even the threat of competition is enough to avoid monopoly pricing behavior. But with regulated substances, entry is expensive given the liability for regulatory test costs incurred upon entry, and entry requires a significant delay to obtain a registration even if relying on others' regulatory data—both of which violate contestable market theory assumptions. While a host of noncooperative game theoretic concepts are useful for understanding behavior with a small number of players, many do not lend themselves well to comparative static analysis, and others approximate the Stackelberg case of price leadership oligopsony examined here.[11]

Suppose market demand is represented by $p = a - bq$ and generic supply is represented by $q_2 = \alpha + \beta p$ where p is price, q is market quantity, and q_2 is the quantity supplied by generic entrants. Then excess supply facing the original entrant after generic entry is obtained by solving $q_1 = q - q_2$ for p,

---

[9] See Federal Insecticide, Fungicide, and Rodenticide Act, Hearings before the Committee on Agriculture, Nutrition, and Forestry, U.S. Senate, 100th Congress, First Session, on S. 1516, Part II, April 29, May 20, and July 30, 1987, p. 182.

[10] I know of only one case where one of three competing firms was driven from a pesticide market where ensuing pricing behavior appeared to follow the implications of a cooperative game. A cooperative approach between two selling firms is to price as a joint monopoly and then divide the gain according to the Nash Bargaining criterion. Of course, overt collusion for this purpose violates the Sherman Antitrust Act, so explicit behavior of this type is rare.

[11] For further discussion, see Rasmusen (1989) or Osborne and Rubinstein (1994), or at a more advanced level Fudenberg and Tirole (1992).

which yields $p = (a - \alpha b - bq_1)/(1 + \beta b)$ where q_1 is the quantity sold by the original entrant. If the original entrant faces constant marginal cost c_1, then the original entrant as a dominant-firm price leader will equate marginal cost c_1 to marginal revenue associated with excess demand, $MR = \partial(pq)/\partial q = (a - \alpha b - 2bq_1)/(1 + \beta b)$, which thus obtains $q_1 = [(a - \alpha b) - c_1(1 + \beta b)]/2b$. Substituting this quantity into the excess demand equation yields

$$p = \frac{a - \alpha b}{2(1 + \beta b)} + \frac{c_1}{2}. \tag{1}$$

Without generic entry, the original entrant equates marginal cost c_1 to the marginal revenue $MR = \partial(pq)/\partial q = a - 2bq$ associated with the market demand, obtaining $p = (a + c_1)/2$. Now, without loss of generality, let the market price and quantity in the case with generic entry be represented by $p = 1$ and $q = 1$ to study relative market effects of generic competition. The generic market share is thus $q_2 = \alpha + \beta$. If the price elasticity of market demand in the case of generic entry is represented by η, then $b = 1/\eta$ and $a = b + 1$. Substituting these results into (1) implies that $c_1 = 1 - (1 - \alpha - \beta)/(\eta + \beta)$. Thus, the monopoly price relative to the price with generic entry is

$$p = 1 + \frac{1}{2\eta} - \frac{1 - \alpha - \beta}{2(\eta + \beta)}. \tag{2}$$

With this model the profit margin under monopoly pricing can be calculated simply as $(p - c_1)/p$.

The dramatic effect of a small amount of competition can thus be calculated as a function of the market demand elasticity (η), the generic market share ($\alpha + \beta$), and the generic supply elasticity (β) as in Table 4. Table 4 expresses the price with generic entry as a percent of the monopoly price in (2) and thus shows how much the price can decline from monopoly levels with small generic market shares. When pesticide demand is highly inelastic, a small competitive market share can bring prices much closer to purely competitive levels than to the monopolistic levels that tend to occur when generic entry is delayed.

The price elasticity of agricultural pesticide demand has been estimated to be -0.25 by Antle (1984). Because of substitution possibilities for individual pesticides, the elasticity for a single pesticide may be somewhat higher. Thus, the table examines a range of pesticide elasticities absolutely larger than -0.25. The table also examines a wide range of supply elasticities, seemingly exhausting realistic possibilities. As the table demonstrates, however, results are not highly sensitive to the supply elasticity.

Table 4. Price Reductions Caused by Generic Entry

Pesticide Demand Elasticity	Generic Market Share	Generic Supply Elasticity						
		0.25	0.50	0.75	1.00	1.25	1.50	1.75
		Price Reduction in Percent						
-0.25	0.05	51.2	57.7	60.4	61.8	62.7	63.4	63.8
-0.25	0.10	52.4	58.3	60.8	62.1	63.0	63.5	64.0
-0.25	0.15	53.5	58.9	61.2	62.4	63.2	63.7	64.1
-0.25	0.20	54.5	59.5	61.5	62.7	63.4	63.9	64.3
-0.25	1.00	66.7	66.7	66.7	66.7	66.7	66.7	66.7
-0.50	0.05	26.8	34.4	38.3	40.6	42.1	43.3	44.1
-0.50	0.10	28.6	35.5	39.0	41.2	42.6	43.7	44.4
-0.50	0.15	30.2	36.5	39.8	41.7	43.1	44.1	44.8
-0.50	0.20	31.8	37.5	40.5	42.3	43.5	44.4	45.1
-0.50	1.00	50.0	50.0	50.0	50.0	50.0	50.0	50.0
-0.75	0.05	16.1	22.3	25.9	28.3	30.0	31.3	32.3
-0.75	0.10	17.8	23.5	26.8	29.1	30.6	31.8	32.7
-0.75	0.15	19.5	24.6	27.7	29.8	31.2	32.3	33.2
-0.75	0.20	21.1	25.7	28.6	30.5	31.8	32.8	33.6
-0.75	1.00	40.0	40.0	40.0	40.0	40.0	40.0	40.0

The results for a demand elasticity of -0.25 show that the price reduction resulting from even a 5 percent generic market share is 51.2–63.8 percent. Comparison of these reductions with price reductions achieved with a 100 percent generic share reveal that most of the price reduction afforded by full competition, 66.7 percent, is achieved by a small competitive generic market share. As the demand elasticity moves up absolutely, the price reductions become smaller but the percentage magnitudes are still large. Even for a demand elasticity of -0.75, the theoretical implications are quite consistent with the price declines from generic competition reported in Table 3. Thus, both the empirical and theoretical evidence imply that even minor generic competition can cause substantial price reductions in pesticide markets under a broad range of market conditions.

5. THE INCENTIVE TO EXTEND MONOPOLY PRICING BY ARTIFICIAL MANIPULATION OF FIFRA

Based on the available information on the price effects of competition and typical general estimates of the input demand elasticity for pesticides, this section calculates the incentive for an original registrant to extend monopoly conditions for an additional year by forestalling generic competition. While the information necessary to evaluate the impacts of generic competition on specific pesticide markets is typically proprietary and thus unavailable for published economic analysis, enough information exists to discuss a few cases based on publicly available data.[12]

Table 5 presents some basic market information on several pesticide markets after generic entry. It evaluates the effects of generic entry on several regulated substances for which minimal information on the price effects of partial competition is suggested by public record or proprietary information made available for this study, and where the necessary market volume could either be determined from the EPA's periodical publication on *Pesticide Industry Sales and Usage* or identified from other sources. The sources of prices with and without partial competition, and market volumes with partial competition, for each substance are explained below. The market volume reported in Table 5 is taken to represent the volume under partial competition in each case because it is reported for a period after generic entrance. The calculations based on these data are straightforward and underestimate the potential effects of full generic competition because in many cases only one or a few generic entrants had entered the markets and, thus, conditions of full competition had not been reached.

To provide a rough assessment of the incentive to extend a monopoly by an additional year, suppose the market volume is observed under partial competition as represented by dominate-firm price leadership. The market volume that would be sold under monopoly is estimated by considering the difference in sales associated with the price difference between partial competition and monopoly on the basis of the input demand elasticity for pesticides as suggested by equations (1) and (2). For this purpose, the demand elasticity of -0.25 for agricultural pesticides estimated by Antle (1984) is used.

The price effect is a simple percentage reduction of the partially competitive price from the monopoly price. In Figure 1, where market demand is D and demand facing the original entrant after generic entry is $D - \overline{q}_2$, the price set by the original entrant equating its marginal cost, $\overline{c}_1$, and marginal revenue is $\tilde{p}_1$ before generic entry and $\overline{p}_1$ after generic entry. The monopoly revenue is found by first determining the monopoly market volume using the price difference, the market volume after entry, and the pesticide demand elasticity, and then multiplying that volume by the monopoly price. The last column of Table 5 provides a first-order

---

[12] Better data are available at substantial cost, for example from Doane Marketing Research, Inc., but the data are proprietary and not available in support of published analyses. Similar calculations based on proprietary data yield quantitative results with implications similar to those presented here and were presented in a confidential report to the EPA (Just 1998).

Table 5. Approximate Incentives for the Original Entrant to Extend Monopoly Pricing

Compound	Monopoly Price	Competitive Price	Unit	Price Effect	Market Volume	Monopoly Revenue	Competitive Revenue	Monopoly Incentive
	($/unit)	($/unit)		(%)	(mil. lbs.)	(mil. $)	(mil. $)	(mil. $)
Linuron	$12.25	$7.75	lbs. a.i.	36.7	4.0	$44.9	$31.0	$16.5
Gibberellic acid	$1.64	$0.85	grams	47.8	14.0	$20.5	$12.0	$9.8
Atrazine	$2.63	$2.00	lbs. a.i.	24.0	70.5	$174.9	$141.0	$41.9
Simazine	$3.88	$2.38	lbs. a.i.	38.7	4.5	$15.9	$10.7	$6.2
Glyphosate	$77.50	$44.52	gallons	42.6	27.5	$481.6	$306.1	$204.9
Trifluralin	$28.50	$21.43	gallons	24.8	25.5	$171.1	$136.6	$42.4

Source: The sources of prices and market volumes are explained in the text.
Regarding market volume as a partial-competition volume, the monopoly volume is estimated using a demand elasticity of -0.25.

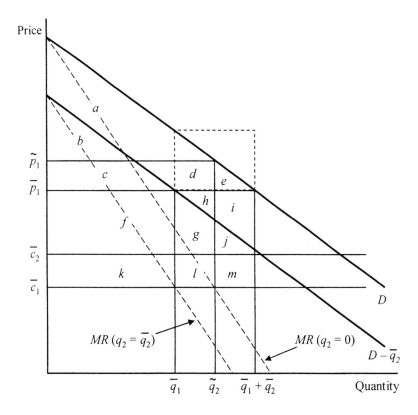

Figure 1. Welfare Implications of Price Leadership

approximation of the incentive to extend a monopoly by one year. This approximation is found by simply multiplying the monopoly market volume by the difference in prices between the monopoly and partial-competition cases (area $c + d$ in Figure 1). This difference also provides a first-order approximation of the benefits received by farmers and consumers from partial competition. It is an underestimate of their benefits from partial competition because the incentive to extend a monopoly also considers the effects on market volume of this change in price. Results later in the paper give a more precise estimate of welfare effects for the original entrant as well as other market groups based on the framework of Figure 1.

5.1. Linuron

Linuron was developed by the original registrant under the brand name Lo-rox. While linuron was once one of the most important herbicides in the

United States, its market has declined substantially in recent years with the development of more effective herbicides. Nevertheless, the linuron market is useful for analysis because the market was larger at the time of generic entry in 1984 and more information regarding generic competition is available due to the explicit nature of the public arbitration decision in *DuPont v. Griffin and Drexel*.

Linuron was patented by Farbwerke Hoechst after first synthesizing it in 1957. DuPont obtained a license including marketing rights in the United States from Hoechst in 1959 and received FIFRA registrations commencing in the early 1960s. Until well after expiration of the patent in 1977, DuPont was the sole supplier of linuron in the United States and thus held a monopoly position. After the linuron patent expired in 1977, Griffin obtained a FIFRA registration in 1979 based on intended use of DuPont-supplied linuron. When linuron could not be purchased from DuPont, Griffin was forced to find an offshore supply and amend its registration accordingly. About the same time, Drexel also received a linuron technical registration. Both firms were first able to start competing as generic entrants in the linuron market beginning in 1984, some 7 years after patent expiration. Another firm, Aceto Chemical Company, had entered the linuron market and gained almost a 10 percent market share before Griffin and Drexel entered the market. However, Aceto apparently left the linuron market to avoid expensive sharing of regulatory test costs and the associated litigation.

Griffin and Drexel were able to gain only small market shares in their first few years of competition. Their combined market share did not exceed about 20 percent prior to 1988.[13] Nevertheless, the price impact was substantial. Price per pound had been increasing rapidly due to inflation in the late 1970s. Yet, the small amount of competition caused the linuron price to decline from $12.25 to $10.75 per pound when it normally would have increased another $.40 to $.50 per pound annually. Thus, price in 1984, the first year of competition, was about 15 percent lower than it would have been without competition. Relative to the price trend that existed prior to generic entry, the effects of generic competition increased in 1985 through 1987 as the price remained almost constant at the reduced level. Furthermore, heavy discounting of prices below the retail prices began to take place. By 1987, linuron was selling for $31 per gallon ($7.75 per pound) after discounts compared to the $49 dollars per gallon implied by the 1983 price of $12.25 per pound.[14]

Table 5 uses an annual market volume of 4 million pounds, which was the approximate market volume when Drexel was still competing. Prices

---

[13] See *E.I. DuPont de Nemours and Company v. Griffin Corporation and Drexel Chemical Corporation*, Docket No. 16 171 0080 86M, December 22, 1988, pp. 37–38.

[14] The prices used here are annual averages of prices reported in the periodical publication, "AgChemPrice" by DPRA Incorporated, Manhattan, Kansas, various issues.

under competition and monopoly are represented by the discounted price of $7.75 per pound in 1987 and the $12.25 per pound price in 1983. These figures reflect an annual monopoly market revenue of $44.9 million compared to a competitive market revenue of $31.0 million. The corresponding reduction in annual profit for the original registrant, at least $16.5 million, represents over half of the total market revenue under competition. Thus, the incentive to extend a monopoly, even for one crop season, is relatively large, as is the effect on farmers and consumers.

5.2. Gibberellic Acid

Gibberellic acid is neither a herbicide, insecticide, nor fungicide but is one of the more important growth regulator pesticides. Imperial Chemical Industries (ICI), a British firm, obtained several patents for gibberellic acid and issued licenses to sell in the United States to five companies, one of which was Abbott Laboratories. The other competitors withdrew, leaving Abbott as the sole producer. From the late 1970s until 1989, only Abbott and Merck sold gibberellic acid, with Merck purchasing all of its supply from Abbott. Thus, the market was monopolized for all intents and purposes. Prices were generally increasing through 1989. After the patent expired, Agtrol Chemical Products applied for and received a technical registration in March 1989 and began to compete as a generic entrant. Initially, Agtrol's market penetration was small, but substantial price impacts began almost immediately. Table 6 shows the time pattern by which the resulting price reductions occurred and how the extent of the price reduction occurred in response to the generic volume sold.

Table 6. Agtrol's Impact on the Gibberellic Acid Market

Year	Sales Volume (1,000 gms.)	Sales Revenue ($1,000)	Rebates ($1,000)	Net Sales ($1,000)	Average Price ($/gram)	Price Impact (percent)	Market Share (percent)
1989	155	264	10	254	1.64	–	1.1
1990	881	1,269	90	1,179	1.34	18.3	6.3
1991	3,067	4,313	334	3,979	1.30	20.7	21.9
1992	4,239	4,770	339	4,431	1.05	36.2	30.3
1993	5,128	5,678	420	5,258	1.03	37.4	36.6
1994	6,183	6,562	463	6,099	0.99	39.7	44.2
1995	6,631	6,830	436	6,394	0.96	41.1	47.4
1996	7,324	7,572	473	7,100	0.97	40.8	52.3
1997	7,017	6,683	693	5,990	0.85	47.9	50.1

Source: Data provided by Graham Stoner, former President of Agtrol Chemical Products.

Note that the market share calculations assume a 14 million gram market.

Agtrol's role in the market in 1989 was minor and late so little price impact occurred. In 1990, price was initially $1.55 per gram to the distributor, but rebates quickly became common as Abbott tried to retain its market share. The actual selling price (after rebates) fell to $1.34, an 18.3 percent reduction in price, as a result of Agtrol gaining just 6.3 percent of the market. In the next year, Abbott reportedly told distributors that it would provide substantial rebates to distributors if 90 percent of their purchases were made from Abbott. Apparently a rumor was also circulated that Agtrol's product was defective, although that rumor was effectively addressed. Distributors evidently declined to buy 90 percent of their product from Abbott. As a result, the effective price which started at $1.40 that year dropped quickly. Most of the 1991 product was sold at $1.25 after rebates. By the end of the marketing year, the price had declined to $1.10. The average selling price for the year was $1.30 per gram after an average rebate of 7–10 percent. As a result of this competition, Merck left the market. However, its departure did not necessarily reduce competition because Merck was relying on supply from Abbott and, thus, likely had limited opportunity to compete with price.

Ultimately, Agtrol was able to achieve a larger market share than typical of generic entrants because of a strong cost position. However, after attaining only a 30.3 percent market share, the resulting price reduction compared to the pre-generic level was already 36.2 percent. After 9 years, Agtrol was able to grow to a 50 percent market share in 1997, which caused a further price decline to $0.85 per gram, for an overall reduction from pre-generic price levels of about 48 percent.

Using a market volume of 14 million grams and the price reduction that occurred with competition as reported in Table 6, the effect of generic competition is estimated to be a reduction in annual market revenue from $20.5 million to $12.0 million, which implies a reduction in profit for the original registrant of at least $9.8 million. This incentive to extend monopoly conditions for one year for the original registrant amounted to over 80 percent of the total market revenue under competition.

5.3. Atrazine

Atrazine was first registered in 1958 by Geigy Chemicals, which later merged into Ciba-Geigy, which eventually merged into Novartis. Ciba-Geigy licensed the patent to 14 companies beginning in 1973 prior to the patent expiration in 1977. In spite of the large number of licensed competitors, the new entrants combined were not able to take more than a minor market share and Novartis remained the dominant seller. While details of the licensing agreements are proprietary, firms receiving licenses to use patents typically must pay royalties that, in effect, impose the monopoly price markup, in which case the market operates as under a monopoly. Thus, major competition typically does not begin until patent expiration.

Atrazine was the top selling pesticide in the United States from 1987 through 1999 according to market volume as reported by the EPA (Kiely, Donaldson, and Grube, p. 14). The EPA estimated the 1994/95 U.S. market volume to be between 68 and 73 million pounds per year in terms of active ingredient. Table 5 uses the midpoint of this range to represent market volume. For the price effects of limited competition, the prices from Table 3 are used, indicating that atrazine prices declined by 24 percent from 2.63/lb. to $2.00/lb. with generic entry. On this basis, the effect of generic competition is estimated to be a decline in market revenue from $174.9 million to $141.0 million. Thus, a lower bound on reduction in profit for the original registrant is $41.9 million per year. The incentive to extend the monopoly for the original registrant for even one year amounts to about 30 percent of the total revenue in the market under competition. Furthermore, because profit margins under partial competition are likely to be much smaller than under monopoly (e.g., 10 to 20 percent rather than 50–80 percent), annual profits for the original registrant under monopoly are likely to be 2 to 4 times larger than received by all producers in the market under competition.

5.4. Simazine

Like atrazine, simazine was first registered by the Novartis predecessor, Geigy Chemicals, in 1958. Ciba-Geigy held the patent when it expired in 1977. The effect of competition was dramatic, as shown in Table 3. Novartis has remained the dominant seller in the post-patent market, illustrating the staying power that is usually possessed by the original registrant.

The annual market volume of 4.5 million pounds active ingredient corresponds to the midpoint of EPA's estimate of 3 to 6 million pounds (Aspelin 1994, p. 22). The price reduction with generic entry in Table 3 amounts to a reduction from $3.88/lb. to $2.38/lb. when converted from gallons to pounds of active ingredient. Based on this information, the annual monopoly market revenue is $15.9 million compared to a competitive market revenue of $10.7 million. The corresponding reduction in annual profit for the original registrant, at least $6.2 million, represents almost 60 percent of the total competitive market revenue.

5.5. Glyphosate

Glyphosate, known as Roundup according to the original registrant's brand name, has been one of the two largest pesticide markets in the United States during the last two decades. Only Metolachlor, also known by the brand name Dual, has had a slightly higher market revenue (Gianessi and Anderson 1995). Only atrazine has had a higher market volume, but even atrazine was displaced by glyphosate as the volume leader by 2001 (Kiely, Donaldson, and Grube 2004, p. 14). The major patent held by Monsanto on glyphosate expired in March 1991. Upon patent expiration, a substantial price decline was realized almost immediately. In the first year, the price declined by over 30 percent. By the second year, the price had declined over 40 percent. Furthermore, this dramatic price reduction occurred with a limited amount of competition. The reason for limited competition was that Monsanto also held some yet unexpired formulation patents in addition to the composition of matter and use patent which expired in 1991. As a result, only two generic firms were able to find a way to register a generic product.[15]

The EPA estimated the 1994/95 U.S. market volume to be between 25 and 30 million pounds per year in terms of active ingredient. Table 5 uses the midpoint of this range to represent market volume. The price fell from $77.50 per pound in 1990 to $44.52 per pound in 1993. The effect of generic competition is estimated to be a decline in annual market revenue from $481.6 million to $306.1 million. This implies a reduction in annual profit for the original registrant of at least $204.9 million per year. Again, the incentive to extend the monopoly period for the original registrant is quite large—in this case, about two-thirds of the total revenue in the market under partial competition. Thus, profits under monopoly may be as much as 3 to 6 times larger than received by all producers in the market under competition.

5.6. Trifluralin

Trifluralin, for which the original registrant's brand name is Treflan, has been the third largest pesticide market in the United States according to market revenue (Gianessi and Anderson 1995). The EPA estimated the 1994/95 U.S. market volume to be between 23 and 28 million pounds per year in terms of active ingredient. Table 5 uses the midpoint of this range to represent market volume, and uses the price effects of generic competition in Table 3 to represent the price effects of switching from monopoly to partial competition. The effect of generic competition is thus estimated to be a decline in annual market revenue from $171.1 million to $136.6 million. This causes a reduction in annual profit for the original registrant of at least $42.4 million per year. Again, the incentive to extend the monopoly market-

---

[15] Glyphosate thus gives another example of how an original entrant can extend the monopoly or restricted competition period. By producing under a basic patent for the product until near patent expiration, and then obtaining a process patent on any unique approach to efficient production near the end of the first patent period, a firm can continue with partial patent protection through a second patent period unless firms can find another process that is economical. However, this possibility is not due to FIFRA manipulation.

ing period for the original registrant is relatively large—a little over 30 percent of the total revenue in the market under partial competition.

5.7. Economic Implications of Incentives to Maintain a Monopoly

The various cases in Table 5 show that the price implications of partial competition on previously monopolized pesticide markets are relatively large. The incentive to continue a pesticide monopoly for the original registrant, even for an additional year, is relatively large and equal to one-third to four-fifths of the total annual revenue of the market under conditions of partial competition. While more accurate proprietary data would be preferable for these calculations, the results provide clear indications of the order of magnitude of results. Because the percentage reductions are so dramatic even when generic competition is small, the economic implications of potential extensions of monopoly marketing due to artificial manipulation of provisions in regulatory statutes are quite large relative to total market benefits.

6. WELFARE EFFECTS OF GENERIC ENTRY ON GENERIC ENTRANTS AND FARMERS/CONSUMERS

Based on the price and quantity effects of partial competition in Table 5, this section calculates conventional producer and consumer welfare effects for original entrants, generic entrants, and farmers/consumers. Calculations follow the framework of Figure 1 and the model associated with equations (1) and (2). These calculations are presented in Table 7. If a market moves from monopoly to a partially competitive equilibrium as a result of patent expiration and generic entry, then the original registrant loses the additional monopoly profit margin on the entire volume that would have been sold under monopoly. Both the original registrant and the generic entrants earn a smaller partially competitive profit margin on their respective sales. The monopoly profit margin (in excess of the partially competitive profit margin) on the entire monopoly market volume is passed on to farmers and consumers. Moreover, farmers and consumers receive an additional benefit because they can adjust by buying more at lower prices.

In addition to dependence on publicly available prices and market volumes as in Table 5, the results in Table 7 depend on data on generic market shares, which are also proprietary and thus largely unavailable in public data. The calculations correspond to the time of generic entry rather than current market conditions. For example, the atrazine and simazine market shares in the second column of Table 5 are taken from the FIFRA data

Table 7. Welfare Impacts of Delaying Generic Entry for One Year

Compound	Generic Market Share	Generic Profit Loss Competitive Margin		Farmer and Consumer Loss	Original Registrant Gain Competitive Margin		Net Social Loss Competitive Margin	
	%	10%	20%		10%	20%	10%	20%
				million $				
Linuron	21.1	$0.7	$1.3	$17.2	$16.9	$17.3	$1.0	$1.3
Gibberellic acid	50.1	$0.6	$1.2	$10.4	$10.3	$10.7	$0.7	$0.8
Atrazine	5.5	$0.8	$1.6	$43.2	$41.9	$41.9	$2.1	$2.8
Simazine	12.8	$0.1	$0.3	$6.5	$6.2	$6.2	$0.4	$0.5
Glyphosate	10.0	$3.1	$6.1	$215.8	$205.1	$205.2	$13.8	$16.8
Trifluralin	10.8	$1.5	$3.0	$43.8	$3.1	$3.8	$2.1	$2.9

Prices and market volumes with and without competition are as given in Table 5. The basis for market shares is given in the text.

compensation award in the arbitration of *Ciba-Geigy v. Drexel*.[16] In the case of atrazine, other generic competitors present were not included in the share in Table 7. Accordingly, the effect on atrazine generic profit is underestimated. The trifluralin market share is based on several assumptions about which technical registrants supply various formulators.[17] The gibberellic acid market share is taken from Table 6. The linuron market share is computed by combining the Griffin and Drexel market shares reported in the FIFRA data compensation award in the arbitration of *DuPont v. Griffin and Drexel*.[18] Finally, the glyphosate generic market share is arbitrarily set to 10 percent simply to illustrate potential effects because proprietary data are lacking.

6.1. Effects of Extending a Monopoly on Generic Pesticide Firms

The second and third numeric columns of Table 7 give the profit lost by generic firms if their access to the market is lost, regardless of how the access is lost. Generic profit is assumed to be zero for the case where market access is lost. The profit of a generic firm in the case of generic competition depends on the average profit margin and the generic firm's share of the market. Because average profit margins are sensitive information and thus generally not available, this study must consider several plausible profit margin levels and perform the calculations conditionally. Table 7 considers two representative profit margin levels which represent typical possibilities for generic entrants after equilibrium price adjustments.

The striking result for every compound represented in Table 7 is that the profit of generic firms is almost inconsequential compared to the effects on farmers and consumers. Obviously, generic competitors are worse off when a monopoly is extended for an additional season, but their losses are small compared to those of farmers and consumers. In linuron, for example, Drexel and Griffin would have foregone profit on about $6 million in sales (20 percent of the $31 million in Table 5) if they had not been able to compete. But at a profit margin of 10 percent, this would amount to only $0.7 million in profit. However, farmers and consumers would have had to pay as much as $4.50 per pound more on the entire market volume of 4 million pounds with a loss of partial competition. This effect alone would make farmers and consumers worse off by $14 million (see Table 5). But, in addi-

---

[16] See *Ciba-Geigy Corporation v. Drexel Chemical Company*, Docket No. 16 171 00321 92G, August 15, 1994, p. 20.

[17] The generic competition is assumed to supply Helena, Wilbur Ellis, Universal Coop, UAP, and Terra. All unaccounted supply is assumed to be attributable to generic supply. The remaining supply is assumed to come from Dintec—the DowElanco/I.Pi.Ci. partnership—which is treated like the original registrant for purposes of analyzing the trifluralin market.

[18] See *E.I. DuPont de Nemours and Company v. Griffin Corporation and Drexel Chemical Corporation*, Docket No. 16 171 0080 86M, December 22, 1988, pp. 37–38.

tion, considering the smaller quantity purchased at the higher price, farmers' and consumers' loss would amount to $17.2 million (Table 7). Thus, farmers and consumers can be big losers when only small marginal competition is prevented.

With a 10 percent profit margin, the generic profit effects are less than $1 million for every product except glyphosate and trifluralin. With a 20 percent profit margin, the profit effects on generic firms are double those at the 10 percent profit margin (and would be triple at a 30 percent profit margin, etc.). The reasons the effects on generic profits are small are twofold: (1) generic firms are typically able to capture only a small market share because of a longstanding brand name reputation held by the original entrant, and (2) profit margins are typically small in the post-patent period because of competition. These results verify that generic firms are losers if original registrants are allowed to extend a monopoly beyond patent expiration, but their losses are small (although all their profit is lost) simply because they are small players.

6.2. Comparison of Generic Profit Potential to FIFRA Compensation Claims

A comparison of the profit potential for generic entrants to the regulatory cost sharing claims of original entrants under FIFRA illuminates the potential for anticompetitive intimidation in absence of a cost sharing standard based on relative market participation. The claimed compensation for linuron from Griffin and Drexel was $12 million (see Table 1). Comparing this claim to the potential of generating an annual profit stream of $0.7 million from generic competition in the linuron market (Table 7 at a 10 percent profit margin) makes clear how an intimidating claim for test cost compensation by an original entrant under FIFRA can not only delay but completely close off generic entry. At standard discount rates, the discounted value of the profit stream would be less than the regulatory cost liability if the claim were granted. Additionally, the linuron market was in a state of dramatic decline at the time of generic entry—a condition often encountered in post-patent markets for substances regulated by FIFRA.

As another example, the claimed compensation for atrazine and simazine from Drexel was $6.67 million (Table 1) while the potential annual generic profit stream was possibly only $0.9 million on atrazine and simazine combined (Table 7, 10 percent profit margin). Again, at discount rates common at the time of this generic entry, the claimed compensation was great enough to entirely extract all potential generic profit. Furthermore, this was a claim against a single generic competitor that was one of three for atrazine and one of two for simazine.

These comparisons, perhaps more than any other in this chapter, underscore the great risk that generic entrants face due to the lack of an explicit regulatory cost sharing standard. The risk that such a claim will be granted is real because FIFRA provides no explicit standard, and because some compensation awards have awarded large shares of claimed compensation based on per capita sharing, royalty, and risk premium principles.

6.3. Effects of Extending a Monopoly on Farmers and Consumers

The fourth numeric column of Table 7 estimates the benefits lost by farmers and consumers combined if a monopoly marketing period is extended by some means for one year. Here, the effects are consistently large relative to the market sizes. Comparing to total market revenues that occur with partial competition (Table 5), the benefits lost to farmers and consumers amount to more than half of total competitive market revenues in 4 of 6 cases. As a percent of competitive market revenue (a generous numeraire since it ignores the cost of production), the benefits for farmers and consumers are 55 percent for linuron, 87 percent for gibberellic acid, 31 percent for atrazine, 61 percent for simazine, 71 percent for glyphosate, and 32 percent for trifluralin. These results suggest that farmers and consumers are the big losers of any regulatory provision that facilitates extending a monopoly beyond the patent period on a FIFRA-regulated substance. Clearly, from Table 7 the effects on farmers and consumers can run into hundreds of millions of dollars per year on major substances.

6.4. Comparison of the Farmer/Consumer Effects to the Original Registrant's Gain

The fifth and sixth numeric columns of Table 7 allow comparison of the benefits and losses among market participants. These columns give the effects on the original registrant and refine the rough calculations in Table 5. Table 7 shows that estimates are not very sensitive to the assumed competitive profit margin. The reason is that the original registrant's profit effects are determined largely by the monopoly pricing premium.

The striking result in Table 7 is that farmers and consumers lose approximately as much as the original registrant gains by extending monopoly conditions (slightly more). For example, a regulatory decision to delay a generic registration for glyphosate long enough to miss one marketing season would have been equivalent to taking a little over $200 million from farmers and consumers and giving it to the original registrant (if the assumed generic market share is appropriate). Thus, the impacts are substantial and take from the many to give to the one.

6.5. Net Benefits for Society Taken as a Whole

The results in Table 7 show that the impact is worse than simply "robbing Peter to pay Paul." As benefits are transferred from farmers and consumers to original registrants, aggregate social benefits are lost in the process. These net losses are reported in the last two columns of Table 7, which sum the effects on generic firms, farmers and consumers, and original registrants. They show, for example, that the cost of transferring $205 million in benefits to the original registrant by extending a monopoly on glyphosate is $13.8 million at a 10 percent competitive profit margin and $16.8 million if the competitive profit margin is 20 percent. This net loss is accounted partially by a loss for generic firms and partially by farmers and consumers having to give up more benefits than the original registrant receives. In these respects, the results are similar across all compounds. The net loss as a percentage of the benefit gained by the original registrant is 4.9–6.9 percent in every case with a 10 percent competitive profit margin, and 6.6–8.2 percent in every case with a 20 percent competitive profit margin.

7. A MODEL OF PRODUCT DEVELOPMENT AND GENERIC ENTRY AFTER PATENT EXPIRATION

A typical claim in arbitrations about regulatory cost sharing is that the regulatory cost is a fixed cost and that therefore economics has nothing to say about how it should be allocated among the original registrant and generic entrants. This section addresses the fallacy of this argument and discusses how regulatory costs should be allocated for economic efficiency. For this purpose and to avoid excessive length in this chapter, the discussion is based a companion paper by Just (2005).

Proper evaluation of the economic effects of how regulatory cost is allocated requires examining both the ex ante incentives for product development and the post-patent incentives for competition in the same framework. This requires modeling both entry decisions as well as subsequent investment and production decisions for both the original registrant and post-patent generic entrants. Taking the patent system as a given means of fostering product development (taking patent law as given in examining pesticide and biotechnology law), Just (2005) demonstrates the relevant issues in a five-stage sequential model of regulation where the stages are characterized as follows:

Stage 1: An original entrant first decides whether to incur the expenses of developing a new product.

Stage 2: The original entrant then decides how much to invest in plant capacity.

Stage 3: The original entrant then produces and sells as a monopolist for n_0 years under patent protection.

Stage 4: Upon patent expiration, potential generic entrants decide whether to incur the fixed expenses of market entry and investment in plant capacity.

Stage 5: Production and sales then take place for another n_1 without patent protection until market termination.

The regulatory cost required to assure society that the product is safe is incurred at the beginning of the second stage. Generic entrants also incur a share of the regulatory cost if they decide to enter in the fourth stage. Stages 2 and 4 take place instantaneously but production takes place under a patent for n_0 years in Stage 3 and for n_1 years in Stage 5.

In this model, annual consumer demand varies stochastically over time and maximizes a representative quasilinear consumer utility subject to a budget constraint where utility is a function of consumption of the regulated good and all other goods. Production takes place under constant marginal cost with a plant capacity constraint. Plant capacity must be decided before production commences and incurs an investment expense proportional to the plant capacity selected. Production costs and demand in future stages are anticipated imperfectly so all investment decisions are made under risk. For example, the product development and plant investment decisions in Stages 1 and 2 are made with risk regarding whether the product's success will be sufficiently great to attract future generic entry and sharing of required regulatory cost upon patent expiration. After product development and plant investment, the original entrant sells monopolistically during Stage 3 based on conditions realized throughout the patent period.

In the post-patent investment period (Stage 4), generic entrants decide whether to invest in plant capacity and incur fixed costs of entry including a regulatory cost share. The original entrant may also decide to adjust plant capacity. These decisions are made with risk regarding future demand for the product and again incur investment costs proportional to the plant capacities selected. As is typical, generic firms may incur lower fixed cost and higher variable cost than the original entrant through off-shore purchases of product manufactured abroad. [19] But the original entrant may be able to continue pro-

---

[19] Generic firms supplied by off-shore producers typically can find only one or two producers whose product quality meets EPA standards. These producers often have substantial capacity constraints. Also, with increasing frequency, original entrants are reportedly imposing requirements on domestic pesticide distributors (there are only about five that matter for U.S. agricultural pesticide sales) such that, say, 90 percent of a specific product must be supplied by the original entrant if the distributor is allowed to sell any of the original entrant's product. This effectively puts a cap on the potential generic market share over which regulatory cost can be spread because few generic firms have the capacity to supply all of a distributor's needs. Thus, the plant capacity constraints of the model are effectively realistic even with

duction with a depreciated plant without incurring further investment expense. As under FIFRA, some additional regulatory cost may also be incurred by the industry through EPA data call-ins at the time of generic entry irrespective of plant capacity or anticipated production. During post-patent competition (Stage 5), production decisions are made where the original entrant is the price leader under price leadership oligopoly and the generic firms are fringe competitors. The resulting prices are anticipated imperfectly at the time of post-patent investment due to demand uncertainty.

Just (2005) examines the mathematical solution to this problem under consumer utility maximization and producer maximization of discounted net profits using the dynamic programming approach of backwardation. The results explain why product price typically declines substantially and market quantity increases upon post-patent competition if generic entry occurs. They also show why consumers and all parties combined receive benefits from generic entry if it occurs and how generic entry depends on the share of regulatory test costs that generic firms incur. Specifically, aggregate social welfare increases as long as the generic firm's marginal production cost, if greater than the original entrant's marginal cost, is greater by no more than half the generic firm's expected profit margin.

Figure 1 provides intuitive understanding of these results and allows intuition about how results are altered under nonlinearity, i.e., under more general consumer utility specifications. In Figure 1, the respective marginal revenue relationships before and after generic entry are $MR(q_2 = 0)$ and $MR(q_2 = \overline{q}_2)$, respectively. Without generic entry, consumer welfare is area $a + b$ and producer profit is area $c + d + f + g + h + k + l$. With generic entry, consumers gain area $c + d + e$, the original entrant loses profit equal to area $c + d + g + h + l$, and the generic entrant gains profit equal to area $g + h + i + j$, for a net social welfare gain of area $e + i + j -$ area l. If generic entry occurs, welfare in the post-patent period is higher by a consumer effect measured by area e (under linearity area e is one-eighth of the dotted box), a profit on the additional quantity sold with generic entry measured by area $i + j$ (which is half of area $g + h + i + j$ under linearity), and a loss to society measured by area l because some of the product previously produced at lower cost $\overline{c}_1$ is produced at the higher cost $\overline{c}_2$ if the generic firm has higher cost. (Any change in fixed cost is not represented in Figure 1.) While nonlinearity would alter the symmetry in Figure 1 (whereby $\overline{q}_1$ and $\overline{q}_1 + \overline{q}_2$ are equidistant from $\tilde{q}_1$), qualitative results are unaltered.

---

offshore purchasing, although they may represent market constraints rather than plant capacity constraints.

7.1. Optimal Sharing of the Fixed Cost of Ensuring Product Safety

The more interesting results from this model are the implications for optimal sharing of test costs required for regulatory purposes. Suppose a one-time regulatory cost K is incurred to test the product and assure society of product safety at the time of original entry. How should this regulatory cost be shared between the original entrant and the later generic entrant(s) assuming, of course, that generic firms incur a share of the regulatory cost conditional upon entry n_0 years later at patent expiration? Suppose the generic entrant is required to compensate the original entrant in the amount αK if entry occurs. At the time of original entry, conditions affecting post-patent competition are subjectively stochastic. To consider the tradeoff of policy incentives for original product innovation versus post-patent competition, the regulatory cost as well as the cost of original product development must be considered random at the time of the Stage 1 product development decision.

Viewing the problem from the standpoint of Stage 1 makes clear how any particular sharing rule represented by α will affect potential future innovations and competition with its consequent implications for social well-being after innovation. The cost of regulation borne by the generic entrant, αK, adds to the fixed cost of generic entry, so that entry becomes less likely as a greater share of regulatory cost is imposed on the generic entrant. The remaining share of regulatory cost adds to the fixed cost ultimately borne by the original entrant so original product development is less likely when the original entrant bears more of the regulatory cost. Of course, if original entry does not occur, then generic entry is precluded. In this framework, the argument that regulatory cost is a fixed cost and therefore its sharing has no economic implications is clearly a very myopic argument made from the standpoint of Stage 5 of the model, thus ignoring the incentives that are created for future innovation and competition.

Just (2005) shows that two stochastic distributions are critical for solving this overall five-stage problem. The first is the distribution function $F(\pi_g)$ of net profit for the generic firm, ignoring regulatory cost, over the post-patent period in the event of entry as viewed from the Stage 1 decision to develop the product. The generic firm will rationally choose to enter, given the regulatory cost allocated to the generic entrant, if and only if its anticipated profit exceeds the allocated regulatory cost αK at the time of patent expiration. Thus, the probability of post-patent competition is $1 - F(\alpha K)$. The second is the distribution function of the original entrant's net profit $G(\pi_o)$ over the entire planning horizon (patent plus post-patent periods), ignoring regulatory cost, as viewed from the date of the Stage 1 product development decision. The original entrant will rationally choose to enter if this profit is greater than the expected regulatory cost borne by the original entrant considering the probability of generic entry, $C = F(\alpha K)K + [1 - F(\alpha K)](1 - \alpha)K$. Thus, the probability of product development and original entry is $1 - G(C)$.

Adding these considerations to the model above, Just (2005) considers the problem of maximizing social welfare over the five-stage horizon with respect to the share α of the regulatory test costs borne by the generic entrant. Aggregate social welfare ignoring discounting is[20]

$$W = G(C)(n_0 + n_1)m + [1 - G(C)]\{n_0 w_0 + n_1 w_1 + [1 - F(\alpha K)]n_1 \Delta w - K\},$$

where m is annual expected consumer income, w_0 is annual expected social welfare benefits from product introduction during the patent period, w_1 is annual expected social welfare benefits from the product during the post patent period if generic entry does not occur, and Δw is the additional annual expected social welfare benefits that occur with competition during the post patent period if generic entry occurs. Whether aggregate social welfare is increasing or decreasing in the share α depends on whether the expected regulatory cost borne by the original entrant is increasing or decreasing in the share. If this cost is increasing in the share, then aggregate social welfare is unambiguously decreasing (assuming producers benefit from entering the market).

However, this cost may be increasing or decreasing in the share. It tends to be increasing if the regulatory cost is high, the marginal effect of the regulatory cost share on the probability of generic entry is substantial, and/or the probability of generic entry is low. All of these conditions appear plausible with typical regulated substances under FIFRA when the share of regulatory costs borne by generic entrants is high, e.g., when test costs are substantial and the share borne by the generic entrant is considerably higher than the share of market sales and profits they receive. The implication is that social welfare is improved by making the share of regulatory test cost borne by the generic entrant as small as possible, at least to the point where these conditions change.

7.2. Sharing of Post-Patent Regulatory Cost

Another interesting problem arises when significant regulatory cost is incurred at the time of patent expiration or generic entry. This may be the case when the EPA issues data call-ins or undertakes a significant re-registration effort. To consider this problem conceptually, suppose the generic entrant does not face liability for original regulatory test cost but must pay a share of

---

[20] Discounting can be simply considered in this model by modifying n_0 and n_1 with additional multiplicative terms representing the effect of the discount factor, so discounting is conceptually inconsequential.

test costs K incurred under a data call-in at the time of patent expiration (Stage 4). No liability for original regulatory test costs may fit the case where original test costs are outside of FIFRA's 15-year window under which compensation is required.

For this case, suppose the additional regulatory test cost is unanticipated even stochastically at the time of the market development decision, in which case the post-patent period can be examined independently. Let $F(\pi_g)$ again denote the distribution function of profit for the generic firm ignoring regulatory cost over the post-patent period in the event of entry but now viewed from the date of the generic entry decision. The generic firm chooses to enter, given the regulatory cost αK, if and only if $\pi_0 > \alpha K$. Thus, the probability of post-patent competition is $1 - F(\alpha K)$. At this point, social welfare over the remaining planning horizon is $W = n_1 w_1 + [1 - F(\alpha K)]n_1 \Delta w - K$. In this case, no social tradeoff between incentives for innovation and competition is generated by the sharing of the regulatory cost. Accordingly, aggregate social welfare over the post-patent period is unambiguously decreasing in α.

This outcome, which seems to unfairly impose all regulatory cost on the original entrant, is quite different if the original entrant behaves competitively in the post-patent market. In this case, the original entrant adjusts plant capacity so that expected price is equal to expected marginal cost. Suppose for simplicity that fixed costs are proportional to plant capacity and that both firms face the same constant marginal cost aside from regulatory cost, $\bar{c}_1 = \bar{c}_2$. Then the generic firm cannot compete in a competitive market if it is allocated a greater share of the regulatory cost than its market share (which is its capacity share in this model). The same statement holds for the original entrant.

Alternative comparisons can be made where marginal costs are not equal for the two firms. If the generic firm has higher cost, $\bar{c}_1 < \bar{c}_2$, then its cost-inferior production may not be driven out of the market if its share of regulatory cost is below its market share. However, if the original entrant has a higher cost, $\bar{c}_2 < \bar{c}_1$, then its cost-inferior production may not be driven out of the market if the generic firm's share of regulatory cost is above its market share. Thus, social optimality is associated with market sharing of the regulatory cost.

7.3. Sharing Test Costs Between Patent and Post-Patent Periods

While optimal sharing of test costs between the patent and post-patent periods obviously depends on the choice of discount rate, the results for sharing post-patent regulatory cost have additional implications for sharing test costs incurred during the patent period. By developing the full market cycle model under the assumption of competitive pricing by the original entrant, the same market-sharing principle is found. That is, test costs incurred during the patent period should be shared according to the market volumes that original and generic entrants sell. In other words, the result that makes the generic share of regulatory test costs zero applies only when the original entrant exercises its market power. If the original entrant prices competitively, then market sharing is optimal. However, when market sharing occurs across the patent and post-patent period, the appropriate sharing is a sharing over the lifetime of the market as opposed to annual market sharing. This type of sharing would thus account for the inequity whereby the generic firm can recover test costs based only on post-patent sales, whereas the original entrant can spread test costs across volumes sold both before and after patent expiration.

7.4. Implications for FIFRA versus TSCA

These results have direct implications related to the test cost sharing provisions of FIFRA and TSCA. Specifically, the market-sharing provision of TSCA is consistent with economic efficiency under competitive behavior when all firms receive benefits of registration over equal time periods. Under FIFRA, the market sharing of post-patent (post-competition) regulatory test costs, which has been awarded in some cases, also appears to be consistent with economic efficiency. However, provisions of FIFRA whereby arbitrators have sometimes awarded test cost compensation exceeding generic market shares and ignored pre-patent time to recover test costs for original entrants are apparently inconsistent with economic efficiency.

Another provision of FIFRA that deserves discussion is the 15-year compensability window. When a generic firm applies for a FIFRA registration by relying on others' test data, FIFRA requires compensation only for the cost of tests submitted to the EPA within the preceding 15 years. This provision has the advantage of reducing the burden of regulatory test cost burden imposed on generic entrants for tests that primarily supported sales by the original entrant. This may hold when much of the testing has been done shortly after the patent was received. However, most pesticides face repeated data "call-ins" that require additional tests from time to time throughout the marketing period as EPA standards are tightened. For example, the 1988 Amendment, which imposed a 9-year schedule for complete re-registration of all FIFRA-regulated substances, required substantial additional test costs over the decade or so following 1988.

These costs according to date of EPA submission typically have a wide distribution across many years but with relative periods of concentration. Arbitration awards to date have considered tests fully compensable until the last day of the 15-year period. For example, when per capita sharing of test costs has been awarded, if a substantial amount of test costs were incurred

13 or 14 years prior to generic entry, then a generic firm could be liable for a full 50 percent per capita share of those costs. But if entrance were postponed another year or two (until the full 15 years had passed), then the generic firm would incur no liability for those regulatory costs. Many arbitrators have concluded that this is legally required because no provision for reduced compensability over time was included in FIFRA (in spite of the fact that no explicit sharing standard was included at all). But this implementation of FIFRA creates a clear incentive for the generic firm to delay competition, which extends the period of monopoly pricing and delays the social and consumer benefits of competition.

8. CONCLUSIONS

The results of this chapter show that the benefits of minor generic competition in the markets for substances regulated under FIFRA can be large—particularly for farmers and consumers. Generic entrants also benefit from entering these markets, but their benefits are typically small compared to the aggregate benefits received by farmers and consumers. Price reductions have been large because patents are highly effective and support high profit margins prior to patent expiration and generic entry. The perhaps surprising result explained by the model in this chapter is how even minor generic competition can have major effects on prices that transfer huge economic surpluses from monopolistic original entrants to farmers and consumers through post-patent generic entry. These conditions will likely be the case for agricultural biotechnologies regulated under FIFRA and TSCA as well.

These conditions imply that the incentives for an original entrant to extend a monopoly through artificial manipulation of FIFRA are great. This chapter has identified a variety of ways that the provisions of FIFRA (or lack thereof) can allow original entrants to delay or discourage generic entry:

1. Preventing generic entrants from relying on previous regulatory tests, which thus forces socially wasteful duplicative regulatory tests. While FIFRA amendments have attempted to remedy this problem, state registration requirements such as in California and Arizona preclude national markets for generic entrants without duplicative testing, even in some cases where equal sharing of federal regulatory test costs has been required.
2. Unwillingness to agree on test cost compensation prior to generic entry (or even to identify the tests for which compensation ultimately will be claimed).
3. Exaggeration of compensation claims with add-ons such as royalties, management fees, interest, early entry compensation, and risk premiums that go far beyond the actual cost of regulatory

tests. With such add-ons, the claimed compensation can exceed the cost of duplicating tests, or claims can exceed the discounted value of all future generic profits.

4. Imposing risk on generic entrants by (i) failure to keep test cost records and then claiming compensation based on questionable ex post estimates of costs, or (ii) claiming royalties, early entry compensation, or risk premiums without quantitative justification.

5. Claiming per capita sharing rather than market sharing of compensation under circumstances where profitability of post-patent competition depends on maintaining low overhead. These circumstances are typical of post-patent competition in unregulated markets but interact adversely with regulation when test costs are not proportionate to market opportunities. For example, (i) only small market shares are generally feasible due to brand name loyalty and marketing schemes that limit generic access to distributors, (ii) generic prices must be discounted below brand name prices to attract sales, and (iii) generic firms must recapture regulatory test costs at competitive rather than monopolistic prices.

6. Unwillingness to share regulatory test costs based on time in the market.

7. Encouraging delay of generic entry by per capita sharing claims that claim greater shares and thus impose greater risk on early entrants (one-half on the first, one-third on the second, etc.).

8. Petitions by the original entrant regarding the generic applicant's product quality, which when carefully timed can cause loss of a full season's sales and effectively delay entry by a year.

Some evidence that substantial delays are being caused in generic entry is provided by prominent pesticides that have not had generic entry for years after patent expiration, e.g., the linuron case where substantive generic competition did not occur until 7 years after patent expiration.

This chapter has further discussed a simple formal model of the tradeoff between incentives for product innovation versus post-patent competition caused by any rule adopted for sharing regulatory test cost between the original entrant and follow-on generic entrants. For the case where the original entrant exercises market power in setting price, the results show that the share of regulatory test costs borne by the generic entrant should be as small as possible when the regulatory cost is high, the marginal effect of the regulatory cost share on the probability of generic entry is substantial, and/or the probability of generic entry is low. The empirical analysis, even though based on limited data available publicly, implies that the regulatory test costs claimed from generic entrants by original entrants are large—so large that they would make entry unprofitable if granted in several cases. Further analysis shows, however, that the result that calls for minimizing generic entrants' share of regulatory test costs is due to the monopolistic (and, after generic entry, dominant-firm leadership) pricing behavior assumed on the part of the original entrant. If the original entrant prices competitively over the period for which test costs support registration, then test costs are best shared on the basis of market share. But the large price declines that typically follow generic entry provide clear evidence that original entrants are not pricing competitively.

A comparison shows that the provisions of TSCA are much more consistent with economic efficiency than are the provisions of FIFRA. Suggestions are generated about how more specific standards for regulatory test cost sharing could be incorporated in FIFRA that would increase economic efficiency and reduce litigation significantly. Given that the various statutes that currently regulate agricultural biotechnology are affected by various entrenched lobbying interests, social welfare may well be improved by developing a new comprehensive policy for biotechnology regulation whereby the regulatory provisions are not subject to strategic manipulation and any regulatory cost sharing is based on an explicit standard.

In conclusion, the reader should note that this chapter is based on market performance and does not consider the externalities that regulated substances may cause for society. One argument that might be advanced for maintaining monopoly pricing is that it reduces use of substances that may have harmful side effects. But this argument has two potential flaws. First, it assumes that the EPA is failing in its mandate to assure society that these substances are being used safely. Second, it ignores the fact that monopoly pricing of newer substances likely leads to greater use of older, more environmentally damaging substances that are now available at more competitive prices. Another qualification that should be borne in mind is that the empirical results in this chapter have necessarily been developed by relying on limited data that are available publicly on profit margins, market shares, and the price effects of generic entry. These matters can likely never be investigated with plentiful data, which permit sound statistical methods, in public reports because of proprietary protection of such data.

REFERENCES

Antle, J.M. 1984. "The Structure of U.S. Agricultural Technology, 1910–78." *American Journal of Agricultural Economics* 66(4): 414–421.

Aspelin, A.L. 1994. "Pesticides Industry Sales and Usage, 1992 and 1993 Market Estimates." Office of Pesticide Programs, U.S. Environmental Protection Agency, Washington, D.C.

_____. 1997. "Pesticides Industry Sales and Usage, 1994 and 1995 Market Estimates." Office of Pesticide Programs, U.S. Environmental Protection Agency, Washington, D.C.

Aspelin, A.L., and A.H. Grube 1999. "Pesticides Industry Sales and Usage, 1996 and 1997 Market Estimates." Office of Pesticide Programs, U.S. Environmental Protection Agency, Washington, D.C.

Aspelin, A.L, A.H. Grube, and R. Torla. 1992. "Pesticides Industry Sales and Usage, 1990 and 1991 Market Estimates." Office of Pesticide Programs, U.S. Environmental Protection Agency, Washington, D.C.

Fudenberg, D., and J. Tirole. 1992. *Game Theory*. Cambridge, MA: MIT Press.

Gianessi, L.P., and J.E. Anderson. 1995. "Pesticide Use in U.S. Crop Production: National Data Report." National Center for Food and Agricultural Policy, Washington, D.C.

Just, R.E. 1998. "Economic Benefits from Registration of Generic Pesticide Products." Report submitted to the U.S. Environmental Protection Agency, Washington, D.C. (confidential).

_____. 2005. "Efficiency in Sharing the Fixed Cost of Assuring Product Safety: The Case of Pesticides." Working paper, University of Maryland, College Park.

Kiely, T., D. Donaldson, and A. Grube. 2004. "Pesticides Industry Sales and Usage, 2000 and 2001 Market Estimates." Office of Pesticide Programs, U.S. Environmental Protection Agency, Washington, D.C.

Levin, R.E., A.K. Klevorick, R.R. Nelson, and S.G. Winter. 1984. "A Survey Research on R&D Appropriability and Technological Opportunity, Part I: Appropriability." Yale University, New Haven, CT.

Osborne, M.J., and A. Rubinstein. 1994. *A Course in Game Theory*. Cambridge, MA: MIT Press.

Rasmusen, E. 1989. *Games and Information: An Introduction to Game Theory*. Oxford, UK: Blackwell Publishers.

U.S. Senate. 1977. Report No. 95-334 of the 95th Congress, 1st Session (1977), pp. 37–38 (prepared by the Office of Pesticide Programs, U.S. Environmental Protection Agency, March 7, 1977).

Chapter 18

REGULATION, TRADE, AND MARKET POWER: AGRICULTURAL CHEMICAL MARKETS AND INCENTIVES FOR BIOTECHNOLOGY

Vincent H. Smith
Montana State University

Abstract: Chemical companies generally support environmental regulatory segregation of Canadian and U.S. agricultural chemical markets, apparently because it enables them to practice third-order price discrimination. This study provides new cross-section evidence that suggests that price discrimination is practiced. We consider the potential implications of chemical market segregation for the innovation and adoption of biotechnologies that are linked to chemical use and chemical prices in the context of a two-sector–two-country model solved as a two-stage game. Biotechnology implications of regulatory harmonization that reintegrates the two chemical markets are then examined.

Key words: price discrimination, agricultural chemicals, biotechnology

1. INTRODUCTION

Incentives for agricultural biotechnology innovation and adoption are in part linked to the markets for other agricultural inputs. In some cases, the structures of the markets for agricultural chemicals may play critical roles in determining whether agricultural biotechnology R&D firms are willing to invest in a specific line of research. These market structures influence the derived demands for agricultural chemicals of agricultural producers and, as a result, the expected profitability of potential agricultural biotechnology innovations. Roundup is a clear-cut example. The derived demands of agricultural producers for Roundup-Ready corn, wheat, and soybeans are transparently related to the price of Roundup because the cost savings obtained from their use are inversely related to the chemical's price. Monsanto's incentives to develop Roundup-Ready varieties of these crops were therefore closely linked to those derived demands and the company's ability both to segment markets and to exploit market power through tied product sales.

Agricultural chemicals are also subject to multiple regulations implemented by several different government agencies, each with their own policy remits and agency objectives, many of which affect market structures and prices. Notwithstanding the existence of numerous interagency task forces and working groups, agencies' objectives are often dissonant, and the regulations and regulatory initiatives through which each agency seeks to accomplish those objectives frequently provide conflicting, incongruent, or, perhaps at best, orthogonal incentives. In the case of agricultural chemicals, the interface of different regulations and proposed regulatory initiatives has implications for agricultural biotechnology. Patents, other property right protections, environmental regulations, and food safety regulations may all affect the prices and availability of biotechnology-related inputs such as pesticides, with consequent implications for investments in the invention of new plant varieties and their adoption by farmers. Trade policy initiatives that affect those prices may also be important.

This chapter provides a case study of these issues by examining the implications of the current agricultural chemical trade policy environment between Canada and the United States for agricultural chemical prices and biotechnology innovation and adoption. In addition, the biotechnology implications of legislative proposals for changes in that environment are examined. Cross-border retail trade in pesticides and herbicides between Canada and the United States is currently illegal. However, several Congressional delegations from individual states have sought to liberalize trade rules by harmonizing pesticide registration policies in response to complaints from farm organizations that prices are lower in Canada than in the United States for identical or almost identical agricultural chemicals. In many instances, identical (or almost identical) herbicides sold on both sides of the border are manufactured by one company or by one company and others under license to that company.

The U.S. agricultural chemical industry, a subset of the U.S. chemical industry, has strongly opposed Congressional harmonization initiatives on several grounds, including the claim that in fact some chemical prices are lower in the United States and that trade liberalization would yield no net gain for U.S. producers. A key member of the U.S. House Committee, with some prompting from the chemical industry, also expressed concerns about trade liberalization between Canada and the United States through harmonized pesticide registration policies because of potential implications for the value of patent rights and incentives for private investments in pesticide R&D.

Thus, this chapter explores the following issues. What are the sources of market segmentation between Canadian and U.S. agricultural chemicals?

Are chemical manufacturers exploiting market segmentation and to what extent? What does market segmentation imply for agricultural biotechnology adoption and innovation in the short run and the long run? And what do regulatory harmonization and trade liberalization imply for agricultural biotechnology innovation and adoption?

2. THE SOURCES OF AGRICULTURAL CHEMICAL MARKET SEGMENTATION

Many of the markets for agricultural chemicals in Canada and the United States are spatially adjacent. For example, some farmers on one side of the Alberta-Montana border (the 49th parallel) are literally a stone's throw away from farmers on the other side of the border but cannot legally purchase the same agricultural chemical from the same chemical dealer. Absent regulatory barriers, arbitrage would almost surely guarantee that prices in those spatially adjacent markets would exhibit very similar patterns and on average be about the same.[1] These markets are segregated because of regulation.

In Canada, at the federal level, pesticide use is regulated by the Pest Management Regulatory Agency (PMRA), established under Canada's 1985 Pest Control Products Act. All pesticides used in Canada must be registered by the PMRA. The terms and scope of use are determined by the Canadian registration and sold under a PMRA-approved label that describes the product and its specific approved uses. A pesticide registered for use in the United States and sold under a U.S. label cannot be used legally in Canada even if its active ingredients are identical to one that has received a Canadian label. Similarly, pesticide use in the United States is regulated under the provisions of the 1947 Federal Insecticide, Fungicide, and Rodenticide Act (FIFRA), as variously amended, and each agricultural chemical is sold under an EPA-approved label. A pesticide registered and sold for use in Canada cannot legally be used in the United States, even if its chemical formulation is identical to that of a product that is registered and sold for use in the United States.[2] Essentially, the pesticide regulatory processes in the two

---

[1] It is well known that prices for homogeneous products are frequently disperse among sellers within well-defined competitive markets. Sorensen (2000), for example, reported substantial price variation for identical drugs among pharmacies in two small towns in upstate New York where each pharmacy was required to post its prices for over 150 commonly used pharmaceuticals. Search models in which information is costly provide explanations of such phenomena (see, for example, Stigler 1961, Salop and Stiglitz 1977, Carlson and McAfee 1983, and McMillan and Morgan 1988). However, they do not imply that average prices from randomly selected samples drawn from the same population in a common market should be statistically significantly different.

[2] Within the United States, under the provisions of FIFRA, individual states may impose stricter (but not weaker) controls over the use of specific pesticides than those required by the

countries prevent legal arbitrage that would otherwise erode most price differences, although some residual differences might persist because of factors
such as transport costs.

As a result of the regulatory processes, price differences for the same
agricultural chemicals have persisted even between components of the markets that are spatially adjacent. That these price differences matter from an
economic perspective is illustrated by the fact that in 2000, a past president
of the Montana Grain Growers Association was found guilty of illegally
conspiring to obtain Roundup from Canada and sell it to other farmers in
Montana and adjacent states.[3] These price differences have led agricultural
commodity organizations on both sides of the border to seek legislation that
would "harmonize" U.S. and Canadian pesticide use regulations.

The U.S. initiative, embodied in Senate Bill 1406, was introduced in
2003 and cosponsored by Senators Dorgan and Conrad of North Dakota,
Daschle and Johnson of South Dakota, and Burns and Baucus of Montana.
The Bill would amend FIFRA to permit the EPA Administrator to register a
Canadian pesticide with identical active ingredients and similar, though not
necessarily identical, formulations to a pesticide already approved by EPA
for use. In hearings on the bill in June of 2004, Mr. Jay Vroom, the president
of CropLife America, the trade association representing manufacturers, distributors, and formulators of agricultural chemicals, testified against the bill.

If, in fact, chemical manufacturers are concerned about losing opportunities for price discrimination, the industry's position vis-à-vis the legislation
is hardly surprising. In addition to voicing several other concerns,[4] the industry's representative claimed that no legislation was needed because Great
Plains producers spent less on pesticides than did Canadian producers. This,
of course, was not an "apples-to-apples" comparison. Canada includes many
agricultural regions with much higher rainfall and much more severe pesticide problems than the semi-arid region of the Northern Great Plains, where
dryland farmers are spending $6 to $10 per acre on pest control (Smith and
Goodwin 1996), compared to $15 or $40 per acre for corn producers in Iowa

---

EPA and may prohibit the use of certain chemicals even though EPA has approved their use.
In Canada, registration of a herbicide is either provided for the whole country or for one or
more of three regions—the Prairie region (Alberta, Manitoba, Saskatchewan, and the Peace
River in British Columbia), the Atlantic Region (consisting of the maritime provinces), and
the region consisting of Ontario, Quebec, and the rest of British Columbia.

[3] Prior to 2002, when Monsanto's U.S. patent on Roundup was still active, McEwan and Daley
(1999), Carlson et al. (1999), and Freshwater (2003), using surveys of agricultural chemical
dealers, reported that the product was substantially cheaper in Manitoba than in North
Dakota. In 2000, Larry Johnson, a farmer and past president of the Montana Grain Growers
Association, was found guilty of over 30 federal counts of conspiring illegally to sell the
Canadian registered version of the product in the United States (Dennison 2000).

[4] One of the industry representative's more profound assertions was that U.S. producers would
be confused by Canadian chemical labels because the same information was provided in both
English and French, although English-speaking Canadians seemed to manage quite well.

or Indiana or, for that matter, the Canadian province of Ontario. In early October of 2004, the government of Canada also issued a memorandum to the provinces indicating its interest in harmonizing regulations because of complaints by Canadian farmers that some important pesticides were cheaper in the United States than in Canada.

Interestingly, these federal initiatives have been introduced even though the U.S. EPA and Canada's PMRA, together with Mexico, established a NAFTA Technical Working Group to address the issue of pesticide regulation harmonization in 1998. One focus of the working group's current five-year agenda, established in November of 2003, is the development of NAFTA labels for pesticides. However, there is no evidence of any substantive progress on the issue. In addition, apparently, a NAFTA label would be issued only at the request of the manufacturer and therefore would not address North American agricultural producers' concerns about price discrimination.

3. PRICE EFFECTS OF AGRICULTURAL CHEMICAL MARKET SEGMENTATION: THE EVIDENCE

Standard single-period models of monopoly pricing predict that when markets are effectively segmented and marginal costs of supplying those markets are identical, the monopolist will "price to market" by offering the commodity at a lower (higher) price in the market in which demand is more (less) price-elastic so as to equate marginal revenue with marginal cost in each market. In addition, such third-order price discrimination will generally increase aggregate economic welfare relative to a monopoly single-price equilibrium when economic welfare is measured by the simple sum of producer and consumer surplus. However, while consumers in the low-price market will gain, consumers in the high-price market will lose. On a net basis, however, the consumer welfare losses in the high-price market will more than offset the gains in the low-price market and consumers as a whole will be worse off. Unambiguously, rents accrued by the monopolist will increase.[5]

Some evidence of price differences in the Canadian and U.S. agricultural chemical markets has been provided in previous studies. More recent evidence for Alberta and Montana is reported by Smith and Johnson (2004), who collected "point in time" price data for 13 agricultural chemicals from a

---

[5] See, for example, Varian (1989), and more recently, Clerides (2004), Stole (2001), and Yoshida (2000). The aggregate effect on consumer welfare depends on the shape of the compensated demand curves in the two-market case, but as market segmentation increases so does the likelihood of reductions in aggregate consumer welfare, and, in the limit, under perfect or first-order price discrimination, all economic surplus accrues to the monopolist.

random survey of agricultural chemical dealerships in northern Montana and a non-random survey of dealerships in southern Alberta where cross-border purchases of agricultural chemicals by farmers would be most likely to occur if trade in retail agricultural chemicals were allowed. The surveys asked respondents on both sides of the border to provide current retail prices for "cash and carry" sales of comparable containers of each agricultural chemical.

In Alberta, the survey was administered by the Pest Risk Management Unit of the Crop Diversification Division of Alberta Agriculture, Food and Rural Development to 14 dealerships whose collective market share in southern Alberta was estimated to be about 80 percent. In northern Montana, a random survey was administered by the Montana State University Agricultural Marketing Policy Center in collaboration with the Montana Department of Agriculture. In Montana, retailers considered likely to serve both Alberta and Montana agricultural producers were located along or near U.S. Route 2, an east-west highway known as "the Highline," which runs parallel to and about 35 miles south of the U.S.–Canadian border. In Alberta, retailers considered likely to serve both Montana and Alberta agricultural producers were located along or near Canada Highway 3, also an east-west highway that in Alberta runs parallel to and about 50 miles north of the U.S.–Canadian border.[6]

The survey was administered to U.S. and Canadian retailers selling agricultural chemicals directly to farmers and ranchers. The survey included 12 herbicides and one pesticide that are widely used both in Alberta and Montana. The agricultural chemicals included in the survey were selected from a list of chemicals identified by the Montana Department of Agriculture in collaboration with representatives of Montana commodity organizations and

---

[6] Information on dealership locations provided by the Montana Department of Agriculture indicated that 120 Montana agricultural chemical dealerships were potential outlets for agricultural producers in both Montana and Alberta. Seventy of these retailers were randomly selected for potential inclusion in the Montana sample and contacted by telephone to ascertain whether the business was an applicator dealing in a very limited number of chemicals or an agricultural chemical dealer selling many of the chemicals of concern at retail. Applicators were excluded from the survey both because of the limited number of chemicals they handled and because many applicators sell agricultural chemicals to agricultural producers only in combined chemical/application packages. The sample selection process was completed when 40 agricultural chemical dealers willing to respond to the survey had been identified. Each of these 40 randomly selected retailers received survey forms within two working days of being contacted. Thirty-two survey forms with usable responses were returned within 14 days by these retailers, an initial response rate of 75 percent. In Alberta, a total of 22 agricultural chemical dealers were identified as potential retail sources of agricultural chemicals for U.S. agricultural producers. All 22 retailers were contacted by Alberta Agriculture, Food and Rural Development, and 14 responded to the survey. Alberta Agriculture, Food and Rural Development estimated that jointly these 14 retailers represent 80 percent of the agricultural chemical market in southern Alberta.

agribusiness organizations that include agricultural chemical dealers. Alberta Agriculture, Food and Rural Development pesticide experts then identified agricultural chemicals on the Montana list that were also registered and used in Alberta.

Agricultural chemical companies market the same or very similar products under different brand names in Canada and the United States. Thus, while herbicide and insecticide product names for the thirteen chemicals differed in the surveys for Alberta and Montana, their active ingredients were the same and their formulations were either identical or very similar (see Table 1). Two of the agricultural chemicals, Mirage and Touchdown, were non-selective herbicides used on fallow and some non-cropland areas. Six agricultural chemicals—Amine 4 and 2, LV6, Bronate Advanced, Clarity, Achieve SC, and Discover—were herbicides used for broadleaf weed control in wheat, barley, and other small grain crops, and on fallow and some non-cropland applications. Four—Everest, Puma 1EC, Ally XP, and Express EP—were selective herbicides used primarily to control grassy weeds, including wild oats in wheat and barley, and one—Warrior (with Zeon)—was a general purpose insecticide.

For each chemical, respondents were given descriptions of active ingredients and formulations obtained from each chemical's U.S. or Canadian product labels. They were then asked (i) to provide prices for either two or three sizes of containers (generally 2.5-gallon, 110-gallon, and 220-gallon containers), (ii) to identify the manufacturer of the chemical, and (iii) to provide any additional pricing information on quantity or manufacturer program discounts to farmers. Respondents in both Alberta and Montana provided very little information on quantity or program discounts, and, with very few exceptions, respondents from Alberta provided price data only for small containers. Thus, the results presented here are restricted to comparisons of prices for agricultural chemicals sold in small containers (2.5-gallon jugs in the United States and 10-liter jugs in Canada).

Every effort was made to compare products with identical active ingredient formulations, but nevertheless some differences in formulations did persist. Data from the labels for each agricultural chemical indicated that the amount of active ingredient per gallon (or ounce) of sales was different in Alberta and Montana for 3 of the 13 chemicals in the survey. These three chemicals are Amine 4 (2, 4-D Amine 500), LV 6 (2, 4-D Ester LV 600), and Bronate Advanced (Buctril M). For these three chemicals, the ratios of the active ingredients in the Canadian products to the active ingredients in the comparable U.S. products were used to adjust the Canadian per unit of sale price (per gallon or per ounce) to an equivalent price for a unit of the product with the same amount of active ingredient as in the U.S. product.

These adjusted prices and their associated estimated standard deviations, reported in Table 2, were used to carry out standard student T comparison of

Table 1. Agricultural Chemicals and Their Major Target Species and Major Uses

Agricultural Chemicals [a]	Target Species/Major Uses
Mirage and Roundup Original	non-selective herbicide for general weed control; fallow and non-cropland areas
Touchdown and Touchdown iQ	non-selective herbicide for general weed control; fallow and non-cropland areas
Amine 4 and 2, 4-D Amine 500	selective herbicide for control of broadleaf weeds; certain crops and non-cropland areas
LV 6 and 2,4-D Ester LV 600	selective herbicide for broadleaf weed control; wheat, barley and non-cropland areas
Bronate Advanced and Buctril M	selective herbicide for certain broadleaf weeds; wheat, barley, oats; rye and flax
Clarity and Banvel II	selective herbicide for broadleaf weeds; CRP, fallow, small grains, and farmstead
Achieve SC and Achieve Liquid	selective herbicide for grassy weeds; wheat and barley
Discover and Horizon 240EC	selective herbicide for grassy weeds; wheat
Everest and Everest	selective herbicide for wild oats, green foxtail and other grassy weeds, and broadleaf weeds; spring, durum, and winter wheat
Puma 1EC and Puma 120 Super	selective herbicide for pigeongrass, wild oats, and millet and barnyardgrass; wheat and barley
Ally XP and Ally Toss and Go	selective herbicide for broadleaf weeds; wheat, barley, and fallow
Express EP and Express Toss and Go	selective herbicide for broadleaf weeds; wheat, barley, and fallow
Warrior (with Zeon) and Matador 120 EC	general insecticide

[a] The first brand name is the chemical's name in the United States and the second is the chemical's name in Canada.

Table 2. Estimated Average Prices and Standard Deviations: Prices Adjusted for Differences in Chemical Formulations (U.S. Dollars)

	Units	Northern Montana			Southern Alberta		
		Average price	Standard deviation	Number of obs.	Average price	Standard deviation	Number of obs.
Mirage and Round Up Original	gallon	20.94	3.92	15	21.68	0.69	12
Touchdown and Touchdown IQ	gallon	31.28	5.06	5	23.22	0.57	14
Amine 4 and 24D Amine 500[a]	gallon	12.16	0.46	28	15.56	0.17	5
LV6 and 24D Ester LV 600[a]	gallon	19.36	0.76	29	19.53	2.04	5
Bronate Advanced and Buctril M	gallon	59.43	3.28	26	42.07	1.36	14
Clarity and Banvel 2[a]	gallon	93.47	3.63	29	96.85	1.89	14
Achieve SC and Achieve Liquid	gallon	220.85	11.1	29	208.37	3.48	12
Discover and Horizon 240 EC	gallon	496.37	23.7	31	448.31	5.54	14
Everest and Everest	ounce	23.45	1.25	29	17.50	0.61	14
Puma 1EC and Puma 120 Super	gallon	181.33	7.57	30	128.97	6.21	14
Ally XP and Ally Toss and Go	ounce	23.27	1.14	23	37.37	1.01	13
Express XP and Express Toss and Go	ounce	18.63	1.14	29	14.82	0.23	7
Warrior and Matador 120 EC	gallon	282.76	8.41	21	361.54	13.15	14

[a] These three chemicals had different formulations in Canada and the United States.

means tests for samples of different sizes. The null hypothesis is that the price of a given chemical in Montana is equal to the price of that chemical in Alberta. Results are presented in Table 3. These results show that for all but one of the 13 agricultural chemicals, Mirage or Roundup Original, average prices were statistically significantly different between northern Montana and southern Alberta. Monsanto's Roundup products now face extensive competition from generic glyphosates as patent rights for the product have expired in both Canada and the United States. Most of the other products are produced by a single chemical company and face no competition from generics.

Average prices of five chemicals were significantly higher in southern Alberta than in northern Montana (Table 3). Four of these chemicals—LV 6 (2, 4-D Ester LV 600), Amine 4 (2, 4-D Amine 500), Clarity (Banvel II), and Ally XP (Ally Toss and Go)—are selective herbicides used to control broadleaf weeds in crops such as wheat and barley, and on fallowed land, on land placed in the Conservation Reserve Program (CRP) (in Montana), and on other "non-cropland" areas. The fifth was the general purpose insecticide Warrior with Zeon (Matador 120 EC).

Average prices for the remaining seven agricultural chemicals were statistically significantly higher in northern Montana than in southern Alberta. Touchdown (and Touchdown iQ) is a non-selective herbicide used for weed control on fallow and on non-cropland areas. Bronate Advanced (and Buctril M) and Express XP (and Express Toss and Go) are selective herbicides used to control broadleaf weeds in wheat and barley and also for fallow, CRP, and non-cropland areas. Achieve SC (and Achieve Liquid), Discover (and Horizon 240 EC), Everest (and Everest), and Puma 1 EC (and Puma 120 Super) are selective herbicides used to control grassy weeds, including wild oats and pigeon grass, in growing crops of wheat and barley.

Although the average prices of a chemical may be statistically significantly different between Montana and Alberta, the economic importance of such differences is what matters. If price differences are small, then Canadian and U.S. farmers may have little to gain from the harmonization of pesticide regulations. The last column of Table 3 shows the average price difference for each chemical as a percentage of the U.S. average price for the chemical. In Table 3, in three cases, the price differences are relatively small and amount to less than 10 percent of the U.S. price, although in one case, Discover, that represents an absolute difference of $52 per gallon. In eight cases, the price differences lie in the range of 19 to 31 percent of the U.S. price. Thus, the prices of all but one of the thirteen agricultural chemicals examined in this study exhibit statistically significant differences and, in most cases, the differences appear to be economically important.

One possible explanation for these differences is that agricultural dealers in Canada may face systematically different costs. Possible sources of such

Table 3. Montana and Alberta Agricultural Chemical: Prices and Price Differences Adjusted for Differences in Formulations

Chemical	Firms	Units	U.S. Price (US $)	Canadian Price (US $)	Price Difference (U.S. Price– Canada Price) (US $)	T-test Value	Percentage Price Difference[a]
Mirage and Roundup Original	generic/Monsanto	gallon	20.94	21.68	-0.74	-0.65	-3.5%
LV 6 and 2, 4-D Ester LV 600	several	gallon	19.36	23.22	-3.86[b]	-7.15	-19.9%
Amine 4 and 2,4-D Amine 500	several/generic	gallon	12.16	15.05	-2.89[b]	-13.71	-23.8%
Clarity and Banvel II	BASF	gallon	93.47	96.85	-3.38[b]	-3.26	-3.6%
Ally XP and Ally Toss and Go	DuPont	ounce	23.27	37.37	-14.10[b]	-37.09	-60.6%
Warrior (Zeon) and Matador 120 EC	Syngenta	gallon	282.76	361.54	-78.78[b]	-21.68	-27.9%
Touchdown and Touchdown Iq	Syngenta	gallon	31.28	23.22	8.06[b]	6.17	25.8%
Bronate Advanced and Buctril M	Bayer	gallon	59.43	44.94	14.49[b]	15.56	24.4%
Achieve SC and Achieve Liquid	Syngenta	gallon	220.85	208.37	12.48[b]	3.79	5.7%
Discover and Horizon 240 EC	Syngenta	gallon	496.37	448.31	48.06[b]	7.45	9.7%
Everest and Everest	Arvesta/Bayer	ounce	23.45	17.50	5.95[b]	16.80	25.4%
Puma 1EC and Puma 120 Super	Bayer/Aventis	gallon	181.33	128.97	52.36[b]	22.54	28.9%
Express XP and Express Toss and Go	DuPont	ounce	18.63	14.82	3.81[b]	8.70	20.5%

[a] The percentage price difference is computed as the ratio of the difference between the U.S. and the Canadian price to the U.S. price. A negative sign implies that the U.S. price is lower and a positive sign implies that the U.S. price is higher.
[b] Difference is statistically significant at the 99 percent confidence level.

differences are differences in product transportation costs, taxes, wages, and other dealership costs. These all seem unlikely as explanations of the above results. First consider transportation costs. Several of these chemicals are purchased in relatively large quantities by individual farmers (for example, Amine 4 is purchased by many Montana farmers in 110-gallon shuttles) and are shipped in bulk to dealers in both Alberta and Montana, in some cases from the same production facility under different labels. Bulk shipping costs by truck from Alberta to Montana for two fertilizers—urea and anhydrous ammonia—were estimated to be about $10 per ton per 100 miles in August 2004 (where a ton is roughly equivalent to about 200 gallons of chemical).[7] Thus, shipping costs amounted to approximately five cents per gallon. Doubling or quadrupling this estimate still implies transportation cost differences of only about 20 cents per gallon, which is small compared with the range of average price differences of between $2.89 and $78.78 per gallon among the twelve chemicals with statistically significant price differences.

Agricultural chemical dealer costs may be systematically different in Alberta and Montana. For example, exchange rate adjusted wages could be lower in Alberta than in Montana. Similarly, there may be systematic differences in tax burdens, energy, and other costs. However, if these cost differences were what mattered, then we would expect retail prices systematically to be either all higher or all lower in one of the two regions (Alberta or Montana). This is not the case. Prices are higher in Alberta for five chemicals and higher in Montana for seven chemicals. This suggests that third-order price discrimination is the real issue.

Table 3 shows the companies that manufacture each chemical, as well as cross-border average prices and price differences. The one chemical for which there is no significant difference in prices is Roundup Original. The product, whose active ingredient is glyphosate, is produced and sold by Monsanto, which held a U.S. patent on the product until 2002, but generic glyphosates are now also sold by several other companies in both Canada and the United States. Among the other twelve chemicals, two face generic competition or are produced by several competing companies (LV6 and Amine 4). The remaining ten are produced either by a single manufacturer (four by Syngenta, two by DuPont, one each by BASF and Bayer) or by Bayer in collaboration with either Aventis or Arvesta. These results suggest that generally the economically and statistically significant price differences are associated with market power and differences in elasticities of demand.

Assuming that this is the case, a natural question is why demand is more own-price elastic in Canada for some chemicals and less own-price elastic for others. Two possibilities spring to mind. The first is that differences in crop mixes in Canada and the United States lead to differences in demand

---

[7] Grain industry sources provided this estimate.

elasticities. While this is a possibility, given that (i) arbitrage is feasible within Canada across large regional markets and also within the United States, and (ii) complex mixes of crops are raised in both countries, it seems unlikely that crop mix is the main issue.[8]

Differences in regulatory regimes may also be important. Suppose, for example, that chemical A is approved for use in Canada but so too are other chemicals that may have different formulations (including different active ingredients) that perform similar functions. In addition, chemical A is also registered in the United States, but no close substitutes are also approved. The difference in the regulatory regimes may well account for the difference in the elasticities of demand (higher in Canada and lower in the United States) and the higher price for chemical A in the United States. The situation may be the exact opposite for chemical B. While, currently, no exhaustive evidence is available to determine which hypothesis is correct, the case of Roundup suggests that differences in regulatory regimes are important. Once Monsanto's U.S. patent for its Roundup product had expired in 2002, as it already had in Canada where close substitutes were then registered for use, similarly close substitutes were registered for use in the United States, and differences in prices in the two markets essentially disappeared.

The question then arises as to why there are differences in the regulatory regimes. In some cases, serendipity may be at work: some firms simply have not sought registration in, say, Canada but have obtained registration in the United States (perhaps because of differences in the sizes of the markets and regulatory approval costs). In others, lobbying of the regulatory authority by producers may be relevant. However, as in the case of Roundup, differences in patents and the timing of patent expirations also may be important. For example, an industry representative suggested to Smith and Johnson (2004) that this was potentially important in explaining why Achieve, a product originally patented in Europe, then in Canada, and only subsequently in the United States, was less expensive in Canada than in the United States.

---

[8] It should be noted that the ability to arbitrage across states in the United States may be restricted because each state must register a chemical for use. However, if a herbicide is approved for a specific use by EPA and registered for that use in one state, then typically other states will also register the herbicide for that use (say, control of wild oats in wheat fields). Moreover, either users (farmers) or manufacturers may initiate the state-level registration process.

4. THE ECONOMIC EFFECTS OF AGRICULTURAL MARKET SEGMENTATION: IMPLICATIONS FOR AGRICULTURAL BIOTECHNOLOGY

There is fairly compelling evidence that agricultural chemical companies practice third-order price discrimination among Canadian and U.S. markets and have done so for many years. A question of considerable interest here concerns the implications of this behavior for investment in the adoption of agricultural biotechnology. A second issue is whether incentives for biotechnology adoption would be increased or decreased if opportunities for price discrimination were obviated by harmonization of pesticide regulations. These issues are examined using a stylized two-country model of the markets for genetically modified crops and the related markets for agricultural chemicals.

4.1. The Model

Consider a simple environment in which the adoption of a new agricultural biotechnology such as a new crop variety is contingent on the price of an agricultural chemical such as glyphosate. The "downstream" firm that develops the innovation will sell the new variety in two markets in country 1 and in country 2 that are segregated because of different regulatory regimes. The "upstream" agricultural chemical firm also is assumed to have a monopoly in both markets, at least initially.

The key elements and parameters of the model are as follows. Adoption of the new variety in each market is contingent on the price of the agricultural chemical in that market. To reflect this possibility, the demand for the new technology in each market is assumed to be a function both of the price of the technology itself, p_i (say the price of genetically modified seed), and the price of the agricultural chemical in that market, r_i. In addition, the biotechnology firm is assumed to have perfect foresight and to know what the demand functions will be in each market once the technology is available.

The biotechnology inverse demand function in each market is assumed to be linear and can be written as

$$p_i = a_i(r_i) - b_i x_i, \tag{1}$$

where x_i is the quantity of the technology demanded by agricultural producers (for example, the amount of seed), p_i is the price of the biotechnology, b_i is a constant, and a_i is an intercept term whose value depends on r_i, itself the price of the agricultural chemical. Thus, for example, if demand for the biotechnology is directly related to the agricultural chemical's price, then $\partial a_i/\partial r_i > 0$, and an increase in r_i shifts the demand curve for the technology vertically upwards, as is the case with Bt corn. It is convenient for the purposes of this analysis to assume that the function $a_i(r_i)$ is linear in r_i[9], that is,

$$a_i = g_i + h_i r_i, \text{ for } i = 1, 2, \tag{2}$$

where g and h are positive constants.[10]

In each market, consumption of the agricultural chemical, y_i, is assumed to be inversely related to its own price, r_i, and the use of the biotechnology; that is, the linear *inverse* demand function for the agricultural chemical may be written as

$$r_i = s_i - t_i x_i - v_i y_i, \text{ for } i = 1, 2, \tag{3}$$

where s, t, and v are positive constants. Thus, there are interesting consequences of adjusting price for the chemical manufacturer. A higher chemical price stimulates adoption of the biotechnology and thereby shifts the demand curve for the chemical inwards, again as with Bt corn. In addition, average and marginal production costs are assumed to be constant for both the downstream and upstream firms, regardless of the market being supplied, equaling c per unit of output for the biotechnology and k per unit of output for the agricultural chemical.

Finally, the biotechnology-innovating firm expects to have a monopoly over the provision of the new technology in each market for a single period. Thus, because it is risk-neutral and operates with certainty, the firm will invest in the expected R&D costs, R, to develop the technology if the joint profits from both markets exceed those R&D costs.

Both the biotechnology firm and the upstream chemical manufacturer are assumed to seek to maximize profits in a single period during which the biotechnology is available.[11] The chemical manufacturer knows the behavior of the downstream biotechnology firm, and therefore the model can be solved in two stages. First, the downstream firm is viewed as taking the upstream firm's pricing decision as parametric to its optimization problem. It

---

[9] This linear function allows for a convenient closed form solution to the model.

[10] Demand for a biotechnology could certainly be decreasing in the price of an agricultural chemical. For example, a genetic innovation could be directed towards controlling for a disease by allowing the use of a pesticide or herbicide, thereby creating a need for the agricultural chemical. This is the Roundup-Ready case. Here, we assume that demand for the biotechnology is increasing in the chemical price, an assumption that reflects the idea that the genetic technology is designed to make a variety resistant to the agricultural chemical, and thereby to permit its less frequent use at more optimal times. Extending the model to deal with the demand-decreasing case is relatively straightforward and simply involves changing the signs of the parameters h in equation (1) and t in equation (2).

[11] This assumption allows the focus to be on market structure issues that are not complicated by intertemporal choices.

then solves its own profit-maximization problem. The upstream firm then uses the downstream firm's sales decision rule, which is parametric in a_i, to solve for its optimum agricultural chemical pricing strategy.

The downstream biotechnology firm's optimization problem is as follows:

$$\underset{x}{Max}\, \Pi_{ds} = \sum_i (a_i(r_i) - b_i x_i)x_i - c\sum_i x_i \,, \tag{4}$$

subject to

$$\Pi_{ds} \geq R, \tag{5}$$

where Π_{ds} denotes the downstream biotechnology firm's profits. Assuming that the minimum profit constraint is non-binding, the downstream firm's optimal production and allocation decision is as follows:

$$x_i^* = \frac{a_i - c}{2b_i}, \quad \text{for i} = 1, 2, \tag{6a}$$

where x_i^* is the optimal level of biotechnology sales in the ith market. Substituting for x in equation (1) from equation (6a), the firm's optimal pricing decision is

$$p_i^* = \frac{a_i + c}{2}, \quad \text{for i} = 1, 2. \tag{6b}$$

These are standard results. From equations (6a) and (6b), it also follows that the downstream firm's optimal profit function is

$$\Pi_{ds} = \sum_i \frac{(a_i - c)^2}{4b_i}. \tag{7}$$

If the biotechnology firm is to produce the technology, then, from equation (4),

$$\Pi_{ds} = \sum_i \frac{(a_i - c)^2}{4b_i} \geq R.$$

It is also worth noting at this point that, transparently, the biotechnology firm's profits are increasing in a_i and decreasing in b_i. Thus, as demand increases and/or pivots and becomes more price-sensitive (b_i decreases), the biotechnology firm's expected profits increase and it is more likely to develop the biotechnology.[12]

---

[12] Recall that for a linear inverse demand function with a slope of $-b$, the formula for the absolute value of the price elasticity of demand is simply $(1/b)(p/x)$, where p is price and x is quantity demanded. Thus, as b decreases, at any given price and quantity, demand becomes more price-sensitive and more price-elastic at any given quantity of sales. Given a fixed mar-

The upstream agricultural chemical producer now seeks to maximize profits, given the production and market allocation decision rules adopted by the downstream biotechnology firm, that is,

$$Max \, \Pi_{us} = \sum_r (r_i - k) y_i \,, \tag{8}$$

where Π_{us} denotes the upstream firm's profits. Recall that the intercept constants in the downstream biotechnology demand functions depend on r_i as defined in equation (2). Substituting for a_i in the downstream firm's optimal decision rule defined in (6) and for the optimal value of x_i in the upstream demand function in equation (3), it follows that r_i may be written as

$$r_i = s_i - t_i \frac{(g_i - h_i r_i - c)}{2b_i} - v_i y_i \,, \tag{9}$$

or, given the upstream firm's optimal rule, solving for r_i:

$$r_i = \frac{\left[s_i - t_i \frac{(g_i - c)}{2b_i} \right]}{1 + \frac{t_i h_i}{2b_i}} - \frac{v_i}{1 + \frac{t_i h_i}{2b_i}} y_i \,. \tag{10}$$

Substituting for each chemical price in equation (8) using equation (9), the optimal solution for sales of the chemical in each market is

$$y* = \frac{(s + k) - [t(g - c + hk)/2b]}{2v} \,, \tag{11}$$

where subscripts have been omitted to simplify notation. Substituting for y from equation (11) in equation (10), it follows that the manufacturer's optimal chemical pricing strategy in each market is

$$r* = \frac{2b(s + k) - t(g - c - hk)}{4b + th} \,. \tag{12}$$

Given the chemical manufacturer's optimal pricing decision in each market, the biotechnology firm's optimal (from the perspective of the chemical manufacturer's perspective) output and pricing decisions can also be identified by substituting for r in equations (6a) and (6b) from equation (12), that is,

---

ginal cost, the firm's optimal level of sales to market i also increases, as is indicated by equation (6).

$$x^{*} = \frac{1}{2}\left[g - c + \frac{h[2b(s+k) - t(g-c-hk)]}{4b+2th}\right] \qquad (13)$$

and

$$p^{*} = \frac{1}{2}\left[g + c + \frac{h[2b(s+k) - t(g-c-hk)]}{4b+2th}\right]. \qquad (14)$$

4.2. Comparative Statics

Equations (11)–(14) provide useful vehicles for examining the economic effects of desegregating the markets for an agricultural chemical through harmonization legislation that allows farmers in one country to import the same chemical from the other country. However, some comparative static results are simply of general interest. Consider, for example, the effects of changes in the marginal cost of producing the agricultural chemical, k, because of R&D in the chemical manufacturing sector. The effects of a change in k on prices and quantities in a country's chemical market are as follows:

$$\frac{\partial r^{*}}{\partial k} = \frac{2b+th}{4b+2th} = 1/2 > 0, \qquad (15a)$$

$$\frac{\partial y^{*}}{\partial k} = \frac{-(2b+th)}{4bv} < 0, \qquad (15b)$$

$$\frac{\partial x^{*}}{\partial k} = \frac{h(b+\frac{hk}{2})}{4b+2th} > 0, \qquad (15c)$$

and

$$\frac{\partial p^{*}}{\partial k} = \frac{h(b+\frac{hk}{2})}{4b+2th} > 0. \qquad (15d)$$

From equations (15), it is clear that a decrease (increase) in the marginal cost of producing the agricultural chemical will reduce (increase) both the biotechnology commodity's price and quantity sold in each market. The reason is straightforward. A decrease in k will unambiguously reduce the price of the agricultural chemical [from equation (15a), $\partial r^{*}/\partial k > 0$]. As shown in equation (3), this decrease in the chemical's price reduces farmers' incentives to adopt the biotechnology. The result is an unambiguous decrease in the profitability of the new technology as p and x have decreased in both markets, and the likelihood that the minimum profit requirement of the biotechnology company will be met has also decreased. Thus, there is a reduction in the likelihood of investment in biotechnology R&D. This is a commonsensical result. A reduction in the potential demand for a new technology will reduce the likelihood that the new technology will be created.

It is also worth noting that the effect of a cost decrease (increase) on sales of the agricultural chemical is positive (negative) in each market. A chemical price reduction (increase) stimulates chemical sales through the effect of r on y in the agricultural chemical demand function, and this direct effect is compounded by the fact that a reduction in r also reduces the demand for the new technology, further increasing the use of the chemical.

The consequences of incomplete harmonization of the use of agricultural chemicals can be linked to the effects of a change in the parameters of the inverse demand function for agricultural chemical prices, that is, the parameters s and v in equation (2). A decrease in s_1, for example, might result from new access to market 1 by manufacturers of substitute chemicals (generics) who already have access to market 2 because of the expiration of a patent in market 2 but not market 1. Similarly, at the same time that s_1 decreases, v_1 might well decrease (that is, in a standard diagram, the demand function for the chemical shifts inward and the slope of the demand function, $1/v$, becomes flatter). The effects of a change in s on prices and quantities in the chemical market are straightforward: $\partial r^*/\partial s = 2b/(2b + 2th) > 0$ and $\partial y^*/\partial s = 1/2v > 0$. Thus a decrease in s in market i reduces both y_i and r_i and, therefore, the profits of the agricultural chemical producers in that market.

The effects of a change in s_i on prices and quantities in the biotechnology market in country i are also unambiguous: $\partial p^*/\partial s = bh/(4b + 2th) > 0$ and $\partial x^*/\partial s = bh/(4b + 2th)$. Thus, a decrease in demand for the agricultural chemical reduces both the price and quantity sold of the biotechnology and, therefore, the expected profits of the biotechnology firm. Harmonization, if it reduced demand for the agricultural chemical, would therefore reduce the probability that the technology would be developed, by lowering the price of the chemical and reducing incentives for the adoption and use of the biotechnology.

An alternative approach to the same issue is as follows. Suppose that the patent for an agricultural chemical such as Roundup has expired in country 2 (say, Canada) but not in country 1 (say, the United States), as was the case with Monsanto's Roundup products in the late 1990s and up to 2002. Thus, in country 2, because of competition from generics, the price of the agricultural chemical is assumed to be equal to its marginal cost, that is, $r_2 = k$.[13] In

---

[13] The assumption is that competition from other companies is sufficient to create a competitive market. This may not be the case. For example, new entrants may be relatively small in

country 1, prior to harmonization, the chemical's price is as defined by equation (12), that is,

$$r_1^* = \frac{2b_1(s_1 + k) - t_1(g_1 - c_1 - h_1 k)}{4b_1 + t_1 h_1}, \tag{16}$$

where r_1^* can generally be shown to be greater than k.[14]

Once harmonization occurs, assuming the chemical company continues to sell in both markets, arbitrage will drive the price of the chemical down to the price of the chemical in country 2 (where $r_2^* = k$) if transportation costs are zero, or to $k + \delta$ if transportation costs equal δ, unless $k + \delta$ exceeds r_1^* as defined in (16). In this case, the effect of harmonization on the biotechnology market is again relatively easy to identify. Substituting for a_i in equations (6a) and (6b),

$$x_1^* = \frac{g_1 + h_1 r_1 - c}{2b_1},$$

and

$$p_1^* = \frac{g_1 + h_1 r_1 + c}{2}.$$

Clearly a decrease in r_1 reduces both p and x in market 1, reducing the profitability of the biotechnology in that market and, therefore, the likelihood that it will be developed in the first place.

However, if harmonization reduces the profits of the chemical company sufficiently, the company may choose to abandon market 2 completely. This could occur if the harmonization legislation were to allow imports of an identical or similar chemical from country 2 into country 1 only when the chemical is sold by the same company in both markets. Under these circumstances, by abandoning country 2's market (where in any case the company is earning zero profits as $r_2^* = k$), the chemical company could protect its profits in country 1 against indirect competition from generic chemicals that are available in country 2 and force the company to price at marginal cost in that market. In this case, where the agricultural chemical company has abandoned country 2, there is no change in the price of the

---

number, resulting in a Cournot oligopoly. The latter case is more complicated to assess but does not obviate the point that price in country 2 will fall.

[14] The algebra is tedious but the economic intuition is straightforward. A monopolist equates marginal revenue with marginal cost where, for a downward-sloping demand curve, marginal revenue is less than price. The situation here is complicated by the fact that a decrease in r lowers x and causes an outward shift in the demand curve for the chemical. This feedback effect is reflected in equation (12) through the terms $t(g - c - hk)$ in the numerator and th in the denominator. These terms disappear if $t = 0$ and $r = 0.5(s + k)/2b$, a standard monopoly pricing result where r is clearly strictly greater than k.

chemical in country 1 and no change in price and quantity in country 1's biotechnology market.

It should be noted that farm lobbies in both Canada and the United States have argued for harmonization rules that permit agricultural chemicals with similar active ingredients to be registered for use in the United States if they are registered for use in Canada, and vice versa. The result would be to allow generic chemicals to be purchased by U.S. farmers from Canadian sellers, effectively undercutting the value of any patent rights. In the context of the above model, both the chemical company and the biotechnology company would be opposed to such a rule.

Assuming there is no competition from generic chemicals and that transportation costs are zero, harmonization would force the chemical company to charge a single price in both markets. This case is not analyzed explicitly here, but the general results of creating a single price in any given market are well known. Price rises in the market with the higher demand elasticity and falls in the market with the lower price elasticity. The complicating factor is that there are feedback effects on the locations of the demand curves in both countries' chemical markets. The implications for the biotechnology firm are ambiguous in this case. A decrease in the price of the chemical lowers demand for the technology, and vice versa. The net effects on total demand for the biotechnology clearly depend on the relative sizes of the two countries' markets for both the chemical and the biotechnology.

5. CONCLUSION

This study has begun an exploration of the links between agricultural chemical markets segregated by regulation and the incentives for the development of agricultural biotechnologies. These links deserve attention for several reasons. First, mergers and subsequent devolutions between chemical companies and biotechnology companies have been prominent. Second, as the new empirical results presented in this chapter suggest, price discrimination seems to be widespread on the part of agricultural chemical manufacturers among country-specific markets. These practices have important implications for biotechnology companies. Many crop biotechnologies are intended to reduce the use of agricultural chemicals, and the demand for those biotechnologies is therefore positively related to chemical prices and vice versa, as is the case with Bt corn. The theoretical results presented in the second part of this chapter suggest that when biotechnology companies offer chemical-saving technologies they probably benefit from regulatory environments that segment agricultural chemical markets on a country-by-country basis. Regulatory harmonization initiatives are likely to reduce agricultural chemical prices in many cases, and especially where competition from generic

chemicals is present in one (or more) but not all of the markets to be integrated by the harmonization initiative. Thus, companies with chemical-saving
biotechnologies and chemical manufacturers have a joint interest in opposing such initiatives.

It should be noted that the analysis presented in the second part of this
chapter is limited in scope. The model has been solved under the assumption
that the upstream chemical producer sets price by utilizing the downstream
biotechnology producer's profit-maximizing decision rule, in which chemical prices are taken to be parametric. The two-stage game could alternatively
be solved by reversing the roles of the biotechnology company and the
chemical manufacturer, or by assuming Cournot-like strategic behavior. The
model could also be extended to examine incentives for mergers between the
upstream and downstream companies. These are all important questions with
substantial implications for aggregate economic welfare and its distribution
that deserve future attention.

REFERENCES

Carlson, G., J. Deal, K. McEwan, and B. Deen. 1999. "Pesticide Price Differentials Between
    Canada and the United States—1999." Unpublished report prepared for the U.S. Depart-
    ment of Agriculture's Economic Research Service and Agricultural and Agri-Food Canada,
    Washington, D.C.
Carlson, J.A., and R. Preston McAfee. 1983. "Discrete Equilibrium Price Dispersion." *The
    Journal of Political Economy* 91(3): 480–493.
Clerides, S.F. 2004. "Price Discrimination with Differentiated Products: Definition and Iden-
    tification." *Economic Inquiry* 42(3): 402–414.
Dennison, M. 2000. "Hi-Line Farmer Convicted." *The Great Falls Tribune*, November 16, p. 1.
Freshwater, D. 2003. "Free Trade, Pesticide Regulation and NAFTA Harmonization." *The
    Estey Center Journal of International Law and Trade Policy* 4(1): 32–57.
McEwan, K., and B. Daley. 1999. "A Review of Agricultural Pesticide Pricing in Canada—
    1997." Unpublished research report, Ridgetown College, Ontario, Canada.
McMillan, J., and P. Morgan. 1988. "Price Dispersion, Price Flexibility, and Repeated Pur-
    chasing." *The Canadian Journal of Economics* 21(4): 883–902.
Salop, S., and G. Stiglitz. 1977. "Bargains and Ripoffs: A Model of Monopolistically Com-
    petitive Rice Dispersion." *The Review of Economic Studies* 44(3): 493–510.
Smith, V.H., and B.K. Goodwin. 1996. "Multiple Peril Crop Insurance, Moral Hazard and
    Agricultural Chemical Use." *American Journal of Agricultural Economics* 78(2): 428–438.
Smith, V.H., and J.B. Johnson. 2004. *Agricultural Chemical Prices in Canada and the United
    States: A Case Study of Alberta and Montana.* Policy Issues Paper No. 4, Montana Sate
    University Agricultural Marketing Policy Center.
Sorensen, A. 2000. "Equilibrium Price Dispersion in Retail Markets for Prescription Drugs."
    The Journal of Political Economy 108(4): 833–850.
Stigler, G.J. 1961. "The Economics of Information." *The Journal of Political Economy* 69(2):
    213–225.

Stole, L.A. 2001. "Price Discrimination in Competitive Environments." Unpublished manuscript, Department of Economics, University of Chicago.

Varian, H. 1989. "Price Discrimination." In R. Schmalensee and R.D. Willig, eds., *The Handbook of Industrial Organization* (Vol. I). Cambridge, MA: Elsevier Science Publishers.

Yoshida, Y. 2000. "Third Degree Price Discrimination in Input Markets: Output and Welfare." *The American Economic Review* 90(1): 240–246.

Chapter 19

REGULATION AND THE STRUCTURE OF BIOTECHNOLOGY INDUSTRIES

Paul Heisey and David Schimmelpfennig
Economic Research Service

Abstract: The agricultural biotechnology industry has experienced consolidation. As a form of sunk costs, increased regulatory costs could contribute to exit by smaller firms and increasing industry concentration. Cost and revenue factors other than regulation, however, are more likely to explain consolidation to date. Regulation may be endogenous, as innovators make greater attempts to influence the regulatory process when marginal benefits of innovation are higher, when environmental/consumer lobbies are less active, and when marginal costs of influence are lower. For large agricultural biotechnology firms, there is a rough positive relationship between firms' R&D or net sales and the amounts they devote to lobbying and campaign contributions.

Key words: agricultural biotechnology, regulation, industry structure, political economy

1. INTRODUCTION

The changing structure of the agricultural biotechnology industry has attracted a great deal of attention both from economists and from the larger segment of the public interested in food policy. Economists may be intrigued by the questions posed both by classic industrial organization theory or by dynamic questions suggested by Schumpeterian models. At the same time, they may be stymied by the relative infancy of the industry and the lack of data on key variables. In the larger public, biotech proponents see the technology as an opportunity for private sector R&D investment to make significant contributions to increased agricultural productivity. Biotech opponents combine fears about potential human health or environmental

The authors gratefully acknowledge suggestions from participants at the NC 1003 conference in Arlington, Virginia, in March 2005. The views expressed here are not necessarily those of the U.S. Department of Agriculture.

externalities with concerns that the technology may lead to multinational corporations exerting ever increasing control over the world's food supply.

Even if attention is restricted to agricultural biotechnology—that is, if the considerably larger amount of activity in medical biotechnology is ignored—and restricted to the first commercially widespread application of agricultural biotechnology, genetically engineered crop varieties, the structure of the industry is complex. This becomes particularly evident when different types of mergers, acquisitions, strategic alliances, and licensing arrangements are considered. If one seed company buys another, this could be a straightforward case of a horizontal merger. Vertical integration might be observed if a trait supplier integrates forward by buying a seed company (or vice versa if the seed company integrates backward). But a great deal of merger activity has been between firms supplying different kinds of inputs— e.g., seed and agricultural chemicals, or more broadly between agricultural input firms and pharmaceutical companies. Economic analysis has been applied to this kind of merger, which is often thought to be motivated by the desire to combine complementary intellectual assets, or to create greater markets for complementary biotechnology and chemical products. In other words, fundamental cost or revenue considerations might lie behind certain observed changes in the agricultural biotechnology industry.[1] Finally, research tools as well as traits might be supplied by small agbiotech firms as well as by more diversified firms, and these firms too might be subject to vertical integration. In fact, given the positions of different types of firms in different parts of the market, any particular merger might combine aspects of horizontal integration, vertical integration, or integration motivated by economies of scope.

How is regulation thought to influence industry structure? The primary mechanism is through the impact of regulatory costs. Regulatory costs have often been regarded as exogenous sunk costs. Sunk costs can be defined as profits foregone if a firm leaves an industry. Ollinger and Fernandez-Cornejo (1998) show in a model of the effects of sunk costs on profitability that increases in sunk costs such as regulatory costs reduce the number of firms in the industry. In other words, increases in regulatory costs induce firm exit, all else equal. Larson and Knudson (1991) argue that exit caused by, among other things, increases in regulatory costs is likely to come primarily from smaller firms. In the case of agricultural biotechnology, however, these predicted effects from regulation are likely to interact with the vertical, horizontal, and complementary integration of the market.

Furthermore, firms can respond to economic regulation in more complex ways as well. And economic regulation comes into being for various reasons, and regulators' objectives might also influence the type and strength

---

[1] Divestitures and spinoffs might be indicators that initial evaluations of firms' potential profitability are not borne out after a merger takes place.

of regulation. Various theories of regulation—normative analysis as a positive theory (NPT), capture theory (CT), and the economic theory of regulation (ET)—have been proposed (Viscusi, Vernon, and Harrington 2000). NPT is based on the hypothesis that regulation occurs most in industries beset with market failures, for example through natural monopoly or externalities (Joskow and Noll 1981). CT began with the empirical observation that regulation often turned out to be pro-producer in the sense of raising industry profit. CT argues that either regulation is supplied in response to an industry's demand for regulation, or that the industry "captures" the regulatory agency over time (Bernstein 1955). ET is based on the ideas that the state's basic power is that of coercion, that an interest group can improve its well-being if it convinces the state to use coercive power on its behalf, and that therefore regulation is supplied in response to interest groups acting to maximize their incomes (Stigler 1971). These theories have primarily been used to analyze traditional regulation of prices or quantities, which may be a complement to or a substitute for antitrust policy.

Regulation of agricultural biotechnology is of a different kind—social regulation based on concerns for health, safety, or the environment. It is fairly straightforward to argue that this regulation does indeed exist in a context of market failure and externalities. Nonetheless, certain aspects of theories of regulation can be applied to the regulation of agricultural biotechnology. In particular, the ET argues that regulation is one mechanism by which an interest group can increase its income by using the state to redistribute wealth from other parts of society to itself. Other things equal, smaller, better organized interest groups are more likely to benefit from influencing regulation than are competing interest groups that are larger and relatively unorganized.

A simple model of the effects of pre-market testing, which is very similar to the broad outlines of regulation in agricultural biotechnology, suggests that there is in fact a socially optimal amount of regulation in the cases of health, safety, or the environment (Grabowski and Vernon 1983). This amount would balance the costs of meeting regulatory standards and costs of delay in the release of beneficial new products against the expected economic damages from the use of a product with negative health or environmental effects. R&D and delay costs are relatively easy to measure; however, reasonable people could well disagree over expected economic damage from potential health or environmental hazards. Therefore, one would expect that different interest groups would likely attempt to influence the level and nature of agricultural biotechnology regulation for informational as well as strategic reasons.

In this chapter we consider all three topics—the structure of the agricultural biotechnology industry, the influence of regulation on that structure, and the political economy of the regulation—in greater detail. The next section briefly describes the regulatory framework for agricultural biotechnology in the United States. The section afterwards outlines changes in the structure of the agricultural biotechnology industry, and how these structural changes might interact with regulation. After that, we present a conceptual model of regulation that includes some political economic considerations. The penultimate section considers a few empirical details that roughly support the theoretical framework for the impacts of regulation on private companies. The last section sketches a few conclusions.

2. THE REGULATORY FRAMEWORK FOR AGRICULTURAL BIOTECHNOLOGY

Agricultural biotechnology in the United States is regulated by three government agencies. The Animal and Plant Health Inspection Service (APHIS) of the U.S. Department of Agriculture regulates plants, plant pests, and veterinary biologics and is charged with determining whether a genetically engineered plant is safe to grow. The U.S. Environmental Protection Agency (EPA) regulates microbial or plant pesticides, and new uses for existing pesticides and novel microorganisms. It determines whether a genetically engineered plant is safe for the environment. The Food and Drug Administration (FDA) of the U.S. Department of Health and Human Services regulates food, food additives, human drugs, feed, and veterinary drugs, and determines whether the product of a genetically engineered plant is safe to consume. Regulation by the three agencies is coordinated through the *Coordinated Framework for the Regulation of Biotechnology*, released by the White House Office of Science and Technology Policy in 1986. When two or more agencies have jurisdiction for a particular issue, the *Coordinated Framework* establishes a lead and secondary agency (Patterson and Josling 2001).

In practice, a new agricultural biotechnology product often encounters APHIS first, then the EPA, and finally the FDA (Belson 2000). At the beginning of the process, APHIS decides whether or not to permit field testing. If plant incorporated protectants (PIPs) are involved—that is biopesticides whose genetic controls have been incorporated into the plant itself—the EPA issues experimental-use permits and then PIP registrations. The FDA issues reports of consultations on bioengineered foods. Finally, APHIS receives and grants petitions for nonregulated status—that is, permission for a plant to be grown commercially.

A cursory glance at immediately available regulatory information suggests that, in fact, companies are often successful in gaining approval for their biotechnology products. This does not take into account products whose development may have been curtailed earlier in the regulatory process but that are not reported in the agency databases. A closer look at the APHIS data, which might roughly represent the beginning of the regulatory process, suggests that very few applications to conduct field trials (under 4 percent) are denied. At the other end of the process, of 104 applications for non-regulated status that APHIS received between 1992 and 2004, 63 have been approved, 27 have been withdrawn, 12 are pending, and 2 received other judgments. Or, of 90 petitions for which decisions have been reached, 63, or 70 percent, have been approved.[2]

There are other ways, however, to measure regulatory effects besides the number of approvals alone. One important variable is the length of time in regulatory review, as regulatory cost increases with review time. Longer review times could be caused by one or more factors, for example a larger number of applications, smaller staff, changes in data handling procedures, or greater scrutiny associated with an increase in regulatory stringency.

In the case of agricultural biotechnology, review time at APHIS for deregulation petitions has more than doubled from an average of 5.9 months, for submissions from 1994 through 1999, to an average of 13.6 months for submissions between 2000 and 2004. Similarly, voluntary consultations with the FDA were reviewed in an average time of 6.4 months from 1994 through 1999, which lengthened to an average time of 13.9 months between 2000 and 2004 (Jaffe 2005). This occurred despite the fact that nearly all the new products reviewed in the latter period were similar to those reviewed earlier, and therefore did not raise fundamentally new questions. Jaffe did not analyze review time for EPA decisions, as PIPs covered by the EPA should be a subset of products regulated by APHIS or the FDA. Judging by public documents produced by the EPA, that agency, too, appears to have become somewhat more deliberate in its review process in recent years. This may be connected in particular to two events—the appearance of StarLink corn in consumer products and public concern over possible harmful effects of Bt on non-target lepidopterans such as the monarch butterfly—that were covered regularly in the popular press.

3. INDUSTRY STRUCTURE IN AGRICULTURAL BIOTECHNOLOGY AND POTENTIAL EFFECTS OF REGULATION

The pattern of changes in agricultural biotechnology industry structure has been addressed both through Schumpeterian analysis and by using transac-

---

[2] Summaries of APHIS decisions can be found at http://www.aphis.usda.gov/brs/database. htm. EPA decisions are recorded at http://www.epa.gov/pesticides/biopesticides/pips/index. htm. FDA consultations are listed at http://www.cfsan.fda.gov/~lrd/biocon.html. Combined information for all three agencies can be found at http://usbiotechreg.nbii.gov/database_pub. asp.

tions cost theory. Kalaitzandonakes and Bjornson (1997) argue that these approaches are complementary. In a Schumpeterian world, radical innovation, as perhaps exemplified by the initial development of agricultural biotechnology, results in lowered barriers to entry in the pre-commercial phase of technology development. In the commercial phase, as products are developed, complementary assets such as past experience, manufacturing capability, and marketing and distribution networks become important. In general, firms that are successful at this stage scale up their activities. Escalation can occur through endogenous investment in R&D or through mergers and acquisitions (Fulton and Giannakas 2001).

This pattern of industry consolidation has been observed in the agricultural biotechnology industry. Count measures of mergers, acquisitions, and other consolidation activities have been provided by Brennan, Pray, and Courtmanche (2000), Kalaitzandonakes and Hayenga (1999), and King (2001). Empirical analyses of the reasons for this consolidation tend to rely on either patents or field trials, as these data are readily available, although they measure only limited points in the research-to-commercialization continuum.[3] Although details differ, analysts support complementarity arguments for consolidation. Graff, Rausser, and Small (2002) argue that if specific agricultural biotechnologies are broadly defined as relating to germplasm, transformation platforms, and traits, mergers tend to combine different technologies. Schimmelpfennig and King (2004) contend further that firms have used mergers to get around intellectual property (IP) licensing holdups. They show that larger firms (characterized as either "chemical" or "multinational" firms) have tended to merge vertically either backwards, by acquiring specialized agricultural biotechnology firms, or forwards, by acquiring seed companies.

In broad economic terms, most of the complementarity arguments are based on cost or revenue considerations, either through economies of scale and scope, or sunk costs. Sunk costs of regulatory approval, advertising, and research are important features of both the seed and chemical markets (Fulton and Giannakis 2001). Analysts of the relationship between changes in agricultural biotechnology industry structure and sunk costs have considered primarily R&D expenditures. Empirically, studies have differed in their accounts of the effects of industry consolidation and R&D investment. Oehmke (2001) finds no relationship between concentration and the level of R&D activity; but Schimmelpfennig, Pray, and Brennan (2004), using a different measure of concentration, find that increased concentration at the R&D level leads to lower R&D intensity.

Exogenous sunk costs such as regulatory costs, on the other hand, may at first glance appear to influence industry structure rather than to be influ-

---

[3] Oehmke and Wolf (2003) propose a modification to the field trial data, based on gene constructs rather than on raw numbers of field trials.

enced by that structure. To date, however, most empirical studies have focused on descriptions of the tri-partite regulation of agricultural biotechnology by APHIS, the EPA, and the FDA, or on calculating the costs of obtaining regulatory approval rather than on the specific effects of regulation on industry structure. Such studies can be found, however, for related industries. As noted, Ollinger and Fernandez-Cornejo (1998) found that in the U.S. pesticide industry, rising product regulation costs led to reduced numbers of firms. Furthermore, they found that smaller firms were affected more by these increased costs than were large firms. Thomas (1990) found that in the pharmaceutical industry, FDA regulations caused large reductions in research productivity for smaller U.S. firms. Large firms, however, benefited, as sales gains due to reduced competition more than offset declines in their research productivity.

Agricultural biotechnology is relatively recent. The Coordinated Framework for regulation was established in 1986; the first approved petition for deregulation of a genetically engineered crop was granted in 1992; and the first notable commercial plantings of genetically engineered crops occurred in 1996. The 10 to 20 year history of investment and product development in agricultural biotechnology is a relatively short time in which to observe regulatory effects on industry consolidation. Regulatory costs could have been one factor reducing the probability that small-scale dedicated agricultural biotechnology firms would scale up through expanded R&D and forward integration. However, the other factors mentioned above—the need to combine genetically engineered traits, transformation platforms, and germplasm, and the need for marketing and distribution networks—were probably more important determinants of industry consolidation.

The economic theory of regulation suggests, however, that in some sense sunk costs associated with regulation might not be completely exogenous. Firms subject to health, safety, or environmental regulation might have incentives to attempt to change the regulatory structure to their benefit. The empirical literature demonstrates that in certain industries, regulation can benefit large firms by increasing their sales through reduced competition, even though it increases their costs, suggesting that these firms might have rather complex motives when they address the regulatory structure. We turn next to a simple conceptual model of regulation in a political economy context.

4. CONCEPTUAL MODEL FOR REGULATION OF AGRICULTURAL BIOTECHNOLOGIES

This highly simplified representation of the supply and demand for regulation focuses on how lobbyists and other interest groups might influence the separate institutions that create the supply and demand for regulation of agricultural biotechnology. The rough framework abstracts quite a bit from the dynamic complexity of evolving benefits and costs of biotechnology innovations, the associated uncertainties, and the risk preferences of the parties that are apparent in the biosafety protocol for agriculture. The model is drawn from transaction cost models of natural resource transfer agreements (Colby 1995) and helps to illustrate how political persuasion can affect both how the regulators operate that oversee the introduction of biotech innovations, and how the private companies develop new technologies.

If we assume that monetary benefits (B) to the innovator increase as the number of their innovations go up, because innovations often extend the market size or share of the innovator, then innovators will usually have incentives to apply political effort (E_B) to accelerate the regulatory approval process for innovations. If the regulatory approval process was ever perceived as unsafely weak, innovators might have incentives to argue for strengthening the regulatory approval process. The mandate of regulators is to protect the public (not improve agricultural productivity), and the fewer innovations coming through the biosafety protocol, the fewer avenues of potential danger they are required to police, the more refinements they can make to existing procedures, and the fewer their losses (L). An accelerated regulatory approval process may not be in their interest of protection, and they might apply political effort (E_R) to strengthen regulations by testifying to legislators about potential dangers. These types of hearings take place regularly in conjunction with budget appropriation decisions.

In addition to direct influence applied to regulatory rule-making by both innovators and regulatory agencies, innovators make campaign contributions and support lobbyists for the same purpose—to influence the form of the regulatory approval process. Campaign contributions are often made through political action committees, in the form of soft money contributions to individual candidates, or to party committees and leadership political action committees. Many biotechnology companies also support either lobbyists or the umbrella Biotechnology Industry Organization (BIO). The level of support given to these groups is denoted by P_B. Lobbyists and political figures often work directly to influence the effectiveness of efforts (E_B) of private companies that support them, but might also choose to counter or refute testimony to legislators given by regulators (E_R).

In addition to choosing their own effort level, E_R, regulators may be influenced by the activities of environmental, ecological, family farm, organic, scenic, and other political interest groups, and the political activities of these groups are denoted by P_R. Unlike innovators who can choose both their own effort and support for lobbyists, regulators can choose only their own effort, E_R, while being influenced exogenously by organizations carrying out P_R both in terms of the impact of P_R on regulator costs, C_R, and on the likely success of innovator efforts E_B. The effects of this asymmetry in the number of choices available to innovators and regulators are explored in their choice problems below. There could be additional feedbacks between strength of biosafety regulation and public opinion concerning food safety, environmental, and other regulations reflected in consumer confidence, but this is difficult to model explicitly, as is modeling the feedback from strength of regulation to the efforts of regulators and innovators.

Innovators are assumed to maximize the net benefits (NB) of a quantity of innovations (Q). Benefits are modeled as a positive, concave function of Q, while costs, C_B, are a function of E_B and P_B. The loss function (L) of regulators is assumed to be downward sloping and convex, and federal regulators are assumed to seek to minimize their losses subject to their costs. Innovator efforts positively influence Q ($\partial Q/\partial E_B > 0$), while efforts of regulators decrease Q ($\partial Q/\partial E_R < 0$), as discussed above. The innovator's objective (net benefit) function with effort E_B and political influence P_B as choice variables can be written as

$$\max NB = B\{Q[E_B, E_R(P_B)]\} - P_R E_B - C_B (E_B, P_B).$$

The innovator's first-order conditions are

$$\frac{\partial NB}{\partial E_B} = 0 \Rightarrow \frac{\partial B}{\partial Q}\frac{\partial Q}{\partial E_B} - P_R - \frac{\partial C_B}{\partial E_B} = 0$$

$$\frac{\partial NB}{\partial P_B} = 0 \Rightarrow \frac{\partial B}{\partial Q}\frac{\partial Q}{\partial E_R}\frac{\partial E_R}{\partial P_B} - \frac{\partial C_B}{\partial P_B} = 0 .$$

These imply that innovator efforts to influence the regulatory approval process are higher when the marginal benefit of an innovation is higher (holding the impact of effort on innovation constant), the more subdued the activities of environmental groups, and the lower the marginal cost of effort. Firms often have to make decisions concerning effort with little more than rough estimates of the impact that innovations might have on market size or of the share of an existing market that they might be able to capture. The second first-order condition shows how pervasive the impact of decisions to support political activism can be by influencing both own marginal costs and the effect of regulator's efforts on the rate of innovation.

The regulator's objective function with effort E_R as its single choice variable is to minimize its net losses and can be written as

$$\min NL = L\{Q[E_B(P_R), E_R]\} - P_B E_R - C_R (E_R, P_R).$$

The regulator's first-order condition is then

$$\frac{\partial L}{\partial E_R} = 0 \Rightarrow \frac{\partial L}{\partial Q}\frac{\partial Q}{\partial E_R} - P_B - \frac{\partial C_R}{\partial E_R} = 0 \,.$$

This condition is analogous to the first first-order condition of the innovator. Regulator efforts to influence the approval process are lower when marginal losses from an additional approval are lower (holding the impact of regulator effort on approvals constant), the greater the political activities of innovators, and the higher the marginal cost of regulator effort. The asymmetry for regulators is reflected in the fact that they are not able to take advantage of the numerous ways that political support can pay off. Regulators are impacted by environmental groups, but there is little these groups can do to impact regulators' agendas. Innovators are able to make campaign contributions and support lobbyists to get the political influences apparent in their second first-order condition working for them. The following section will further attempt to characterize the relationships between agbiotech companies and their use of lobbyists and contributions to political campaigns.

5. SUNK COSTS AND INNOVATOR SUPPORT FOR POLITICAL ACTION

In order to examine empirical evidence on agricultural biotechnology company support for lobbyists and contributions to political campaigns, we first examine some relevant characteristics of the other variables from the firm's decision problem in the previous section. The quantity of innovations, Q, discussed previously as monotonically related to firm benefits, market share, sales, and company output variables, has a much more complicated discontinuous relationship to the inputs for Q. Research and development is a notoriously risky business involving substantial sunk costs in terms of the infrastructure, human, and intellectual capital involved. Company headquarters, a sales force, and marketing team all involve sunk costs of course, but are not product-, or innovation-, specific. Mergers and acquisitions that are widespread in the agbiotech industry have been partially explained in terms of motivations to gain control of specific R&D assets like germplasm or transformation technologies for the production of individual products as discussed. Licensing is so infrequent among agbiotech firms that mergers and acquisitions may have been the easiest way to obtain access to these technologies for many firms, which incidentally introduced additional sunk costs. To acquire all or part of their R&D capital, companies have to be acquired lock, stock, and barrel, even if layoffs and spinoffs take place in the future.

To ameliorate some of the bias that might be introduced by looking at data on campaign contributions and lobbyist support related to some continuous response variables and others with significant sunk costs, we consider some of both types of company data. For a continuous output variable we examine net sales, or gross sales, minus returns, discounts, and allowances. This output target is under constant scrutiny by the company, and by stock analysts outside the firm. Sales forecasts are made and revised within some firms on a weekly basis, but certainly quarterly sales figures are viewed by the entire industry for changes in market size and share of the market. For a sunk cost variable we consider R&D expenditures, keeping in mind the difficulties in analyzing data that mixes several program areas together with product development pipelines of varying lengths.

The data on lobbying expenditures are from federal lobbying reports filed with Congress, summarized by the Center for Responsive Politics, and referenced on BIO's website (http://www.BIO.org, accessed February 2005). Figure 1 shows that even though Monsanto had twice the agbiotech budget of the next most R&D intensive firm, it also had twice the amount of lobbying expenditures of the next biggest lobby supporter (Dow). In 2000, at the height of the consumer acceptance debate for genetically modified food, Monsanto's lobbying was off the figure at over 13 times the spending levels of BASF, Pioneer, and Syngenta.

Campaign contributions are based on data released by the Federal Election Commission (through May 27, 2003) and include contributions to political action committees (PACs), and soft money and individual contributions to federal candidates, party committees, and leadership PACs as discussed earlier. Contributions in Figure 2 are for biannual election cycles (two years) and were summarized by the Center for Responsive Politics, and referenced on BIO's website (accessed February 2005). The figure shows that until 2000, when Monsanto substantially increased contributions, Pioneer was the leader in contributions per dollar of net sales. Dow's use of this instrument had declined steadily since 1992.

To attempt to determine if any time-invariant relationships might exist between the variables in the first two figures, two additional representations were considered in Figures 3 and 4 where each scatter plot point represents one year for one of the companies. As might be predicted from the previous discussion, there is a point to the far right in each figure (representing Monsanto in 2000), but otherwise there appear to be positive relationships (at least more positive than negative) between political persuasion and company characteristics.

To interpret the scatter plot in Figure 3, we first note that even after disregarding the outlier to the right, Monsanto's ratio of lobbying to R&D is still tops in the industry, with the next two rightmost data points belonging to them. All of the points at the top of the graph are associated with DuPont/

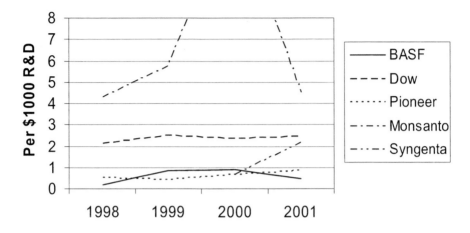

Figure 1. Lobbying Expenditures per $1,000 R&D

Sources: http://www.capitaleye.org/bio-lobbying.asp and Compustat (Standard & Poor's).

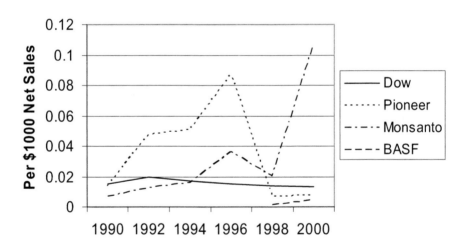

Figure 2. Campaign Contributions per $1,000 Net Sales

Sources: http://www.capitaleye.org/bio-lobbying.asp and Compustat (Standard & Poor's).

Pioneer, which indicates the size of their R&D program relative to other companies, but these data include their research into high-performance materials, synthetic fibers, electronics, and specialty chemicals in addition to agbiotech. Insiders put DuPont's agricultural program at less than half of Monsanto's (Cavalieri 2004). Taking into account that several agbiotech

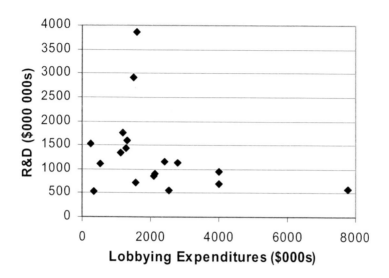

Figure 3. Lobbying and R&D Expenditures (1998–2001)
Source: Figure 1.

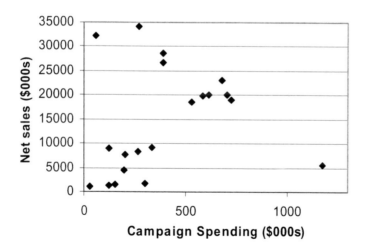

Figure 4. Campaign Contributions and Net Sales (1990–2000)
Source: Figure 2.

companies do not invest in lobbyists and that there are a number of points along the vertical axis that do not appear in the Figure, the relationship between lobbying and R&D is probably positive.

Figure 4 shows the same sort of positive relationship between contributions and net sales. Disregarding the outlier at the far right again, and considering that the points in the middle of the scatter are all Dow, and five of the points near the origin are Monsanto between 1990 and 1998, when these two loci are joined up there is clearly a positive association between net sales and campaign spending.

By combining the direct empirical evidence on agricultural biotechnology company support for lobbyists and contributions to political campaigns [in relation to both sales (market size or share) and R&D (sunk costs)] with relationships implied by the innovator's first-order conditions, several regularities are apparent. Empirically we see that even though sales and R&D are fundamentally different kinds of company information, both are positively related to political activism. Second, if the marginal benefit of an innovation (the result of R&D) or an increase in market share/size (measured by sales) is increasing (as predicted theoretically), then innovator efforts to influence the regulatory process (by themselves, through lobbyists, or using politicians) should be lower when sales and R&D are lower and the (roughly) positive relationship observed agrees with the theoretical framework. In fact, the case of Monsanto in 2000 agrees with this summary in that it was precisely at the point that the company had the largest market shares, the strongest pipeline of new products, and the most at stake that they invested heavily in political support.

6. CONCLUSIONS

Environmental and health regulation is likely to have complicated effects on industry structure in agricultural biotechnology. First, theoretical considerations and empirical results from other industries suggest that the net effect of regulation on industry structure is likely to be a reduction of firm numbers, with exit more likely for smaller firms. It is likely, however, that other cost and revenue factors have been more important to consolidation in the agricultural biotechnology industry to date than regulation. Second, our analysis suggests that innovators will make greater efforts to influence the regulatory approval process when the marginal benefit of an innovation is higher, when environmental/consumer lobbies are less active, and the lower the marginal cost of achieving this influence (all else equal). Empirically, for relatively large players in the agricultural biotechnology industry, we find a rough positive relationship between firms' level of R&D and net sales and the amounts they devote to lobbying and campaign contributions. Third, the time required to complete the regulatory process does appear to have lengthened, and this raises regulatory costs to firms. This could be due to increased stringency, or other administrative factors could be contributing to regulatory delays. It remains to be seen whether this contributes in a major way to further reductions in firm numbers in the core agricultural biotechnology industry, or whether the firms that remain have sunk costs that keep them in the game under a wide range of possible future regulatory conditions.

REFERENCES

Belson, N.A. 2000. "U.S. Regulation of Agricultural Biotechnology: An Overview." *AgBio-Forum* 3(4): 268–280.

Bernstein, M.H. 1955. *Regulating Business by Independent Commission*. Princeton, NJ: Princeton University Press.

Brennan, M.F., C.E. Pray, and A. Courtmanche. 2000. "Impact of Industry Concentration on Innovation in the U.S. Plant Biotech Industry." In W.H. Lesser, ed., *Transitions in Agbiotech: Economics of Strategy and Policy*. Food Marketing Policy Center, University of Connecticut, Storrs.

Cavalieri, A. 2004. "Commercialization of Agricultural Biotechnology Products." Presentation made at the conference on Frontier Plant Biotechnology: Advancing Crop Productivity and Market Potential, hosted by the Center for Strategic and International Studies and the Howard Hughes Medical Institute. Available online at http://www.csis.org/tech/Bio[-] tech/events/index.htm (accessed May 2005).

Colby, B.G. 1995. "Regulation, Imperfect Markets, and Transaction Costs: The Elusive Quest for Efficiency in Water Allocation." In D.W. Bromley, ed., *The Handbook of Environmental Economics*. Cambridge, MA: Blackwell Publishers.

Fulton, M., and K. Giannakas. 2001. "Agricultural Biotechnology and Industry Structure." *AgBioForum* 4(2): 137–151.

Grabowski, H.G., and J.M. Vernon. 1983. *The Regulation of Pharmaceuticals: Balancing the Benefits and Risks*. Washington, D.C.: American Enterprise Institute.

Graff, G.D., G.C. Rausser, and A.A. Small. 2002. "Agricultural Biotechnology's Complementary Intellectual Assets." *Review of Economics and Statistics* 85(2): 349–363.

Jaffe, G. 2005. "Withering on the Vine: Will Agricultural Biotech's Promises Bear Fruit?" Center for Science in the Public Interest, Washington, D.C. Available online at http://cspi[-] net.org/new/pdf/withering_on_the_vine.pdf (accessed May 2005).

Joskow, P.L., and R.G. Noll. 1981. "Regulation in Theory and Practice: An Overview." In G. Fromm, ed., *Studies in Public Regulation*. Cambridge, MA: MIT Press.

Kalaitzandonakes, N, and B. Bjornson. 1997. "Vertical and Horizontal Coordination in the Agro-Biotechnology Industry: Evidence and Implications." *Journal of Agricultural and Applied Economics* 29(1): 129–139.

Kalaitzandonakes, N., and M. Hayenga. 1999. "Structural Change in the Biotechnology and Seed Industrial Complex: Theory and Evidence." Presented at the NE-165 conference "Transitions in Agbiotech: Economics of Strategy and Policy," Washington, D.C., June 24–25.

King, J.L. 2001. "Concentration and Technology in Agricultural Input Industries." Agricultural Information Bulletin Number No. 763, Economic Research Service, U.S. Department of Agriculture. Available online at http://www.ers.usda.gov/publications/aib763/ (accessed May 2005).

Larson, B.A., and M.K. Knudson. 1991. "Public Regulation of Agricultural Biotechnology Field Tests: Economic Implications of Alternative Approaches." *American Journal of*

Agricultural Economics 73(4): 1074–1082.

Oehmke, J.F. 2001. "Biotechnology R&D Races, Industry Structure, and Public and Private Sector Research Orientation." *AgBioForum* 4(2): 105–114.

Oehmke, J.F., and C.A. Wolf. 2003. "Measuring Concentration in the Biotechnology Industry: Adjusting for Interfirm Transfer of Genetic Materials." *AgBioForum* 6(3): 134–140.

Ollinger, M., and J. Fernandez-Cornejo. 1998. "Sunk Costs and Regulation in the U.S. Pesticide Industry." *International Journal of Industrial Organization* 16(2): 139–168.

Patterson, L.A., and T. Josling. 2001. "Regulating Biotechnology: Comparing EU and US Approaches." Paper presented at the 76th Annual Conference of the Western Economic Association International, San Francisco, July 8.

Schimmelpfennig, D., and J. King. 2004. "Mergers, Acquisitions, and Flows of Agbiotech Intellectual Property." Paper presented at the 8th International Consortium on Agricultural Biotechnology Research (ICABR) International Conference on Agricultural Biotechnology: International Trade and Domestic Production, Ravello, Italy, July 8–11.

Schimmelpfennig, D.E., C.E. Pray, and M.F. Brennan. 2004. "The Impact of Seed Industry Concentration on Innovation: A Study of U.S. Biotech Market Leaders." *Agricultural Economics* 30(2): 157–167.

Stigler, G.J. 1971. "The Theory of Economic Regulation." *Bell Journal of Economics and Management Science* 2(1): 3–21.

Thomas, L.G. 1990. "Regulation and Firm Size: FDA Impacts on Innovation." *RAND Journal of Economics* 21(4): 497–517.

Viscusi, W.K., J.M. Vernon, and J.E. Harrington, Jr. 2000. *Economics of Regulation and Antitrust* (3rd edition). Cambridge, MA, and London: MIT Press.

Chapter 20

THE SOCIAL WELFARE IMPLICATIONS OF INTELLECTUAL PROPERTY PROTECTION: IMITATION AND GOING OFF PATENT

James F. Oehmke
Michigan State University

Abstract: Successful biotechnology innovations have captured significant market shares and are thought to lead to non-competitive input markets. These market structures are determined not just by the current innovation, but also by the history of prior innovation and the potential for future innovation or imitation. This paper examines the relationships among future innovation or imitation, current market structure, and the social welfare. Model results suggest that the social welfare effects of going off patent can be larger than those generated by the discovery itself, especially when a private-sector firm extracts a majority of the generated social surplus in the form of rents.

Key words: agricultural biotechnology, intellectual property, patent, social welfare

1. INTRODUCTION

The heart of regulatory approaches to intellectual property protection and regulation of temporary monopoly or market power is the conflict between static and dynamic efficiency. This conflict arises because the discovery/ invention of new products or processes almost always requires the investment of resources, and sometimes leads to an accrual of monopoly or market power by the successful innovator (or innovating firm). The new product or process itself is associated with economic growth and thus welfare improvement and dynamic efficiency in the context of secular growth. The temporary monopoly power that the successful innovator obtains while she is the sole provider of the innovative product/process leads to the usual static inefficiency measured by deadweight loss in a partial-equilibrium setting. However, it is the excess profit from this market power that entices firms to invest resources in the inventive activity in the first place. Thus, the temporary market power and associated "excess" profits seem to be necessary to balance the required investment in inventive activity and generate an acceptable long-run growth trajectory.

The ability to acquire temporary market power depends on the legal infrastructure available to protect intellectual property, among other factors. In relation to agricultural biotechnology, the primary methods of intellectual property protection are utility patents, plant variety protection certificates, and trade secrets. For example, transgenic varieties of corn and soybeans are protected by a complex of utility patents covering not just the plant itself, but the genes inserted, the insertion mechanism, gene promoters, markers, and a variety of other components (Lesser 2005). Transgenic cotton varieties, for reasons that are not entirely clear, are most often protected by plant variety protection certificates (Oehmke, Naseem, and Schimmelpfennig 2005). Traditionally bred hybrid varieties of various crops have often been protected by trade secrets in the form of complex inter-crosses that are not discernible from the commercialized variety.

Much of the literature analyzing the tradeoffs between dynamic and static efficiency focuses on the protection of intellectual property as new innovations are introduced. Certainly this is the time at which the innovating firm's rents start to accrue. However, at an aggregate level, there are reasons to believe that from an ex-ante social welfare perspective, the change in social welfare from the introduction of a new innovation may be much lower than the change in the firm's profits. Indeed, a zero-profit condition would all but ensure that for the industry as a whole, the winning firm's rents are dissipated by R&D costs of "losing" firms.

Moreover, there is substantial evidence from both the agricultural and non-agricultural industries that substantial welfare gains occur when there is competition to produce the innovative product, e.g., when a pharmaceutical comes "off patent." As agriculture moves into the biotechnology era, in which firms are using utility patents to protect their intellectual property, it becomes more likely that innovating firms will grab increasingly large shares of the rents from innovation. Thus there will be social benefits when these innovations are imitated or go off patent.

The purpose of this paper is to present a model of innovation and industry structure that enables calculation of the social welfare effects of innovation and intellectual property protection in a fully dynamic environment. We are particularly interested in the dynamics associated with innovations coming "off patent." The next section provides a brief summary of selected literature. The third section develops a dynamic model of endogenous R&D. This model represents R&D activity as an investment with an uncertain payoff. It provides specific functional forms and equilibrium conditions to determine the level of investment in R&D, but is general enough to allow for a variety of output market structures. The fourth section provides an example of how to marry different output market structures with the dynamic model of R&D. The final section draws conclusions.

2. THE STATE OF THE LITERATURE

Recent literature has focused on measuring single-period profits from successful innovation in static, non-competitive markets (e.g., Huang and Sexton 1996, Alston, Sexton, and Zhang 1997, and Alston, Sexton, and Zhang 1999). One motivation for examining non-competitive markets is that technological advances can generate competitive advantages, so that firms owning the latest technology gain market power, and the resulting market structure is non-competitive. For example, only a few firms own the intellectual property underlying glyphosate-tolerant soybeans, leading to non-competitive forces in the soybean complex (Moschini, Lapan, and Sobolevsky 1999). Non-competitive market structures can cause standard social surplus calculations based on competitive-market models to be incorrect (Moschini and Lapan 1997).

However, the static nature of the current literature is limiting. This is especially true in the agricultural context, where *continual* innovation and technological improvements are the rule, not the exception. Static models allow no mechanism for future innovations to affect market structure, profits derived from current-generation technology, or social benefits.

For example, consider the sequence of papers by Voon (1994), Sexton and Sexton (1996), and Voon (1996), discussing whether research benefits are greater under perfect competition than under monopoly. Taking a static perspective, Sexton and Sexton argue that "it is well known that the aggregate social loss from monopoly is the so-called deadweight loss, DWL....it is straightforward to see that the necessary and sufficient condition for the benefits of the innovation to be greater (less) under monopoly than under perfect competition is that DWL decline (increase) as a consequence of the innovation" (p. 205). Voon (1996) agrees: "Voon (1994) supports ... [Sexton and Sexton's 1996] conclusion that the research benefit is greater under perfect competition that [sic] under monopoly. An implication for this is that public research aimed to raise national welfare may be more optimally allocated in favor of less concentrated enterprises" (p. 205, parentheses in original). However, this tells only part of the story: the part associated with a single innovation. A systemic perspective would entertain the possibility that the existence of monopoly profits or other rents associated with imperfect competition provide the incentive for industry to innovate (cf. Schumpeter 1968). The analysis of an "efficient" market structure must include not just the static effects of the market structure (DWL), but the dynamic effects of market structure on the rate of future innovation (cf. Davidson and Segerstrom 1998), and may lead to policy implications very different from those of Voon (1994). Consequently, welfare and policy analysis based on a static model are inappropriate.

Moschini and Lapan (1997) focus on non-competitive input market structures that arise due to successful innovation. This is an important step in understanding the relationships among innovation, market power, and social welfare. Their paper uses a partial equilibrium approach and emphasizes the time at which the innovation is commercialized. There are two related issues with the Moschini and Lapan (1997) approach. The first is the lack of an equilibrium condition(s) that makes endogenous the level of private-sector R&D investment at both the firm and industry levels. To the extent that Moschini and Lapan (1997) are focusing on innovating firms that accrue market power sufficient to effect a non-competitive market structure, the firm likely will also capture a significant portion of the social surplus from the innovation as rents.[1] This does not mean that the industry accrues the same level of rents. Particularly in cases where the innovation is subject to intellectual property protection, it is expected that industry R&D expenditures in a particular area (herbicide-tolerant soybeans) will be greater than the R&D expenditures by the winning firm, as multiple firms may engage in R&D to discover this product.[2] More importantly, in the private sector, we would expect there to be some sort of equilibrium relationship between the expected level of rents (determined in part by the market structure) and the amount that firms individually and the industry as a whole are willing to invest in pursuit of these rents.[3] The presence of such a relationship is an additional piece of information that can be brought to bear on the welfare analysis of the innovation.

The second issue is that the partial-equilibrium analysis of social welfare gains from innovation is not recalculated when the innovation goes off patent. However, anecdotal evidence from the pharmaceutical industry would suggest that going off patent seems to generate significant changes in industry structure and important welfare benefits by making the innovation more widely available at a lower price (e.g., a name drug becomes available through generic substitutes). Nothing in the agricultural economics literature

---

[1] Certainly, in some cases, both private and social gains (apart from the private gains) appear to be generated at the time the innovation is introduced to the market (e.g., Bt cotton; c.f. Falck-Zepeda, Traxler, and Nelson 2000, and Oehmke and Wolf 2004).

[2] Good data on expenditures by R&D orientation are not available, but an examination of field trial data shows that the Animal and Plant Health Inspection service has issued 375 permits for field trials of herbicide-tolerant transgenic soybeans. Monsanto, whose Roundup Ready® varieties dominate the market, conducts 123 or just fewer than one-third of these field trials.

[3] Perhaps because agricultural economists have traditionally examined returns to public research, we typically do not assume an equilibrium relationship between the level of rents and the level of investment in R&D to obtain those rents.

seems to deal with the social welfare implications of going off patent. Several major plant biotechnologies either have gone off patent very recently or will go off patent in the near future (e.g., 35S, some Agrobacterium applications, the gene gun, several CRY sequences). Thus it is timely to consider the market structure and welfare effects of going off patent.

In this paper, we provide a general dynamic context for the analysis of R&D in non-competitive markets. In particular, the dynamic model addresses two issues raised by Alston, Sexton, and Zhang (1997)—research investment and long-run benefits: "Understanding the implications of market power of agribusiness firms for the magnitude and distribution of benefits from given research-induced supply shifts is just the first step toward understanding the full consequences of market power for research investment patterns and long-run research benefits" (p. 1264). To address the issue of research investment patterns we endogenize the research investment decision at both the firm and industry level. The endogeneity comes in the forms of intertemporal arbitrage and first-order conditions at the firm level and an R&D industry zero-profit condition. We address the issue of long-run research benefits by examining transmission mechanisms that result in consumers eventually capturing the bulk of research benefits when innovations go off patent, and how non-competitive market structures influence these mechanisms. Finally, we explore behavior and policy implications related to innovations coming off patent.

The paper proceeds with a description of the dynamic model.

3. THE MODEL

3.1. Assumptions

The model rests on four basic assumptions, which are standard in neo-Schumpeterian models (see, e.g., Dinopoulos 1994 or Aghion and Howitt 1998):

- *Innovative activity takes the form of races to discover the next-generation technology.* For example, consider varietal development. Innovative activity takes the form of racing to discover the next-generation variety, whether this be a higher-yielding variety, or one that is more resistant, or one that has better processing characteristics. The first firm to discover such a variety has a competitive advantage in the marketplace.

- *There is always an innovation race taking place.* Even before one variety is released, breeders are hard at working discovering the next improvement.

- *Each new technology is based upon and improves on the previous technology.* For example, each new high-yielding variety has to have yields even higher than the previously best variety (or some other advantage). Similarly, animals with high lean muscle mass must be even leaner than those produced from the previous genotype or environment. With biotechnology, the ability to "stack" more than one bioengineered trait in a single variety will determine the leaders in the next generation of corn, cotton, and soybean varieties. Finally, current advances also rest on previous advances in a scientific sense. For example, it is difficult to stack biotechnology traits without having first introduced at least one trait.

- *Old technologies are available at marginal cost (even if there are no buyers).* It is usually pretty easy to find someone selling last year's seed variety, or lower quality dairy or beef cattle. Because these products no longer hold a competitive advantage, it is intuitively plausible that they do not generate quasi-rents from technical advantages and thus are priced at marginal cost. Formal models of this type of pricing decision are summarized in Aghion and Howitt (1998) and Barro and Sala-i-Martin (1995).

These four assumptions allow characterization of the dynamic process of accruing social benefits over time as a consequence of R&D and innovative activity.

3.2. Innovation and R&D

An innovation is a technological advance that improves agricultural production or products. The innovation may act to reduce production costs, or to increase the quality of the agricultural product. The innovations potentially available to the world are indexed by $j \in \{0, 1, 2, ...\}$, with innovation j being superior to innovation $j - 1$ in the sense of cost reduction or quality improvement. Innovative activity (R&D) consists of a sequence of races; firms enter the race by investing resources in R&D. The winner of race j discovers and is the sole owner of technology j. Immediately upon the discovery of j, the race to discover innovation $j + 1$ begins.

R&D is inherently an uncertain process. A greater investment in R&D means that the firm is more likely to win the race. Let $R_{i,j}$ denote the level of R&D activity by firm i in the race to discover technology j, and R_j denote

R&D activity by the industry in race j. Let $\varnothing(R_{i,j}, R_j)dt$ and $\Phi(R_j)dt$ denote the probabilities that firm i and the industry, respectively, discover technology j in the time interval $(t, t + dt)$. We assume that these probabilities are independent of t, but that the probability of discovering innovation $j + 1$ before innovation j is zero. We also assume that as a stochastic process the innovation process is memory-less; that is, the probability of innovation at any instant is independent of the time lapse since the previous innovation.[4]

The firm that wins race j may choose to protect its intellectual property (IP) using one or more types of IP protection: plant variety patent (PVP) protection, as used for most hirsute cotton varieties; utility patent (UP) protection, as used for most transgenic crops; or trade secrets, as used for many traditionally bred hybrids. For simplicity we represent the patent protection as giving the innovator sole rights to the innovation until such time as the patent expires. However, we model a dynamic situation in which the dominant form of change is when a new innovation is discovered.[5] The discovery of the new innovation makes the prior innovation obsolete, even if the patent or trade secret remains intact. We will consider innovation j to go off patent when the patent on innovation j expires, when a rival discovers the trade secret, or when innovation $j + 1$ is discovered. In the model we focus on the discovery of innovation $j + 1$, but the example of the next section shows how to accommodate patent expiration or erosion of the trade secret.

To maintain generality, we impose as little specification as possible on the market structure for the innovative product. The innovating firm earns profits by selling its innovation, e.g., transgenic seeds. The profit function for the innovating firm at each instant in time, when the jth innovation has been discovered but the $j + 1^{st}$ has not, is

$$\Pi_j = \Pi_j\left(p, q;\ \omega,\ \lambda,\ \zeta,\ \mu,\ \sigma, \{(j + k, t);\ k = 1, 2, 3, \ldots\}\right). \qquad (1)$$

The variables p and q represent the price and quantity of the innovation produced (or licensed for production). Depending on the structure of the pricing strategy or game, the innovative firm may choose p and/or q. ω is a vector of other prices such as R&D inputs or competing innovations, λ is a vector of

---

[4] The Poisson process is one example of a memory-less process. Graf, Rausser, and Small (2003) and Oehmke, Wolf, and Raper (2005) have used the Poisson process in empirical and theoretical (respectively) models of agricultural biotech R&D. Despite its memory-less characteristic, the Poisson process retains the characteristic that the probability of instant innovation is zero, as is the probability of never innovating.

[5] In the example of the next section we examine a case representing trade secrets as intellectual property protection that erodes over time as other firms discover pertinent aspects of the secrets, although in reality any type of IP protection could have combinations of these characteristics.

innovation characteristics such as measures of quality, ζ is a vector of demand characteristics such as elasticity, μ is a vector of market structure characteristics such as degree of market power or type of market structure, and σ is a vector of cost parameters that influence the marginal cost of producing the innovation once discovered. The last term in the expression for Π_j recognizes that the rents earned by innovation j may depend on the future innovations $j + k$, and the times t at which they become available. This is because the discovery of a "better" innovation will tend to reduce the profits to be earned from the prior innovation. This is a general way of accommodating Moschini and Lapan's (1997) point that new innovations can affect market structure.

Similarly, we specify a general social welfare function for each instant in time:

$$S_j = S_j\left(p,\, q;\, \omega,\, \lambda, \zeta, \mu,\, \sigma,\, \{(j+k,t); k = 1,2,3,\ldots\}\right). \tag{2}$$

Equation (2) is general enough to allow specifications in which S is social surplus, the typical partial equilibrium measure of welfare gains from innovation (Alston, Norton, and Pardey 1995), or other social welfare measures such as weighted average of the gains to different political constituent groups. The inclusion of μ, the vector of market structure characteristics, explicitly allows for market structure to influence the welfare gains from innovation à la Moschini and Lapan (1997) and Alston, Sexton, and Zhang (1997, 1999). Even though we have indexed S by the innovation, the social welfare gains from an innovation may persist even after that innovation has been replaced (standing on the shoulders of giants). That is, society may realize gains from innovation j even after innovation $j + 1$ has been discovered. As with firm profits, the level of social welfare gain may be affected by future innovations.

R&D is a costly activity. Let $C_R(R_{i,j};\, \rho_i)$ denote the cost to firm i of conducting an amount $R_{i,j}$ of R&D activity directed toward discovering the jth innovation; ρ_i is a vector of R&D cost-related parameters, some or all of which may be firm-specific.[6] Let

$$R_j = \sum_i R_{i,j}$$

---

[6] At this point, we do not distinguish whether the "firm" is a private-sector entity or a public-sector organization such as a university or agricultural experiment station. The cost formulation is general enough to allow the different types of organizations to have different cost structures. Specific applications may focus on instances in which primarily public-sector organizations are involved (basic research, open-pollinated varieties), in which primarily private-sector organizations are involved (glyphosate-tolerant hybrids), or in which both public and private organizations are involved (improved pesticide sprayers).

denote the amount of R&D conducted by the industry. Conducting an amount $R_{i,j}$ of R&D gives the firm a probability $\phi(R_{i,j}; R_j)dt$ of innovating successfully in any time period of duration dt, conditional on the prior discovery of innovation $j-1$ and on innovation j not yet having been discovered. The corresponding probability that some firm in the industry will innovate in a time period of duration dt is $\Phi(R_j)dt$.

Let V_j denote the value of the firm that has discovered innovation j, that is, the value of the firm after the R&D has been completed successfully. In anticipation of this discovery, firms will engage in R&D to be the first to discover innovation j and "win" the "prize" of having their firm value become V_j. Firms will engage in R&D until the expected marginal benefit from the R&D equals the marginal cost of R&D:

$$\phi'(R_{i,j}; R_j)V_j = C_R'(R_{i,j}; \rho_i).$$ (3)

The left side is the expected marginal benefit: the marginal change in probability that firm i will innovate successfully times the value of being the innovator.[7] The right side is the marginal cost of R&D.

The two key model equations follow. The first is the intertemporal arbitrage condition that relates firm value to the rents earned from successful innovation. In the analysis of publicly funded research, there has been little reason to believe that universities, for example, gain rents that are related to the value of their innovations—in fact, part of the university ethos (at least until recently) has been to make important discoveries widely available at low cost. The private sector R&D firm, however, counts on innovations to drive the profit statements. The value of the private sector firm will depend on the rents that the firm is deriving from its innovation. In particular, private sector firms will be subject to the arbitrage condition:

$$(r + \Phi(R_{j+1}))V_j = \Pi_j.$$ (4)

Investors purchase the firm in order to get a stream of returns, namely firm profits as shown on the right side of equation (4). In order to own this profit stream investors will bid up (down) the value of the firm until their return on investments consists of the risk-free return, r, plus an idiosyncratic risk premium. The idiosyncratic risk comes from the fact that the profit stream may not last forever—in this version of the model the profit stream vanishes when the $j + 1^{st}$ innovation is discovered and makes the jth innovation obsolete. The probability that the $j + 1^{st}$ innovation is discovered at any instant in

---

[7] More formally we would need to introduce dt terms into each side of the equation and take limits as dt goes to zero; we shall omit such formality in this paper.

time is $\Phi(R_{j+1})$, assuming that the innovation was not previously discovered. In equilibrium the firm profits suffice to exactly cover the risk-free return plus the risk premium. Another way of interpreting the arbitrage equation is that it shows that the value of the firm that discovers innovation j is the discounted flow of expected future profits from the innovation. Momentarily we will look at alternative formulations of the arbitrage equation (4).

The arbitrage equation (4) evinces a subtle but critical change for the measurement of innovation benefits. Equation (4) describes a relationship between profits derived from innovation j and R&D conducted on innovation $j + 1$. In other words, this equation shows the links between different R&D races. Thus, when equation (4) applies, it is inappropriate to examine the effects of any single innovation in isolation from the stream of innovations that surrounds the discovery under observation.

The second key model equation is the zero-profit condition. Firms will enter the R&D competition until the expected profits from firm entry are zero:

$$\phi(R_{i,j}; R_j)V_j = C_R(R_{i,j}; \rho_i). \tag{5}$$

The left side of the zero-profit condition (5) is the expected return to the firm from carrying out an amount $R_{i,j}$ of R&D. The right side is the cost of conducting that amount of R&D. For generality, we allow this cost function to vary by innovation, so that the cost function for R&D on innovation j may be different from that for innovation $j + 1$. Firms enter (or exit) the industry, changing the industry level of R&D, R_j. This changes the probability that any firm i successfully innovates $\phi(R_{i,j}; R_j)dt$. In equilibrium, firms enter until the expected profits from entry fall to zero, i.e., until the expected returns from entry equal the costs of being in the hunt to discover the innovation.

As with the arbitrage equation (4), the zero-profit condition (5) appears to be harmless but has important implications for the valuation of innovation benefits. Intuitively, the zero-profit condition can be thought of as a condition stating that on average firms bid away the expected rents from successful innovation. Even though the winning firm will obtain rents from the innovation, on average (and therefore in the industry as a whole) firms earn zero profits from innovation. That is, the losses from those firms that lose the innovation race and thus have nothing to show for their R&D investment offset the rents of the winning firm.

The zero-profit condition thus contrasts with the traditional experience of agricultural economists in evaluation of public-sector R&D activities. There is no reason to expect that the public sector would be subjected to a zero-profit constraint. The usual assumption in benefit-cost analysis is that public research expenditures are exogenously determined [although there is a literature on the political economy of public research expenditures—e.g., see the models of Oehmke and Yao (1990), Rausser and Foster (1990), or de Gorter, Nielson, and Rausser (1992)]. When examining private-sector bio-technology research the assumption that the level of R&D expenditures is exogenous to the level of expected profits is untenable, and should be re-placed with a zero-profit condition such as equation (5) (or some other man-ner of making endogenous the amount of R&D conducted at the industry level if the investigator so desires).

3.3. The Social Welfare Benefits of Innovation

In its most general form, the net present value of the social benefits from innovation j is

$$\int_0^\infty e^{-rt} S_j(t) dt , \qquad (6)$$

where for simplicity we have let the time index t stand for the values at time t of the parameters specified in the social welfare equation (2). There are two major issues with a general specification such as equation (6). First, the generic form provides no guidance as to how the investigator might actually implement this calculation; second, equation (6) implicitly assumes that the investigator has perfect knowledge of future social benefits as there is no accounting for uncertainty in equation (6).

A more practical approach is to specify a more specific form of equation (6) that accounts for model predictions of how social benefits will evolve as technology and industry structure evolves. This approach should be consis-tent with the arbitrage equation (4) that links innovations j and $j + 1$. One in-tertemporal model that is consistent with equation (4) is when the probability of discovering innovation $j + 1$ follows a Poisson distribution with parameter $\Phi(R_{j+1})$. This results in a conditional (on not having been previously discov-ered) probability of discovery during any time period of length dt of $\Phi(R_{j+1})dt$. From the information set available at time 0, the unconditional probability of the innovation being discovered during a specified time period $(t, t+dt)$, $t > 0$, is $(1 - e^{-\Phi(R_j)})\Phi(R_j)$. Let the time $t = 0$ denote the time at which the innovation is discovered. Assume further that the social welfare benefits from innovation j occur at one of two levels: from the time of discovering j until the time that $j + 1$ is discovered the social welfare benefits per period are $S_j(t|j)$, i.e., the social welfare benefits from innovation j conditional on j being the last innovation discovered. Beginning at the point at which innova-tion $j + 1$ is discovered, and continuing thereafter, the social welfare is given by $S_j(t|j + 1)$. The expected social welfare benefits from innovation j at time $t = 0$ are thus

$$\int_0^\infty [[\int_0^t e^{-rs} S_j(s \mid j) ds + e^{-rt} \int_t^\infty e^{-rs} S_j(s \mid j+1) ds] \Phi(R_j)(1 - e^{-\Phi(R_j)t}) dt . \quad (7)$$

The interior integrals take as given that innovation $j+1$ will be discovered at time t. The first interior integral gives the present value of the social welfare from innovation j until innovation $j+1$ is discovered. The second interior integral gives the present value of the social welfare from innovation j after innovation $j+1$ is discovered. The exterior integral accounts for the probability that innovation $j+1$ is actually first discovered at time t, for all $t \geq 0$.

3.4. Implementation of and Extensions to the Basic Model

Implementation of the basic model is eased considerably by specification of an appropriate and simple functional form for the probability of innovation $\Phi(R_{j+1})$. For example, Tolley et al. (1985) model trade secrets and patents as deteriorating over time at a constant rate [i.e., $\Phi(R_{j+1}) \equiv \gamma$] as the rest of the industry learns what makes the innovation successful and/or how to imitate this success without infringing on the patent. Oehmke, Wolf, and Raper (2005) model biotechnology industry evolution using the firm and industry functional forms $\varphi(R_{i,j}; R_j) \equiv R_{i,j}/(R_j+K)$ and $\varphi(R_j) \equiv R_j/(R_j+K)$, respectively, where K is a parameter that determines the curvature of the function, with $K > 0$ guaranteeing decreasing returns to scale. Mathematically, these forms turn out to be very tractable and make the probability of success depend on the level of R&D activity. Dinopoulos (1994) and Segerstrom (1998) have comparable but different functional forms in which the probability of innovation depends on R&D.

The basic model presented in this paper can also be consistent with various models of industry structure and evolution. The key equations are the specifications of the profit and social welfare functions, so that they are compatible with the industry structure at any point in time, and the arbitrage equation, which is affected by changes in industry structure. For example, Oehmke, Wolf, and Raper (2005), using basically this same framework, show how an unanticipated decline in profits leads to a cyclically out-of-steady-state adjustment path, and apply the model to R&D in transgenic soybeans.

Segerstrom (1991) endogeneized the decision to imitate by requiring that firms invest in imitative R&D in order to copy the innovation, leading to an innovation-imitation-innovation cycle. In this situation the profit function is such that after innovation j is discovered the next R&D activity is to imitate this innovation (call it the R&D race to discover innovation j'). Upon discovery of innovation j' the two innovators act as cooperative duopolists and split the monopoly profits [Segerstrom (1991) gives conditions under which this is a subgame perfect equilibrium], until innovation $j+1$ is discovered

and destroys the duopoly. Translating Segerstrom's model into our framework results in two arbitrage equations, one describing the value of the duopoly firms and one describing the value of the first firm to innovate before
imitation:

$$\left(r + \Phi(R_{j+1})\right)V_{j'} = \Pi_{j'} \tag{4'}$$

$$rV_j + \Phi(R_{j'})\left(V_j - V_{j'}\right) = \Pi_j. \tag{4''}$$

Equation (4) describes the value of a duopoly firm that faces the threat of
losing its duopoly profits, $\Pi_{j'}$, from the discovery of innovation $j+1$. It is
analogous to equation (4). Equation (4''), describing the value of the innovative firm in this context, is also analogous to (4). The difference between (4)
and (4'') is that in the latter equation, the risk premium is associated with the
probability $\Phi(R_{j'})$ that a firm successfully imitates, and the loss in firm value
if that imitation occurs is the difference between the value of the firm as the
monopoly owner of the innovation and the value of the firm as a duopolist.

Moschini and Lapan (2006) have modeled the profit functions for an
industry that evolves from a single innovator with a patent to competition
from "generic" imitators that enter once the original innovation goes off
patent. By changing the nature of the Segerstrom's (1991) duopoly game
[Moschini and Lapan (2006) use a Stackleberg price leader] one can consider putting these into an innovation-imitation (with additional innovation if
desired) framework such as equations (4') and (4'').

Additional specifications that may be of interest are those of continued
monopoly power on the part of one firm. The motivation for this interest is
the persistent dominant position of Monsanto with respect to bST and to
transgenic crop technologies. One interesting model is that of the persistent
monopolist (Harris and Vickers 1985). In this model, the firm that discovers
innovation j obtains both some monopoly power in the market for the input
that embodies innovation j (the transgenic seed market) and some advantage
in discovering the next innovation, $j+1$. This monopolist invests enough in
R&D to perpetuate its competitive and innovative advantage by discovering
and commercializing a sequence of innovations (e.g., stacking insect resistance and herbicide tolerance). An alternative model is that the monopolist
does only enough R&D on innovation $j+1$ to indicate to other firms that it
could be the first to discover and patent this innovation, if it so desired, and
use this as a barrier to entry into the R&D race to discover the next innovation. This acts to perpetuate the firm's market power as sole purveyor of innovation j.

4. EXAMPLE

This section presents a simple graphical application of the model. The point is to show how the dynamic approach modifies the partial equilibrium approach traditionally used in agricultural economics.

The economy is closed and the demand for the agricultural product is constant elasticity with elasticity $-\varepsilon < 0$. This demand curve is depicted in Figure 1. The inelastic demand for the agricultural output translates into an inelastic demand for the innovation.

There are a large number of agricultural producers behaving as price takers. For the agricultural good, the unit cost of production conditional on access to technology j is constant. C_j is the unit cost of producing the agricultural output exclusive of costs specific to accessing technology j (i.e., technology fees). At the outset of the example the only technology available is $j = 0$, which we shall call the traditional technology. For concreteness we think of this technology as a traditionally bred crop variety. This variety is available free of technology fees to all producers (although there will obviously be seed costs). The initial equilibrium in the agricultural product market is depicted by the intersection of the supply and demand curves in Figure 1.

A firm discovers and commercializes innovation $j = 1$. For concreteness we think of this innovation as a genetically modified plant variety that lowers production costs (e.g., Bt cotton). Assume that the cost of producing Bt

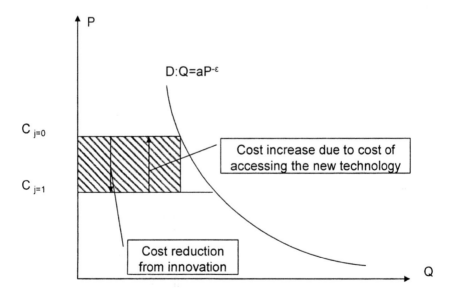

Figure 1. The Effects of Innovation $j=1$ and Monopoly Pricing of the Innovation $j=1$ as Seen in the Agricultural Product Market

seeds is the same as producing other hybrid seeds (i.e., once product development is completed the cost of embodying the Bt in the seed is negligible—
once the gene is in the seed the Bt varieties can be reproduced with the same
techniques used for producing seed of traditional varieties).

We use the agricultural product market to derive the price (technology
fee) that the firm will charge producers to access this innovation. We represent the effects of the Bt technology on agricultural production as lowering
the unit agricultural production costs to $C_1 < C_0$.

The technology fee depends on the level of cost reduction and the structure of the R&D industry. For expositional simplicity, calibrate the technology fee or price of accessing the technology so that it is a price per unit of
agricultural output. That is, the units of the innovative input are scaled so
that a change of one unit of agricultural output results in a change of one unit
of purchased innovative input (e.g., we are talking about a technology fee
per acre rather than per bag of seed). This allows us to interpret the demand
curve for the agricultural output as also being the derived demand curve for
the innovative input. Exercising its market power as the only firm with the
Bt seeds, the innovating firm would like to move up the inelastic portion of
the derived demand curve by restricting the quantity (raising the price) of
available Bt seed. This movement up the demand curve is limited by the
presence of the traditional technology. Thus, the optimal technology fee
(from the innovator's perspective) will be the difference in unit costs of the
Bt technology from the traditional technology: $\tau_1 = C_0 - C_1$. To see this, if the
innovating firm charges more than the difference between unit costs of production with the different technologies, then suppliers of traditional seeds
will undercut it and take away the market. If the innovating firm charges less
than the difference between unit production costs for the two technologies,
then it could increase profits by raising the price. The rents earned by the
firm are thus τ_1 times the quantity of seed sold, depicted by the diagonally
shaded area in Figure 1.

In a traditional, static framework the innovation rents earned by the firm
are considered to be an increase in social surplus (albeit acquired totally by
the firm). These would be the per-period gross benefits from research, to be
compared with the costs of the resources devoted to the R&D effort.

In the dynamic approach, these rents are still considered to be rents, but
there is explicit recognition that they are offset by R&D costs. From an expected present value perspective, the zero-profit condition means that at the
outset of an R&D race, expected future rents are exactly equal to the expected value of industry R&D costs. That is, on average, the profits shown
by the shaded area in Figure 1 are just sufficient to offset the R&D costs plus
interest. Thus, on average, society does not see a positive net present value

of the R&D activity until after the innovating firm's monopoly profits have dissipated.[8]

The innovating firm's monopoly profits dissipate when innovation $j = 2$ is discovered. The effects of innovation 2 are represented in Figure 2. With Bertrand price competition, innovator 2 will charge a technology fee equal to $\tau_2 = C_1 - C_2$ and take all the business away from innovator 1.[9] The rents earned by innovator 2 are equal to the technology fee times the quantity sold. As with the first innovation, these rents are represented by a rectangle between the lines C_1 and C_2, the demand curve, and the axis, and are depicted by the diagonally shaded area in Figure 2.

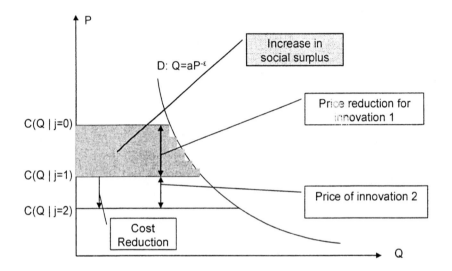

Figure 2. The Effects of Innovation 2 and Limit-Pricing in the Agricultural Product Market

---

[8] Another way to think of this is in the context of an infinitely lived representative consumer. With a non-changing instantaneous utility function, utility in any instant increases only if consumption increases. The question becomes: Does the discovery of innovation j increase consumption? With the technology fee for innovation j set at $\tau_1 = C_0 - C_1$, there is no decrease in the cost of production including technology fee, and no change in the quantity of the agricultural output produced. Thus there is no gain in utility from additional consumption of the agricultural good. Moreover, the zero-profit condition ensures that investment in the industry portfolio yields no excess profits, so there is no additional income with which to buy other consumer products. Hence consumption does not change, and so neither does utility.

[9] For simplicity assume that if the farmer is indifferent between two technologies, he buys the newer one.

The industry structure now becomes very interesting. Innovator 1 is not actually selling any product, but the threat of entry from this innovator prevents innovator 2 from raising the technology fee above the cost differential $C_1 - C_2$, or else innovator 1 would still have a market for the first innovation. Thus, the limit on the size of the technology fee τ_2 arises from the reduction in the price at which innovation 1 is (would be) offered, namely any first-innovation technology fee τ_1 greater than 0 (recall that we have assumed that the cost of producing and distributing Bt seeds is no greater than the cost for any other seed and is separate from the technology fee). Thus, the industry structure is one in which a firm with monopoly power over innovation 2 is limited in its ability to extract rents by the threat of competition from innovator 1.

Analogously to the first innovation, at the industry level the rents earned by innovator 2 are dissipated in the form of lower values for those firms conducting R&D on this innovation but that failed to discover it. Hence there is no change in social welfare represented by the shift from C_1 to C_2 *per se*. However, there is an increase in social welfare due to the change in technology fee for innovation 1. Specifically, innovation 1 is now available at an implicit technology fee of zero. Thus society gains at the time innovation 2 is introduced, this gain being the social surplus from innovation 1 that was previously captured by the first innovator. (Another way of thinking about this is that some of the rents from innovator 2 are captured by the innovator, but limit pricing means that some of the rents are dissipated to society in the form of social surplus.) The change in social surplus is depicted by the fully shaded area in Figure 2, which is equivalent to the diagonally shaded area in Figure 1.[10]

This example shows that the gains to society from going off patent can be very large relative to the gains to society at the time of the innovation. The measurement of actual gains in any specific case is an empirical issue.

5. CONCLUSIONS

This paper develops a dynamic model of innovation and industry structure that is appropriate to the measurement of welfare effects from innovation. The dynamics are introduced via an intertemporal arbitrage equation. An

---

[10] Since the limit on the rents obtained by the innovating firm is expected at the time that firms are making their R&D investment decisions, the zero-profit condition ensures that in the sense of expected value, the rents (under limit pricing) equal the R&D costs. Similarly, it is expected at the time that firms are investing in R&D for the first innovation that eventually innovation 2 will eliminate any rents from innovation 1. This is implicitly incorporated into the zero-profit condition for innovation 1 *via* the risk premium in the arbitrage condition.

additional novel feature of the approach is the use of a zero-profit condition on industry level R&D expenditures by the private sector.

Examination of the model and a graphical example leads to three conclusions:

- The model captures issues of dynamic market structure and industry evolution that may have significant influence on the welfare effects of technological innovation.
- The measurements needed to implement the model are essentially social surplus measurements in non-competitive markets; the key difference is that in the dynamic model these measurements should account for dynamic changes in market structure.
- The social welfare gains from a technological innovation with intellectual property protection may be largest when the innovation no longer enjoys intellectual property protection and society can access the benefits of the innovation at low or marginal cost.

REFERENCES

Aghion, P., and R. Howitt. 1998. *Endogenous Growth Theory*. Cambridge, MA: MIT Press.

Alston, J., G.W. Norton, and P. Pardey. 1995. *Science Under Scarcity*. Ithaca, NY: Cornell University Press.

Alston, J., R. Sexton, and M. Zhang. 1997. "The Effects of Imperfect Competition on the Size and Distribution of Research Benefits." *American Journal of Agricultural Economics* 79(4): 1252–1265.

_____. 1999. "Imperfect Competition, Functional Forms, and the Size and Distribution of Research Benefits." *Agricultural Economics* 21(2): 155–172.

Barro, R., and X. Sala-i-Martin. 1995. *Economic Growth*. New York: McGraw-Hill, Inc.

Davidson, C., and P. Segerstrom. 1998. "R&D Subsidies and Economic Growth." *RAND Journal of Economics* 29(3): 548–577.

de Gorter, H., D. Nielson, and G. Rausser. 1992. "Productive and Predatory Public Policies: Research Expenditures and Producer Subsidies in Agriculture." *American Journal of Agricultural Economics* 74(1): 27–37.

Dinopoulos, E. 1994. "Schumpeterian Growth Theory: An Overview." *Osaka City University Economic Review* 29(1–2): 1–21.

Falck-Zepeda, J.B., G. Traxler, and R.G. Nelson. 2000. "Surplus Distribution from the Introduction of a Biotechnology Innovation." *American Journal of Agricultural Economics* 82(2): 360–369.

Graf, G.D., G.C. Rausser, and A.A. Small. 2003. "Agricultural Biotechnology's Complementary Intellectual Assets." *Review of Economics and Statistics* 85(2): 349–363.

Harris, C.J., and J.S. Vickers. 1985. "Patent Races and the Persistence of Monopoly." *Journal of Industrial Economics* 33(4): 461–481.

Huang, S., and R. Sexton. 1996. "Measuring Returns to an Innovation in an Imperfectly Competitive Market: Application to Mechanical Harvesting of Processing Tomatoes in Taiwan." *American Journal of Agricultural Economics* 78(3): 558–571.

Lesser, W. 2005. "Intellectual Property Rights in a Changing Political Environment: Perspectives on the Types and Administration of Protection." *AgBioForum* 8(1&2): 64–72.

Moschini, G., and H. Lapan. 1997. "Intellectual Property Rights and the Welfare Effects of Agricultural Research." *American Journal of Agricultural Economics* 79(4): 1229–1242.

_____. 2006. "Labeling Regulations and Segregation of First- and Second-Generation GM Products: Innovation Incentives and Welfare Effects." In R.E. Just, J.M. Alston, and and D. Zilberman, eds., *Regulating Agricultural Biotechnology: Economics and Policy.* New York: Springer.

Moschini, G., H. Lapan, and A. Sobolevsky. 1999. "Roundup Ready® Soybeans and Welfare Effects in the Soybean Complex." Staff Paper No. 324, Department of Economics, Iowa State University, Ames.

Naseem, A., J.F. Oehmke, and D.E. Schimmelpfennig. 2005. "Does Plant Variety Intellectual Property Protection Improve Farm Productivity? Evidence from Cotton Varieties." *AgBioForum* 8(2&3): 100–107.

Oehmke, J.F., and C.A. Wolf. 2004. "Why Is Monsanto Leaving Money on the Table? Monopoly Pricing and Technology Valuation Distributions with Heterogeneous Adopters." *Journal of Agricultural and Applied Economics* 36(3): 705–718.

Oehmke, J.F., C.A. Wolf, and K.C. Raper. 2005. "On Cyclical Industry Evolution in Agricultural Biotechnology R&D." *Journal of Agricultural & Food Industrial Organization* 3(2): Article 1. Available online at http://www.bepress.com/jafio/vol3/iss2/art1 (accessed March 2005).

Oehmke, J.F., and X. Yao. 1990. "A Policy Preference Function for Government Intervention in the U.S. Wheat Market." *American Journal of Agricultural Economics* 72(3): 631–640.

Rausser, G., and W. Foster. 1990. "Political Preference Functions and Public Policy Reform." *American Journal of Agricultural Economics* 72(3): 641–652.

Schumpeter, J.A. 1968. *The Theory of Economic Development* (8th printing). Cambridge: Harvard University Press.

Segerstrom, P. 1991. "Innovation, Imitation and Economic Growth." *Journal of Political Economy* 99(4): 807–827.

Segerstrom, P.S. 1998. "Endogenous Growth Without Scale Effects." *The American Economic Review* 88(5): 1290–1310.

Sexton, R.J., and T.A. Sexton. 1996. "Comment: Measuring Research Benefits in an Imperfect Market." *Agricultural Economics* 13(23): 201–204.

Tolley, G.S., J.H. Hodge, W.N. Thurman, and J.F. Oehmke. 1985. "A Framework for Evaluation of R&D Policies." In G.S. Tolley, J.H. Hodge, and J.F. Oehmke, eds., *The Economics of R&D Policy.* New York: Praeger Press.

Voon, T.J. 1994. "Measuring Research Benefits in an Imperfect Market." *Agricultural Economics* 10(21): 89–93.

_____. 1996. "Reply: Measuring Research Benefits in an Imperfect Market." *Agricultural Economics* 13(23): 205.

Part III

CASE STUDIES ON THE ECONOMICS OF REGULATING AGRICULTURAL BIOTECHNOLOGY

Chapter 21

INTERNATIONAL APPROVAL AND LABELING REGULATIONS OF GENETICALLY MODIFIED FOOD IN MAJOR TRADING COUNTRIES

Colin A. Carter[*] and Guillaume P. Gruère[†]

University of California, Davis, [*] *and International Food Policy Research Institute*[†]

Abstract: We review the approval and labeling regulations covering genetically modified (GM) foods in the United States, the European Union (EU), Japan, Canada, Australia, and New Zealand. We divide these countries into three groups according to their regulatory approach. At one extreme, the United States and Canada use pragmatic and science-based regulations, and at the other extreme the EU uses stringent and precautionary regulations. Finally, Japan and Australia/New Zealand have intermediate regulatory approaches. We argue that labeling requirements in importing nations have affected international trade, and that approval regulations are more likely than labeling regulations to be harmonized in the future.

Key words: genetically modified (GM) food, labeling, international trade

1. INTRODUCTION

Agricultural biotechnology has been under increasing regulatory scrutiny. Today, food produced with agricultural biotechnology is typically subject to pre-market safety approval and to labeling regulations. Ironically, as research in crop bioengineering in different countries seemed to be converging towards common goals at the beginning of the 1990s, approval and labeling regulations covering genetically modified (GM) food started to quickly diverge across nations. In fact, the current approval procedures and labeling regulations covering GM foods differ widely across OECD countries, with a large and visible gap between North America and European Union (EU) approaches. The difference between the EU and the U.S. regulatory approaches has become a textbook example of international policy tension and has be-

---

The contents of this paper reflect the views of the authors alone.

come the focus of a number of political and economic studies (e.g., Isaac 2002). At the same time, regulatory developments in many other countries have combined portions of the U.S. and EU approaches, as we explain below.

In this chapter, we analyze international approval and labeling regulations in major food trading countries. We limit the country-by-country description to the United States, the EU, Japan, Canada, Australia, and New Zealand. In the next section we review the general regulatory framework common to these countries and highlight the main differences. In Section 3, we summarize the approval and labeling regulations in each of these countries. Section 4 discusses the effects of labeling regulations on international trade and discusses the prospects for international harmonization of labeling policies. The last section concludes with some implications of GM food labeling for the future of agricultural biotechnology.

2. APPROVAL AND LABELING REGULATIONS OF GM FOOD

2.1. The Regulatory Process from Farm to Fork

The approval of any new GM food follows a regulatory path from the field to the consumer. Figure 1 shows the regulations of the EU (top boxes), the U.S. (bottom boxes), and the intermediary regulations of the other major trading countries. Each regulatory path starts on the left side of the figure, with a company applying for authorization to commercialize the use of a new GM food or feed. At the end of the process (on the right side of Figure 1), the product is eventually sold to the end users.

As part of the safety assessment and food labeling procedures it is important to note that the definition of "novel food" differs significantly across countries. It can be based on the product itself and the specific characteristics of the product, relative to traditional food products (the so-called *product-related standard*). Alternatively, it can be driven by the novelty characteristics of the production technology *(process-related standard)*. In the first case, a new GM food product is considered novel only if the genetic modification enhances or changes the product's chemical composition in significant ways, whereas in the second case, all food derived from GM crops is considered novel.

The *novelty criterion* is important in that it triggers a series of administrative steps required to obtain pre-marketing food safety approval. The authorization to commercialize GM food products is generally obtained by passing a food safety assessment test. In most countries, biotech companies applying for pre-marketing approval must submit a scientific assessment of safety risks. As a second step after pre-marketing approval, in some countries, the application is then submitted to political authorities for approval.

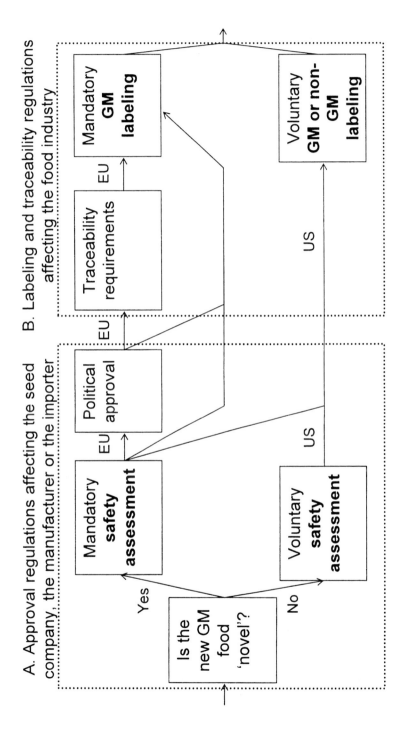

Figure 1. Regulatory Trajectory of Approved GM Food from the Field to the Consumer

The scientific assessment procedures tend to have similar features across developed countries: they follow the general approach defined in the OECD consensus report written in 2000, in collaboration with experts from the World Health Organization (WHO) and the United Nations Food and Agriculture Organization (FAO). Safety assessment is commonly based on the concept of *substantial equivalence*, as defined by the OECD in 1993, and endorsed by the FAO/WHO joint consultation in 1996. The establishment of substantial equivalence for any GM food is not an absolute safety assessment, but it guarantees that the food is no less safe than traditional food under similar consumption patterns and processing procedures. The general steps are the following:[1] (i) description of host organism, (ii) description of donor organism, (iii) molecular characteristics of genetic modification, (iv) identification of primary and secondary gene products, (v) evaluation of safety of expected novel substances (toxins) in the food, (vi) assessment of novel food potential allergenicity, and (vii) evaluation of unintended effects on food composition (change of the concentration of nutrients or natural toxins, identification of anti-nutrients altered, and evaluation of safety of compounds that show a change in concentration).

In addition, food processing and consumption patterns are accounted for at every step of the process. The applicant has to identify the likely consumers and evaluate the usual amount consumed, and must assess the effects of consuming the food at different intake levels. At the end of the process, there are three possible outcomes: (i) the GM food is considered substantially equivalent to traditional food, (ii) the GM food is substantially equivalent to the traditional counterpart except for defined differences (which will require more study), and (iii) the GM food is not substantially equivalent to the traditional food (which then leads to a reassessment with other methods, leading to an absolute risk assessment). Until 2004, all commercial varieties of GM crops were designed to produce agronomic input traits, and therefore there were no major differences in the final food product; thus all of the approved varieties were routinely considered substantially equivalent in most countries.

Although scientific risk assessments are similar across countries in many ways, the risk management steps differ in important ways. In some countries, all GM food varieties have to be approved by different bodies of the legislation; in some others the system does not require legislative approval.

Once the approval procedure is satisfactorily completed, the GM crop or ingredient may be sold to food companies. The seed companies do not face any additional food safety requirements. However, in many developed countries, food companies (i.e., food processors and retailers) have to separate and label those products that contain or are derived from GM ingredi-

---

[1] Source: http://www.agbios.com/ (accessed December 2004).

ents. In countries with traceability requirements, the purchaser of any GM crop must track, separate, and conduct analyses at several points of the production chain to guarantee its integrity of their GM or non-GM product.[2] In these countries, food companies are subject to inspections from public agencies, and are required to demonstrate their ability to establish the origin of GM or non-GM ingredients in any final product.

2.2. Divergence in Approaches

Figure 1 also shows the major cross-country divergent regulatory approaches. The EU regulations follow the regulatory path linking all of the boxes on the top of Figure 1. In contrast, the U.S. approach (for substantially equivalent products) follows the path that goes through the bottom boxes. All other countries that we consider follow an intermediate path connecting both top and/or bottom boxes. We will now discuss rules that are "product-based" versus "process-based," and rules that are "voluntary" versus "mandatory."

The first regulatory dichotomy depends on whether the rules are "product-based" or "process-based." The United States bases its regulations on end products, and thus substantially equivalent GM food is subject to the same basic regulations as other foods. Canada's regulations are also product-based, but Canada also treats all GM food as "novel." It therefore requires GM foods to pass a pre-marketing safety assessment, but then approved GM food is considered conventional and subject to the same laws as traditional foods. Most other countries base their regulations on the production process and regard biotech food as novel, requiring a specific set of approval and labeling regulations.

The regulatory framework in Canada and the United States respects the scientific consensus that was in place at the beginning of the 1990s, which states that foods produced with biotechnology do not necessarily present any new or greater risk than traditional foods. This approach is based on the concept of *substantial equivalence*. The safety reviews are for the purposes of determining that the GM food is equivalent to traditional foods in terms of risk.

On the other hand, the EU has built its safety system under the premise that GM food may present specific risks and thus should be treated separately from conventional foods. The EU approval and labeling procedures for GM food are based on the *precautionary* principle. This approach postulates that biotech food may contain unknown risks, and EU countries believe they should take appropriate measures to limit the development of future

---

[2] For instance, the EU requires traceability for products consisting of or containing GMOs to keep track of the origin and destination of GM shipments pursuant to Regulation (EC) No. 1830/2003.

unknown risks to human health and the environment. The Cartagena Protocol on Biosafety, which counts more than 100 ratifying countries, is also based on process-based standards (it aims to regulate the movement of living modified organisms, which include GM crops, GM seeds, and GM food) and the precautionary principle.

The second major regulatory division among countries relates to whether or not the regulations are voluntary or mandatory. To date, voluntary labeling and mandatory labeling have had opposite effects in the food industry, the choice of technology, consumer choice, and consumer information. Mandatory labeling requires food processors and retailers to separate their ingredients, which can be costly, especially for small or less integrated firms. Labeling regulations also affect the perceived quality of the final product to consumers. Current food products derived from biotechnology are not providing any visible benefits to consumers, and thus labeling products as "GM" is equivalent to posting a negative sticker or a hazard warning on the product, whereas labeling a product as "non-GM" confers a new quality attribute.

In theory, labeling requirements should let consumers decide whether or not to purchase GM food products according to their own preferences. However, because of processor and retailer refusal to buy GM ingredients for fear of losing market share and reputation, mandatory labeling in rich countries has acted as a *de facto* market ban (Carter and Gruère 2003). Thus labeling regulations may be the key to market access. These regulations have effectively prevented approved GM food products, recognized as safe by the public authorities, from accessing the market in countries where there is a share of consumers willing to buy GM products (Noussair, Robin, and Ruffieux 2004).

In contrast, voluntary versus mandatory approval processes have not resulted in different outcomes. The United States has a voluntary consultation process for safety assessment, but to date all GM food has completed the consultation process.

3. THREE DISTINCT REGULATORY APPROACHES

Approval and labeling procedures are heterogeneous at the international level, as shown in Table 1. The five columns in Table 1 correspond to general characteristics of the GM food approval and labeling regulations; the first two columns apply to risk regulations and the three right-most columns apply to the nature of the labeling regulations. These five characteristics are sufficient to categorize each country's regulations, and to help us classify our group of countries according to their regulatory approaches.

These regulations result in different outcomes in terms of varieties approved and GM food availability at the consumer level. For instance,

Table 1. GM Approval and Labeling Regulations in Major Trading Countries

	Mandatory or voluntary safety assessment (M/V)	Political approval required	Mandatory or voluntary labeling (M/V)	Labeling based on process (P) or end product (E)	Labeling tolerance level
Australia/New Zealand	M	Y	M	E	1%
Brazil	M	Y	M	P	1%
Canada	M	N	V	E	5%
China	M	Y	M	E	0%
EU	M	Y	M	P	0.9%
Japan	M	Y	M&V	E	5%
Korea	M	Y	M	E	3%
Russia	M	Y	M	P	1%
United States	V	N	V	E	5%

Source: National regulations, U.S. Department of Agriculture's Foreign Agricultural Service attaché reports.

Table 2 shows the number of varieties approved for food or feed in a number of major trading countries as of 2004. This table confirms our division of major trading countries: the United States and Canada have each approved over 60 varieties, whereas the EU has only approved a total of 14 varieties, and Japan and Australia have approved 34 and 24 varieties, respectively. We will now describe the regulatory systems of these three regions.

Table 2. Number of Commercial GM Varieties Approved for Food, Feed, and/or Marketing Use as of 2004

	Food only	Feed only	Food + Feed	Marketing[†]	Total
Australia	10		14		24
Brazil			1		1
Canada	15	2	46		63
China			1	5	6
EU	1	1	4*	8	14
Japan	2	4	28**		34
Korea	3				3
Russia	1***				1
United States	3	2	56		61

Source: AGBIOS database, http://www.agbios.com/ (accessed January 2005).

[†] Products listed as approved for marketing are those which have been approved for placing on the market.
* Including three varieties approved for food, feed, and marketing.
** Including one variety approved for food, feed, and marketing.
*** Variety approved for food and marketing.

3.1. The United States and Canada

The U.S./Canada approach to biotech risk management is product-related, and labeling is voluntary. Both these countries have adopted centralized federal policies regarding GM food approval and labeling.

Regulations of agricultural biotechnology in the United States were first introduced in 1986 under the "Coordinated Framework for the Regulation of Biotechnology." This framework—still in use today—is based on four principles: (i) existing laws are adequate, (ii) regulations should be based on product not process, (iii) GMOs are not fundamentally different from conventional crops, and (iv) oversight authority is exercised only when there is evidence of risk that would make it unreasonable to introduce the product.[3] The Coordinated Framework also outlines regulatory oversight of GM

---

[3] Source: http://www.agbios.com/ (accessed December 2004).

plants, dividing the responsibilities between governmental agencies. The U.S. Food and Drug Administration (FDA) is in charge of GM foods, food additives, processing aids, and biotech medical products under the Federal Food Drugs and Cosmetics Act (FFDCA). The FDA is also responsible for food labeling, with the exception of meat and poultry products.[4]

The FDA's approval procedure is based on consultations with seed producers and food companies. In 1992, the FDA published a "Statement of Policy on GM food," including a risk-based decision tree, under the assumption that GM food is no more risky than its traditional counterpart. In 1994, in accordance with its 1992 interpretation of the FFDCA, the FDA introduced a simplified approval procedure for GM foods. Under this procedure, applications for approval of GM foods are not subject to comprehensive scientific review because there is no evidence that they pose significant risks. GM foods are "generally recognized as safe" (GRAS) unless, in the judgment of the manufacturer, there is a reason for concern (e.g., allergy problems). Foods considered GRAS are not subject to pre-market risk assessment by the FDA. In 1997, the FDA published guidelines for the voluntary consultation process. The FDA registers and publishes the consultation dates. Although the consultation process is voluntary, seed companies have subjected all GM foods and feeds marketed in the United States to the consultation process since its introduction in 1994.[5]

In May 2000, the U.S. government announced a new initiative, which included a plan for FDA to implement a mandatory safety assessment system, requiring firms to submit their product for official approval 120 days before the introduction of a new GM food. But the FDA dropped this plan in 2003 (Miller and Conko 2004, p. 55). The 2000 initiative also aimed at producing voluntary labeling guidelines for GM and non-GM food. Under the 1992 FDA statement, labeling is not required for GM food "substantially equivalent" to the corresponding conventional food in terms of composition, nutrition, and safety. But following the 2000 initiative, in January 2001, the FDA issued "guidelines" for voluntary labeling of GM/non-GM foods.

Interestingly, the U.S. food and biotechnology industry is actively lobbying for mandatory safety assessment,[6] but is definitely opposed to a mandatory labeling system (Grocery Manufacturers of America 2001a, 2003; Muth, Mancini, and Viator 2003). U.S. consumers trust the FDA; there-

---

[4] Meat and poultry products are under the responsibility of the U.S. Department of Agriculture.

[5] Miller and Conko (2004) explain the incentives for seed companies to submit their products to the consultation process: first, they benefit from an implicit seal of approval from the FDA, and second, they also prefer to reduce the risk of future litigation.

[6] See Miller and Conko 2004 (p. 54), Grocery Manufacturers of America 2001b, and a letter by Carl B. Feldbaum, President of the Biotechnology Industry Organization, to President Bush on January 15, 2004, available at http://www.bio.org/news/features/2004Letter_to_PresBush. pdf (accessed March 2005).

fore it may be more acceptable for consumers to have a mandatory safety assessment consistent with a voluntary labeling system. This idea is consistent with the principle that labeling is not required for any purpose other than providing information related to food safety, and there is no risk associated with approved GM food. But the FDA approach is based on science, and there is no scientific evidence of inherent risk with GM crops, so there is no need to single out the approval process for GM foods that are substantially equivalent to other foods.

The Canadian approach is similar to that of the United States in principle, but differs in practice. It is based on an agreement between the Canadian federal agencies in 1993, which was renewed in 1998. Three agencies are involved: the Canadian Food Inspection Agency (CFIA), Health Canada (HC), and Environment Canada. For food regulations, HC is responsible for assessing the human health safety aspects of food derived through biotechnology. The CFIA shares responsibility for the regulation of products derived from biotechnology including plants, animal feeds and animal feed ingredients.

Even if the regulatory approach is based on product rather than process, any GM food is considered a novel food by definition, as a "product modified by genetic manipulation." HC is in charge of food safety assessments for all regulated products, including novel food products under the Novel Food Regulation of October 27, 1999. A manufacturer or importer must notify HC forty-five days prior to first marketing a novel food. HC provides a response within forty-five days or, if more data are required, within ninety days. The safety assessment follows 1994 HC guidelines (under review in 2003), which are based on the joint OECD/FAO/WHO concept of substantial equivalence.

Like the FDA, HC has no labeling mandate for substantially equivalent food. Canada adopted a voluntary labeling system in April 2004. The regulation defines which foods are eligible for a GM or non-GM label. The system will be enforced by the CFIA, but is based on *understandable, truthful and not misleading* labeling. The CFIA conducts inspections only in cases of civil claims of mislabeling.

3.2. The European Approach

In contrast to North American regulations, the EU approach is precautionary, process-related, and includes mandatory labeling. Most non-EU countries in Europe (e.g., the Eastern European nations) are adopting regulations similar to those in the EU. And Switzerland is changing its threshold level to become compatible with EU standards. Therefore we will limit our discussion in this section to EU regulations.

The EU regulatory system for GM foods has become increasingly more stringent. In 1990, the European Council adopted Directive 90/220 on the deliberate release of GM organisms into the environment. The directive regulated approval of GM crops for field trials and cultivation, and it also governed the approval of GM food. This first regulation did not define any specific approval procedures or labeling regulations. In 1997, the EU Parliament and the EU Council adopted Regulation 258/97, entitled the Novel Foods Regulation. This regulation applied to new food products including GM foods, and it defined approval procedures requiring proof that any GM food is safe for human consumption. Later, the EU commission and the Council published Regulations 1813/97 and 1139/98, which required the labeling of food products containing approved GM soybeans and GM corn. These regulations were augmented by Regulation 49/2000, introducing mandatory labeling of GM food and GM ingredients at the one percent level, and Regulation 50/2000, extending the labeling requirements to food ingredients containing GM additives and flavorings.

The EU's most recent laws on GM food authorization (Regulation 1829/2003 and Regulation 1830/2003) took effect on April 18, 2004. These regulations established procedures for evaluating potential risks from GM food, and laid down rules on labeling of GM food and feed. Approvals are now granted for a period of 10 years, renewable. There is a zero percent threshold for unapproved GM crops.[7] Labeling is extended to animal feed, food sold by caterers, and food derived from GM ingredients, even if the end product has no significant traces of transgenic DNA or proteins. The threshold for labeling is 0.9 percent. One major addition is the traceability requirements for GM and non-GM food: any food potentially containing GM material has to be tracked all the way from the farm to the consumer. This requires food companies to keep track of all shipments and to conduct DNA or protein tests at different stages. There is no labeling requirement for products such as meat, milk, or eggs produced from animals fed with GM feed.

Under the new EU authorization system, as described in Commission Regulation 641/2004 of April 6, 2004, a company that intends to market a GM food product in the EU must follow four successive steps (Reuters Factbox 2004). First, the company must apply to the relevant authority of the EU member state where the product is first to be marketed, and provide a full risk assessment, a monitoring plan, a labeling proposal, and a detection method. Second, if the authority gives a favorable opinion, the member state informs other member states via the European Commission. Third, if there are no objections by other member states, the notifying state or its national food safety authority may authorize the product for marketing throughout

---

[7] Clearly, a zero percent threshold could lead to problems—say, if unapproved GM canola is found in a cargo of grain shipped from North America to the EU.

the EU. Fourth, if there are no objections, a decision is required at the EU level and the following procedure is initiated. The Commission asks the independent European Food Safety Authority (EFSA) for an opinion based on a risk assessment procedure. The EFSA must give an opinion within 6 months. If the opinion is favorable, the Commission submits a draft decision to the Standing Committee on the Food Chain and Animal Health, made up of scientific experts from the member states. If the committee approves the authorization, the Commission adopts the decision and authorizes the new GM food product. If the committee does not agree, the Commission sends its draft approval to the Council of Ministers (agricultural or environmental ministers), who have three months to reject or adopt it. If they do not act within this time, the Commission may adopt its own decision and authorize the new GM food product.

Globally, the EU has the most comprehensive regulations on GM food. The new labeling and traceability regulations were introduced to force member states to end the *de facto* 4-year moratorium on new GM crops and to respond to the pressure imposed by the United States and other countries when they launched a WTO dispute on the moratorium. Since then, the EU has approved a few new GM varieties, and it is now the labeling regulations that have become the new *de facto* trade barrier. Although labeling was introduced to provide consumer choice, the mandatory labeling system has encouraged all EU food processors and retailers to avoid GM ingredients entirely.

According to surveys and polls, a majority of EU consumers claim that they do not want to eat GM food, but at the same time some empirical studies (Noussair, Robin, and Ruffieux 2004) have shown that a positive share of consumers would be willing to purchase GM products in the EU. Carter and Gruère (2003) argued that the EU mandatory labeling system has acted as a majority voting system where the winner takes all.

Currently, it is almost impossible to find GM food products in the EU. Of course GM animal feed is available because animal products do not need to be labeled. It is unlikely that the positions of the retailers and food processors will change, unless there is an abrupt shift in consumer acceptance. If some new GM products are developed that provide significant consumer benefits, or if a new GM product offers significant cost savings, then the EU retailers and processors might budge from their current position.

3.3. The Japanese and Australian/New Zealand Approaches

Japan's regulations and the common regulations of Australia and New Zealand (ANZ) have similar approval and labeling procedures, including mandatory safety assessment and mandatory labeling based on differences in products and with a number of exemptions. Labeling is based on the end products.

In Australia and New Zealand, Food Standards Australia New Zealand (FSANZ) is responsible for the safety and commercialization of GM foods. The regulation is defined in Food Standard 1.5.2, which was incorporated into the Food Standards Code on May 13, 1999, and in an amended form on July 12, 2000. The standard has two provisions: the first component addresses the mandatory safety assessment requirement and the second the mandatory labeling requirement.

Approval is based on a scientific assessment by FSANZ and by the formal approval of the ANZ Food Regulation Ministerial Council. The FSANZ has 12 months to make recommendations to the Ministerial Council. Applicants must submit a number of documents and test results, which follow a procedure similar to that described in Section 2.1 above. The safety approval procedure applies to GM additive and processing aids. Applicants can pay to accelerate the approval procedure. More importantly, applicants have to pay an application fee if FSANZ determines that the authorization of their products would result in an "*exclusive capturable commercial benefit.*"

On July 28, 2000, the Health Ministers of the ANZ Food Standards Council resolved to amend the labeling provision of the Food Standards Code to include GM food where (i) novel DNA and/or protein is present in the final food, or (ii) the food has altered characteristics. Interestingly, the second option triggers the labeling requirement without considering the presence of novel DNA or protein in the final product. Moreover, the legal definition of "altered characteristics" includes foods with "significant ethical, cultural and religious concerns" (Division 2, Clause 7e, of Food Standard 1.5.2), which could in theory be applied to any GM food in a country where consumers are largely opposed to agricultural biotechnology. At the same time, exemptions to the labeling requirements are granted to highly refined food, GM processing aids, food additives without transgenic proteins or DNA, GM flavors that represent less than 0.1 percent of the final food, food sold in restaurants or by caterers, and ingredients that are less than one percent GM.

The FSANZ system has granted approval to 24 types of GM food in Australia as of 2004 (see Table 2). At the same time, the mandatory labeling system has provided an incentive for processors and retailers to avoid using GM ingredients. It is hard to find GM-labeled products in Australia's retail shops (Foster, Berry, and Hogan 2003). However, approved GM varieties of food exempted from labeling (e.g., soy oil) are common in the food chain.

In 2000, Japan introduced regulations defining the authorization procedure. The Ministry of Health Labor and Welfare (MHLW) is in charge of the approval procedure for GM food. All GM food, GM processing aids, and GM food additives are subject to pre-marketing safety assessment. The safety assessment includes information regarding the host, the vector, the inserted gene, the recombinants, and the toxicity levels. If the application to MHLW is complete, it is then submitted to the Expert Panel of the Biotechnology Subcommittee within the Food Sanitation Committee. The Panel reviews and makes recommendations to the Biotechnology Subcommittee, which then passes its judgment on to the Food Sanitation Committee. This committee makes a recommendation to MHLW's minister, and if approved the new variety is announced in the Japanese Gazette. It usually takes about one year to go through the regulatory process.

The MHLW enforces standards under the Food Sanitation Law (FSL), and it samples and tests imported foodstuffs at ports of entry. The testing focuses on GM foods approved abroad but not in Japan. There is a zero percent tolerance for unapproved GM material. After the Starlink corn food scare, Japan increased the frequency of food safety inspections on corn from 5 percent to 50 percent of all cargoes.

The Ministry of Agriculture, Fisheries and Forestry (MAFF) is responsible for environmental safety approval, feed safety assessment, and biotech labeling rules. The MAFF's environmental assessment is voluntary but all companies comply. The MAFF's feed safety assessment is mandatory, from April 1, 2003. All applications for feed approval are reviewed by the Feed Division of MAFF, and then sent to the Expert Committee of the Agricultural Materials Council. There is a one percent tolerance level for the unintentional presence of GM feed that has been approved in other countries, under the condition that the exporting country's safety assessments are deemed equivalent to Japan's.

Japan's mandatory labeling scheme was introduced on April 1, 2001, under the Law on Standardization and Proper Labeling of Agricultural and Forestry Products, which was introduced into the Japanese Agricultural Standards (JAS). Labeling is required for all GM food if DNA/protein can be detected in the finished food products and if the GM ingredient is one of the top three ingredients and accounts for more than 5 percent of the total weight.[8] The MAFF list of products subject to mandatory labeling included 30 foods in 2003. Importantly, there are no labeling requirements for soy oil or corn oil, except if the oil has special properties (such as high oleic soy oil). The labeling regulations are enforced jointly by MAFF and MHLW under the JAS and the FSL, respectively. In addition to the mandatory GM labeling requirements, there is a voluntary labeling option for non-GM, subject to identity preservation procedures.

Overall, the Japanese policy can be described as pragmatic, in the sense that it requires the labeling of GM food but the regulations do not cover all products and the tolerance levels are reasonable. Food processors and retail-

---

[8] This 5 percent tolerance level is informal but currently applied.

ers in Japan have typically avoided products that require GM labels. As in the EU, most GM products are used for animal feed, but unlike in the EU, many highly processed products derived from GM ingredients (e.g., soy oil) are sold without labels.

4. INTERNATIONAL TRADE AND REGULATORY HARMONIZATION

Labeling and approval regulations in major developed countries have affected international trade in all GM crops except cotton (because most cotton products are not edible and thus not subject to food safety and labeling requirements). In particular, regulations in the EU and Japan—two large importers of these crops—have impacted trade. Furthermore, the choice of regulations in the EU and Japan has discouraged the development of new GM food varieties and at the same time encouraged third countries to adopt labeling requirements similar to those in the EU.

The EU regulations managed to halt almost all corn imports from the United States during the *de facto* moratorium (between 1998 and 2004). U.S. exports of corn byproducts to the EU also fell. But at the same time—even during the moratorium—the EU still imported a relatively large volume of GM soybeans for feed, mainly because meat from animals fed with GM crops is exempt from EU labeling requirements. Since the bovine spongiform encephalopathy (BSE) crisis, EU farmers are required to use only vegetable feed, and the EU is incapable of producing enough soybeans for its animals. Thus EU farmers import GM soybeans from Brazil, the United States, and Argentina.

In the EU, food processors have switched to non-GM ingredients and so farmers have been discouraged from adopting potentially beneficial GM crops. For instance, Dewar et al. (2003) showed that the use of GM sugar beets would provide significant yield and environmental benefits, but no EU farmer will adopt GM sugar beets because nobody would buy his or her product. Spain is the only EU country with significant GM crop production. Cohen and Paarlberg (2002) argued that many developing countries have delayed or rejected crop biotechnology because of the stringent regulations in the EU.

In contrast, Japanese regulations have not had a dramatic impact on imports of soybeans, corn, and. canola because of pragmatic regulations and exemptions to the labeling requirements granted to soy and canola oil and other processed products. Although there are no official data, up to 20 percent of Japan's annual soybean imports (of 5 mmt) may be non-GM. The non-GM soybeans are mainly used for tofu, which unlike soybean oil, is subject to Japan's GM labeling regulations. In addition, Japan imports non-GM corn each year from the United States and China. This corn is used in food products (such as snack foods) that are subject to the GM labeling rules. Most of Japan's imported corn is GM, which is used for animal feed, and the final meat product does not have to be labeled there.

It is difficult to find GM-labeled products at the retail level in Japan, but many products labeled as non-GM are available to consumers. The selective mandatory labeling regulations have acted as an intermediate set of rules— between voluntary labeling of non-GM (like in Canada) and a mandatory labeling of GM (like in the EU). At the same time, a substantial amount of food eaten in Japan contains GM but doesn't have to be labeled. These products include cheese, soy sauce, some baked goods, and numerous manufactured foods.

The Japanese labeling requirements may have influenced other Asian countries in their choice of GM labeling regulations. Thailand and South Korea have similar requirements to Japan's; Vietnam, Indonesia, and Malaysia are considering similar labeling policies.

At the international level, two main organizations have worked towards a harmonized regulatory approach on agricultural biotechnology. First, the Codex Alimentarius has been working on finding a common terminology, a common food safety approval procedure, and a common position on labeling, since the beginning of the 1990s. The Codex has failed thus far to reach any agreement on the labeling question. Second, the United Nations Cartagena Protocol on Biosafety was created in an effort to conserve biodiversity. It was ratified in September 2003. The Protocol will allow importers of GM crops to request information regarding GM content and GM varieties and if necessary to ban imports of GM crops as a precautionary measure. Kalaitzandonakes (2004) shows that these regulations could impose a substantial cost on exporters and importers of the main GM and non-GM crops. The Protocol clearly uses a precautionary approach, and was supported by the EU as a necessary procedure to protect human health and the environment from GM crops with unknown risks.

In addition, the United States, Argentina, and Canada have filed a WTO dispute over the EU moratorium on new GM crops, in place from 1998 to 2004. This dispute is important in the sense that it will provide a precedent regarding GM food and standards based on process and production methods.

To sum up, the harmonization effort is led by the two transatlantic powers through the Cartagena Protocol on one side—supported by the EU—and the WTO dispute on the other, which was launched by the United States. Discussions at the Codex may balance the two powers, but it is doubtful that it will generate a consensus. In the meantime, regional factors have pushed the globe towards local harmonization of labeling and approval procedures in Asia, Europe, and North America.

5. IMPLICATIONS OF GM FOOD LABELING FOR THE FUTURE OF PLANT BIOTECHNOLOGY

The present set of international regulations covering GM crops and associated production patterns is unstable. Will more GM crops be planted and cover an increasing area, while at the same time more stringent approval and labeling regulations are being introduced in many countries? Will political considerations continue to discourage the spread of a technology that has the potential to vastly improve the plight of farmers in the developing world? There are two main factors that will determine whether biotech crops will rise or decline in the next few years. The first issue is the stability of the transatlantic regulations. The second relates to adoption decisions and regulatory changes in Asia. We will address each of these two points below.

5.1. Sustaining the Transatlantic Status Quo

Several authors (e.g., Tothova and Oehmke 2004) have argued that the international landscape on GM food is bipolar—the United States and the EU at the two extremes. At the same time, most of these studies assume that the EU and U.S. regulations are rigid. Can we rule out meaningful changes in either the EU and U.S. regulations, or both?

On the EU side, regulations will not be relaxed in the short term, for two reasons. First, the European approach is based on the *precautionary principle*, which means that any deregulation would be acceptable only if there are a significant number of research reports by trusted institutions showing that there are no risks associated with crop biotechnology. This is highly unlikely to happen, because unknown risks cannot be quantified, by definition, and most European scientific institutions are not trusted in matters of food safety.

Second, in the EU, opposition to GM food is strong in many member countries. The current regulations are deemed necessary by a majority of voters, food producers, farmers, ecological groups, and politicians, so there are no obvious benefits from deregulation. Thus, deregulation may come about only from external pressure, and a WTO dispute is probably the only way to put foreign pressure on the EU in this area. However, even if there was a WTO ruling against the EU's mandatory labeling regime, it is unlikely that this would result in deregulation, in view of the outcome of the U.S.–EU hormone-beef dispute.

The EU labeling regulations may become more stringent with extensions to now exempted products (such as meat or processing aids) in the next few years. In particular, there is increasing pressure from green political parties and non-governmental organizations to extend the labeling requirements to include meat, milk, and eggs produced with GM grains and oilseeds. The inclusion of these animal products in the EU labeling regulations would have a much larger effect on global commodity markets than current policies do. If meat labeling became mandatory, EU ranchers would have to choose whether to sell their product labeled as GM or switch to non-GM feed. In the current market, the largest retailers have enforced bans of GM-labeled products, so there would be no meat from animals fed with GM sold in supermarkets. If EU consumers continue to be opposed to GM, if the media continues to publicize labeling exemptions, and if the green pressure groups continue to influence political parties, this labeling extension will remain a distinct possibility. Any move in this direction would be viewed by the United States as an effort by the EU to erect another non-tariff trade barrier.

In the United States, there is some pressure for tighter GM regulations. In November 1999, forty-nine members of Congress sent a letter to the FDA requesting mandatory labeling of GM food. Two bills on the labeling of GM food were introduced in the U.S. Senate and House in 1999 and 2000 (Bernauer 2003, p. 58). In May 2002, Congressman Dennis Kucinich introduced a proposed bill for mandatory labeling of GM food (H.R. 4814).[9] At the state level, Oregon had a referendum in 2002 (Measure 27) on the mandatory labeling of GM food. It was rejected by 73 percent of the voters. Two other ballot initiatives were proposed in Florida and Washington, although neither attracted a sufficient number of signatures to be put on the ballot. Furthermore, twelve other states introduced a total of thirty-two propositions or bills related to GM food labeling requirements between 2001 and 2003 (Pew Initiative on Food and Biotechnology 2003). In April 2004, the State of Vermont passed a law (entitled Act 97) requiring the labeling of all GM seeds.

Thus far, all serious attempts to introduce mandatory labeling in the United States have failed. The media frequently reports that most Americans would support mandatory labeling requirements for GM food.[10] In this context, it is reasonable to expect that one or more states may pass a law or approve a new ballot initiative on mandatory GM labeling in the next few years.

Moreover, things may change quickly in Canada, with the recent push for mandatory labeling in Quebec. If this province were to adopt its own labeling requirements, given there is considerable opposition towards GM food there, large food processors may choose to purchase only non-GM ingredients in order to avoid any risk of losing market share. Moreover, the processors may prefer to harmonize their processing chains by opting for non-GM ingredients for all products sold in Canada, if not the entire North

---

[9] It was referred to the House subcommittee on Farm Commodities and Risk Management on June 4, 2002.

[10] For instance, a recent poll by the Pew Initiative on Food and Biotechnology showed that 92 percent of respondents support mandatory labeling of GM food or GM ingredients (see Pew Initiative on Food and Biotechnology 2004).

American continent. Although the adoption of provincial regulations will not force the federal authorities to follow, the market will speak, and companies may start to avoid the purchasing of GM ingredients.

5.2. The Importance of Asia

According to Feffer (2004), Asia holds the key to the future of plant biotechnology, and we agree. Indeed, several factors suggest that the future of plant biotechnology will depend primarily on developments in Asian countries. First, Asia has a very large population base; second, this region has dynamic agricultural sectors; third, most Asian countries have strong research capacities in plant biotechnology; and fourth, the adoption of Bt cotton has generated large benefits to small-scale farmers in China and India (Pray et al. 2002, Bennett et al. 2004), which may encourage governments to push towards the adoption of other GM crops.

China will have the principal role. Since the end of the 1990s, the Chinese government has invested heavily in research on plant biotechnology (Huang et al. 2002). Notwithstanding the fact that it has banned the production of GM soybeans and corn, the Chinese government is considering the commercialization of several varieties of GM rice in 2005 or 2006 (Jia 2004). China has over 20 percent of the world's rice area, and at the present time China's rice farmers spend about US$150 per hectare for chemical control of insects and diseases. Despite this significant expenditure on chemical control, China still loses up to 10 percent of its potential rice crop each year due to insects and diseases. So China will forgo a significant payoff if it does not introduce GM rice, and it will have first-mover advantage. China has the proven GM rice technology ready to be released, and it is now simply a political decision. Furthermore, China also has a strong research program in GM wheat and corn, and commercialization of these GM crops will likely follow GM rice.

China has also implemented a very stringent mandatory labeling policy for GM food, although it has not been enforced in all provinces. Unlike Japan, China requires the labeling of soy oil produced from GM soybeans. But unlike in developed countries, many GM products are sold to consumers in food markets and may or may not be labeled. In all likelihood, GM rice will also be labeled as such in China as soon as it is commercialized.

The adoption of GM rice by China would change the world's biotech political landscape dramatically, as it would encourage India and other rice producers and exporters to follow China's lead.[11] The global rice market will be impacted once China realizes a yield boost and enjoys reduced costs. We

---

[11] Following reports that China is close to commercializing GM rice, Australia has announced plans to begin experimental crop trials with GM rice.

do note that the EU is not a large consumer of rice, and Japan is importing rice only because it was forced to as part of the Uruguay Round of the GATT agreement. So China's introduction of GM rice will not disrupt the world market for rice, except to lower the marginal cost of production.

Because GM rice seems to promise significant producer and environmental benefits, and because the introduction of GM rice would not seriously jeopardize its trade interests, it is likely that China will commercialize GM rice soon. If China introduces GM rice successfully and if labeling continues to have no significant effect on consumers there, crop biotechnology will continue to spread rapidly in China and throughout Asia.

In both Japan and South Korea[12] consumer surveys show wide opposition to GM food and GM crops (McCluskey et al. 2003, Finke and Kim 2003). This is somewhat ironic, because they now consume a relatively large amount of GM products that do not have to be labeled. If for whatever reason Japan or Korea decides to extend their labeling regulations to include soy oil, this would have an important effect on trade.

Other Asian countries lie between China and Japan in terms of their attitudes towards GM crop production and regulations. As Asian countries grow larger and richer, their choice of technology will be crucial for plant biotechnology. So these countries are the center stage of a soft war between EU and U.S. agricultural interests. On the one hand, EU regulations implicitly act as a threat to countries adopting GM crops: any country choosing to do so may have trouble exporting to the EU and other European countries. For example, the recent announcement of the introduction of GM papaya in Thailand provoked an immediate reaction from food companies importing papayas in Germany and France, and they declared that they would not purchase any more papayas grown in Thailand. On the other hand, the United States is implicitly pushing for the adoption and approval of new GM crops in Asia.

To sum up, we have argued that the EU and the United States will either keep their current regulations in place or adopt stricter labeling regulations in the near future. At the same time, Asia is at a crossroads, where it can either embrace or reject plant biotechnology. Other countries also play a role, but they will not affect the future of crop biotechnology as much as these three regions.

In addition, we made a case that labeling regulations will play an important role in the future of GM food, and labeling rules will not likely be harmonized across large developed countries in the near future. In contrast, ap-

---

[12] South Korea has mandatory labeling regulations for GM food that are similar to Japan's, although the tolerance level in Korea is 3 percent and it applies to the top five ingredients. In Korea, products such as soybean oil, corn oil, soy sauce, and corn syrup are exempt from GM labeling requirements.

proval procedures for new GM crops may be partially harmonized, but agreement on these rules will not affect trade as much as labeling will.

Our chapter underscores the fact that public and private stakeholders should move forward to design new crops targeted either at consumers in developed countries (i.e., changing the image of GM food) or targeted at developing countries (i.e., focusing on other markets). Given the intransigence of rich consumers, we believe that the next surge in crop biotechnology uptake will originate in the developing world.

REFERENCES

Bennett, R.M., Y. Ismael, U. Kambhampati, and S. Morse. 2004. "Economic Impact of Genetically Modified Cotton in India." *AgBioForum* 7(3): 96–100.

Bernauer, T. 2003. *Genes, Trade, and Regulation: The Seeds of Conflict in Food Biotechnology*. Princeton, NJ: Princeton University Press.

Carter, C.A., and G.P. Gruère. 2003. "Mandatory Labeling of Genetically Modified Foods: Does It Really Provide Consumer Choice?" *AgBioForum* 6(1–2): 65–67.

Cohen, J.I., and R. Paarlberg. 2002. "Explaining Restricted Approval and Availability of GM Crops in Developing Countries." *AgBiotechNet* 4: Article ABN 097.

Dewar, A.M., M. May, I. Woiwod, L. Haylock, G. Champion, B.H. Garner, R.J.N. Sands, A. Qi, and J. Pidgeon. 2003. "A Novel Approach to the Use of Genetically Modified Herbicide Tolerant Crops for Environmental Benefit." Proceedings of the Royal Society, *Biological Sciences* (B) 270(1513): 335–340.

Feffer, J. 2004. "Asia Holds the Key to the Future of GM Food." *YaleGlobal* (December 2).

Finke, M.S., and H. Kim. 2003. "Attitudes about Genetically Modified Foods Among Korean and American College Students." *AgBioForum* 6(4): 191–197.

Foster, M., P. Berry, and J. Hogan. 2003. "Market Access Issues for GM Products." Report No. 03.13, Australian Bureau of Agricultural and Resource Economics, Canberra (July).

Grocery Manufacturers of America (GMA). 2001a. "FDA Proposal for Biotech Labeling and Product Approval: A Victory for Consumers." Press release, GMA, Washington, D.C. (January 17).

_____. 2001b. "FDA Should Require Testing Methods for New Biotech Crops, Says GMA." Press release, GMA, Washington, D.C. (May 3).

_____. 2003. "European Union Biotech Labeling Regulation a Barrier to Trade, Says GMA." Press release, GMA, Washington, D.C. (November 25).

Huang, J., S. Rozelle, C. Pray, and Q. Wang. 2002. "Plant Biotechnology in China." *Science* 295: 674–677.

Isaac, G.E. 2002. *Agricultural Biotechnology and Transatlantic Trade: Regulatory Barriers to GM Crops*. Oxon, UK: CABI Publishing Inc.

Jia, H. 2004. "China Ramps Efforts to Commercialize GM Rice." *Nature Biotechnology* 22(6): 642.

Kalaitzandonakes, N. 2004. "The Potential Impacts of the Biosafety Protocol on Agricultural Commodity Trade." IPC Technology Issue Brief, International Food and Agricultural Trade Policy Council (IPC), Washington, D.C. (December 26).

McCluskey, J.J., K.M. Grimsrud, H. Ouchi, and T.I. Wahl. 2003. "Consumer Response to Genetically Modified Food Products in Japan." *Agriculture and Resource Economics Review* 32(2): 222–231.

Miller, H.I., and G. Conko. 2004. *The Frankenfood Myth*. Westport, CT: Praeger Publishers.

Muth, M.K., D. Mancini, and C. Viator. 2003. "U.S. Food Manufacturer Assessment of and Responses to Bioengineered Foods." *AgBioForum* 5(3): 90–100.

Noussair, C., S. Robin, and B. Ruffieux. 2004. "Do Consumers Really Refuse to Buy Genetically Modified Food?" *The Economic Journal* 114(492): 102–120.

Pew Initiative on Food and Biotechnology. 2003. Legislative tracker, available online at http://pewagbiotech.org/resources/factsheets/legislation/index.php (accessed December 2004).

_____. 2004. "Overview of Findings: 2004 Focus Groups and Polls." Available online at http://pewagbiotech.org/research/2004update/overview.pdf (accessed December 2004).

Pray, C., J. Huang, R. Hu, and S. Rozelle. 2002. "Five Years of Bt Cotton in China: The Benefits Continue." *The Plant Journal* 31(4): 423–430.

Reuters Factbox. 2004. "EU's Legal Labyrinth on GMO." Reuters, Brussels, Belgium (November 11).

Tothova, M., and J.F. Oehmke. 2004. "Genetically Modified Food Standards as Trade Barriers: Harmonization, Compromise, and Sub-Global Agreements." *Journal of Agricultural & Food Industrial Organization* 2(2): Article 5.

Chapter 22

BENEFITS AND COSTS OF BIOSAFETY REGULATION IN INDIA AND CHINA

Carl E. Pray,[*] Jikun Huang,[†] Ruifa Hu,[†] Qihuai Wang,[‡] Bharat Ramaswami,[**] and Prajakta Bengali[††]
Rutgers University,[*] Chinese Academy of Science,[†] Chinese Ministry of Agriculture,[‡] Indian Statistical Institute,[**] and Marketics Technologies India Pvt. Ltd.[††]

Abstract: Developing countries are rushing to build effective biosafety regulatory systems in response to the spread of genetically modified organisms. So far, few of them have considered the costs and benefits of building and implementing such a system. This chapter takes a modest step in that direction by listing some of the possible costs and benefits of regulation and then giving some examples of costs and benefits to regulation in India and China. It finds that private firms' costs of complying with biosafety regulations for Bt cotton have been substantial in India but much less in China. On the benefits side the paper provides a detailed example of economic benefits from restricting unapproved genes in China and evidence that a transparent regulatory system might increase the demand for GM products.

Key words: regulation, biotechnology costs, India, China

1. INTRODUCTION

Most countries have or are developing a system to regulate transgenic crops in response to concerns of environmentalists and consumers about the crops' safety and pressures from the biotechnology companies and government scientists who want regulations that assure people of their products' safety and value. So far most of the regulations and enforcement mechanisms that have been put in place have focused on reducing the possibility of health and environmental problems of new products with no consideration of the costs

The authors acknowledge the financial support of the National Science Foundation of China (70021001 and 70333001) for surveys in China, the USDA Economic Research Service for support of the India component of this project, and the International Food Policy Research Institute for its financial support.

that these regulations impose on society or the risks inherent in current crop production practices. In addition there has been even less empirical work on attempting to measure the size of the benefits from incremental reductions of risk. As a result many countries are putting in place regulations that impose large costs and risks on groups like farmers and biotechnology companies without assessing whether there are substantial benefits to society from incurring these costs.

The objective of this chapter is to start quantifying some of the costs and benefits of biosafety regulations in two developing countries with the most advanced biosafety regulations: India and China. There is no attempt to be comprehensive in our documentation of costs and benefits; rather, we attempt to give some examples of costs and benefits that can be documented and quantified. The examples of costs include costs of complying with the regulations by firms and government research institutions, and the foregone income of farmers who cannot get access to safe technology. Examples of benefits include increased income for farmers from replacing less effective technology with more effective technology, keeping current pest control effective for longer periods of time, and increasing the demand for biotechnology through more effective and transparent regulations.

This chapter is organized as follows. Section 2 provides an overview of the activities of the regulatory systems and the spread of transgenic crops. Section 3 describes briefly the evolution and structures of the biosafety regulatory systems of the two countries. Section 4 looks at the costs of compliance in both countries and recent changes. Section 5 discusses some preliminary evidence on benefits of biosafety regulations. The concluding section attempts to pull out the main lessons from the Indian and Chinese evidence.

2. THE SPREAD OF REGULATED TRANSGENIC CROPS

2.1. India

In India the biosafety regulatory system has tested genetically modified (GM) cotton, rice, mustard, maize, potatoes, eggplant, tomatoes, pigeon pea, and cabbage, but the government has not revealed the precise numbers of field trials. By mid-2003 at least 34 events (genes introduced into a specific background crop variety) from the private sector were being tested in the nine crops just listed (Sharma, Charak, and Ramanaiah 2003). Most of the genes were *Bacillus thuringiensis* (Bt) genes for pesticide resistance, followed by some genes for herbicide tolerance, one set of genes for improving yield through better hybrids, and finally some genes for disease resistance. GM rice varieties for improved nutrition had not started trials in 2003 but now are at the early stages of field trials. So far only two crops have been put forward for commercialization: Bt cotton and hybrid mustard.

The transgenic hybrid mustard program was started by the Indian seed company Proagro in collaboration with the Belgium biotech company PGS. The multinational seed company Aventis purchased both of these companies, and then, in 2001, Bayer, the German chemical company, purchased Aventis. The genes that were used to produce hybrid mustard have been used in canola to produce hybrid canola cultivars in Canada and the United States. Proagro started working towards biosafety approval in the mid-1990s. Government regulators asked for another set of trials in 2003, but Bayer officials in India decided that they would not continue trying to commercialize hybrid mustard in India.

Monsanto's Bt gene in cotton is the only one that has been approved for cultivation. Monsanto formed a joint venture with Maharashtra Hybrids Company (MAHYCO), called Monsanto MAHYCO Biotechnology (MMB), to commercialize transgenic cotton. Three varieties of Bt cotton from MAHYCO were approved for cultivation in 2002 in central and southern India, and one Bt cotton variety from Rasi Seed Company was approved in 2004, also for south India. In 2005, ten more Bt cotton varieties of several different companies were approved. The approved GM cotton has spread quite rapidly in India—in 2004 it was planted on about 400,000 hectares (Monsanto India 2004).

In addition, a small company named NavBharat started growing a Bt cotton variety, NB-151, in the western Indian state of Gujarat in 2000 after registering this variety with the government as an insect-resistant variety. In 2001 it was discovered that NB-151 was being grown on between 6,000 and 12,000 acres. The government forced NavBharat to stop selling it after 2001 because it had not gone through the biosafety regulatory process, but smaller seed companies, former contract seed producers for NavBharat, and large farmers had the inbred lines needed to produce this seed. These small companies were able to keep producing the hybrid NB-151. In addition, many farmers saved their seeds and planted the second generation of the hybrid, known as the F2 generation. Officials of the Seed Association of India estimated in December 2003 that about 400,000 acres of F2s were planted and that the total Bt cotton acreage was between 700,000 and one million acres in 2003. The seed industry officials also estimated (in December 2003) that in 2004 about 2 million acres would be planted.

Both types of Bt cotton in India have reduced pesticide use, but the cost savings from reducing pesticide use have been offset by the higher cost of MMB seeds. The cost of NB-151 seed and its relatives is considerably less, so farmers using those varieties are gaining income through reduced pesticide costs. The main economic benefit for both types of Bt cotton has been to increase output per unit of land between 45 to 63 percent, which has led to a large increase in net income (Bennett et al. 2004).

2.2. China

More money has been invested in biotechnology research and technology transfer in China than in India (Huang et al. 2002), which accounts in part for the fact that there have been many more field trials of transgenic crops in China. From 1997 to July 2003, the Chinese government received 1,044 applications for field trials, environmental release, pre-production, or commercialization. Of these, 777 were approved, covering more than 60 crops and several animals, as well as numerous microorganisms. The GM crops that have been approved for commercial cultivation are cotton, tomatoes, sweet and chili peppers, and petunias. A total of 30 transgenic cotton varieties by 2003 and more than 140 transgenic cotton varieties by 2004 that use the Chinese Academy of Agricultural Sciences' (CAAS) Bt or the stacked Bt and Cowpea Trypsin Inhibitor (CpTi) or Monsanto's Bt have been approved for 14 provinces. Most of these varieties use the CAAS gene. Only six transgenic cotton varieties in nine provinces using Monsanto's Bt had regulatory approval for commercial production in 2003. In five provinces, varieties with Monsanto Bt were approved in 2004. No new transgenic crops have been approved since 1999.

In addition to cotton, the crop which has attracted the most interest among scientists and regulators is rice, which is China's major food crop. Many types of transgenic rice varieties and hybrids have reached and passed field trial and environmental trial phases since the late 1990s. Transgenic Bt rice varieties and hybrids that are resistant to rice stemborer and leaf roller were approved for environmental trials in 1997 and 1998 (Zhang, Liu, and Zhao 1999). Other scientists introduced the CpTi gene into rice, creating rice varieties with another type of resistance to stemborers. This product was approved for environmental trials in 1999 (Chen 2000). Transgenic rice with Xa21 and Xa7 genes for resistance to bacterial blight has been approved for environmental trials since 1997 (Chen 2000). In interviews with scientists, we also found that field trials have been underway since 1998 for transgenic rice with herbicide tolerance (using the Bar gene) as well as varieties expressing drought and salinity tolerance.

Four transgenic rice hybrids have advanced to the pre-production trials stage—the earliest pre-production trials started in 2001. Two insect-resistant hybrids—GM Xianyou 63 and Kemingdao—contain stemborer-resistant Bt genes. The hybrid GM II Youming 86 contains the CpTi gene, which provides resistance to stem borers also. A fourth hybrid contains the Xa21 genes, which provide resistance to bacterial blight, one of the most prevalent diseases in rice production areas in central China (Huang et al. 2005).

The transgenic cotton varieties with Bt genes from CAAS and Monsanto have spread rapidly since 1997. By 2004 they covered almost 3.7 million hectares, or about 65 percent of the cotton area of China. A survey by the Center for Chinese Agricultural Policy of the Chinese Academy of Science found that about 60 percent of the Bt area contains the CAAS insect-resistant genes, while the rest contain the Monsanto Bt gene.

Most of the area is covered by approved varieties, but a portion of the area—at most about 20 percent—may be covered by varieties containing unapproved seed. In China a government cotton research institute developed some bollworm-resistant pureline varieties and hybrids containing an unapproved Bt gene. It entered these insect-resistant varieties in the provincial variety registration trials and got approval for sale in 1998. It started selling seed in Henan and Shandong provinces through an already-established joint venture with the Shandong Provincial Seed Companies. In 2001 the Center for Chinese Agricultural Policy surveyed cotton farmers in north China (Hebei, Henan, Shandong, Anhui, and Jiangsu provinces) and found that approximately 20 percent of the Bt cotton fields were planted with the illegal Bt (CCAP unpublished data), but the area seems to have declined recently.

The adoption of Bt varieties has reduced pesticide use to control bollworm and increased cotton yields (Pray et al. 2002). The widespread adoption of these varieties has also pushed cotton prices down, but the net result is higher net income for people who adopt Bt cotton (Pray et al. 2002).

3. BIOSAFETY REGULATIONS IN INDIA AND CHINA

3.1. Indian Biosafety Regulations

The goal of the Indian regulatory system is to ensure that India's GM crops pose no major risk to food safety, environmental safety, or agricultural production, and that there are no adverse economic impacts on farmers. This last goal is one that many developed countries do not include in their biosafety regulatory systems, but one that most developing countries have included. The biosafety regulatory system was established in its current form in 1990 by guidelines issued through the Ministry of Science's Department of Biotechnology ("Recombinant DNA Guidelines"), with some modifications in August 1998 ("Revised Guidelines for Research in Transgenic Plants and Guidelines for Toxicity, Allergenicity Evaluation of Transgenic Seeds, Plants and Plant Parts").

The Indian biotechnology regulatory system has three layers. At the bottom, institutional biosafety committees must be established in any public or private institute using rDNA in its research. These committees contain scientists from their respective institute and a member from the Ministry of Science's Department of Biotechnology. There are more than 230 of these committees in India, of which 70 deal with agricultural biotechnology. They can approve contained research at institutes unless the research uses a particularly hazardous gene or technique. That type of research must be approved by the Review Committee on Genetic Manipulation (RCGM), which is the next layer of the system.

The RCGM is in the Ministry of Science's Department of Biotechnology and regulates agricultural biotechnology research up to large-scale field trials. It requests food biosafety, environmental impact, and agronomic data from applicants who wish to do research or conduct field trials. It gives permits to import GM material for research. The RCGM is made up primarily of scientists (including agricultural scientists) and can request people with specialized knowledge to review cases. It has a monitoring-cum-evaluation committee that monitors limited and large-scale field trials of GM crops and which is made up primarily of agricultural scientists.

The Genetic Engineering Approval Committee (GEAC) is under the Ministry of Environment and Forestry. It is the agency that gives permits for commercial production of GM crops, large-scale field trials of GM crops, and the imports of GM commercial products. The committee members are primarily bureaucrats representing different ministries, and they draw on the scientific expertise of each ministry.

The main steps in the biosafety regulatory process for a new GM event (new gene inserted into one location on a specific variety's genome) is shown in column 1 of Table 1. Columns 2 and 3 show the data generated for regulators and which committees regulate each step. If little is known about the event or it is thought to be risky, then the committee at the next level has to sign off on the experiment or trial. For example, an insitutional biosafety committee could not approve at its institute a greenhouse experiment with risk category III events. The approval of RCGM would also be required. After the event in a specific variety proves that it is safe for food, the environment, and agriculture, and that it will be economically beneficial for farmers, the GEAC approves it for commercial use. It will also have to go through several years of testing by the state and national variety trials to prove its agronomic superiority over the current varieties.

If an approved GM event is backcrossed into a new plant variety, the developers of the new variety do not have to produce new food safety and environmental data. However, they do have to put it through at least two years of agronomic trials to obtain GEAC clearance, and then it has to go through several more years of the variety trials.

India has experienced difficulties in enforcing biosafety regulations. As mentioned above, a local entrepreneur—NavBharat Seed Company—acquired

Table 1. Comparison of Indian and Chinese Biosafety Laws, Institutions, and Impacts

	India	China
Policy objectives	Ensure that GM crops pose no major risks to food safety, environmental safety, and agricultural production, and no adverse economic impacts on farmers.	Promote biotechnology R&D; tighten safety control of genetic engineering work; guarantee public health; prevent environmental pollution; maintain ecological balance
Legislative history	▪ 1986: Environmental protection law ▪ 1990: "Recombinant DNA Guidelines" (Ministry of Science, Department of Biotechnology) ▪ 1998: "Revised Guidelines for Research in Transgenic Plants" and "Guidelines for Toxicity, Allergenicity Evaluation of Transgenic Seeds, Plants and Plant Parts" (Ministry of Science, Department of Biotechnology)	▪ 1993: "Safety Administration Regulation on Genetic Engineering" (MOST) ▪ 1996: "Safety Administration Implementation Regulation on Agricultural Biological Genetic Engineering" (MOA) ▪ 2001: "Regulation on the Safety Administration of Agricultural GMOs" (State Council) ▪ 2002: Implementing regulations: safety evaluation administration of agricultural GMOs; safety administration of ag GMO imports; ag GMO labeling administration (MOA)
Institutional structure	▪ Ministry of Environment and Forestry's multi-ministry GEAC (for commercial production, large field trials, GM product imports) ▪ Ministry of Science's Department of Biotechnology's Review Committee on Genetic Manipulation (RCGM) (for contained research on hazardous gene or technique, and all research up to large-scale field trials) ▪ State biosafety committees ▪ Institutional Biosafety Committees (IBSC) at each research institute	▪ Allied Ministerial Meeting (MOA, MOST, State Development and Planning Commission, MPH, Ministry of Foreign Economy and Trade, SEPA) ▪ Office of Agricultural Genetic Engineering Biosafety Administration (OGEBA), within MOA ▪ National Agricultural GMO Biosafety Committee ▪ Provincial Biosafety Management Offices ▪ Institutional Biosafety Committees

cont'd.

Table 1. (cont'd.)

Submissions and approvals:		
a. cases considered	a. Unknown	a. 1,044 applications
b. approved – field testing	b. Nine crops	b. 60 crops
c. approved – commercialization	c. one crop (cotton), 4 varieties in 2004	c. four crops (cotton, tomatoes, peppers, petunias), 181 varieties in 2004
Diffusion – commercial production	Cotton: 1.2 million ha in 2004	Cotton – 3.7 million ha (60 percent CAAS, 40 percent Monsanto) in 2004
Enforcement – institutions that are active	▪ Environmental ministry GEAC ▪ State Departments of Agriculture ▪ Courts ▪ Local NGOs and international NGOs such as Greenpeace	▪ Ministry of Agriculture – National Agricultural GMO Biosafety Committee OGEB ▪ Provincial and local agricultural bureaus ▪ No local NGOs; only Greenpeace allowed
Enforcement – percent of area illegal	66 percent illegal Bt cotton in 2004	20 percent was illegal in 2000; now less

or developed transgenic cotton varieties and then started marketing them without putting them through the biosafety regulatory system. When MMB brought this to the attention of the GEAC, the GEAC declared that production of untested transgenic varieties should be stopped. The local government did stop NavBharat from selling these varieties, but did not stop their spread by smaller seed companies and farmers. The illegal varieties now cover two-thirds of the Bt cotton acreage in India (Singh 2004).

3.2. China's Biosafety Regulations

In response to the emerging progress in China's agricultural biotechnology, the first biosafety regulation, "Safety Administration Regulation on Genetic Engineering," was issued by the Ministry of Science and Technology (MOST) in 1993 (see Table 1 for a summary of regulations). This regulation consisted of general principles, safety categories, risk evaluation, application and approval, safety control measures, and legal responsibilities. After this regulation was decreed, MOST required relevant ministries to draft and issue corresponding biosafety regulations on biological engineering [i.e., the Ministry of Agriculture (MOA) for agriculture and the Ministry of Public Health for food safety]. The MOA issued the "Implementation Regulation on Agricultural Biological Genetic Engineering" in 1996. Labeling was not part of this regulation, nor was any restriction imposed on imports or exports of GM products. The regulation did control genetically modified organisms (GMOs) for research and commercial production. Under this regulation, the National Agricultural GMO Biosafety Committee (Biosafety Committee) was established in 1997 to provide MOA with expert advice on biosafety regulation.

In May 2001 the State Council decreed a new and general rule on biosafety called the "Regulation on the Safety Administration of Agricultural GMOs." This new regulation replaced the 1993 regulation issued by MOST. Under the new regulation, the MOA issued three new implementing regulations on biosafety management, trade, and labeling of GM farm products. The implementing regulations were to take effect after March 20, 2002. They included several important changes to existing procedures, and details of regulatory responsibilities after commercialization. The changes included an extra preproduction trial stage prior to commercial approval, new processing regulations for GM products, labeling requirements for marketing, new export and import regulations for GMOs and GMO products, and local- and provincial-level GMO monitoring guidelines. In the meantime, the Ministry of Public Health also promulgated its first regulation on GMO food safety in April 2002, to take effect after July 2002.

The MOA is the primary institution in charge of the formulation and implementation of biosafety regulations on agricultural GMOs and their commercialization. In order to incorporate representation of stakeholders from different ministries, the State Council established an Allied Ministerial Meeting comprising leaders from the MOA, the State Development Planning Commission (SDPC), the MOST, the Ministry of Public Health, the Ministry of Foreign Economy and Trade (MOFET), the Inspection and Quarantine Agency, and the State Environmental Protection Authority (SEPA). This Allied Ministerial Meeting coordinates key issues related to biosafety of agricultural GMOs, examines and approves the applications for GMO commercialization, determines the list of GMOs for labeling, and establishes import or export policies for agricultural GMOs and their products. The routine work and daily operations are handled by the Office of Agricultural Genetic Engineering Biosafety Administration (OGEBA) under MOA.

The Biosafety Committee remains the major player in the process of biosafety management. Currently, the Biosafety Committee is composed of 56 members, most of whom are agricultural scientists. The committee meets twice each year to evaluate all biosafety assessment applications related to experimental research, field trials, environmental release, pre-production trials, and commercialization of agricultural GMOs. It makes recommendations to OGEBA based on the results of its biosafety assessments. OGEBA is responsible for the final approval of decisions.

The Ministry of Public Health (MPH) is responsible for food safety management of biotechnology products (processed products based on GMOs). The Appraisal Committee, consisting of food health, nutrition, and toxicology experts nominated by MPH, is responsible for reviewing and assessing GM foods, as they have been designated a novel food. SEPA participates in GMO biosafety management through the Allied Ministerial Meeting and through its members on the Biosafety Committee. Although SEPA has taken the responsibility for international biosafety protocol, its focus on biotechnology in China is limited to biodiversity.

In 2005, all 31 provinces in China established provincial biosafety management offices under provincial agricultural bureaus. These biosafety management offices collect local statistics on and monitor the performance of research and commercialization of agricultural biotechnology in their provinces and assess and approve (or not approve) all applications of GM-related research, field trials, and commercialization in their provinces. Only those cases that are approved by provincial biosafety management offices are submitted to the Biosafety Committee for further assessment.

Part of the reason for strengthening these provincial biosafety offices is that China's difficulties in enforcing biosafety regulations have been similar to those experienced in India. A local enterprise—in this case a government research institute—acquired a Bt gene and developed Bt cotton varieties and then started marketing them without obtaining approval from the national biosafety committee. The biosafety committee found out about them and declared that the production of untested transgenic varieties should be stopped, but its declaration was not implemented. In the Chinese case the government seems to have been able to induce the research institute to substitute the approved Bt gene from CAAS for the unapproved gene (see Pray et al. 2006).

3.3. Comparison of Chinese and Indian Biosafety Regulatory Systems

Indian and Chinese regulations can be compared in Table 1. The goals, history, and structure of the regulations are quite similar. The major difference in the structure of these systems is that in China the regulatory and enforcement machinery is all under the Ministry of Agriculture, while in India regulation is implemented by the Ministry of Environment and Forestry and by the Department of Biotechnology, which is part of the Ministry of Science and Technology.

4. COST OF REGULATIONS

The U.S. Environmental Protection Agency (2000) has developed a framework for evaluating the costs of regulations. The main categories are listed in Table 2. The first is the cost of compliance with the regulations, which falls on the biotech firms that are trying to get transgenic products approved for commercial production or consumption and then keep them in the market. The second category of costs is the budgetary costs to the government of running the biosafety system and enforcing it. The third category is social welfare costs. These include the costs to agricultural producers who have to plant refugia to prevent the development of pests that are resistant to the pesticides in the plants. An additional social cost is the producer's surplus that farmers lose because they are prevented from using a safe technology, and the consumer surplus that consumers lose because farmers are producing less than they otherwise would. Another cost of biosafety regulations is that they reduce incentives for agricultural research, which could have long-term impacts on agricultural productivity growth and incomes.

This chapter looks at three categories of costs: (i) the compliance costs faced by biotechnology companies and research institutes, (ii) the costs to farmers of foregone income because they are trying to comply with refuge requirements in India or are forced to use conventional maize, which acts as a refuge, in China, and (iii) the costs to farmers of foregone income because they were not able to adopt as quickly.

Table 2. Examples of Social Cost Categories from Biosafety Regulations

Social Cost Category	Examples
Real-resource compliance costs	▪ Cost of generating data to obtain and maintain permission to sell transgenic technology ▪ Foregone income of farmers from growing insect refugia ▪ Operation and maintenance of new labs and equipment
Government sector regulatory costs	▪ Monitoring/reporting ▪ Training/administration ▪ Enforcement/litigation ▪ Permitting
Social welfare losses	▪ Higher consumer and producer prices of food and fiber due to slower adoption of cost-saving technology ▪ Additional legal/administrative costs
Transitional social costs	▪ Unemployment ▪ Firm closing ▪ Public sector institution abandoning research projects of public interest ▪ Transaction costs ▪ Disrupted production

Source: Taken from U.S. Environmental Protection Agency (2000) and modified by authors.

4.1. Compliance Costs—India

As discussed above, only two genes—the Bt gene Cry1Ac in cotton and the genes for hybrid mustard—have been proposed to the GEAC for commercial release. In addition, Bt eggplant and high-protein potato have been tested fairly extensively. Our cost estimates are based on these crops. We collected the cost data through extensive interviews with the companies or research institutions in December 2003.

The first events approved (in 2002) were Monsanto's Bt gene in three cotton hybrid cultivars from MAHYCO. The government gave MMB permission to commercialize Bt cotton for three years, after which time the application had to be reviewed again. Thus, there are post-approval regulatory costs as well as pre-approval costs.

The pre-approval compliance costs without salaries were about $900,000 and total pre-approval cost was perhaps $1.8 million.[1] It took six years of trials to get approval. A certain amount of this time would have been required to produce and obtain approval for a new conventional (non-transgenic) variety anyway, but the biosafety requirements added three to four years of trials and tests to the commercialization process. In addition, MMB estimated that it would have between $100,000 and $200,000 of further ex-

---

[1] A detailed breakdown of this cost is available in Pray, Bengali, and Ramaswami (2005).

penses in order to meet the three-year renewal requirement. Monsanto expects that in the future, GM events such as Bt maize (primarily an animal feed in India) will cost about $500,000 in regulatory costs, excluding salaries, to bring to market. In contrast, in December 2003 the compliance cost for a new chemical pesticide in India was about $200,000, according to Monsanto India's government affairs officer.

The genes that Bayer Agrosciences and its predecessors used to produce hybrid mustard have been used in canola (which is closely related to mustard) to produce hybrid canola cultivars in Canada and the United States. They have cleared the biosafety regulations in those countries. However, these genes have not been commercialized in mustard anywhere in the world. Therefore, Bayer and its predecessors decided that they would not commercialize transgenic mustard in India before conducting research to show India's neighbors and trading partners that this particular genetic event was safe. They started working towards biosafety approval in the mid-1990s. Since then, between US$3 and 4 million was spent in the United States and Europe to ensure that it met the international food safety requirements. In addition, $1 to 1.5 million was spent in India on environmental trials and nutritional standards.

After Bayer and its predecessors spent between $4 and $5 million on meeting regulatory requirements in India and elsewhere, GEAC asked for another set of trials in 2003. Bayer officials in New Delhi informed us in December 2003 that, because of the continued costs and the uncertainty about whether this product would ever be approved and the potential market size for GM mustard, they have decided not to continue trying to commercialize hybrid mustard in India.

If the plant variety containing a specific event has been approved for commercial use by the GEAC and then is backcrossed into another variety, the GEAC requires two to three years of agronomic trials before the new variety is approved for commercialization. If the number of biosafety trials are 15 the first season and 40 the next two seasons, and each ICAR trial costs private companies about $1,000, as they do at present, then the cost of introducing a new variety will be almost $100,000 (Pray, Bengali, and Ramaswami 2005).

In contrast to private sector scientists, who estimated high compliance costs, most public sector scientists felt that costs were not a major constraint on their research or commercialization efforts. For them, the main problem is the years lost in the regulatory process. One of the few institutes that have a long experience with the biosafety process is the National Research Center for Biotechnology at the Indian Agricultural Research Institute (IARI). It started Bt research in 1996, and its vegetable program has had Bt varieties in the regulatory system for a number of years. It developed a Bt eggplant using a Cry1Ab gene that controls 70 percent of the fruit borer attack. The Institute had agronomic trials in a controlled environment in 1998/99, 1999/2000, and 2000/2001. In 2003 it was permitted to conduct field trials in five locations. The cost of controlled environment trials and field trials so far has been about $10,000. If the Institute is required to do two more years of field trials in 10 locations, this will add another $10,000. Late in 2003 the Institute asked for and received estimates of the cost of meeting the food safety requirements. The estimates ranged from 1 to 1.5 million rupees (Rs) (US$22,000 to $33,000) for everything needed (Indian Agricultural Research Institute 2003). Thus, with no more food safety and two more years of field trials, the total cost would come to $53,000. This is a large sum for the Indian research system, but IARI has grant money to cover it.

The high-protein potato research at the Center for Plant Genomics Research in New Delhi was often used by the Department of Biotechnology as an example of the consumer benefits from GM technology. The lead scientist on this project, Dr. Niranjan Chakraborty, reported in early 2004 that the Center's costs of meeting regulatory requirements have been negligible— some costs of the allergenicity/toxicity and the costs of labor and fertilizer for three years of field trials. The institutes that conducted the tests absorbed all other costs. They still have not submitted their data to the GEAC to get approval for wide-scale testing of the technology. So there will be more costs at the next level and at least two more years of testing, but at the moment the total costs and the time required has been limited.

4.2. Compliance Costs—China

The data on the costs of complying with biosafety regulations in China suggests a pattern somewhat different from India's system—the cost of compliance paid by the government institutes and private companies is low relative to India, and the Chinese system for approving new varieties containing approved events has evolved into a fairly speedy one, at least for cotton. However, getting other food crops approved—such as maize or rice—has been a long process for both public and private institutions.

For foreign companies that are operating through joint ventures in China we have some data, but so far we do not have a complete accounting of compliance costs. We know that it took much less time and money to get approval for Bt cotton in the late 1990s in China than it did in India. The Chinese biosafety committee was still working out what data was required and assumed that since Bt cotton was approved in the United States, it presented no food safety or agronomic problems. Thus, for Bt cotton Monsanto had to produce its international dossier, which had evidence of food safety, environmental impact, and efficacy. The only trials that were required in China were two years of environmental impact trials and some mouse-feeding trials to check the safety of the oil and seed cake. Similarly, when the Biotechnology Research Institute of CAAS wanted to get permission to commercialize its Bt cotton, it had to produce scientific data on the characteristics of the gene construct and had to show several years of field trial data indicating the efficacy of the technology in Chinese conditions. The Chinese government brought together a large group of scientists to several meetings in the mid-1990s to discuss pest resistance management. They decided that there was no need to require farmers to grow conventional cotton on part of their cotton field as a breeding ground for susceptible pests because farmers already grow other crops such as maize where susceptible pests can breed (see discussion on refugia on page 503 below).

After the biosafety committee decided to allow commercialization of Bt cotton in 1997, the international controversies about the safety of biotech started to cause concern in the press and the public. As a result the government decided that it needed to know more about the possible health risks to humans of eating Bt cotton seed oil and the risks to animals from eating cotton seed cake. The Chinese government, after the release of Bt cotton, did conduct more studies in 2000. MOST provided the Ministry of Health's Nutrition and Food Safety Institute about 5 million renminbi (RMB) (US$604,000) for food safety testing, first of cotton seed oil and cake and then of Xa21 rice. An additional RMB 5 million was given to the Institute of Plant Protection, which is part of the Ministry of Agriculture, for research on the environmental impact of GM crops (Peng 2000).

To find out how much research institutes' and seed companies' costs of compliance with the regulations were in 2004, the authors interviewed a number of government research institutes and commercial seed companies that have been pushing varieties of Bt cotton through the regulatory system. Unfortunately, none of these companies had new varieties of Bt or CpTi; they had simply new varieties of cotton that used the CAAS or Monsanto Bt events as one of the parents. The results of these interviews are shown in Table 3. The total costs to research institutes of developing and bringing to market a new Bt cotton variety was about $88,000. Of that about $75,000 went to plant breeding, while $13,000 went to biosafety regulations.

Private firms did not give us precise numbers on their cost of getting the first Bt cottons through the biosafety process, but it clearly was small because the Chinese government did not require that food safety testing be done on the cotton seed oil or cotton seed cake in China and required only limited environmental testing in China. To calculate what it would cost now to comply with the biosafety requirements, we relied on the cost of tests that are mandated by the Chinese government. Table 4 shows the costs of these tests for the private firms and for government institutes. The cost per year is the cost per trial times the number of trials that are typically required for each type of test. The cost per trial for private firms is typically about three times more than the cost for government research institutes. The minimum

Table 3. Costs of Compliance with the Biosafety Requirements for Chinese Cotton Varieties (2000 prices, in U.S. dollars)

	Jimian 38	Zhong mian 41	GKz8	GKz23	Nannong 98-7	Average
Breeding	110,775	58,717	52,906	104,116	46,247	74,576
Biosafety trials						
Small-scale	1,816	4,237	242	3,995	2,421	2,542
Medium-scale	3,632	4,964	605	4,843	3,269	3,511
Pre-production	5,448	8,232	969	8,111	5,085	5,569
Safety certificate	1,453	2,785	121	1,211	1,574	1,453
Subtotal	*12,349*	*20,218*	*3,390*	*18,160*	*12,349*	*13,317*
Total	123,123	78,935	56,295	122,276	58,596	87,893

Source: Survey of 5 Chinese research institutes.

Table 4. Cost in China for Approval of a New GM Field Crop Event: Private and Public Sector (in U.S. dollars)

	Years required	Companies			Government		
		Cost/ year	Cost min. years	Cost max. years	Cost/ year	Cost min. years	Cost max. years
Environmental safety			32,800	32,800		32,800	32,800
Food safety							
Anti-nutrients			120	120		120	120
90-day rat feeding			14,500	14,500		14,500	14,500
Pilot field trial	1 – 2	1,816	1,816	3,632	581	581	1,162
Environmental field trial	2 – 4	5,085	10,170	20,340	1,695	3,390	6,780
Pre-production field trial	1 – 3	6053	6,053	18,159	1,896	1,896	5,688
Total			65,459	89,551		53,287	61,150

Sources: Years are best guess from companies. Costs are from Chinese Ministry of Agriculture (http://www. stee.agri.gov.cn/ biosafety/jcjy/bz_jcjy/t20031029_131210.htm).

number of years is the minimum required according to government regulations. The maximum number of years is a guess because all crops have been stuck in either environmental trials since 1999 or have been in pre-production trials. For example, Bt maize has been at the environmental trial stage since 1999, and Bt rice has been in preproduction trials for at least three years.

The total cost of compliance for private firms, excluding the salaries of the company's regulatory staff in Beijing, is at least $65,000 (Table 4). Using a more realistic guess at the number of years, the total would be about $90,000. The big cost may be the salaries of the regulatory people in Beijing and elsewhere. There are several people in firms' regulatory offices, and they have to work on these genes for 8 to 10 years. This mounts up. At today's salaries for highly trained professionals in Beijing, assuming that one person would take care of four new gene events over a number of years, the cost of personnel could easily add up to $100,000. Government institutes would pay less to meet biosafety regulations, somewhere between $53,000 and $61,000 without salaries.

In response to concerns of government research institutes and private companies about the slowness and costs of regulations for Bt cotton varieties, the Chinese government changed the way it conducted trials to reduce these costs. New GM varieties can now be simultaneously tested in the field trials and the variety registration trials. In addition, in September 2004, MOA implemented measures for Bt cotton that allow (i) developers of a Bt cotton variety that received a production safety certificate from one province to directly apply for the safety certificates for all provinces within the same cotton-ecological region without going back to the National Biosafety Committee in Beijing, (ii) developers of an approved variety to apply production safety certificates directly to other provinces in other cotton ecological regions, and (iii) a Bt variety that was developed from approved varieties with safety approval to also be entered directly to provincial biosafety committees for approval.

4.3. Comparing India and China

The costs of compliance in India and China are summarized in Table 5. The table suggests that the costs of compliance are higher in India than in China for private firms, while the costs for government research institutes are less than for private firms in both countries. The lower costs of compliance in China than in India may be due in part to the fact that the government absorbed much of the cost of testing the food safety of cotton seed oil and cake and transgenic rice, as well as studies on the gene flow and other environmental issues of concern for transgenic rice and maize. Of course, the actual cost for GM events from the public sector in India is unknown since no GM varieties from the public sector have made their way through the regulatory system. In both systems, the real problem is not money, but the time required and the uncertainty about whether any major food grains will be approved for commercial production.

The Chinese system was much faster than the Indian system in approving Bt cotton and a few minor crops. Now, however, both systems are undecided about food crops that have had extensive testing—mustard in India and rice in China. Thus, there are major uncertainties for the developers of

Table 5. Comparing Biosafety Compliance Costs in India and China Without Personnel Costs

	India – Actual Costs or Industry Estimates	China – Current Official Costs of Trials
New Bt cotton gene, private corporation	$900,000	$89,500
New Bt cotton, government research institute	Not available	$53,000–61,000
Approval of Bt in new variety	$100,000 (estimated by multinational companies)	$13,300 (from survey of government research institutes)
Bt eggplant, Indian government	$53,000	Not available
Bt maize, private corporation	$500,000	$89,500

Source: Interviews by authors.

transgenic food crops about whether they will ever be approved. The Chinese system does appear to be faster in approving varieties of crops which already have transgenic cultivars in production. In addition, in China the biosafety regulations' field trials (small trial, middle or environmental release trial, pre-production trial, and varietal agronomy trial/registration) are allowed to overlap with the previous stage, so the time to go through all stages could be shortened by nearly half.

5. BENEFITS FROM BIOSAFETY REGULATION

The benefits from biosafety regulation are supposed to be reduced risk of human disease, allergic reactions, antibiotic resistance, and nutrition problems; reduced risk of negative environmental impacts such as reduced biodiversity, in particular endangered species; and reduced risk of damaging agriculture through crops that are particularly susceptible to disease or pests or that lead to rapid development of pesticide-resistant insects and weeds. As mentioned above, Indian and to a certain extent Chinese regulators also try to protect their small farmers from the possibility of economic losses from the adoption of unprofitable crops. These benefits and ways of measuring the costs and benefits of these regulations are listed in Table 6. The problem with measuring these benefits is that it is difficult to find any reliable numbers to base them on. For example, what is the probability that biotechnology companies or government institutes will attempt to commercialize a

Table 6. Social Costs and Benefits of Regulation

Benefit Category	Examples of Service Flows	Costs	Valuation Methods of Benefits
Human Health • mortality risks • morbidity risks	Reduced risks of fatality Reduced risks of • allergic reactions • sickness from malnutrition	Food safety trials on mice, humans ???	Averting behavior Hedonics Stated preference Cost of illness
Environmental health • biodiversity risk • endangered species	Reduced risk of • losing economically valuable species • losing aesthetically pleasing species	Costs of • environmental field trials • measurement of pollen flows • assessing impact of pollen flow • foregone income due to buffer crop or variety	Expected value of biodiversity in pharmaceuticals and agriculture Willingness to pay
Agricultural production and costs Risk of • crop losses from resistant pests • crop losses from disease/pest epidemic because no agrodiversity • sickness or death of farm animals	Reduced risk that • plant pesticide will lose efficacy • new pest or disease could destroy crop • animals will be allergic or will miss nutrients	Field trials to assess agricultural impact Foregone income from refugia Animal feeding trials	Crop, disease, and insect modeling Costs of increased pesticide use Animal disease models Cost of treating animals
Economic risk that • small farmer will adopt an unprofitable technology • small farmer will adopt a technology with no/limited market • pollen flow will make organic crops non-organic		Agronomic trials plus social economic studies Studies of potential markets	Lost income from unprofitable variety Cost of finding new markets Lose premium price for organic crop

product that will end up causing an allergic attack in humans or that they will inadvertently insert a gene that increases the possibility of cancer? And how much will the regulatory system be able to reduce that probability? Where could an investigator get some reasonable numbers? The closest analogy might be the pesticide industry and pesticide regulation. However, even there it has been virtually impossible to statistically link actual pesticide in farmers' fields to sickness of farmers or consumers (Roberts, Buzby, and Lichtenberg 2003).

Since it is so difficult to measure benefits, we next consider an in-depth case study of one measurable benefit of regulation—the benefits to farmers and consumers from limiting the use of an inferior Bt gene—and then discuss some preliminary evidence on possible benefits of regulatory activities: mandatory refugia and increased demand for biotech products.

5.1. Restricting Inferior Genes

The primary case study of the benefits of regulation comes from China. This is the case mentioned above in which Chinese farmers were being sold a Bt cotton variety containing a Bt gene that had not been approved by the biosafety committee because a research institute had commercialized it without sending it to the biosafety committee. Scientists tested this variety against other Bt varieties for the biosafety committee and found that it was not as effective in protecting cotton because its toxicity declined early in the season (Wu 2002). The biosafety committee was also concerned that the low levels of toxicity would lead to more rapid development of bollworms that were resistant to Bt.

In a paper that focused on a number of different seed industry policies, Hu et al. (2005) developed an econometric model of the impact of the different types of insect-resistant genes and insect-resistant varieties, using various sources on yields and pesticide use. One component of the model estimated the effectiveness of the unapproved and the approved Bt genes. The model was estimated using data from approximately 400 cotton-growing farmers in northern China in the years 1999–2001. We can use the coefficients of the seeds with an approved Bt gene and the *unapproved* Bt genes to calculate the impact of these genes on yields and pesticide use. The pesticide equation shows little difference in the reduction of pesticide use. The seeds that did not go through the biosafety regulation process were almost as effective (38.53 kg reduction) as the legitimate Bt genes from CAAS or Monsanto (41.45 kg or 39.77 kg reduction). However, the yields of unapproved varieties were lower than those of the approved varieties. The yields of unapproved varieties were not statistically different from conventional varieties of cotton, while the varieties with approved genes yielded 19.2 to 25.7 percent more than conventional varieties.

We then calculate the effect of increased regulation of biosafety by developing scenarios about what would have happened had regulations been enforced and the unapproved gene kept off the market (see Table 7). The baseline shows the benefits from Bt cotton that actually took place, using figures from Hu et al. (2005). We then keep all assumptions the same as the baseline and assume one change: instead of allowing varieties with unapproved genes in the fields of farmers due to weak enforcement, we assume that biosafety regulations have successfully prevented the use of varieties with this gene.

The impact of suppressing these varieties would depend on whether the companies providing the approved gene would have been able to supply enough seed of varieties to replace those with the unapproved genes. In scenario 1, we assume that the companies cannot supply the seed and that farmers have to be satisfied with non-Bt cotton. This is a possible scenario because Jidai, the company selling the Monsanto Bt varieties, did not have permission from the biosafety committee to sell seed in most provinces where the unapproved varieties were grown, and Biocentury, the company that sold the CAAS Bt varieties, was still having management problems that made it difficult for it to meet farmers' actual demand in 2001. In scenario 2, we assume that Jidai and Biocentury were able to supply farmers with enough legitimate seed with approved genetic material.

The basic objective of the scenarios, whether there is zero replacement of varieties with unapproved and approved material (scenario 1) or complete replacement (scenario 2), is to simulate the impact on society if the biosafety committee and the Ministry of Agriculture are able to enforce their biosafety regulations. In order to carry out the simulation, we assume that the return to the unapproved varieties from our sample for 2001, which accounted for about 20 percent of China's Bt cotton area, either reverted to the returns of conventional cotton [which means the returns would fall by the size of the coefficient on the unapproved variable in the regression found in the tables in Hu et al. (2005)] or rose by an amount based on the differences between the size of the coefficiencies on the Jidai/Biocentury legitimate varieties and the coefficient on the unapproved variety.

When there is no substitution between the unapproved variety and the varieties with the approved Bt genes, we can see that enforcing biosafety regulation can actually impose a cost on farmers (compare farmers' income in the baseline and scenario 1 of Table 7). In scenario 1, the enforcement of biosafety regulations eliminates unapproved varieties, but farmers have to revert to conventional cotton varieties. Although the yields of unapproved varieties were lower and levels of pesticide use higher than for varieties with approved genes (that is, legitimate varieties), the yields were higher and pesticide use levels much lower than for conventional cotton. This means that, without being able to substitute in better alternatives, biosafety

Table 7. Annual Benefit for 2001 Under Different Policy Scenarios (in million RMBs)

	Farmer Welfare	Consumer Welfare	Seed Company Profits	Monsanto Royalties	CAAS Royalties
Baseline	2,586	1,302	69	27	17
Scenarios					
1. Biosafety enforcement, approved Bt genes only in approved provinces, and no unidentified Bt genes plus no replacement	1,916	957	69	27	17
2. Biosafety enforcement, approved Bt genes only in approved provinces, and no unidentified Bt genes plus complete replacement	2,884	1,441	206	80	51

Source: Hu et al. 2005.

management can reduce producer welfare [by RMB 670 million (RMB 2,586 million minus RMB 1,916 million)]. With less cotton available, prices rise and consumers also get hurt slightly. Since in scenario 1 the suppliers of the unapproved seed disappear and Jidai and Biocentury do not sell any replacement seed, their profits and the royalties of Monsanto and CAAS are the same.

When seed from Jidai and Biocentury replaces the seed that contained the unapproved gene (scenario 2), farmers gain about RMB 300 million over the baseline because they produce yields that are higher and use less pesticide than when they use seed with the unapproved gene. The gains to consumers are RMB 139 million. These gains are substantial. The gains to the seed and gene supply industry are substantial—they triple their profits, seed companies gaining RMB 137 million and gene providers gaining RMB 87 million (RMB 53 million for Monsanto plus RMB 34 million for CAAS). The supply industries' gains in profits (RMB 234 million) are close to the producer gains (though farmers are the largest single gainer).

In fact, what seems to have happened is that the government gradually pushed the inferior varieties out so that by 2004, when the next survey was conducted, the area of the unapproved gene had dropped to near zero in the surveyed areas (Pray et al. 2006). The process was slow enough to allow the other varieties time to move in. As a result, there was no overall decline in Bt varieties, and farmers and consumers captured the benefits from the regulations.

5.2. Benefits from Refugia

Refugia are parts of a crop field that are left without the protection of a plant pesticide like Bt so that the few insects that are resistant will be able to mate with susceptible pests in the refuge area instead of other resistant insects in the protected area. Since the susceptibility trait is the dominant trait and the resistance trait recessive, the offspring of the resistant-susceptible match will be susceptible. If the resistant insects crossed only with other resistant insects, their offspring would be resistant. Thus a sufficiently large refuge area will reduce the number of resistant offspring and slow the buildup of resistance. The economic benefits are that farmers will be able to continue to use this plant pesticide for a longer time to control the pests. Farmers will not have to suffer losses from the pests or switch to new, more expensive pesticides as quickly.

There are, however, costs to slowing the development of these resistant pests. The first is that farmers have to plant part of their field with a less profitable variety. So, their lost profits are a cost. The second cost is the cost to seed companies of educating farmers to plant non-Bt seed, providing them with non-Bt seed, and monitoring their compliance. A third cost would be the budgetary costs of government efforts to educate farmers and to enforce refugia.

In India, farmers are required to plant a refuge of four rows of non-Bt cotton around each Bt cotton field. The farmers' costs will include the lost production on the four rows of non-Bt cotton and some lost production on the neighboring Bt fields that are damaged by the susceptible pests before they die. Seed companies are required to include a packet of non-Bt cotton seed with each packet of Bt seed, and then are supposed to help enforce the refugia requirement. Their costs include the cost of providing free packets of non-Bt cotton seed with every packet of Bt seed, the cost of educating farmers about the value of planting a refuge, and the cost of then monitoring and enforcing the refugia. The government also has a role in the education of farmers, in monitoring refuge use, and in enforcing the refugia, all of which have costs.

In China a refuge in cotton fields is not required because there are a number of alternative host crops such as corn, peanuts, and some vegetables, where the resistant bollworms can breed with the susceptible insects. An attempt to determine whether there are any benefits from cotton refugia in China has been carried out by Fangbin Qiao (2005) at the University of California, Davis. He based his model of the optimal size of refugia in China on a model of the development of resistant pests that was developed for Bt cotton at North Carolina State University. Then data on farmers' fields was collected in a survey in northern China to calibrate the model for Chinese agriculture. In addition, information on Chinese pests, which are different from the pests that attack cotton in the United States, was also included to calibrate the model. However, in the China case, he finds that the benefits from cotton refugia were smaller than the costs, and that Chinese policymakers seem to have made the economically efficient decision.

5.3. Increasing Demand for Biotech Products

If a country can establish a biosafety regulatory system that actually provides people with assurance that the risks from new technology have been examined by regulators and that these risks are at an acceptably low level, the demand for these technologies and products made from these technologies will shift outward. This shift would increase both the consumer welfare, which could be measured by an increase in consumer surplus, and producer welfare, which could be measured by an increase in producer surplus.

There is some evidence from China that such a shift would take place if consumers were more aware of the fact that the regulatory machinery was actually in place. A recent study (Zhang 2005) of consumer attitudes toward biotechnology by the Center for Chinese Agricultural Policy and Rutgers University was based on a survey of a random sample of households conducted in 11 cities from 5 provinces in the east part of China in 2003. That study found that only 20 percent of respondents knew that the food and environmental safety of GM food products was tested in China. Ten percent of consumers thought that the government had not tested food or environmental safety, and 70 percent did not know whether there were safety tests or not (see Figure 1). Knowledge that the government has conducted food and environmental safety tests of GM products can significantly increase consumer trust in the safety of GM food products. Figure 2 shows that more than 80 percent of respondents strongly agreed or somewhat agreed that their trust in the safety of GM products would increase if they knew that these products had been examined for food safety and environmental safety.

In addition, there is evidence that consumers would buy more GM foods if they knew that there was some oversight going on. Of the consumers who knew that the food and environment tests had been performed in China, 78 percent said that they would be likely to buy the disease- and pest-resistant fruit or vegetable if its price were the same as the non-GM one (Table 8). This percentage is about 5 percent higher than that in the consumer group of those who answered "no" or "don't know" in response to the question of whether the government had tested the food safety or environmental safety of GM foods. Similar results can be found for four other selected products (see Table 8). Regression analysis confirmed that these differences in the means were associated with greater acceptance of GM soybean oil and pork fed with GM feed.

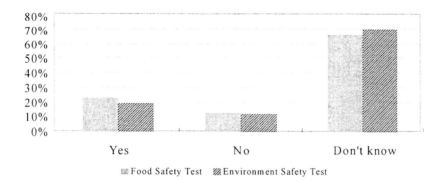

Figure 1. Response to Survey Question—Has the Chinese Government Tested for Food Safety and Environmental Safety of GM Food Products?

Source: Center for Chinese Agricultural Policy, 2003 survey.

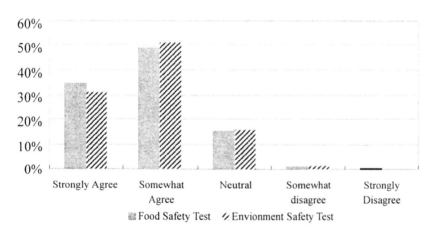

Figure 2. Response to Survey Question—Will Food and Environmental Tests Increase Your Trust in the Safety of GM Products?

Source: Center for Chinese Agricultural Policy, 2003 survey.

5.4. Summary of Benefits

Our examination of three possible cases of benefits from biosafety regulations for GM crops suggests that some of the benefits from biosafety regulations are amenable to measurement with standard economic tools. The benefits clearly are all very sensitive to the assumptions made about what the true counterfactuals would be. However, it is certainly conceivable that in

Table 8. Acceptance of GM Food by Chinese Consumers If They Think it Is Tested by the Government

	Food Safety		Environmental Safety	
	"Yes" – testing	"No" and "Don't know"	"Yes" – testing	"No" and "Don't know"
Disease- or pest-resistant GM vegetable	77.8	74.3	80.9	73.8
Oil from disease- or pest-resistant GM soybean	69.8	64.7	72.3	64.3
Disease- or pest-resistant GM rice	75.6	75.1	77.1	74.8
Nutritionally enhanced GM rice	77.3	73.6	78.2	73.5
Livestock fed by GM soybean or corn	65.0	61.3	68.3	60.7

Source: Zhang (2005).

many cases these benefits would be sufficient to justify a modest investment in the regulatory system.

6. CONCLUSIONS

Developing countries around the world are rushing to build effective biosafety regulatory systems in response to the spread of GMOs and international agreements such as the Convention on Biodiversity and the Cartagena Protocol. So far, few of them have had the time to think about the costs and benefits of building and implementing such a system, and few economists have tried to measure their costs and benefits. This chapter takes a modest step in that direction by listing some of the possible costs and benefits of regulation and then giving some examples of costs and benefits to regulation in India and China, which have two of the largest and most developed regulatory systems in the developing world.

The main costs that this chapter concentrates on are the costs that research institutes and companies must incur to comply with the data needs of the regulators. The example of Bt cotton shows that these costs can be quite substantial, particularly for foreign firms in India. In China these costs have been considerably less. It is not clear how much it will cost to get approval for commercial production of a GM food crop in either country, since none have yet been approved.

The main benefits that people believe they will achieve from regulations are reductions in the risks of health or environmental problems from GMOs. So far there is little scientific evidence of actual health risks from GMOs and less evidence that regulation in developing countries has reduced these risks. However, there are some other examples of benefits that can come from biosafety regulation. In one case, it was possible to measure the benefits that regulators could have achieved by limiting the spread of a gene that was inferior to other genes that were available. In a market with perfect information, markets would have quickly driven out the inferior product, but in China, markets for new technology are far from perfect. Assuming that when the government removed the inferior good, the better varieties would quickly move in to replace it, farmers and consumers would benefit from the regulators' actions. In fact, what seems to have happened is that the government gradually pushed the inferior varieties out, allowing the other variety time to move in, and farmers and consumers captured the benefits.

The other two examples of the benefit to regulation that could be measured were the refuge requirements in Bt cotton and the shift in the demand curve for GM food. The one empirical attempt to measure the benefits from refugia in Bt cotton in China suggested that there were no benefits from this particular regulation in China in this particular situation of many millions of small farmers with mixed cropping patterns. The example of the impact on demand for GM products of biosafety regulations suggests not only that the regulations themselves are important but that it is also important for people to know about them (only 20 percent of the Chinese population did know). When asked, Chinese survey respondents said that regulatory oversight would give them more confidence in GM food, and this was supported by the fact that those who knew about these regulations were more likely to approve of GM food.

These results suggest that more empirical research needs to be done on the costs and benefits of biosafety regulation in developing countries and that policymakers should then use this information as part of their decision making process when they consider implementing new regulatory legislation.

REFERENCES

Bennett, R.M., Y. Ismael, U. Kambhampati, and S. Morse. 2004. "Economic Impact of Genetically Modified Cotton in India." *AgBioForum* 7(3): 96–100. Available at http://www.agbioforum.org.

Chen, J. 2000. "Transgenic Plant Program." *Biotechnology Engineering Progress* 20 (special issue): 14–16.

Hu, R., C.E. Pray, J. Huang, S. Rozelle, C. Fan, and C. Zhang. 2005. "Intellectual Property, Seed and Bio-safety in China: Who Benefits from Policy Reform?" Discussion paper, Center for Chinese Agricultural Policy, Chinese Academy of Science, Beijing.

Huang, J., S. Rozelle, C.E. Pray, and Q. Wang. 2002. "Plant Biotechnology in China." *Science* 295(5555): 674–677.

Huang, J., R. Hu, S. Rozelle, and C. Pray. 2005. "Insect-Resistant GM Rice in Farmers' Fields: Assessing Productivity and Health Effects in China." *Science* 308(5722): 688–690.

Indian Agricultural Research Institute. 2003. New Delhi, India. Personal communication (December 12).

Monsanto India. 2004. Personal communication.

Peng, Y. 2000. Scientist, Institute of Plant Breeding, Chinese Academy of Agricultural Science, Beijing, China. Personal communication (July).

Pray, C.E., P. Bengali, and B. Ramaswami. 2005. "Costs and Benefits of Biosafety Regulation in India: A Preliminary Assessment." *Quarterly Journal of International Agriculture* 44(3): 267–289.

Pray, C.E., J. Huang, R. Hu, and S. Rozelle. 2002. "Five Years of Bt Cotton in China: The Benefits Continue." *The Plant Journal* 31(4): 423–430.

Pray, C.E., B. Ramaswami, J. Huang, P. Bengali, R. Hu, and H. Zhang. 2006. "Costs and Enforcement of Biosafety Regulation in India and China." In *International Journal of Technology and Globalization* (forthcoming).

Qiao, F. 2005. "Optimal Refuge Strategies to Manage the Resistance Evolution of Pests: A Case Study of Bt Cotton in China." Unpublished Ph.D. thesis chapter, Department of Agricultural and Resource Economics, University of California, Davis.

Roberts, T., J.C. Buzby, and E. Lichtenberg. 2003. "Economic Consequences of Foodborne Hazards." In R.H. Schmidt, ed., *Food Safety Handbook*. New York: John Wiley and Sons.

Sharma, M., K.S. Charak, and T.V. Ramanaiah. 2003. "Agricultural Biotechnology Research in India: Status and Policies." *Current Science* 84(3): 297–302.

Singh, S.K. 2004. "India Cotton and Products." Annual GAIN Report No. IN4047, U.S. Embassy, New Delhi, India.

U.S. Environmental Protection Agency. 2000. "Guidelines for Preparing Economic Analysis." Report No. EPA 240-R-00-003, U.S. Environmental Protection Agency, Washington, D.C. (available online at www.epa.gov/economics).

Wu, K.M. 2002. Impacts of Bt Cotton on Status of Insect Pests and Resistance Risk of Cotton Bollworm in China." Presentation at the Conference on Resistance Management for Bt-Crops in China: Economic and Biological Considerations, North Carolina State University, Raleigh, April 28.

Zhang, C. 2005. "Measuring the Impact of Biosafety Regulations on Consumer Acceptance of Genetically Modified Food in China." Master's thesis, Department of Food and Business Economics, Rutgers University, New Brunswick, NJ.

Zhang, X., J. Liu, and Q. Zhao. 1999. "Transfer of High Lysine-Rich Gene into Maize by Microprojectile Bombardment and Detection of Transgenic Plants." *Journal of Agricultural Biotechnology* 7(4): 363–367.

Chapter 23

BIOSAFETY REGULATION OF GENETICALLY MODIFIED ORPHAN CROPS IN DEVELOPING COUNTRIES: A WAY FORWARD

José Falck Zepeda and Joel I. Cohen
International Food Policy Research Institute

Abstract: Orphan crops are critical to developing country strategies for poverty alleviation. However, some productivity constraints of orphan crops cannot be addressed by conventional research, but potentially through genetically modified
(GM) crops. These undergo biosafety regulatory assessment in-country.
Significant scientific and regulatory gaps exist for orphan crops, potential
genes, and transformation protocols. Biosafety assessments of orphan crops
may require generation of new information. This becomes difficult if extensive
new data is asked for during research, especially for public sector institutions in
developing countries investing in GM technologies. We propose alternative approaches to help orphan crops move forward in their testing and development,
achieving safety goals, while taking into account the limited resources available
for this effort.

Key words: biosafety, regulations, orphan crops, developing countries

1. INTRODUCTION

Many developing countries increasingly include in their R&D agendas responses to the needs of smallholder producers. These responses are becoming an integral part of the overall strategy to address poverty alleviation.

The authors would like to acknowledge that this chapter was made possible through the
support provided for the activities of the Program for Biosafety Systems (PBS) by the Office
of the Administrator, Bureau for Economic Growth, Agriculture and Trade/ Environment and
Science Policy, U.S. Agency for International Development, under the terms of Award No.
EEM-A-00-03-00001-00. The opinions expressed herein are those of the author(s) and do not
necessarily reflect the views of the U.S. Agency for International Development. We would
also like to thank Patricia Zambrano from IFPRI for her invaluable compilation of data used
in this chapter.

Strategies may include orphan crops, as they are invaluable in helping secure the livelihood of poor producers. However, R&D policies and strategies have to be redressed in light of newer considerations of biosafety analysis mandated to signatory countries of the Cartagena Protocol, the implementation arm of the Convention on Biological Diversity. The biosafety approval process of genetically modified (GM) orphan crops is of particular interest to national agricultural research systems and international agricultural research centers. Involvement in the development of GM technologies implies compliance with biosafety regulatory requirements. Critically it is important to highlight the distinction between conventional and modern biotechnology from a regulatory point of view, because only the products of genetically modified techniques are subject to biosafety regulations. Other biotechnology techniques such as tissue culture, marker assisted selection and breeding, and mutageneses are exempt from undergoing the biosafety regulations. In the case of orphan crops, all conventional and modern biotechnology techniques have been used by the scientific community with different degrees of success.

Researchers and institutions have underscored the need to increase investments in R&D and knowledge creation for orphan crops (Naylor et al. 2004). From a regulatory and a biotechnology innovation perspective, this implies a significant knowledge gap with regard to orphan crops in general, but also with regard to the potential protocols and processes used in manipulation and derivation of useful genes to address specific needs of developing countries, including those located in tropical climates (Naylor et al. 2002). Knowledge gaps for GM orphan crops include information of parent crop and inserted gene constructs, transformation methods, and other information necessary to make a regulatory decision. This will increase the amount of time required for the biosafety assessment and decision making, but will also increase the cost of regulating orphan crops. These two developments may motivate national and international research centers to decrease or not make investments in GM orphan crops. There are a large number of productivity constraints not being resolved—in spite of years of investments by developing countries' research systems—by conventional plant breeding. Therefore, investments in GM biotechnologies for orphan crops may be critical for many countries.

GM orphan crops may contribute by reducing or resolving these productivity constraints not addressed by conventional means. However, only a few GM crops (all commercial) are presently approved for use in developing countries. This scenario is slowly changing as new GM crops and livestock products have been developed by research conducted by developing countries and their collaborating partners (Cohen 2005, Atanassov et al. 2004). All of these products require biosafety regulatory review and approval. Building regulatory capacity is underway across countries and sub-regions, in part related to compliance needs for the implementation of the Cartagena Protocol for Biosafety, supported widely by the United Nations Environment Program (UNEP) Global Environmental Facility (GEF) project. Such regulatory capacity ensures both human and environmental safety while balancing opportunities and perceived risks from biotechnology.

This chapter discusses issues that orphan crops are likely to face in a regulatory evaluation. We contend that the biosafety analysis to determine the safety (risk) attributes of these crops will not be any different than with commercial crops. However, we also argue that because of the potentially high social returns to investments that orphan crops have, scientists, regulators, and decision makers in the developing world and in the international community will need to propose innovative ways to help these crops through the biosafety regulatory process. This includes addressing the lack of familiarity and the current knowledge gap of GM orphan crops and paying closer attention to biosafety regulatory design and implementation. In addition, we propose that the overall decision making process for approval of the crop incorporates, in addition to the biosafety considerations, all the potential benefits and costs of the adoption and use of these crops. A final decision to use a particular biotechnology for orphan crops should be the result of carefully made time-limited biosafety evaluation efforts. These efforts should be based on a flexible biosafety regulatory framework that strikes a socially accepted and negotiated balance between precaution, technology use and promotion, and the overall economic growth and poverty-alleviation goals. The resulting biosafety system will need compliance with the spirit of the Cartagena Protocol on Biosafety[1] (see Section 4) and will supply technologies contributing to the livelihood of smallholder producers in developing countries.

2. BACKGROUND

Roughly half of the world's requirement for energy and protein is provided by just three crops: maize, wheat, and rice. About 30 crops provide most human nutritional needs, with 100 species composing roughly 90 percent of the food crop supply (Global Forum on Agricultural Research 2001). However, thousands of species are utilized worldwide to supply macro- and micro-nutrients, and to provide fibers, spices, medicines, and other needs. Padulosi et al. (2002) compiled a portfolio of studies that show that humans use 2,450 species for agronomic uses. Studies compiled by Padulosi et al.

---

[1] The Cartagena Protocol is our point of departure for the discussion of issues surrounding regulation of orphan crops, as the biosafety design and implementation process in most countries will be couched in this treaty.

also show that humans utilize about 100,000 plant species for all uses. These plants have been described in different forums as *underutilized, underdeveloped, neglected, minor, promising,* or *orphan* crops. All of these terms are very ambiguous, however; for convenience we will refer to them collectively in this chapter simply as orphan crops.

Because of the ongoing confusion over the definitions of underutilized and neglected crops, we will use the definitions used by the International Plant Genetic Resources Institute (IPGRI) for these categories of crops (as described in Padulosi et al. 2002 and Eyzaguirre, Padulosi, and Hodgkin 1999). IPGRI defines underutilized crops as those that were "once more widely grown but are today falling into disuse for a variety of agronomic, genetic, economic and cultural factors." In turn, neglected crops are defined as those grown primarily by farmers in their centers of origin or diversity. Farmers are critical in IPGRI's definition because they are viewed as custodians and main actors in preserving and promoting crop diversity in centers of origin and/or diversity. "Centers of origin" are defined as those regions containing an extraordinary range of wild counterparts of cultivated species, useful plants, or the geographic origins of non-cultured genera. "Centers of diversity" refer to those circumscribed regions where a particular crop shows significant genetic/phenotypic variation.

The original definitions by the Russian scientist Vavilov did not differentiate between origin and diversity; they were both one and the same. As later research has shown, a center of diversity may be located near or in a center of origin, but in many cases significantly away from a center of origin. Modern definitions open the possibility of two separate concepts. Some neglected crops may be distributed globally, but they tend to occupy restricted socio-agro-ecological niches. These crops occupy restricted agroecosystem niches, and this complicates the ability to exchange products and knowledge. It increases the complexity of activities destined for germplasm enhancement and use. Although in their restricted niche there may be significant use of these crops, they continue to be inadequately characterized and researched. Often a crop may be underutilized in some regions but not in others. In some cases, a crop may be an important component of the daily diet of millions of people, but poor marketing conditions makes it largely underutilized (and thus underrepresented) in economic terms.

Naylor et al. (2004) documents the case of 27 orphan crops that occupy areas between 0.5 and 38 million hectares each in developing countries. The total area for the 27 orphan crops described by Naylor et al. is roughly 250 million hectares. In addition, Naylor et al. estimate that there are approximately 70 million hectares planted to a wide variety of fruits and vegetables. The authors used representative yields and prices for these crops to estimate total value for these crops, which reaches almost US$100 billion in developing countries. These estimates do not include bananas, plantain, and some other major fruits and vegetables. According to Naylor et al. (2004), these crops can have an even more significant role and value within a region. In sub-Saharan Africa, sorghum and millet are planted over a larger area than are rice and wheat—41 million hectares versus 9 million hectares, respectively. The authors also make the point that roots and tubers play a dominant role in addressing food security, providing more than 400 kcal of energy per person per day, an amount significant in a region with significant nutritional deficits overall.

The benefits of utilizing orphan crops in developing countries should be considered in relation to the production constraints these crops face, including biotic and abiotic stresses. These stresses include pest and disease, and vulnerability to salt, drought, and aluminum. Many of these stresses have proved to be intractable for conventional plant breeding efforts in spite of years of investment. The advent of modern biotechnology,[2] including the genetic identification and manipulation of individual (and groups of) genes within and from outside a species, may provide opportunities to address some of these problems in a timely manner. Examples of possible developments include virus resistance in papaya, fungal resistance in bananas, and insect resistance in cotton.

The situation of orphan crops is complicated because of their existence within complex agro-ecological systems usually rich in biodiversity. Farmers in these areas manage systems that in many cases contain crops and livestock that are either underutilized, under-invested in, or in many cases neglected, especially by the private sector, but also by national and international agricultural research centers. The private sector has pointed out that orphan crops are a public good and that therefore there are limitations to their ability to recover investments. Because of the high initial costs of closing the gap in terms of knowledge of orphan crops, the total cost of doing R&D in these crops would likely be high. The small-scale use of these crops may limit the feasibility of dedicating plant breeding and biotechnology programs to this type of research, given the severely limited economies of scale that the innovator would likely capture. This may help explain the private sector's low level of investment in these types of crops (Wu and Butz 2004).

This background leads to critical biosafety questions for all stakeholders. How do these countries reconcile the need to evaluate biotechnologies for their safety, the realities of technology as an engine of growth, and their

---

[2] Cohen (1999) defines biotechnology as products arising from cellular or molecular biology and the resulting techniques coming from these disciplines for improving the genetic makeup and agronomic management of crops and animals. This definition allows us to focus on products arising from both traditional and modern biotechnology. Traditional biotechnology includes products of tissue culture, micro-propagation, and those used to eliminate diseases. Modern approaches will consider use of DNA diagnostic probes, recombinant vaccines, and products of genetic modification.

stated development goals? How is the international research community going to develop its biotechnology research agenda within the scope of nationally based biosafety systems? Developing countries' answers to these questions are complex and will be done mostly on a country-by-country and perhaps case-by-case basis. In this chapter we examine the special case of the biosafety regulations of GM orphan crops.

3. STATUS OF BIOTECHNOLOGY R&D AND REGULATIONS IN DEVELOPING COUNTRIES

Developing countries have invested human and financial resources in developing agricultural biotechnologies. Figure 1 presents results from a survey done by the International Service for National Agricultural Research (ISNAR) in six countries: China, Colombia, Indonesia, Kenya, Mexico, and Zimbabwe. In total these countries invested $193 million (in 1985 US$ purchasing power parity). These amounts are small in comparison to what multinational corporations and larger national agricultural research systems are spending in biotechnology research. Yet they represent a commitment that is likely to have increased since the survey was completed and that is reflected in the current level of outputs in the public sector.

Figure 2 presents data on GM crop biotechnology capacities in 16 developing countries. The data was compiled by the International Food Policy Research Institute (IFPRI) under the "Next Harvest" study. The Next Harvest study documents 209 unique GM events in 24 crops. These events are all products of the public sector in the 16 countries. Events documented cover distinct phenotypical categories including virus, insect, fungal, and bacterial resistance.

To obtain a proxy of the relative distribution of efforts on GM crop research performed by the international research community in the Next Harvest study, we disaggregated the data into those crops "studied" and those "not studied" by the CGIAR. This comparison is important because those crops studied by the CGIAR represent the bulk of investments in plant genetic resources improvement in the public and private sector globally. Efforts destined to those crops not served by the CGIAR thus become a (very rough) proxy for investments in GM orphan crop research. Results of this exercise are presented in Table 1.

The column labeled "Non-CGIAR Crops" in Table 1 shows 86 events grouped in 4 distinct stages of the biosafety/R&D process: laboratory/greenhouse, confined field trial, scale-up, and commercialization. Data compiled in this table show that a significant number of events are in the laboratory/greenhouse and confined field trials (78 events), and that very few are in the scale-up (3) and commercialization stages (5 events). In fact, the only country

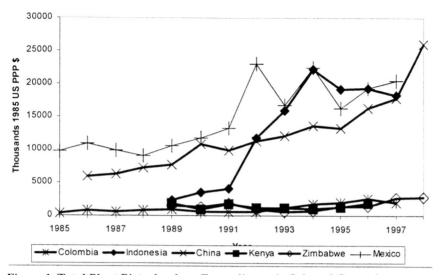

Figure 1. Total Plant Biotechnology Expenditures in Selected Countries

Notes: Compiled by authors of the International Service for National Agricultural Research (ISNAR) plant biotechnology surveys as described in Falconi (1999), Torres and Falconi (2000), and Huang et al. (2001).

with commercialization of a public sector GM crop is China. In the list of crops in this table, we can see a significant number that may qualify truly as orphan crops. A complete quantitative analysis of these crops, to determine whether to classify them as orphan or not, will be undertaken in the near future. For completeness, we have included data for all the events in the Next Harvest study studied by the CGIAR in the column labeled "CGIAR Crops."

Figure 3 groups the different crops according to continent and regulatory status. The regulatory situation is similar across continents in the sense that there is a higher proportion of GM technologies in the laboratory/greenhouse stages, with a much smaller number of technologies advancing to the confined field trial, scale-up, and commercialization phases. This is somewhat the standard distribution of a research pipeline where technologies are eliminated as they move to the commercialization phase. Although the "Next Harvest" study captured a snapshot in time of technologies in different regulatory stages, all regulators and scientists that contributed to the assembly of the data set expressed a sense of paralysis—that is, that technologies were not moving to the next stages of the regulatory process.

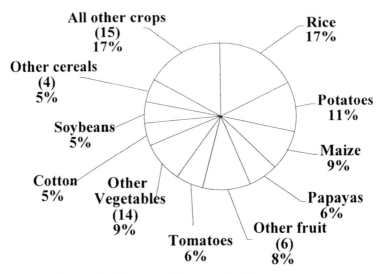

Figure 2. Distribution by Crops and Phenotypical Groups

Notes: The "Next Harvest" study conducted by the International Food Policy Research Institute (IFPRI) included 16 countries, 46 crops, and 209 transformations destined to address own problems. Phenotypical traits included in the "Next Harvest" study included insect resistance, product quality, virus resistance, fungal resistance, herbicide tolerance, agronomic properties, bacterial resistance, and others.

Source: Compiled by Patricia Zambrano from IFPRI's "Next Harvest" study. This study is described in the paper by Atanassov et al. (2004).

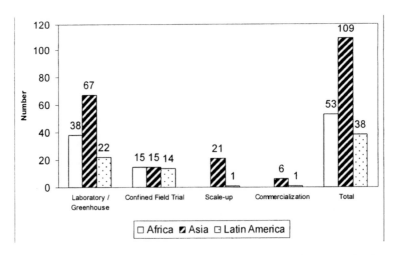

Figure 3. Regulatory Status of GM Crop Technologies by Continent in the "Next Harvest" Data Set

Source: Compiled by Patricia Zambrano from IFPRI's "Next Harvest" project as described in Cohen (2005) and Atanassov et al. (2004).

Table 1. Distribution by Country, Crop, and Regulatory Status of GM Crop Events in IFPRI's "Next Harvest" (NH) Data Set for Both CGIAR and non-CGIAR Crops

Country	Area of NH crops over total area harvested (%)	Value of production of NH crops over total value of production (%)	Food and fiber crops	CGIAR Crops					Non-CGIAR Crops	Grand Total
				Lab. / green.	Confined field trial	Scale-up	Comm. release	Total	Total	
Argentina	25.1	37.9	alfalfa, citrus, strawberry, sunflowers	13				13	8	21
Brazil	21.5	22.5	papayas		1			1	8	9
China	17.0	29.3	cabbage, chili, cotton, melons, papayas, tomatoes		3	3	5	11	19	30
Egypt	16.9	31.8	cotton, cucumber, melons, squash and marrow, tomatoes, watermelons	6	2			8	9	17
India	10.7	29.6	cabbage, cauliflower, citrus, eggplant, mung beans, muskmelon, mustard/rapeseed, tomatoes	7	4			11	10	21

cont'd.

Note: Costa Rica has 5, Kenya 3, and Mexico 3 CGIAR crop events not included in this table.

Source: Author's and Patricia Zambrano's compilation from IFPRI's "Next Harvest" and FAOStat data sets.

Table 1 (cont'd.)

Country	Area of NH crops over total area harvested (%)	Value of production of NH crops over total value of production (%)	Food and fiber crops	CGIAR Crops					Non-CGIAR Crops	Grand Total
				Lab. / green.	Confined field trial	Scale-up	Comm. release	Total	Total	
Indonesia	24.0	47.3	cacao, chili pepper, coffee, mung beans, papayas, shallot, sugar cane	8				8	16	24
Malaysia	27.2	51.7	oil palms, papayas	1	2			3	2	5
Pakistan	3.8	15.2	cotton		2			2	3	5
Philippines	25.5	34.5	mangoes, papayas, tomatoes	5				5	12	17
South Africa	19.3	30.1	apples, grapes, lupin, melons, strawberry, sugar cane, tomatoes, vegetables–indigenous	12	4			16	12	28
Thailand	21.7	31.0	cotton, papayas, pepper	4	2			6	1	7
Zimbabwe	13.5	16.7	cotton, tomatoes	2				2	3	5
Grand total				58	20	3	5	86	114	200

Note: Costa Rica has 5, Kenya 3, and Mexico 3 CGIAR crop events not included in this table.

Source: Author's and Patricia Zambrano's compilation from IFPRI's "Next Harvest" and FAOStat data sets.

4. CONCEPTS AND DEFINITIONS OF BIOSAFETY REGULATIONS AND REGULATORY DESIGN

4.1. Background

Adopted in January 2000 as a supplement to the Convention on Biological Diversity, the Cartagena Protocol on Biosafety (the Protocol) addresses the safe transfer, handling, and use of living modified organisms (LMOs) that may have an adverse effect on biodiversity, taking into account risks to human health and focusing specifically on transboundary movements (Secretariat of the Convention on Biological Diversity 2000). The Protocol allows governments to indicate their willingness to accept imports of agricultural commodities that include LMOs by communicating their decision to the world community via the Biosafety Clearing House, a mechanism set up to facilitate the exchange of information on, and experience with, LMOs. The aim is to ensure that recipient countries have the capacity to assess risks involving modern biotechnology.

National, regional, and international agencies have recognized that successful implementation of the Protocol is contingent on the development of national biosafety capacity in countries that have yet to establish, or are in the process of establishing, biosafety systems. Parties to the Protocol must develop or have access to "the necessary capacities to act on and respond to their rights and obligations," with considerable flexibility as to how importing countries may meet their obligations with respect to risk-management decision making and to the implementation of these decisions. As stated in Article 16, which deals with risk management, each party has an obligation to "establish and maintain appropriate mechanisms, measures, and strategies to regulate, manage, and control risks identified in the risk assessment provisions." Finally, the need for capacity-building should be addressed, including "the enhancement of technological and institutional capacity in biosafety" (Secretariat of the Convention on Biological Diversity 2000).

An alternative definition is that biosafety regulation is both the regulatory system and other risk analysis measures designed to ensure that applications of modern biotechnology are *safe* for human health, agriculture, and the environment. According to this definition, biosafety is a principle that tempers the adoption of a new technology with careful consideration of its potential effects on all stakeholders and on the environment.

4.2. Risk Analysis and Biosafety Regulatory Systems

Risk analysis was developed originally by academics and regulatory agencies in industrialized countries to address the need for a consistent and logical methodology framework to meet statutory mandates of protecting consumers and the environment. Regulators needed a systematic (and science-based) approach because they had to balance several conflicting issues.[3] A 100-percent safety requirement would mean that regulators would not approve any new products or activities. However, since the amount of risk associated with not approving technologies might actually be greater than the risk of approving a technology, regulators would not be truly fulfilling their responsibility to protect consumers and the environment if they failed to approve a technology whose benefits outweighed its risks.

The regulators' task is then to decide, once a risk has been identified for each product or activity, if the risk of approval is greater or lesser than the risk of non-approval. Their goal is always to choose the path that minimizes the risk to society to the greatest extent possible, tempered by the potential benefits of the technology. They accomplish this by conducting a risk assessment and then, if a product or activity is approved, putting into place risk management and communication procedures (recognizing that zero risk is unattainable). In fact, we propose in this chapter that this decision making process needs to be augmented to include the consideration of not only risks, but also of benefits and costs in the decision making process.

A standard requiring that a product be absolutely safe is unreasonable as it requires proving something that is impossible to prove. Such a standard is the result of many stakeholders conditioned to think in dichotomous terms: "A product is safe or it is not safe." We posit that this line of thinking is not correct. Clearly, the risk or safety profile of a technology is based on probabilities and its respective distribution of the effects, both positive and negative, of using technologies. This is an inherent characteristic of technology innovations and their use.

However, when the regulator performs a risk analysis to determine safety, it is done in the context of an arbitrary standard set by society that establishes a boundary between safe and unsafe use of the technology. Establishing an arbitrary level of safety puts the regulator in the middle of a logical paradox: the "Verites paradox."[4] At the heart of this paradox lies the

---

[3] The first issue is an insistence on the part of some stakeholders that GM technologies be absolutely safe before releasing them into the environment. A second is the realization that all technological innovations carry a level of risk, measured often by frequencies. A third is the societal need of technological innovation as a necessary building block for economic growth. A fourth is the lack of clarity about what constitutes a reasonable safety standard by which risk can be assessed and managed accordingly. A fifth is the lack of clarity about what is a sensible decision making process that is acceptable to all members of society. Finally there is risk associated with *not* developing new products or carrying out innovative activities. Implicit in these issues is that products and activities can have benefits, costs, and risk trade-offs.

[4] The "Verites paradox" is usually explained by using the analogy of a pile of sand. If one takes one grain of sand from the pile, does the pile stop being a pile? For the first grain of sand the answer is usually no. If one continues to take individual grains of sand from the pile, at what point does the pile stop being a pile? Removing which grain of sand makes the pile

concept of the existence of vagueness ("fuzziness") in describing the boundary between being and not-being (safety and non-safety). If an arbitrary limit for a particular substance (i.e., allergenic proteins) is set, then it can be argued that small departures above and/or below the standard do not result in a significant change in the level of safety, and therefore the standard loses its relevance. In addition, any proposed safety standard has to be tempered with the toxicology principle that indicates that all substances can be deadly (or damaging) depending on the dose and the exposure. The mere presence of a "hazardous" substance in a product does not make the product unsafe per se. Therefore, careful consideration is given to the hazard, dose, and response levels of using the substance, and to the inherent possibility that the biosafety regulatory analysis may be wrong in its conclusion.[5]

4.3. General Approaches to Biosafety Regulations

All regulatory processes are by their nature precautionary to a degree. Controversy can arise when these basic principles are misunderstood or misconstrued by actors in the process and extreme interpretations of a particular risk and/or implementation are made. We can identify two general approaches to biosafety regulation implemented globally. These are the substantial equivalence and the precautionary approaches. Both approaches are not necessarily incompatible with each other; controversy generally arises when strict interpretation and implementation of either approach is used by the regulatory framework as its foundation.[6]

Substantial equivalence. One general approach is called "substantial equivalence." This is a standard employed by OECD (Organization for Economic Cooperation and Development), WHO/FAO (World Health Organization/Food and Agriculture Organization), and different agencies in the United States. Biotech products need to be proven to be *as safe* as products pro-

---

no longer a pile? The fuzzy boundary between being and not-being is a direct result of Aristotle's dichotomous thinking processes.

[5] To address this issue, the standard is typically raised by a specified multiple to ensure that susceptible individuals (e.g., the elderly and children) are not harmed by the presence of a toxic substance in food. This safety margin is generally arbitrary; very rarely is there a good scientific base for setting such a value.

[6] The differences between both processes are directly linked to the difference between the social and scientific rationalities of risk analysis frameworks as described in Isaac (2004). The precautionary approach is linked to the social rationality, whereas the substantial equivalence is linked to the scientific rationality. Our paper proposes a pathway that considers both approaches jointly for biosafety regulatory design and implementation that will lead to the desired income of GM technologies that are determined to be safe and effective. Most importantly, ones that are developed to address the needs of resource-poor farmers in developing countries.

duced by their conventional counterparts. The focus is on considering the types and levels of risk associated with biotechnology in light of the "proven" safety of the counterpart. This approach embodies the recognition that to hold biotechnology to a higher standard than its conventional counterpart may result in opportunity costs or loss of broader benefits to society. For example, discouraging the option of using pest-resistant biotechnology crops (through over-regulation or bans) may result in the lost opportunity to reduce pesticide use on certain crops because the predominant alternative currently used is pesticides.

The substantial equivalence approach is closely linked to the concepts of product (or products) deemed to be GRASS ("generally recognized as safe") and with "familiarity of use." The concept held by U.S. and other countries' regulators is that a product derived from another product (or products) that have been GRASS and whose use is familiar is itself safe. Therefore, substantial equivalence, GRASS, and familiarity are interconnected in that a biotechnology innovation has to be proven to be at least as safe (substantial equivalence) as its best conventional counterpart. Proving a GM biotechnology to be as safe as its conventional counterpart is facilitated if the parent crop and the gene utilized in the transformation are themselves considered to be GRASS and to have a history of safe use (familiarity). If there is a significant change in the product so that it is no longer "substantially equivalent," then the product continues to undergo additional tests to better characterize its safety profile. The regulation of GM orphan crops and their biosafety analysis is further complicated by the higher probability that the crop has not been characterized with regard to nutrients and anti-nutrients and amino acid and protein profiles compared to commercial crops.

Precautionary approach. The Cartagena Protocol makes reference to a "precautionary approach" and reaffirms the precaution language in Principle 15 of the Rio Declaration on Environment and Development. The principle states that, "In order to protect the environment, the precautionary approach shall be widely applied by States according to their capabilities. Where there are threats of serious or irreversible damage, lack of full scientific certainty shall not be used as a reason for postponing cost-effective measures to prevent environmental degradation." This approach can halt importation of LMOs, or genetically modified seeds, if determined that sufficient scientific information is not available as regards their environmental impact. This approach has also been referenced in different treaties and publications, yet there is not a consensus of what exactly is involved in determining an acceptable level of safety for society in general.

4.4. Converging Towards New Models of Biosafety Regulation Design and Implementation

Differences of perception between substantial equivalence and precautionary approaches emphasize the need for new models of biosafety design and implementation that empower developing countries to address their own local situation and capacity. One potential area of exploration is new models that seek to streamline regulations for GM applications.[7] Of particular interest are those models that work with technologies considered low risk and with which sufficient familiarity has been gained through extensive use. In regulatory design, the level of regulatory oversight should match the presumed level of risk. This scientifically sound approach is a long-established paradigm that applies not only to GMOs but to other regulated technologies as well. It does not in any way advocate less than a full regulatory review, but what it does tell us is that what constitutes a full regulatory review is not the same in all cases. We contend that following this approach is scientifically sound, appropriate to the risk involved, and consistent with the Cartagena Protocol.[8]

4.5. Steps in the R&D/Biosafety Implementation Process

Scientists and regulators make an initial assessment of the GM constructs or GM events (a combination of a specific gene construct and a crop) before starting a research project. Those research projects that are deemed to have the potential to affect the environment or public health have to be conducted in fully contained facilities such as a greenhouse (for example see India's Department of Biotechnology guidelines in Table 2). Otherwise the research may be conducted under confinement conditions using different isolation techniques such as erecting physical barriers or otherwise eliminating the possibility of dispersion of vegetative material or, for sexually reproducible plants, the removal of flowers. The initial evaluation of a new innovation's

---

[7] It is worthwhile to note that this is a common element in most functional national biosafety systems (see, for example, Article 7 and Annex V of the EU Directive 2001/18/EC on the deliberate release of GMOs, and Article 13 of the Cartagena Protocol).

[8] For example, Article 15 of the Cartagena Protocol ("Risk Assessment") states that "Risk assessments undertaken pursuant to this Protocol shall be carried out in a scientifically sound manner, in accordance with Annex III and taking into account recognized risk assessment techniques. Such risk assessments shall be based, at a minimum, on information provided in accordance with Article 8 and other available scientific evidence in order to identify and evaluate the possible adverse effects of living modified organisms on the conservation and sustainable use of biological diversity, taking also into account risks to human health."

In addition, in Annex III ("Risk Assessment") it states that "Risk assessment should be carried out on a case-by-case basis. The required information may vary in nature and level of detail from case to case, depending on the living modified organism concerned, its intended use and the likely potential receiving environment."

Table 2. Categories of Genetic Engineering Experiments and Containment Requirements on Plants and Notifications in India's Biosafety Approval Process

Category	Description	Containment requirement
I.	- Routine cloning of defined genes - Defined non-coding stretches of DNA - Open reading frames in defined genes of *E. Coli* or other bacterial and fungal hosts that are generally considered as safe - This category of experiment needs communication only with IBSC	Routine good laboratory practices
II.	- Includes laboratory, greenhouse, and net house experiments in contained environment - Defines DNA fragments non-pathogenic to humans and animals used for genetic transformations - Category requires permission by IBSC to perform experiments. Decision by IBSC would be communicated to the RCGM before execution, and RCGM would put this information on record	Experiments need to be performed in green house / net house facilities whose design ensures arrest of transgenes within the facility
III.	- High-risk experiments where escape of transgenic trait into the open environment could cause significant alterations to biosphere, ecosystem, and plants and animals, the effects of which cannot be judged precisely - Category requires permission by the RCGM and notification to the Department of Biotechnology	Experiments need to be performed in greenhouse / net house facilities whose design ensures near complete isolation from the open environment and prevention of the entry of insects

Note: "IBSC" stands for Institutional Biosafety Committee and "RCGM" is the Review Committee for Genetic Manipulation.

Source: Department of Biotechnology, India (1998).

risk is based on familiarity, accumulated knowledge of the parent crop, event, and the method used to transform the product.

Regulatory processes include a sequential set of stages that require applications to institutional, regional, and/or national biosafety committees for initiating research in a stage. Regulators examine these applications, considering the parent crop, the transformation method, the gene construct, and the GM crop in relation to food, feed, and environmental safety issues (see Table 3). A typical set of stages may include the following:

- laboratory
- glasshouse/greenhouse (contained)
- confined field trials
- scale up (extended) field trials
- commercialization

As each of these stages requires the innovator to submit an application for advancement to a later stage as part of the compliance process, biosafety applications become a tool that supports risk analysis of GM technologies. If not enough consideration is given to the design of the application forms, the applicant may end up needing to submit unnecessary data to the regulators. For example, the Kenya Biosafety Committee and the Program for Biosafety Systems (a biosafety capacity-building program coordinated by IFPRI) contributed significantly to the evaluation of application forms in Kenya. After successful revisions of the forms with the active participation of Kenyan regulators and decision makers, the forms were reduced from 30 pages to approximately 5 pages, by sharpening focus and factoring in questions of local safety importance. All of the issues above, if not addressed accordingly, may increase the cost of the biosafety regulatory process, on top of the R&D necessary to close the gap on knowledge on orphan crops.

5. BIOSAFETY REGULATORY COSTS

There are growing concerns that insufficient funding (together with insufficient expertise) may hamper the implementation of effective biosafety systems in many if not most developing countries. Likewise, there are also concerns that regulatory compliance costs may pose an entry barrier to biotechnology products coming from public research institutions, universities, and local private firms. In particular, excessive requirements and overly stringent conditions imposed by biosafety committees may inhibit technology transfer and adoption in developing countries. An important question that emerges concerns the trade-off between investments to reach additional safety or reduced risk, and benefits to society from such investments.

Table 3. Issues Potentially Considered During a Biosafety Evaluation Process

What is examined?	Characterization/familiarity	Food/feed safety	Environmental safety
parent crop	• history of safe use • toxins/allergens • asexual/sexual reproduction characteristics	• nutritional composition • anti-nutrients	• existence of wild relatives • center of diversity and/or origin
transformation process	• history of safe use • sources of genes • insertion process • gene construct • markers	• existence of markers	not done
gene product	• history of safe use • protein purification • mechanism of action/specificity • homology	• potein characterization, structure, and expression • digestibility • bioavailability • acute toxicity • allergenicity	• vertical and horizontal gene transfer potential

the GM crop/food/feed	• history of safe use
	• inheritance and phenotypic stability after transformation
	• DNA sequence of the insert in the plant genome
	• border sequence
	• expression data
	• compositional analysis – protein and amino acid profile
	• nutritional equivalence
		• in vitro digestibility assay
		• skin allergenicity
		• dermal
		• heat stability and amino acid homology
		• toxicology assessment
		  - acute and repeat-dose oral, inhaled and dermal toxicity
		  - mutagenicity
		  - sub-chronic and chronic toxicity
		  - oncogenicity
		  - effects on the immune and endocrine systems
		• dietary risk characterization
			• weediness of crop plant
			• out-crossing/gene flow: potential
			• out-crossing/gene flow impacts
			• effects on non-target organisms
			• effects on biodiversity

There are few estimates of the cost of compliance with the biosafety regulatory process. Table 4 presents a compilation of some of the estimates presented in different venues. These estimates for the cost of regulation vary from US$0.9–4.0 million. These estimates do not consider the cost of R&D to develop the GM technology or the opportunity cost of capital investments. The range of regulatory costs presented here varies according to country, accumulated knowledge, information about the transformation, type of event, and the inherent cost differences between countries. In many cases it is very difficult (if not impossible) to separate R&D costs from those of the regulatory process, as the regulatory process may have been done by the R&D process in any case. In most cases, the separation between the two types of costs will be subjective and arbitrary.

As indicated above, these costs do not include the (opportunity) cost of capital. DiMasi (2003), in an estimate of the cost of pharmaceutical R&D and regulations, showed that because of the time lag due to R&D and regulations, the opportunity cost of capital can be as high as 50 percent of the total cost of development. For pharmaceuticals, the total cost of R&D and regulations can be as high as $800 million dollars, of which half is the opportunity cost of capital.

Development of a GM product is the first step in the adoption/diffusion process. Additional and probably significant expenses need to be added to the cost of development. These may include costs of intellectual property rights, extension, knowledge sharing, segregation, traceability identity preservation, labeling, and international trade effects. From the standpoint of orphan crops, the main lesson from the review of these preliminary studies is that the cost of doing R&D for orphan crops will be increased significantly by the cost of regulation. Cost of regulation may run in the millions of dollars depending on the complexity of the process, which in turn is based on the degree of familiarity-accumulated knowledge. Since accumulated knowledge on regulatory issues of orphan crops is not as significant as with cereals and other large-scale crops, obtaining this knowledge will probably require significant investments. In some cases the cost of regulation may be higher than the cost of R&D, thus aggravating the investment problems that developing countries face with underutilized and orphan crops.

6. STRATEGIES FOR REGULATORY ACTION

So far, experience with food and environmental safety of GM crops has been positive. Yet, as more technologies are being developed and evaluated, there will be increased pressures on existing biosafety systems. Implementation of the Cartagena Protocol, with the assistance of UNEP-GEF and other projects, including the Program for Biosafety Systems (PBS) by IFPRI and

Table 4. Estimates Cost of Biosafety Regulations

Type of crop (example)	Crop	Country	Event approved in developed countries	Estimated costs of biosafety regulations
Food crop	maize	India	yes	$500,000–1,500,000
	maize	Kenya	yes	980,000
	rice	India	no	$1,500,000–2,000,000
	rice	Costa Rica	no	2,800,000
	beans	Brazil	no	700,000
	vegetable	India	no	$4,000,000
	soybeans	Brazil	yes	4,000,000
	potatoes	South Africa	yes	980,000
	potatoes	Brazil	yes	980,000
	papaya	Brazil	yes	
Non-food crop	cotton	India	yes	$500,000–1,000,000
	jute	India	no	$1,000,000–1,500,000

Source: Compilation by authors of estimates by Quemada (2003), Odhiambo (2002), Sampaio (2002), Sittenfeld (2002), and in a study by Pray, Bengali, and Ramaswami (2004).

other partners, has meant significant review and revisions of the regulatory system. This has meant a strong movement away from interim processes that utilize existing laws and guidelines for conducting initial biosafety evaluations.

A significant gap in the current systems is the lack of appropriate open use or commercialization approval mechanisms, as well as post-approval monitoring systems. In the next section we propose two distinct avenues to help move GM orphan crops through the regulatory system while at the same time addressing meaningfully the issue of ensuring a socially determined level of safety. One avenue addresses issues pertaining to orphan crops specifically. The second avenue addresses improving the overall effectiveness of the general biosafety process, which in turn will help GM orphan crops move forward. In both cases, the main theme is to overcome the fundamental problems of compliance with biosafety regulations, while dealing with the issues of lack of familiarity and accumulated knowledge on these crops, and the well-known problems of R&D in the developing world.

6.1. Orphan Crop Specific Recommendations

Increase in resources dedicated to orphan crop R&D and regulatory issues. A significant increase in the amount of resources dedicated to the conservation *in situ* and *ex situ* of orphan crops is required for ensuring preservation of global biodiversity. The United Nations' Global Diversity Trust Fund is one of the responses of the international community in this respect. However, there is no equivalent fund for the sustainable improvement and use of global biodiversity. In particular there is not enough support for struggling plant breeding programs that could conduct the research necessary to close the knowledge gap on orphan crops. Investments in R&D and the genetic characterization of key orphan crops should help close the current knowledge gap of these crops.

Closing the gap in knowledge and the use of innovative models. Investments in orphan crops should be consistently directed to address the issue of familiarity. One potential avenue is to explore models that would consider orphan crops as family or "related" groups and therefore use data from related crops as the basis for R&D and regulation processes. This process is akin to what is proposed for R&D in Naylor et al. (2002) and Taylor, Kent, and Fauquet (2004). A second avenue for evaluating regulatory testing is the use of risk modeling techniques. This could help avoid unnecessary data collection and analysis.

6.2. General Biosafety Processes Recommendations

Flexibility in the design of regulatory frameworks. Regulatory frameworks should be designed to have sufficient flexibility to ensure their appropriate implementation. Issues to consider are the following:

- Ability to recognize and critically accept data generated elsewhere, in particular for characterization and food safety. Although there may be quite a bit of latitude in utilizing information for food safety characterization, we believe that there is less scope for this in the assessment of environmental safety. However, there are some proposals to utilize a diverse set of databases and methodologies for the initial safety evaluation of environmental issues (Hancock 2003).

- Exploration of the possibility of implementing a single-window–single-regulator mechanism as in Australia and now proposed for India. The rationale behind this approach is to facilitate the process for applicants and at the same time to centralize decision making under one roof.

- Exploration of the continued utilization of formal frameworks such as the ISNAR/IFPRI-FAO biosafety conceptual framework. This framework was designed to establish a thought process behind the design of the biosafety regulatory system in developing countries by examining issues such as the inventory of human, physical, and financial resources, as well the policy and legislative status of the country. Application of this framework clearly spells out the different trade-offs and options that may be allowed for in the design, implementation, and revision of the biosafety system.

Rationalizing the application process. Taylor, Kent, and Fauquet (2004) document the case of a proposed GM cassava submitted for Nigeria's review. The application form at the time consisted of 150 questions; the Donald Danforth Plant Sciences Center's application totaled 60 pages. Completing the application required a significant amount of human and financial resources that many public sector institutions may not have. The regulatory process must ensure that the relevant questions are being asked at each stage of the biosafety regulatory process and that no unnecessary information is requested.

Prescriptive versus performance-based regulations. Most regulatory systems establish a set of regulatory processes to be complied with by the applicant. In some cases, there is very little flexibility contained within the regulatory system to adapt to accumulation of knowledge on and experience with the specific crop and gene construct, and with GM crops in general. In contrast, performance-based regulations give applicants the flexibility to comply with a set of standards through innovative ways and thus respect the goals of the regulatory process.

Understanding the effects of regulation. Regulators and policymakers need to understand regulatory policies in terms of costs, benefits, and safety trade-offs between approval and non-approval of a technology. It is important to understand the cumulative and in some cases unanticipated impacts of regulation. This process should help foster a greater sensitivity to the constraints that applicants and technology innovators are likely to face when dealing with regulations, while at the same time contribute to an environment where collaboration with all stakeholders is enabled and thus contributes to ensure the success of regulation.

Greater use of risk analysis and decision making processes. As described in this chapter, regulatory systems in general have been characterized by a marked emphasis on minimizing the possibility of a catastrophic event. In many cases, regulatory systems consider safety issues only, without considering the potential cost of not approving a technology. We contend that there has to be a more comprehensive use of risk analysis techniques that would expand the decision making process to include consideration of all the relevant information, including the safety profile, the potential benefits and magnitude of the constraint, and livelihood issues.

7. CONCLUDING REMARKS

In this chapter we have discussed several issues related to the regulation of orphan crops. It is our contention that there will be few differences between the evaluation process for orphan crops and for commercial crops. Although the design of regulations cannot address specific orphan crops, there is considerable scope to facilitate the movement of GM innovations deemed to be of value through the biosafety regulatory process. The main issues to be addressed are the lack of familiarity with the crop, the transformation process of the crop, and the paucity of appropriate genes to address the needs of developing countries, particularly those located in tropical climates. We propose a distinct set of orphan crop specific and general strategies to streamline and improve the efficiency of the implementation of the biosafety process.

REFERENCES

Atanassov, A., A. Bahieldin, J. Brink, M. Burachik, J.I. Cohen, V. Dhawan, R.V. Ebora, J. Falck-Zepeda, L. Herrera-Estrella, J. Komen, F.C. Low, E. Omaliko, B. Odhiambo, H. Quemada, Y. Peng, M.J. Sampaio, I. Sithole-Niang, A. Sittenfeld, M. Smale, Sutrisno, R. Valyasevi, Y. Zafar, and P. Zambrano. 2004. "To Reach the Poor: Results from the ISNAR-IFPRI Next Harvest Study on Genetically Modified Crops, Public Research, and Policy Implications." EPTD Discussion Paper No. 116, Environment and Production Technology Division, International Food Policy Research Institute, Washington, D.C.

Cohen, J.I. (ed.) 1999. "Managing Agricultural Biotechnology." Wallingford, UK: CABI Publishing.

_____. 2005. "Poorer Nations Turn to Publicly Developed GM Crops." *Nature* 23(1): 27–33.

DiMasi, J.A., R.W. Hansen, and H.G. Grabowski. 2003. "The Price of Innovation: New Estimates of Drug Development Costs." *Journal of Health Economics* 22(2): 151–185.

Department of Biotechnology – India. 1998. "Revised Guidelines for Research in Transgenic Plants and Guidelines for Toxicity and Allergenicity of Transgenic Seeds, Plants and Plant Parts." Department of Biotechnology, Ministry of Science and Technology, Government of India, New Delhi.

Eyzaguirre, P., S. Padulosi, and T. Hodgkin. 1999. "IPGRI's Strategy for Neglected and Underutilized Species and the Human Dimension of Agrobiodiversity." In S. Padulosi, ed., *Priority Setting for Underutilized and Neglected Plant Species of the Mediterranean Region* (report of the IPGRI Conference, February 9–11, 1998, in Aleppo, Syria). International Plant Genetic Resources Institute, Rome, Italy.

Falconi, C. 1999. "Agricultural Biotechnology Research Capacity in Four Countries." ISNAR Discussion Paper No. 42, International Service for National Agricultural Research, The Hague, the Netherlands.

Food and Agricultural Organization. 2004. "State of Food and Agriculture 2003–2004." FAO Agriculture Series No. 35, Food and Agriculture Organization of the United Nations, Rome, Italy.

Global Forum on Agricultural Research (GFAR). 2001. "Under-utilized and Orphan Species and Commodities – A Global Framework for Action." A position paper of the GFAR-UOC Secretariat, Rome, Italy.

Hancock, J.F. 2003. "A Framework for Assessing the Risk of Transgenic Crops." *BioScience* 53(5): 512–519.

Huang, J., Q. Wang, Y. Zhang, and J. Falck-Zepeda. 2001. "Agricultural Biotechnology Research Indicators: China." ISNAR Discussion Paper, No. 01-5, International Service for National Agricultural Research, The Hague, the Netherlands.

Isaac, G. E. 2004. "The Interaction Between Levels of Rule Making in International Trade and Investment: The Case of Sanitary and Phytosanitary Measures." Discussion paper prepared for the "Workshop on the Interaction Between Levels of Rule Making in International Trade and Investment UNU CRIS/LSE ITPU Project" in Brussels, Belgium, December.

Naylor, R.L., W.P. Falcon, R.M. Goodman, M.M. Jahn, T. Sengooba, H. Tefera, and R.J. Nelson. 2004. "Biotechnology in the Developing World: A Case for Increased Investments in Orphan Crops." *Food Policy* 29(1): 15–44.

Naylor, R.L., R. Nelson, W. Falcon, R. Goodman, M. Jahn, J. Kalazicgh, T. Sengooba, and H. Tefera. 2002. "Integrating New Genetic Technologies Into the Improvement of Orphan Crops in Least Developed Countries" Paper presented at the 6th International ICABR (International Consortium on Agricultural Biotechnology Research) Conference on Agricultural Biotechnologies: New Avenues for Production, Consumption and Technology Transfer, July 11–14, in Ravello, Italy.

Odhiambo, B. 2002. "Products of Modern Biotechnology Arising from Public Research Collaboration in Kenya." Paper presented at the conference "Next Harvest – Advancing Biotechnology's Public Good: Technology Assessment, Regulation and Dissemination," October 7–9, International Service for National Agricultural Research, The Hague, the Netherlands.

Padulosi, S., T. Hodgkin, J.T. Williams, and N. Haq. 2002. "Underutilized Crops: Trends, Challenges and Opportunities in the 21st Century." In J.M.M. Engels, V.R. Rao, A.H.D. Brown, and M.T. Jackson, eds., *Managing Plant Genetic Diversity*. Oxon, UK: CABI Publishing.

Pray, C.E., P. Bengali, and B. Ramaswami. 2004. "Costs and Benefits of Biosafety Regulation in India: A Preliminary Assessment." Paper presented at the 8th ICABR (International Consortium on Agricultural Biotechnology Research) Conference, July 8–11, Ravello, Italy.

Quemada, H. 2003. "Developing a Regulatory Package for Insect Tolerant Potatoes for African Farmers: Projected Data Requirements for Regulatory Approval in South Africa." Paper presented at the symposium "Strengthening Biosafety Capacity for Development," June 9–11, Dikhololo, South Africa.

Sampaio, M.J. 2002. "Ag-Biotechnology GMO Regulations/IP Progresses and Constraints." Paper presented at the conference "Next Harvest – Advancing Biotechnology's Public Good: Technology Assessment, Regulation and Dissemination," October 7–9, International Service for National Agricultural Research, The Hague, the Netherlands.

Secretariat of the Convention on Biological Diversity. 2000. "Cartagena Protocol on Biosafety to the Convention on Biological Diversity: Text and Annexes." Secretariat of the Convention on Biological Diversity, Montreal, Québec. Available online at http://www.biodiv.org/doc/legal/cartagena-protocol-en.pdf (accessed July 28, 2005).

Sittenfeld, A. 2002. "Agricultural Biotechnology in Costa Rica: Status of Transgenic Crops." Paper presented at the conference "Next Harvest – Advancing Biotechnology's Public Good: Technology Assessment, Regulation and Dissemination," October 7–9, International Service for National Agricultural Research, The Hague, the Netherlands.

Taylor, N., L. Kent, and C. Fauquet. 2000. "Progress and Challenges for the Deployment of Transgenic Technologies in Cassava." *AgBioForum* 7(1&2): 51–56.

Torres, R., and C. Falconi. 2000. "Agricultural Biotechnology Research Indicators: Colombia." ISNAR Discussion Paper No. 00-5, International Service for National Agricultural Research, The Hague, the Netherlands.

Williams, J.T., and N. Haq. 2002. "Global Research on Underutilized Crops: An Assessment of Current Activities and Proposals for Enhanced Cooperation." International Centre for Underutilised Crops, Southampton, UK.

Wu, F., and W.P. Butz. 2004. "The Future of Genetically Modified Crops: Lessons from the Green Revolution." RAND Corporation Monograph Series MG-161, RAND Corporation, Santa Monica, California.

Chapter 24

Bt RESISTANCE MANAGEMENT: THE ECONOMICS OF REFUGES

George B. Frisvold
University of Arizona

Abstract: Refuge requirements have been the primary regulatory tool used to delay pest resistance to Bt crops. This chapter presents a simple method to estimate the annual cost of refuges to producers, applying it to Bt cotton. It also examines broader welfare impacts, estimating how Bt cotton acreage restrictions affect producer surplus, consumer surplus, seed supplier profits, and commodity program outlays. The implications of grower adoption behavior—partial adoption, aggregate adoption, and refuge choice—for regulatory costs are examined. Empirical examples illustrate how providing multiple refuge options significantly reduces regulatory costs.

Key words: resistance management, refuges, technology adoption, Bt cotton

1. INTRODUCTION

A growing literature reports that transgenic cotton varieties producing toxins from *Bacillus thuringiensis* (Bt) have led to significant economic benefits from yield gains, reductions in conventional insecticide sprays, or both throughout the world (Cotton Research and Development Council 2002; Doyle, Reeve, and Barclay 2002; Falck-Zepeda, Traxler, and Nelson 2000a, 2000b; Frisvold and Tronstad 2002; Frisvold, Tronstad, and Reeves 2006; Gianessi et al. 2002; Huang, Hu, Rozelle, Qiao, and Pray 2002; Huang, Hu, and van Tongeren 2002; Huang, Rozelle, Pray, and Wang 2002; Ismael, Bennett, and Morse 2002; Marra 2001; Pray et al. 2001; Pray et al. 2002; Price et al. 2003; Qaim, Cap, and de Janvry 2003; Qaim and Zilberman 2003; Traxler et al. 2002). Estimates of U.S. domestic welfare benefits from Bt cotton adoption have been in the range of $150–$250 million annually (Falck-Zepeda, Traxler, and Nelson 2000a, 2000b; Frisvold and Tronstad 2002: Frisvold, Tronstad, and Reeves 2006; Price et al. 2003).

The benefits of Bt cotton could be short-lived, however, if there were rapid evolution of pest resistance. The main targets of the first generation of Bt cotton are tobacco budworm (*Heliothis virescens*), cotton bollworm (*Helicoverpa zea*), and pink bollworm (*Pectinophora gossypiella*) in the United States, and *Helicoverpa armigera* in China. In the United States, resistance evolution in budworm and cotton bollworm to organochlorine, organophosphate, and carbamate insecticides greatly reduced the effectiveness of these compounds by the late 1970s (Livingston, Carlson, and Fackler 2004). Resistance evolution in budworm to pyrethroid insecticides, a replacement for these other compounds, led to field failures and large yield losses (29 percent) in Alabama in 1995 (Williams, various years; Gianessi et al. 2002). Because transgenic Bt crops express the toxin continuously, many entomologists believed that pests would quickly evolve resistance to Bt crops. This belief was based on prior experience with other pesticides, laboratory selected resistance to Bt toxins in insects, and the development of in-field resistance to Bt sprays by diamondback moth (Tabashnik et al. 2003).

Foliar Bt sprays, long used in pest control, are less toxic to non-target insect species, birds, and mammals than broad-spectrum pesticides and are the most widely used insecticides in U.S. organic agriculture (Hutcheson 2003). Environmental and organic farming groups have raised concerns that widespread planting of Bt crops could speed evolution of pest resistance to the Bt toxin and potentially undermine IPM and organic pest control strategies.

The U.S. Environmental Protection Agency (EPA) has authority to manage pesticide resistance under FIFRA (the Federal Insecticide, Fungicide, and Rodenticide Act). To delay the evolution of resistance, EPA requires growers who plant Bt cotton to also plant non-Bt cotton on a minimum percentage of their total cotton acreage. These non-Bt acres serve as a refuge for susceptible pests, allowing them to survive and mate with adults that have become resistant to the Bt toxin and thereby delay the development of resistance in the pest population. Experimental evidence and results from entomological simulation models suggest that refuges can significantly delay the onset of resistance (Carrière and Tabashnik 2001, Tabashnik et al. 2003).

For agricultural producers, refuges imply an intertemporal economic trade-off. Refuge requirements mean that growers must forego the annual, short-run benefits of Bt crops on a portion of their planted acres. Refuges, however, preserve the efficacy of Bt varieties over a longer period of time. Both economic theory and experience with conventional insecticides suggest that individual growers may over-use pest control technologies relative to the social optimum, bringing about resistance prematurely (Carlson and Wetzstein 1993). For growers the trade-off is between receiving annual benefits, B, of the Bt technology for T years versus the benefits net of the opportunity cost of the refuge $B - C$ for $T + N$ years. The overall impact of refuge requirements on the present value of grower returns will depend on B,

C, T, and N, as well as the discount rate, where $C = C(N)$ and $C'(N) > 0$. Larger refuges are more costly, but purchase greater longevity of the Bt technology. The long-run impacts of a given refuge policy will also depend on the availability, timing, and pricing of backstop technologies. A backstop exists in the form of crop varieties with different or multiple Bt toxins that could replace first-generation Bt crop varieties.

1.1. Aims and Scope of Chapter

Previous economic studies of refuges have focused on normative analysis, asking, what would the optimal refuge strategy be? Management of resistance via refuges is treated as an optimal control or dynamic programming problem, where the refuge configuration that maximizes the present value of grower returns is estimated and compared to existing refuge requirements (Hurley, Babcock, and Hellmich 2001; Hurley, Secchi, Babcock, and Hellmich 2002; Laxminarayan and Simpson 2002; Livingston, Carlson, and Fackler 2004; Secchi, Hurley, and Hellmich 2001; Secchi and Babcock 2003). Studies have examined potential grower gains from altering refuge size and type and from allowing refuge sizes to adjust over time. Some model results suggest that the present value of grower returns would be higher with smaller refuges than are currently required, while others provide justification for current policies. Studies generally find that grower returns are higher with some refuge than with no refuge at all. Laxminarayan and Simpson (2002) examine the conditions under which it might be desirable to delay introduction of refuges. Other simulation results suggest that gains from dynamic refuges would be quite modest (Livingston, Carlson, and Fackler 2004; Secchi, Hurley, and Hellmich 2001).

In contrast to these studies, this chapter focuses on positive analysis— assessment of actual (rather than optimal) refuge policies. What are the short-run, annual costs of Bt cotton refuge requirements in the United States? How are the costs of current regulations distributed between different groups in the economy? How do the costs and benefits of refuge regulation compare to a case with no regulations at all? Under what conditions are long-run producer returns greater under current refuges than without them?

The remainder of the chapter proceeds as follows. Section 2 addresses the short-run annual costs of refuge requirement in U.S. cotton production. A simple method is introduced to estimate the annual cost of refuges. An important advantage of this method is its reliance on data that is reported in numerous available studies of the economic benefits of Bt crops. We next demonstrate why this simple refuge cost calculation represents an upper-bound cost estimate. We then use this method to develop upper-bound estimates of Bt cotton refuge costs in the United States. A more complete regulatory impact analysis of refuge requirements would account for the affects on cotton purchasers, the seed supply industry, and the federal budget deficit. Section 2 concludes by presenting estimates of the distribution of costs of Bt cotton acreage restrictions across different segments of the U.S. economy.

Section 3 develops trade-off curves illustrating how different factors affect the longer-term benefits and costs of refuge policies. An iso-cost or trade-off surface is derived that represents the impacts of refuges on producer returns as a function of time to resistance with and without refuges. While modest data requirements define the shape of the iso-cost surface, entomological data is needed to determine where growers in a region are on that surface. A major result from this exercise is that, if current U.S. refuge requirements prevented resistance from developing by 2002, then those refuge requirements "paid for themselves." In other words, grower returns would be higher with the current refuge policies than with no refuges.

Section 4 considers an empirical example of refuge costs and pink bollworm resistance in Arizona. The trade-off curve approach is applied to estimate a range of estimates for refuge costs. The Arizona example also illustrates the importance of grower adoption decisions and the availability of multiple refuge options in lowering overall regulatory costs.

2. Bt COTTON REFUGE REQUIREMENTS

Beginning in 1996 the EPA required growers planting Bt cotton to adopt one of two refuge options. Based on entomological simulation models, the EPA believed that the two options would produce comparable levels of susceptible lepidopteran pests. First, the 80:20 external sprayed refuge required growers to plant 25 acres of non-Bt cotton for every 100 acres of Bt cotton (i.e., 20 percent of total acreage in non-Bt cotton). The refuge could be sprayed for lepidopteran pests, but not with foliar Bt sprays. The second, 96:4 external unsprayed refuge option required growers to plant four acres of non-Bt cotton for every 100 acres of Bt cotton planted (3.85 percent of total acreage in non-Bt cotton). This refuge could not be sprayed for lepidoptera. In counties where Bt cotton accounted for 75 percent or more of cotton acres, refuges had to be planted within one mile of Bt cotton.

In 2001, EPA altered and expanded refuge options (Matten and Reynolds 2003, EPA 2001). The 80:20 sprayed option was maintained, but refuges were required to be planted within one mile of Bt cotton. The 96:4 external unsprayed option was changed to a 95:5 requirement (5 percent non-Bt cotton), with the distance limit set at one-half mile. The EPA also permitted a 95:5 embedded refuge option that allows a 5 percent block of non-Bt cotton to be planted within a Bt cotton field. If the field is treated for lepidoptera, the refuge may be treated with the same pesticide (except Bt sprays) at the same rate within 24 hours. For areas affected only by pink bollworm, growers could plant refuge cotton as one single non-Bt cotton row for every 6–10 rows of Bt cotton, amounting to a minimum 10 percent refuge. The new rules introduced community refuges, allowing multiple growers to configure their acreage to comply with refuge requirements in aggregate.

2.1. A Simple Model of Refuge Cost and Choice

Consider now how refuge requirements and choice affect producer returns. Profits per acre from growing non-Bt cotton are

$$\pi_n = py\,(1 - \delta_n) - c_n,\tag{1}$$

where p is the cotton price, y is potential yield with no pest damage, c_n is insecticide costs on non-Bt cotton, and δ_n is percentage yield damage on non-Bt cotton. Profits per acre from growing Bt cotton are

$$\pi_b = py\,(1 - \delta_b) - c_b - t,\tag{2}$$

where δ_b is percentage yield damage, c_b is insecticide costs on Bt cotton, and t is the per-acre technology fee paid as a price premium on Bt seed.

Under the 80:20 option, growers must plant at least 20 percent of their cotton acreage to non-Bt cotton. A grower's cost of complying with the 80:20 refuge requirement is the economic gain foregone by not planting Bt cotton on 100 percent of total cotton acreage. This refuge cost R_s is

$$R_s = \pi_b - [0.8\,\pi_b + 0.2\,\pi_n] = 0.2\,[\,py\,(\delta_n - \delta_b) + (c_n - c_b) - t\,].\tag{3}$$

The refuge cost R_s is just 20 percent of the per-acre profit advantage of Bt cotton. The term $py\,(\delta_n - \delta_b)$ is the revenue advantage of Bt cotton, while $(c_n - c_b) - t$ represents the net change in pest control costs, counting the technology fee, from adopting Bt cotton.

This 20 percent figure is an upper bound of refuge costs. EPA allows other refuge options, but profit-maximizing growers will choose these other options only if they yield higher profits than the 80:20 option. This occurs when their refuge costs are lower than R_s. To see this, consider the 95:5 un-sprayed refuge option. Here growers must plant at least a 5 percent non-Bt cotton refuge. Further, this refuge may not be sprayed to control bollworm, budworm, or pink bollworm. Profits on the unsprayed refuge are

$$\pi_u = py\,(1 - \delta_u),\tag{4}$$

where δ_u is yield damage on the unsprayed refuge and $\delta_u > \delta_n > \delta_b$. The cost of compliance with the 95:5 refuge requirement R_u is

$$R_u = 0.05 \left[py \left(\delta_u - \delta_b \right) - \left(c_b + t \right) \right]. \qquad (5)$$

The term $py \left(\delta_u - \delta_b \right)$ is the lost revenue from lower yield on the refuge, while $\left(c_b + t \right)$ is the cost savings from not spraying for target pests on the refuge or paying the technology fee. Growers will choose the 95:5 unsprayed refuge over the 80:20 sprayed refuge if it returns higher profits,

$$\left[0.95 \, \pi_b + 0.05 \, \pi_u \right] - \left[0.80 \, \pi_b + 0.20 \, \pi_n \right] > 0. \qquad (6)$$

But (6) may be rearranged:

$$\left[0.95 \, \pi_b + 0.05 \, \pi_u \right] - \left[0.80 \, \pi_b + 0.20 \, \pi_n \right] \qquad (7)$$
$$= 0.2 \left[py \left(\delta_n - \delta_b \right) + \left(c_c - c_b - t \right) \right] - 0.05 \left[py \left(\delta_u - \delta_b \right) - \left(c_b + t \right) \right]$$
$$= R_s - R_u.$$

Equation (7) simply means that when the 95:5 unsprayed refuge is chosen, the refuge cost must be less than 20 percent of the gain from Bt cotton ($R_s - R_u > 0$). The same logic applies to the other refuge options, so the 20 percent figure is an upper bound for cost estimates of refuges.

2.2. Upper Bound Cost Estimates of Refuges

The previous section illustrates how upper-bound refuge costs may be derived based on estimates of the per-acre profit gain from Bt cotton adoption. Table 1 reports estimates of Bt profit gains from two literature surveys, one conducted by the National Center for Agricultural Policy (NCFAP) (Gianessi et al. 2002) and the other conducted for the International Food Policy Research Institute (IFPRI) (Marra 2001). The NCFAP study focuses on results for the year 2000, while the IFPRI estimates are derived from a survey of studies covering 1996–1999. Estimates of the impact of Bt cotton vary widely across states and years, ranging from large gains to large losses.

Table 2 reports imputed upper-bound refuge cost estimates based on the NCFAP study and the mean impact estimates from the IFPRI study. Average refuge costs weighted by state Bt acreage in 2000 range between $4 per acre based on NCFAP benefit estimates to about $10 per acre based on IFPRI mean estimates. These estimates are comparable to industry estimates of Bt cotton gains and (implied) refuge costs. Hudson, Mullins, and Mills' (2003) summary of Bt cotton returns based on studies from 1995–2002 reports an unweighted average Bt cotton profit gain of just under $40 per acre, implying

Table 1. Estimated Profit Gains Per Acre from Bt Cotton Adoption from Two Literature Surveys

State	NCFAP	IFPRI (mean)	IFPRI (low)	IFPRI (high)
Alabama	$ 2.64	$ 77.60	$ 38.70	$ 116.50
Arizona	$ 53.29	$ 57.50	$ (104.00)	$ 465.00
Arkansas	$ 12.28	$ 44.21	$ (26.95)	$ 86.74
California[a]	$ 53.28			
Florida[b]	$ 2.64			
Georgia	$ 2.64	$ 92.00	$ 38.70	$ 169.20
Louisiana	$ 12.20	$ 16.50	$ (3.10)	$ 36.00
Mississippi	$ 29.06	$ 34.50	$ (3.10)	$ 79.50
Missouri[c]	$ 9.12			
New Mexico[a]	$ 53.28			
North Carolina	$ 9.31	$ 20.50	$ (25.30)	$ 95.10
Oklahoma	$ 62.73	$ 53.80	$ 25.50	$ 85.50
South Carolina	$ 9.17	$ 51.80	$ 17.10	$ 80.10
Tennessee	$ 9.12	$ 67.50	$ 60.70	$ 74.30
Texas	$ 62.57	$ 46.00	$ 46.00	$ 46.00
Virginia	$ 9.32	$ 41.70	$ 41.70	$ 41.70

[a] Extrapolated from Arizona.
[b] Extrapolated from Georgia.
[c] Extrapolated from Tennessee.

an upper bound refuge cost of nearly $8 per acre. In comments to EPA and the FIFRA Scientific Advisory Panel, the National Cotton Council (NCC) reported an estimated average cost (in a typical pest year) of $4.46 for a 10 percentage-point increase of sprayed refuge (Isbell 2000). This would imply an upper bound refuge cost of $8.92 per acre.

The last two columns of Table 2 extrapolate upper bound refuge costs to the state and national level. The calculation assumes that growers would receive 100/80 = 1.25 of their actual returns absent Bt cotton planting restrictions. This places upper bound estimates of the cost of refuge requirements at between $25 to $60 million.

This upper bound refuge cost estimate assumes that refuge requirements are a binding constraint on Bt adoption and grower profits. This may not be the case for many growers, however. In its comments to the EPA, the NCC stated:

> According to Monsanto, approximately 40 percent of the growers planting Bt cotton in 1999 employed the 96-4 refuge option on their entire farm. Another 20 percent utilized the 80-20 option on their entire farm. The remaining 40 percent of growers employed the 80-20 option, but only on a portion of their cotton acres. On average these growers planted Bt cotton on approximately half of their total cotton acres [Isbell 2000, p. 6].

Table 2. Imputed Upper Bound Costs of Refuge Requirements Based on Benefit Estimates

State	Bt cotton acres	Net Gain from Bt Cotton ($/acre)		Imputed Refuge Cost ($/acre)		Imputed Refuge Cost ($000)	
		NCFAP	IFPRI (mean)	NCFAP	IFPRI (mean)	NCFAP	IFPRI (mean)
Alabama	344	2.64	77.60	0.53	15.52	227	6,674
Arizona	162	53.29	57.50	10.66	11.50	2,158	2,329
Arkansas	570	12.28	44.21	2.46	8.84	1,750	6,300
California	46	53.28	NR	10.66	NR	613	NR
Florida	80	2.64	NR	0.53	NR	53	NR
Georgia	648	2.64	92.00	0.53	18.40	428	14,904
Louisiana	563	12.20	16.50	2.44	3.30	1,717	2,322
Mississippi	1024	29.06	34.50	5.81	6.90	7,439	8,832
Missouri	85	9.12	NR	1.82	NR	194	NR
New Mexico	25	53.28	NR	10.66	NR	333	NR
North Carolina	379	9.31	20.50	1.86	4.10	882	1,942
Oklahoma	94	62.73	53.80	12.55	10.76	1,474	1,264
South Carolina	209	9.17	51.80	1.83	10.36	479	2,707
Tennessee	429	9.12	67.50	1.82	13.50	978	7,239
Texas	442	62.57	46.00	12.51	9.20	6,914	5,083
Virginia	44	9.32	41.70	1.86	8.34	103	459
Total	5,144	20.02	48.94	4.00	9.79	25,742	60,055

Notes: NR = no benefit estimates reported. Bt acreage data comes from Gianessi et al. (2002).

This suggests that for 40 percent of Bt cotton adopters in 1999, the refuge requirements were not a binding constraint.

The NCC comments go on to estimate average per-acre costs of increasing refuge sizes from 5–10 percent for the unsprayed refuge option and from 20–30 percent for the sprayed refuge option. Further, the analysis extrapolates up to national cost estimates. The extrapolation overstates the costs of larger refuge requirements because it does not allow for producer substitution between refuge options. In their base case, grower returns under the 95:5 unsprayed and 80:20 sprayed options are essentially equal, so growers would be indifferent between each option. Grower losses moving to the 90:10 option are estimated to be $8.23 per acre, while grower losses moving to the 70:30 option are $4.46 per acre. National losses are estimated to be $25 million. This appears to come from a calculation of

$$[\$8.23 \, / \, \text{acre} \times 2.2 \text{ million Bt acres under the 95:5 option} \times 100/95] \quad (8)$$

$$+ \, [\$4.46 \, / \, \text{acre} \times 1.1 \text{ million Bt acres under the 80:20 option} \times 100/80]$$

$$= \$25 \text{ million.}$$

This exercise, however, begs the question, why would growers with this cost and return profile choose the 90:10 unsprayed option over the 70:30 sprayed option? Or, to put it another way, why would growers choose a refuge option that yielded $3.77 less in profits per acre? If all growers switched to the 70:30 option, total costs of more stringent refuge requirements would be closer to $16.5 million. The NCC estimates can also be used to impute an upper bound estimate of the cost of the existing refuge requirements of $33 million.

What lessons can we draw from these various refuge cost estimates? First, refuge costs vary widely. This means that—for the purpose of cost-benefit analysis—single-year estimates may be poor substitutes for longer-term averages. It also means that costs of refuges for particular growers in certain years can be significant, relative to their net farm income. Second, growers will substitute between refuge options to maximize returns. Failure to account for such substitution will overstate refuge cost estimates. Third, following from this, providing growers with multiple refuge options can significantly reduce refuge compliance costs.

2.3. Adoption Behavior and Refuge Costs

The costs of refuge requirements to individual growers will depend both on individual adoption and more aggregate, area-wide adoption choices. First, consider individual adoption. There is evidence that many cotton growers have planted *less* Bt cotton than legally permitted under current refuge requirements (Isbell 2000, Carrière et al. 2005). This implies that current refuge regulations impose no costs on these growers (aside from any compliance reporting or verification costs). If Bt cotton is more profitable to grow than conventional cotton, though, why don't growers plant as much of it as allowed? One reason for this partial adoption might be learning by doing associated with the initial phase of technology introduction. Growers may test out the new technology on a smaller portion of their acres before committing to full adoption.

Another explanation for partial adoption of agricultural technologies may be risk aversion (Just and Zilberman 1983). Alexander, Fernandez-Cornejo, and Goodhue (2003) and Hurley, Mitchell, and Rice (2004) explore adoption of Bt corn under different types of risk. Although they do not include refuge requirements in their acreage allocation decisions, the results of both these studies suggest that these requirements may not be binding constraints. While there is some evidence that Bt cotton reduces the variance of target pest damage (Frisvold and Pochat 2004), Banerjee, Martin, and Wetzstein (2005) have found evidence that both the mean and the variance of Bt cotton profits are higher than for non-Bt cotton. While risk aversion may provide an explanation of partial adoption of Bt cotton, it raises other questions. To what extent should EPA include risk premiums in assessments of Bt crops and, in turn, the cost of refuge requirements? Work along the lines of Alexander, Fernandez-Cornejo, and Goodhue (2003), Banerjee, Martin, and Wetzstein (2005), and Hurley, Mitchell, and Rice (2004) may prove useful in answering these questions.

Patterns of aggregate (area-wide) adoption may affect both the costs and the benefits of refuges. On the benefits side, some authors have examined how aggregate adoption can affect benefits via its effect on the time to resistance. Secchi and Babcock (2003) and Hurley (2005) have noted that time to resistance depends both on mandated, "structured" refuges and the presence of natural refuges that may be provided by alternative pest hosts or by non-adopters of Bt varieties. Incomplete and slow diffusion of Bt varieties at a regional level can delay the onset of resistance.

Bt cotton adoption has been fairly rapid in many areas, however. Introduced in 1996, Bt cotton was planted on less than half of U.S. cotton in 2004—46 percent according to U.S. Department of Agriculture survey data (U.S. Department of Agriculture 2004), and 48 percent according to National Cotton Council survey data (Williams, various years). Yet this average masks large regional variations. Bt cotton adoption remains below 20 percent in Texas and California, which account for 48 percent of U.S. planted cotton acres, but is much higher in other parts of the country.

It is also possible that very high rates of aggregate adoption may reduce the annual opportunity cost of refuges. Table 3 shows differences in Bt vs. non-Bt cotton performance in three areas with high adoption rates reported in Williams (various years). Revenue losses from cotton bollworm and tobacco budworm damage are reported along with costs to control those pests. In the three areas reported, the costs of pest damage and control costs on non-Bt acres are lower than the additional price premium (technology fee) for Bt cotton.

This does *not* necessarily suggest that Bt cotton is not profitable to adopt on average. A simple comparison of means does not account for potential sample selection bias. Areas with greater pest damage and control costs are more likely to adopt Bt cotton (Frisvold and Pochat 2004). Also, insect suppression of the Bt cotton may spill over to the non-Bt acres. The observed yield and control cost differences between Bt and non-Bt cotton may be biased downward as a result (Marra 2001).

However, the numbers in Table 3 raise the question of whether there would be any economic gain to growers from planting Bt cotton on these remaining non-Bt acres. Figure 1 shows a relationship—at this point hypothetical—between the profit advantage of Bt cotton to an individual grower and aggregate, regional adoption. For lower levels of regional adoption, the advantage of Bt cotton, in terms of higher revenue and lower insecticide costs, remains relatively constant. Once aggregate adoption reaches a threshold

Table 3. Performance of Bt versus Non-Bt Cotton in Three High Adoption Areas

	Tennessee	Louisiana	Mississippi Delta
Percentage Bt cotton[a]	93%	93%	90%
	– $ / acre –	– $ / acre –	– $ / acre –
Revenue loss[b]			
Non-Bt	$ 18.71	$ 2.70	$ 29.12
Bt	$ 6.17	$ 1.95	$ 14.82
Bt advantage	$ 12.54	$ 0.75	$ 14.30
Pest control cost[a]			
Non-Bt	$ 9.67	$ 2.97	$ 10.71
Bt	$ 9.24	$ 1.01	$ 6.08
Bt advantage	$ 0.43	$ 1.96	$ 4.63
Overall Bt advantage	$ 12.97	$ 2.71	$ 18.93
Bt technology fee[a]	$ 23.00	$ 27.63	$ 28.00

[a] From Williams (various years [2004]).
[b] Yield loss from Williams (various years [2004]) multiplied by $0.65 per pound to represent market price plus loan deficiency payment received per pound.

level A_1, however, the pest population is suppressed sufficiently around non-Bt cotton that Bt cotton's advantage erodes. When adoption reaches A_2, the gross gain from adoption does not exceed the technology fee.

The hypothesis put forth here is that very high adoption areas such as Tennessee, Louisiana, or the Mississippi Delta may represent cases where aggregate adoption exceeds this second threshold level A_2. As one Mississippi grower has noted, "When we have a lot of Bt cotton out there [on surrounding farms] we haven't had the budworm problems" (Robinson 2001).

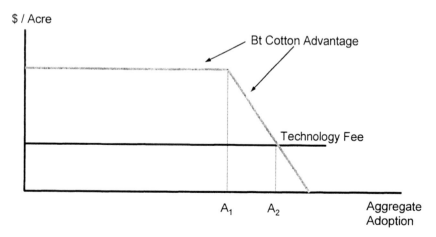

Figure 1. Hypothesized Gross Advantage of Bt Cotton to an Individual Grower as a Function of Regional Aggregate Adoption

Carrière et al. (2003) found significant reductions in Arizona pink bollworm populations in all regions where Bt cotton adoption rates exceeded 65 percent. At lower adoption rates, changes in population density varied. The region with the highest rate of Bt cotton adoption (87 percent) had a 62 percent decline in pink bollworm population.

If there is no gain from expanding Bt cotton acreage, then the costs of current refuge requirements in high adoption areas may be negligible. In discussing this type of halo effect with respect to Bt corn, Alstad and Andow (1996) suggest that

> ... refuges are areas where growers derive some degree of free protection from the adjacent Bt plots. The halo is a positive externality, or indirect benefit that leverages a grower's investment in Bt seed and provides a short-term economic incentive for implementation of resistance management [p. 180].

This does not mean that larger refuge requirements would not impose costs on growers. In Figure 1, greater restrictions could involve a movement from the right of A_2 to the left of A_2.

2.4. Welfare Impacts of Restricting Bt Cotton Acres

A complete regulatory impact analysis of refuge requirements would consider more than just impacts on producers. By limiting production, refuges affect the price of cotton, consumer surplus, and commodity program payments for cotton. Lichtenberg and Zilberman (1986) have demonstrated how failure to account for commodity program effects can seriously bias estimates of pesticide regulation costs. Refuge requirements also reduce sales and monopoly rents from the seed supply sector. Moschini and Lapan (1997) argue that assessments of welfare impacts of privately developed agricultural technologies should include impacts on rents to monopolist-innovators. Following this line of reasoning, we consider how seed supplier rents are affected by regulation of technology.

Table 4 reports estimates of the domestic welfare impacts of restrictions on Bt cotton acreage based on a simulation model of the world cotton market. The model embeds a multi-state mathematical programming model of the U.S. cotton sector in a three-region trade model that includes the United States, China, and the Rest of World (Frisvold, Tronstad, and Reeves 2006). The model is calibrated to 2001 economic data. Yield and pest control cost differences are derived from Falck-Zepeda, Traxler, and Nelson (2000a, 2000b), Gianessi et al. (2002), Hudson, Mullins, and Mills (2003), Marra (2001), Williams (various years), and other sources. In the baseline, U.S. state-level Bt cotton acreage is calibrated to match 2001 adoption rates. Adoption of Bt cotton is restricted by equal percentages in each state, with restrictions

Table 4. Net U.S. Domestic Cost of Reducing U.S. Bt Cotton Acreage[a]

	Percentage Reduction in Bt Cotton Acreage			
	10%	25%	50%	100%
Reduction in:				
consumer surplus (CS)	2.5	6.1	12.3	24.5
producer surplus (PS)	16.5	41.2	82.4	164.6
government payments (GP)	13.6	33.9	67.7	134.4
monopolist rents (MR)	14.3	35.7	71.3	142.7
Net cost of acreage reduction[b]	19.6	49.1	98.4	197.4
Lost government payments as a share of producer loss	82.4%	82.3%	82.2%	81.7%
Lost monopolist-innovator rents as a share of domestic costs	73.0%	72.7%	72.5%	72.3%

[a] Based on the simulation model used in Frisvold, Tronstad, and Reeves (2006).
[b] Change in (CS + PS + MR) – change in GP.

ranging from reductions of 10 percent to 100 percent, the latter representing a complete ban on Bt cotton.

The U.S. domestic costs of Bt cotton acreage restrictions range from $20 million for a 10 percent reduction to $198 million for a complete ban on Bt cotton (Table 4). For growers, 82 percent of the cost of Bt acreage restrictions is from loss of cotton program payments. Restricting Bt cotton acreage lowers national yields and production, but raises the price of cotton. Lower production and higher prices reduce government outlays for loan deficiency payments. The welfare impact is similar to that found by Lichtenberg and Zilberman (1986). The presence of coupled program payments makes regulation more costly for producers, while reducing the net economic cost of regulation. Lost monopolist-innovator rents (from lost seed sales and profits) represent over 72 percent of the domestic cost of Bt cotton acreage restrictions.

3. BENEFITS AND COSTS OF REFUGES: INTERTEMPORAL TRADE-OFF CURVES

This section introduces simple methods to weigh the annual costs of refuges against their potential longer-term benefits. The focus here is narrow, considering costs and benefits from the perspective of an individual grower or growers in a multi-county area. Ignored are the impacts of refuge requirements on seed supplier profits, consumer surplus, or commodity program payments. Also ignored are potential negative impacts of resistance on users of foliar Bt sprays. While these factors are important, the focus here is to address a specific question: how costly, if costly at all, are refuges to growers in the long run?

With no refuge requirement, the present value of the profit gain from Bt cotton over non-Bt cotton, Y_0, is

$$Y_0 = \sum_{t=0}^{t_r} y_t /(1+i)^t, \tag{9}$$

where y_t is the per-acre profit advantage of Bt cotton over non-Bt cotton in year t, i is the discount rate, and t_r is the year that resistance makes Bt cotton no more profitable to grow than non-Bt cotton. Given a fixed profit advantage and assuming profit maximization, without a refuge requirement growers will plant only Bt cotton. The present value of the gain from Bt cotton adoption under a given refuge requirement Y_R is

$$Y_R = \sum_{t=0}^{t_r} y_t (1-R_t)/(1+i)^t + \sum_{t=t_r}^{T} y_t (1-R_t)/(1+i)^t, \tag{10}$$

where R_t is the annual opportunity cost of the refuge requirement and T is the year of resistance onset under the given refuge. Both R_t and T will depend on refuge size s and a vector of requirements θ such as configuration requirements, distance requirements, and insecticide spraying restrictions ($R=R(s, \theta)$, $T = T(s, \theta)$). Functions R and T each are increasing in s and θ, implying that more stringent refuge requirements increase annual regulatory costs, but prolong the efficacy of the Bt cotton for more years. The terms s and θ could measure actual rather than mandated requirements, allowing the costs and benefits of noncompliance to be measured. Time T may be truncated $T \leq T_b$, where T_b is the arrival date of a backstop technology.

The net cost of the refuge requirement $Y_R - Y_0$ can be written as

$$Y_R - Y_0 = \sum_{t=t_r}^{T(s,\theta,T_b)} y_t (1-R(s,\theta))/(1+i)^t - \sum_{t=0}^{t_r} y_t R(s,\theta)/(1+i)^t. \tag{11}$$

A refuge will be more costly if the annual opportunity cost is greater (larger R) or if it takes longer for resistance to develop without a refuge (larger t_r).

The net cost or benefit of a given refuge relative to no refuge, $Y_R - Y_0$, can be expressed in terms of the time to resistance without a refuge t_r and the number of additional years of Bt efficacy that the given refuge provides, $T - t_r$ (Figure 2). The refuge is more costly to producers when resistance is slow to evolve. If resistance develops quickly the refuge is more likely to benefit producers. The introduction of a backstop technology will truncate the trade-

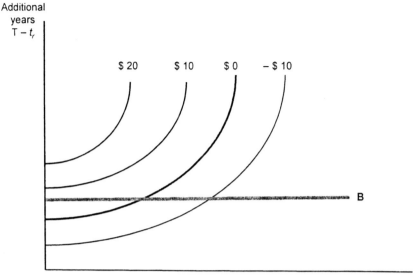

Figure 2. Impact of Refuge Requirement on Grower Returns as a Function of Time to Resistance Without a Refuge and Additional Years of Bt Efficacy Provided by Refuge

off surface. In Figure 2, introduction of a backstop means that only outcomes below line *B* are relevant.

While basic economic data are needed to construct the trade-off surface, one needs entomological data to determine where producers are situated on the surface—specifically, time to resistance with and without refuges. Unfortunately, there remains a high degree of scientific uncertainty about time to resistance. EPA reports simulation model results estimating years to resistance for cotton bollworm of 2.2 years with a 0 percent sprayed refuge, 7.25 years for a 10 percent refuge, and 10.5 years for a 20 percent refuge (EPA 2001). For a 4 percent unsprayed refuge, EPA reports simulation results estimating years to resistance ranging from 4 to 12 years for cotton bollworm and budworm (EPA 2001). In both cases, the number of years to resistance is the number of years it takes the resistant allele frequency in the pest population to reach 0.5. Qaim, Cap, and de Janvry (2003) cite simulation model results for cotton bollworm where, with a 0 percent refuge, the frequency of the resistant allele reaches 0.5 in four years and 1.0 in 6–7 years. In the same model the resistant allele frequency falls over time under both 20 percent and 5 percent refuges. All these results suggest that, from a grower's perspective, some refuge would be preferable to no refuge at all.

They provide less clear insights into the optimal refuge size, however. Livingston, Carlson, and Fackler (2004) report an optimal refuge size of greater than 0 percent but smaller than current refuge requirements for a time horizon of six or more years. The difference in the annualized present value (APV) of profits between current and optimal refuge size was on the order of $3 per acre.

3.1. Costs and Benefits of Refuges: Alternative Scenarios

Table 5 shows the annualized present value of net refuge costs, $(Y_R - Y_0) / T$, under different assumptions for time to resistance without a refuge and regulatory response to resistance onset. A number of common assumptions were made in all cases. First, the annual profit gain from Bt cotton over non-

Table 5. Net Cost of Refuge Requirements as a Function of Annual Opportunity Costs, Time to Resistance, and Regulatory Response

Year of resistance without refuge	Percent opportunity cost of refuge (R)	R two years (and beyond) after resistance	Extra years of Bt efficacy from refuge[a]	Annualized PV of refuge cost ($/acre)[b, c]
> 2004	20%	n.a.	n.a.	6.20
2004	20%	n.a.	1	3.61
2003	20%	100%	2	0.84
2002	20%	100%	3	(2.12)
2002	20%	50%	3	(0.83)
2001	20%	100%	4	(5.29)
2001	20%	50%	4	(2.61)
> 2004	30%	n.a.	n.a.	9.30
2004	30%	n.a.	1	6.71
2003	30%	100%	2	3.94
2002	30%	100%	3	0.98
2002	30%	50%	3	2.27
2001	30%	100%	4	(2.19)
2001	30%	50%	4	0.49
> 2004	10%	n.a.	n.a.	3.10
2004	10%	n.a.	1	0.51
2003	10%	100%	2	(2.26)
2002	10%	100%	3	(5.22)
2002	10%	50%	3	(3.92)

Notes: "n.a." is not applicable.

[a] Backstop technology assumed available by 2005.

[b] Negative value means refuges increase grower returns.

[c] Seventy percent discount rate with annual profit gain from Bt cotton over non-Bt cotton of $40/acre.

Bt cotton is $40 per acre. This is comparable to average returns reported in Hudson, Mullins, and Mills (2003) and intermediate between average estimates reported by Gianessi et al. (2002) and by Marra (2001). The discount rate i is 7 percent, the initial time period ($t = 0$) is 1996, and time T is 2004. A backstop technology—in the form of new Bt varieties—becomes available by 2005. In the year of resistance onset, Bt cotton is no more profitable than non-Bt cotton. This assumption may actually understate the initial cost of a resistance event if growers must make unanticipated insecticide applications after paying the higher price for Bt cotton seed. Growers in the affected region are not allowed to plant Bt cotton in the year immediately following a resistance outbreak. The opportunity cost of refuges, R, is varied from 10–30 percent. The 20 percent case corresponds to the current 80:20 sprayed refuge option.

Two regulatory responses to resistance are considered. In one case, growers may not plant Bt cotton at all once resistance occurs until the backstop varieties become available in 2005. In the second case, growers are permitted to plant Bt cotton two years after the resistance outbreak, but are required to plant 50 percent of their crop in refuges. These regulatory responses are purely hypothetical, but provide a plausible range of responses. EPA required the development of remedial action plans to respond to resistance events (EPA 2001). These had some specific monitoring, notification, and reporting requirements, but flexibility was maintained for producer response in areas with a verified resistance event. To address potential pink bollworm resistance, the Arizona Bt Working Group, comprised of university, cooperative extension, cotton grower, and seed industry representatives, recommended designating a remedial action zone falling within 6 miles of sections of land where resistance occurred (Carrière et al. 2001, EPA 2001). If fields with verified resistance were detected in more than three different townships within a particular cotton-growing region, the entire region could be designated as an action zone. The working group also recommended additional chemical control of pink bollworm, early cotton harvest, release of sterile pink bollworms, and parasitic nematode treatments as possible immediate responses. For the following year, it recommended that no Bt cotton be sold within the action zone. Reintroduction of Bt cotton would depend on bioassays to determine whether the resistant allele frequency had declined sufficiently. Remedial action plans for tobacco budworm and cotton bollworm have followed the same basic structure of the Arizona model, with EPA listing as possible responses suspension of Bt cotton acreage and alteration of future refuge sizes (EPA 2001). Because there have been no actual resistance events, it remains unknown how long Bt sales would have been restricted or how refuge size would have been adjusted in the event of an actual resistance event.

The net costs of a refuge decline the earlier resistance occurs without a refuge (Table 5). Negative costs mean that grower returns are greater with the refuge than without it. Consider first the case where $R = 20$ percent. If resistance had occurred by 2002 without refuges, then refuges actually would have increased grower returns relative to the no-refuge case. This result holds both when Bt cotton planting is discontinued completely after resistance and when planting with 50 percent refuges is allowed after two years. It also holds under a higher discount rate of 9 percent. The result is also insensitive to assumptions about the level and time path of Bt cotton profit gain. For example, Hudson, Mullins, and Mills (2003) and Gianessi et al. (2002) found higher returns to Bt cotton adoption in earlier periods. However, when using annual averages from Hudson, Mullins, and Mills (2003) (allowing y_t to vary by year), refuges with $R = 20$ percent are still profitable relative to no refuges if they prevented resistance onset by or before 2002.

Table 5 suggests that the annualized impact of current refuge requirements ($R = 20$ percent) ranges from a cost of $6.20 per acre in annualized present value (APV) to a gain of $5.29 APV. This latter gain would have come if refuges prevented resistance from occurring in 2001 and also prevented the halt of subsequent Bt cotton planting until the arrival of a backstop in 2005. Could one assign probabilities to the range of outcomes? Perhaps, but as Tabashnik et al. (2003) point out, our ability to predict time to resistance in the field remains quite limited. The range of impacts of the type presented in Table 5 could, however, provide a useful starting point for a more comprehensive regulatory impact assessment of refuge policies.

4. AN EXAMPLE FROM ARIZONA

Arizona is a state where, *ex ante*, both the annual opportunity cost of refuges and the potential for resistance were high. In Arizona, pink bollworm (PBW) is the primary lepidopteran cotton pest. Statewide, the mean (median) percentage of acres treated for PBW was 69 percent (98 percent) from 1986–1995, before Bt cotton was introduced, and 39 percent (23 percent) from 1996–2003. Growers have virtually ceased spraying for PBW on Bt cotton acres since 1997 (Williams, various years). Bt cotton confers an average yield gain of 5.5 percent, while saving three insecticide applications per year (Frisvold, Tronstad, and Reeves 2006).

In the 1990s, however, the risk of evolution of PBW resistance to Bt cotton also appeared high. Alternate plant hosts that could serve as a natural refuge are virtually non-existent in the state and laboratory selection produced strains of PBW able to survive and reproduce on Bt cotton (Carrière et al. 2001). The mean frequency of a recessive allele conferring resistance to

Bt toxin was estimated to be 0.16 in strains of PBW from Arizona cotton fields collected in 1997 (Tabashnik et al. 2000). By 1998, adoption rates had reached 75 percent or more in major cotton-producing areas of central Arizona (Carrière et al. 2001, Carrière et al. 2005).

Widespread adoption of Bt cotton has led to area-wide suppression of PBW populations (Carrière et al. 2003). Despite early fears, estimated resistant allele frequencies have fallen since 1997. Based on simulation models, Carrière and Tabashnik (2001) suggest that refuges, fitness costs, and incomplete resistance may all have contributed to delaying resistance.

While the risk of resistance may be lower than initially feared, annual refuge costs have also declined over time. Results of a detailed survey of cotton planting in Arizona found evidence of over-compliance with embedded and in-field refuge size requirements (Carrière et al. 2005). The survey used global positioning systems and micro-surveyors to map cotton fields in counties accounting for 96 percent of state cotton production. Embedded refuges are blocks of non-Bt cotton at least 150 feet wide planted within fields of Bt cotton. Refuge area must be at least 5 percent of total cotton area. Average embedded refuge size was 35 percent in 2002 and 2003— seven times the minimum requirement. For in-field refuges, EPA requires growers to plant at least one row of non-Bt cotton for every 6 to 10 Bt acres in a field, amounting to a minimum 10 percent refuge requirement. For in-field refuges, average refuge size was 28 percent in 2002 and 25 percent in 2003, more than double the minimum requirement.

Average refuge sizes much larger than the minimum requirements indicate that refuge requirements are not binding constraints on many growers' planting decisions. Moreover, roughly a quarter of Arizona's 2003 Bt acreage was planted with embedded or in-field refuges. Introduction of these options in 2001 has lowered regulatory costs on a significant scale.

Why do growers over-comply? One reason may be that for risk-averse growers, the refuge size that maximizes expected utility is greater than the minimum requirement. Another possibility is that there are halo effects on embedded and in-field refuges such that pest damage is significantly reduced on non-Bt cotton. Growers can avoid $32 per acre in Bt technology fees for every additional acre of non-Bt cotton they plant. Moreover, regional suppression of the pink bollworm population in Arizona would reduce the advantage of planting Bt cotton.

To quantify the long-run gain to producers from the new refuge options, the APV of Arizona cotton refuge costs, $(Y_R - Y_0) / T$, was calculated. The APV with $R = 20$ percent was based on an average yield gain of 5.5 percent and a savings of three insecticide sprays per year (Frisvold, Tronstad, and Reeves 2006) and based on actual Bt cotton adoption rates, state cotton yields, prices received, and coupled program payments received. The APV of refuge costs over the 1996–2004 period was $7.92. For growers who switched to an embedded or in-field refuge in 2001 and planted more than the minimum required refuge size, there would be no shadow cost of the refuge requirement from 2001 to 2004. For these growers, the APV of refuge costs from 1996 to 2004 would be $4.90, a reduction of 38 percent.

5. CONCLUSIONS

This chapter introduced a simple method to estimate an upper bound of annual costs of Bt cotton refuges. Weighted by state Bt acreage, these costs range from $4 to $10 per acre where refuge requirements are binding. For many growers, though, refuge requirements may not be binding, implying zero costs. Moreover, these cost estimates are average, not marginal costs. In high-adoption areas, pest pressure may become low enough that there would be limited grower gains to relaxing current requirements.

Based on programming model simulations, the U.S. costs of Bt cotton acreage restrictions range from $20 million for a 10 percent reduction to $198 million for a complete ban. For growers, 82 percent of the cost of the restrictions is from loss of cotton program payments. Lost seed supplier profits represent 72 percent of the domestic cost of Bt cotton acreage restrictions.

To obtain better estimates of refuge costs (and benefits), more research is needed that examines different aspects of grower adoption behavior, namely partial adoption, aggregate adoption, risk aversion, and refuge choice. Empirical examples illustrated how providing multiple refuge options significantly reduces regulatory costs.

REFERENCES

Alexander, C., J. Fernandez-Cornejo, and R.E. Goodhue. 2003. "Effects of the GM Controversy on Iowa Corn-Soybean Farmers' Acreage Allocation Decisions." *Journal of Agricultural and Resource Economics* 28(3): 580–595.

Alstad, D.N., and D.A. Andow. 1996. "Implementing Management of Insect Resistance to Transgenic Crops." *AgBiotech News and Information* 8(10): 177–181.

Banerjee, S., S.W. Martin, and M.E. Wetzstein. 2005. "The Influence of the Value of Bt Cotton on Risk and Welfare." Paper presented at the Southern Agricultural Economics Association 37th Annual Meetings, Little Rock, AK, February 5–9.

Carlson, G.A., and M.E. Wetzstein. 1993. "Pesticides and Pest Management." In G. Carlson, D. Zilberman, and J. Miranowski, eds., *Agricultural and Environmental Resource Economics.* Oxford: Oxford University Press.

Carrière, Y., T.J. Dennehy, B. Pedersen, S. Haller, C. Ellers-Kirk, L. Antilla, Y. Liu, E. Willott, and B.E. Tabashnik. 2001. "Large-scale Management of Insect Resistance to Transgenic Cotton in Arizona: Can Transgenic Insecticidal Crops be Sustained?" *Journal of Economic Entomology* 94(2): 315–325.

Carrière, Y., C. Ellers-Kirk, K. Kumar, S. Heuberger, M. Whitlow, L. Antilla, T.J. Dennehy, and B.E. Tabashnik. 2005. "Long-term Evaluation of Compliance with Refuge Requirements for Bt Cotton." *Pest Management Science* 61(4): 327–330.

Carrière, Y., C. Ellers-Kirk, M. Sisterson, L. Antilla, M. Whitlow, T.J. Dennehy, and B.E. Tabashnik. 2003. "Long-term Regional Suppression of Pink Bollworm by *Bacillus thuringiensis* Cotton." *Proceedings of the National Academy of Sciences* 100(4): 1519–1523.

Carrière, Y., and B.E. Tabashnik. 2001. "Reversing Insect Adaptations to Transgenic Insecticidal Plants." *Proceedings of the Royal Society London.* Series B 268: 1475–1480.

Cotton Research and Development Council (CRDC). 2002. *Annual Report 2001–2002*. Narrabri New South Wales, Australia.

Doyle, B., I. Reeve, and E. Barclay. 2002. "The Performance of Ingard Cotton in Australia During the 2000/2001 Season." Institute for Rural Futures, University of New England, Armidale, New South Wales, Australia.

EPA [see U.S. Environmental Protection Agency].

Falck-Zepeda, J.B., G. Traxler, and R.G. Nelson. 2000a. "Rent Creation and Distribution from Biotechnology Innovations: The Case of Bt Cotton and Herbicide-Tolerant Soybeans in 1997." *Agribusiness* 16(1): 21–32.

———. 2000b. "Surplus Distribution from the Introduction of a Biotechnology Introduction." *American Journal of Agricultural Economics* 82(2): 360–369.

Frisvold, G., and R. Pochat. 2004. "Impacts of Bt Cotton Adoption: What State-level Data Can and Can't Tell Us." In *Proceedings of the Beltwide Cotton Conferences*. National Cotton Council, Memphis, TN.

Frisvold, G., and R. Tronstad. 2002. "Economic Impacts of Bt Cotton Adoption in the United States." In N. Kalaitzandonakes, ed., *The Economic and Environmental Impacts of Agbiotech: A Global Perspective*. Norwell, MA: Kluwer-Plenam.

Frisvold, G.B., R. Tronstad, and J.M. Reeves. 2006. "International Impacts of Bt Cotton Adoption." In R.E. Evenson and V. Santaniello, eds., *International Trade and Policies for Genetically Modified Products*. Wallingford, UK: CABI Publishing.

Gianessi, L.P., C.S. Silvers, S. Sankula, and J.E. Carpenter. 2002. "Plant Biotechnology: Current and Potential Impact for Improving Pest Management in U.S. Agriculture: An Analysis of 40 Case Studies." National Center for Food and Agricultural Policy, Washington, D.C. (June).

Huang, J., R. Hu, S. Rozelle, F. Qiao, and C.E. Pray. 2002. "Transgenic Varieties and Productivity of Smallholder Cotton Farmers in China." *Australian Journal of Agricultural and Resource Economics* 46(3): 367–387.

Huang, J.R., H. Hu, and F. van Tongeren. 2002. "Biotechnology Boosts to Crop Productivity in China and Its Impact on Global Trade." Working Paper No. 02-E7, Chinese Center for Agricultural Policy, Beijing.

Huang, J., S. Rozelle, C. Pray, and Q. Wang. 2002. "Plant Biotechnology in China." *Science* 295(25): 674–677.

Hudson, J., W. Mullins, and J.M. Mills. 2003. "Eight Years of Economic Comparisons of Bollgard® Cotton." In *Proceedings of the Beltwide Cotton Conferences*. National Cotton Council, Memphis, TN.

Hurley, T.M. 2005. "Bt Resistance Management: Experiences From the U.S." In J. Wessler, ed., *Environmental Costs and Benefits of Transgenic Crops in Europe* (Wageningen UR Frontis Series Vol. 7). Dordrecht: Springer.

Hurley, T.M., B. Babcock, and R.L. Hellmich. 2001. "Bt Corn and Insect Resistance: An Economic Assessment of Refuges." *Journal of Agricultural and Resource Economics* 26(1): 176–194.

Hurley, T.M., P.D. Mitchell, and M.E. Rice. 2004. "Risk and the Value of Bt Corn." *American Journal of Agricultural Economics* 86(2): 345–358.

Hurley, T.M., S. Secchi, B.A. Babcock, and R.L. Hellmich. 2002. "Managing the Risk of European Corn Borer Resistance to Bt Corn." *Environmental and Resource Economics* 22(4): 537–558.

Hutcheson, T. 2003. "EPA: Bt Cotton Harmful to Organic." Organic Trade Association, Greenfield, MA (November 14).

Isbell, H. 2000. "Comments of the National Cotton Council on Bt Resistance Management and Benefits of Bollgard Cotton." Public comment submitted to the U.S. Environmental Protection Agency and the FIFRA Scientific Advisory Panel. National Cotton Council, Memphis, TN (October 4).

Ismael, Y., R. Bennett, and S. Morse. 2002. "Benefits from Bt Cotton Use by Smallholder Farmers in South Africa." *AgBioForum* 5(1): 1–5.

Just, R.E., and D. Zilberman. 1983. "Stochastic Structure, Farm Size, and Technology Adoption in Developing Agriculture." *Oxford Economics Papers* 35(2): 307–328.

Laxminarayan, R., and R.D. Simpson. 2002. "Refuge Strategies for Managing Pest Resistance in Transgenic Agriculture." *Environmental and Resource Economics* 22(4): 521–536.

Lichtenberg, E., and D. Zilberman. 1986. "The Welfare Economics of Price Supports in U.S. Agriculture." *American Economic Review* 76(5): 1135–1141.

Livingston, M., G.A. Carlson, and P.A. Fackler. 2004. "Managing Resistance Evolution in Two Pests to Two Toxins with Refugia." *American Journal of Agricultural Economics* 86(1): 1–13.

Marra, M.C. 2001. "The Farm Level Impacts of Transgenic Crops: A Critical Review of the Evidence." In P.G. Pardey, ed., *The Future of Food: Biotechnology Markets in an International Setting.* Baltimore: Johns Hopkins Press and International Food Policy Research Institute.

Matten, S.R., and A.H. Reynolds. 2003. "Current Resistance Management Requirements for Bt Cotton in the United States." *Journal of New Seeds* 5(2–3): 137–178.

Moschini, G., and H. Lapan. 1997. "Intellectual Property Rights and the Welfare Effects of Agricultural R&D." *American Journal of Agricultural Economics* 79(4): 1229–1242.

Pray, C., J. Huang, R. Hu, S. Rozelle. 2002. "Five Years of Bt cotton in China: The Benefits Continue." *The Plant Journal* 31(4): 423–430.

Pray, C., D. Ma, J. Huang, and F. Qiao. 2001. "Impact of Bt cotton in China." *World Development* 29(5): 813–825.

Price, G.K., W. Lin, J. Falck-Zepeda, and J. Fernandez-Cornejo. 2003. "Size and Distribution of Market Benefits From Adopting Biotech Crops." Technical Bulletin No. TB-1906, Economic Research Service, U.S. Department of Agriculture, Washington, D.C. (November).

Qaim, M., E. Cap, A. de Janvry. 2003. "Agronomics and Sustainability of Transgenic Cotton in Argentina." *AgBioForum* 6(1–2): 41–47.

Qaim, M., and D. Zilberman. 2003. "Yield Effects of Genetically Modified Crops in Developing Countries." *Science* 299(5608): 900–902.

Robinson, E. 2001. "Producer Hedge: Half Bt, Half Conventional Cotton." *Delta Farm Press* (Clarksdale, MS), May 11, p. 19.

Secchi, S., and B. Babcock. 2003. "Pest Mobility, Market Share, and the Efficacy of Using Refuge Requirements for Resistance Management." In R. Laxminarayan, ed., *Battling Resistance to Antibiotics and Pesticides: An Economic Approach.* Washington, D.C.: Resources for the Future Press.

Secchi, S., T.M. Hurley, and R.L. Hellmich. 2001. "Managing European Corn Borer Resistance to Bt Corn with Dynamic Refuges." Working Paper No. 01-287, Center for Agricultural and Rural Development, Iowa State University, Ames.

Tabashnik, B.E., Y. Carrière, T.J. Dennehy, S. Morin, M. Sisterton, R. Roush, A. Shelton, and J. Zhao. 2003. "Insect Resistance to Transgenic Bt Crops: Lessons from the Laboratory and Field." *Journal of Economic Entomology* 96(4): 1031–1038.

Tabashnik, B.E., A.L. Patin, T.J. Dennehy, Y. Liu, Y. Carrière, M.A. Sims, and L. Antilla. 2000. "Frequency of Resistance to *Bacillus thuringiensis* in Field Populations of Pink Bollworm." *Proceedings of the National Academy of Sciences* 97(24): 12980–12984.

Traxler, G., S. Godoy-Avila, J. Falck-Zepeda, and J. Espinoza-Arellano. 2002. "Transgenic Cotton in Mexico: Economic and Environmental Impacts." In N. Kalaitzandonakes, ed., *The Economic and Environmental Impacts of Agbiotech: A Global Perspective.* Norwell, MA: Kluwer-Plenam.

U.S. Department of Agriculture. 2004. "Acreage." Report No. CR PR 2-5 (6-04), National Agricultural Statistics Service. Washington, D.C. (June 30).

U.S. Environmental Protection Agency. 2001. "Bt Plant-Incorporated Protectants: Biopesticides Registration Action Document." Biopesticides and Pollution Prevention Division, U.S. Environmental Protection Agency, Washington, D.C.

Williams, M.R. Various years. "Cotton Insect Losses." In *Proceedings of the Beltwide Cotton Conferences.* National Cotton Council, Memphis, TN.

Chapter 25

MANAGING EUROPEAN CORN BORER RESISTANCE TO Bt CORN WITH DYNAMIC REFUGES

Silvia Secchi,[*] Terrance M. Hurley,[†] Bruce A. Babcock,[*] and Richard L. Hellmich[*†]
Iowa State University,[*] University of Minnesota,[†] USDA Agricultural Research Service[‡]

Abstract: Genetically engineered Bt (*Bacillus thuringiensis*) corn provides farmers with a new tool for controlling the European corn borer (ECB). The high efficacy and potential rapid adoption of Bt corn has raised concerns that the ECB will develop resistance to Bt. The Environmental Protection Agency has responded to these concerns by requiring farmers to plant refuge corn. Current refuge requirements are based on models that do not consider the value of dynamically varying refuge in response to increased scarcity and diminished control over time or the importance of backstop technologies currently being developed. The purpose of this chapter is to evaluate dynamically optimal refuge requirements with the arrival of alternative backstop technologies and to compare the results to an optimal static refuge policy. The results show that a dynamically optimal refuge requirement provides only modest benefits above a static optimum. The results also show how the type of backstop technology and characteristics of ECB population dynamics affect the optimal refuge requirement.

Key words: Bt corn, refuge strategy, optimal control, pesticide resistance

1. INTRODUCTION

Bt corn is genetically engineered to produce a protein found in the soil bacterium *Bacillus thuringiensis* (Bt). The protein is toxic when consumed by lepidopteran insects such as the European corn borer (ECB), which is estimated to cost U.S. farmers over $1 billion annually in yield loss and control costs (Mason et al. 1996). The high efficacy and full season control provided by Bt corn has resulted in its rapid adoption by farmers. Between 1996 and 2004, the percentage of Bt corn acreage in the United States increased from less than one to 27 percent (U.S. Department of Agriculture 2004). The rapid adoption of Bt corn raises concerns that the ECB will develop resistance to it. The U.S. Environmental Protection Agency (EPA) (EPA 1998a) has responded to these concerns by requiring farmers to plant refuge corn. Refuge slows the proliferation of resistance by allowing susceptible pests to thrive and mate with resistant ones (EPA 1998a).

Previous studies provide rationale for the EPA's resistance management requirement. Pests are a detrimental renewable resource because they propagate and damage crops (Hueth and Regev 1974; Regev, Gutierrez, and Feder 1976; and Regev, Shalit, and Gutierrez 1983). Pest susceptibility (the converse of resistance) is a valuable resource because susceptible pests are controllable (Hueth and Regev 1974, and Regev, Shalit, and Gutierrez 1983). The use of pesticides reduces the biological capital of susceptibility as it increases resistance through natural selection, thereby making pests less controllable in the future. The ECB is a mobile pest that farmers will treat as common property (Clark and Carlson 1990). Thus, farmers are unlikely to privately manage resistance.

Early literature characterized the dynamic optimal dose of pesticides for managing resistance. Recent policy work for Bt crops explores the value of static refuges, that is, refuges whose size is fixed throughout the period in which the pesticide is used (e.g., Alstad and Andow 1995; Roush and Osmond 1996; Gould 1998; Onstad and Gould 1998a, 1998b; Hurley et al. 1999; and Hurley, Babcock, and Hellmich 2001). Along the same lines, Livingston, Carlson, and Fackler (2000) discuss the case of a static refuge policy that maximizes the sum of discounted profits for the producer when the producer uses a decision rule to choose the optimal amount of refuge subject to the static refuge policy cap. Static refuge has been the focus of this research because it is consistent with early and current EPA policy. However, static refuge is unlikely to be first-best policy in resource management terms because varying the size of refuge is similar to varying pesticide dose.

Recent policy research for Bt corn has also focused on commercialized varieties of Bt corn that rely on one of two toxins (EPA 2005), while ignoring effort currently underway to develop new varieties with multiple toxins "stacked" in the same plant. Resistance is thought to evolve more slowly when pests must overcome multiple toxins. These toxins are designed with different modes of action, so cross resistance is unlikely. Therefore, the introduction of multiple toxin Bt corn may make less refuge optimal because it can eventually replace existing single toxin varieties of Bt corn.

The purpose of this chapter is to (i) determine the difference between the optimal dynamic policy and the second-best static policy for refuge in the case of Bt corn and (ii) consider how the introduction of a new technology changes the optimal and second-best policies. The sensitivity of these results to ECB population characteristics is also explored. Specifically, we consider how the optimal and second-best policy change based on whether it is possible to eradicate ECB using Bt corn—a scenario predicted by many simulation models, but viewed skeptically by many entomologists.

2. THE MODEL

Even with parsimonious models, it is difficult to analytically characterize the optimal path of refuge because increasing refuge leaves more of the crop unprotected and increases future pest pressure, but also slows resistance, improving future crop protection and decreasing future pest pressure. Ultimately, which of these effects dominates is an empirical question. We explore this question by evaluating the optimal path of refuge for a typical continuous corn region in the North Central United States assuming Bt corn is planted to control European corn borer (ECB).

Following Alstad and Andow (1995, 1996), Roush and Osmond (1996), Gould (1998), Onstad and Gould (1998a, 1998b), we consider a simplified production region with a single crop and pest. The region is closed to migration. While there is a single crop, there are two different varieties. The first is a toxic Bt variety. The second is a non-toxic refuge variety. The proportion of the refuge planted in season t is denoted by $1 \geq \phi_t \geq 0$. The proportion of resistant pests in season t is $1 \geq R_t \geq 0$ and the number of pests is $N_t \geq 0$. Π_t is the value of agricultural production, which determines the value of pests and pest susceptibility in season t, while Ω_T is the salvage value of pests and pest susceptibility for all $t \geq T$, the season when a new technology is introduced.

The ECB is a mobile diploid[1] pest that reproduces sexually with as many as four generations a season. ECBs in warmer southern climates produce three to four generations, while ECBs in more temperate northern climates produce one to two generations. A bivoltine (two-generation) population is typical for most of the North Central United States (Mason et al. 1996).[2]

The development of resistance is a function of natural selection caused by the use of Bt. Bt corn currently uses toxins with a single mode of action, so while there are two different toxins, cross-resistance is likely. Therefore, we assume resistance is conferred by a single allele that is not sex linked. An allele can be either resistant (r) or susceptible (s). Each parent contributes an allele, so the offspring's gene, the combination of alleles contributed by its parents, will be determined according to Table 1. The frequency at which parents are homozygote resistant (rr), heterozygote (rs), or homozygote

---

[1] A diploid organism carries in the nucleus of each cell two sets of chromosomes, one from each parent.

[2] In some areas, farmers can face two different strains of European corn borer. For instance, a farmer may face both a univoltine and bivoltine population. While this situation is not considered here, the model can be readily extended to such scenarios.

Table 1. Possible Offspring Genotypes Given Mother's and Father's Genotypes

	Father	rr		rs		ss	
Mother	alleles	r	r	r	s	s	s
rr	r	rr	rr	rr	rs	rs	rs
	r	rr	rr	rr	rs	rs	rs
rs	r	rr	rr	rr	rs	rs	rs
	s	rs	rs	rs	ss	ss	ss
ss	s	rs	rs	rs	ss	ss	ss
	s	rs	rs	rs	ss	ss	ss

susceptible (ss) determines the probabilities of the offspring's genotype. Bt corn produces a high dose and is believed to kill all the ss and almost all the rs pests throughout the season. The evolution of resistance depends on the initial frequency of resistant alleles and on the genotypic survival rates, which, in turn, depend on whether the crop is Bt or refuge.

The backstop technology we model uses two toxins with different modes of action. Therefore, we assume that resistance to each toxin is conferred by two separate genes, a and b. We explore two possible scenarios. In the first, one of the toxins is the original toxin, while the second is novel. In the second, two novel toxins are introduced. This allows us to quantify the effect of a positive value of susceptibility to the original toxin in the salvage function. We define $\rho_{\gamma i} = [\rho^{rr}_{\gamma i}, \rho^{ss}_{\gamma i}, \rho^{rs}_{\gamma i}]$ as the survival rate of resistant and susceptible homozygotes and heterozygotes for gene $\gamma = a$, b and crop i where $i = 0$ for the Bt and 1 for the refuge crop. Following Hurley, Babcock, and Hellmich (2001), we assume $\rho_{a0} = \rho_{b0} = [1.0, 0.0, 0.02]$ and $\rho_{a1} = \rho_{b1} = [1.0, 1.0, 1.0]$. This implies that the two toxins are equally effective in the elimination of pests. Hurley, Babcock, and Hellmich (2001) consider a single gene model and assume that the initial frequency of resistant alleles is 3.2×10^{-4}. We assume that the initial frequency of resistant alleles is the same for both genes and equal to 3.2×10^{-4}.

A gamete represents the combination of alleles a parent contributes to its offspring for each gene. With a single gene there are two possible gametes: r and s. With two genes, there are four: $r|r$, $r|s$, $s|r$, and $s|s$. Therefore, we define R_g as a 1×4 vector of the proportion of each type of gamete at the beginning of generation g: $[R^{r|r}_g, R^{r|s}_g, R^{s|r}_g, R^{s|s}_g]$. The initial gamete proportions are $R_0 = [1.0 \times 10^{-7}, 3.2 \times 10^{-4}, 3.2 \times 10^{-4}, 0.9993]$. The initial gamete proportions at T when two novel toxins are introduced are $R_T = [1.0 \times 10^{-7}, 3.2 \times 10^{-4}, 3.2 \times 10^{-4}, 0.9993]$. When a novel toxin is added to supplement an existing toxin, R_T will depend on how much resistance remains for the original toxin. The dynamics of resistance with two genes are detailed in the Appendix.

To capture the change in ECB from one generation to the next, we adopt the modified logistic growth model,

$$N_{g+1} = \beta_{0g} + \beta_{1g}\rho_g N_g + \beta_{2g}(\rho_g N_g)^2 + \rho_g N_g, \tag{1}$$

used by Hurley, Babcock, and Hellmich (2001), where β_{0g}, β_{1g}, and β_{2g} are parameters to estimate, ρ_g is the survival rate of ECB in generation g, and N_g is the number of pests at the beginning of generation g. Note that $\rho_g N_g$ reflects the number of pests that survive in generation g. The traditional logistic growth model is modified with β_{0g}, which, when positive, eliminates the possibility of eradicating ECB. The reason for the choice of this growth function is, as Hurley, Babcock, and Hellmich (2001) show, that the high efficacy of Bt corn results in near eradication or "heavy" suppression of ECB with a conventional logistic growth function. Many entomologists express skepticism about such a result because of the potential for weeds and other plants to serve as alternative hosts for ECB. Therefore the modified growth function is used to test the sensitivity of results to the degree of pest suppression.

We define the current value of agricultural production between period T_1 and T_2 as the average annualized net revenues per acre for Bt and refuge corn:

$$\Pi(T_1, T_2) = \frac{\sum_{t=T_1}^{T_2} \delta^{t-T_1}\left\{(1-\phi_t)[pY(1-D_t^0)-C^0]+\phi_t[pY(1-D_t^1)-C^1]\right\}}{\sum_{t=T_1}^{T_2} \delta^{t-T_1}} \tag{2}$$

where Y is equal to the pest-free yield, p is equal to the real price of corn, D_t^i is the proportion of pest-free yield lost to the ECB on crop i in season t, and C^i is the cost of production for crop i. The proportion of yield loss is defined explicitly as $D_t^i = \text{Min}[1.0, \rho_{2t+1}N_{2t+1}d^2 + \rho_{2t}N_{2t}d^1]$, where d^1 and d^2 are the constant yield loss per pest for first and second generation ECB, respectively.

Using equation (2), we define the value function as $\Pi_t = \Pi(0, T-1)$ and the salvage value as

$$\Omega_T = \frac{1}{1-\delta}\underset{\phi_T}{Max}\left\{\Pi(T,T')\middle|1.0\geq\phi_t=\phi_T\geq 0.0, N_{t+1}=n(\phi_t,N_t,R_t)+N_t,\right.$$

and

$$R_{t+1} = r(\phi_t, R_t) + R_t \ \forall \ t \in [T, T']\}.$$

Thus, the value function reflects the annualized present value of production between the initial season and season $T-1$. The salvage value reflects the value of a stream of income equal to the annualized value of the new technology for $T'-T$ years when an optimal static refuge is used to manage resistance. Our salvage value assumes that a new technology arrives every $T'-T$ seasons to restore the efficacy of pest control as resistance develops to the current pest control technology. We use an optimal static refuge to calculate the salvage value of the new technology to reduce the computational burden of solving the model and because simulations using static and dynamic refuges with the new technology (available on request) indicate that an optimal static refuge provides a good approximation to the optimal dynamic refuge.

Note that the use of the technology necessarily reduces its efficacy. That is, it increases resistance. The new technology, be it dependent or independent from the old one, is superior in the sense that it mines a new stock of susceptibility, and is therefore more effective. The problem is equivalent to discovering a new mine with lower extraction costs. Since the new technology considered is better than the existing technology, it is optimal to introduce it immediately.

Having parametrically specified the evolution of resistance, the ECB population dynamics, the value function, and the salvage function, benchmark parameters are now chosen. Table 2 presents the benchmark configuration for all but the population dynamics. Table 3 presents estimated parameters for two alternative population models.

National Agricultural Statistical Service (NASS) and Economic Research Service (ERS) data provide values for the real price, pest-free yield, and production cost of refuge corn. The real price of corn, $2.35, is the monthly average from 1991 to 1996 deflated to 1992.[3] The average Iowa yield from 1991 to 1996 was about 123 bushels per acre. Assuming an average annual ECB yield loss of 6.4 percent (Calvin 1995) implies that the pest-free yield is 130 bushels per acre. Excluding returns to management, the average production cost, $185, comes from 1995 ERS corn budgets deflated to 1992 prices. The interest rate used for discounting is 4 percent.

The pest-free yield and production cost of Bt corn is the same as refuge. We are unaware of studies showing a significant difference in Bt and conventional corn yields in the absence of ECB. While farmers typically pay a $7 to $10 per acre technology fee for Bt seed, this premium does not reflect an increase in the marginal cost of growing Bt corn from a social perspective. Once Bt is introduced into corn, the cost of producing Bt and

---

[3] Depending on the rate of adoption of Bt corn and the refuge size, there could be supply-side price effects that are not included here.

Table 2. Benchmark Parameter Values

Parameter	Existing Technology	New Technology
Economic Parameters		
years	15	15
discount rate	$1/(1+0.04)$	$1/(1+0.04)$
price of corn ($/bushel)	$2.35	$2.35
pest-free yield (bushels/acre)	130	130
production cost ($/acre)	$185	$185
1st generation constant marginal yield loss (pests/plant)	0.055	0.055
2nd generation constant marginal yield loss (pests/plant)	0.028	0.028
Biological Parameters		
initial pest population (pests/plant)	0.23	N_{15}
recombination factor	0.5	0.5
initial gamete proportions (R_0')	$\begin{bmatrix} 1.0 \times 10^{-7} \\ 3.2 \times 10^{-4} \\ 3.2 \times 10^{-4} \\ 0.9993 \end{bmatrix}$	R_{15}
Gene a		
refuge survival rates for all genotypes	1.00	1.00
survival rate of resistant homozygotes on Bt corn	1.00	1.00
survival rate of susceptible homozygotes on Bt corn	0.00	0.00
survival rate of heterozygotes on Bt corn	0.02	0.02
Gene b		
refuge survival rates for all genotypes	1.00	1.00
survival rate of resistant homozygotes on Bt corn	1.00	1.00
survival rate of susceptible homozygotes on Bt corn	1.00	0.00
survival rate of heterozygotes on Bt corn	1.00	0.02

conventional corn seed stock is essentially identical. Since research and development costs for Bt corn are sunk, the technology fee reflects an economic rent transferred from growers to Bt corn registrants for using the Bt technology.

Hurley, Babcock, and Hellmich (2001) consider two specifications for the population model and find very different results. We explore the same two specifications. The first assumes that population growth follows a logistic curve with no intercept: $\beta_{0g} = 0$. In this case, in the absence of pest resistance, eradication is possible. When pest resistance develops, heavy ECB suppression results instead of eradication. The second specification estimates a positive intercept for the growth curve: $\beta_{0g} > 0$. Therefore,

Table 3. European Corn Borer Population Model Parameters[a]

Parameters	First Generation		Second Generation	
	Heavy Suppression	*Light* Suppression	*Heavy* Suppression	*Light* Suppression
constant	0.000	0.028	0.00	0.26
previous population	-0.757	-0.802	7.76	5.96
previous population squared	-0.053	-0.040	-10.30	-8.13
equilibrium population without Bt corn (pest/plant)	0.248	0.227	1.54	1.43
calibration factor	1.01		0.97	

[a] Population parameters adopted from Hurley, Babcock, and Hellmich (2001).

eradication is not possible even with susceptible pests, while when resistance has developed, ECB suppression is "light." The biological difference between heavy and light suppression is the amount of time it takes low ECB populations to return to carrying capacity. This amount of time is longer with heavy suppression.

Hurley, Babcock, and Hellmich (2001) estimate different parameters for first and second generation ECB using data reported in Calvin (1995). These parameter estimates are reproduced in Table 3. The calibration factors that are also reported ensure that the steady state ECB population is comparable across specifications when no pest control is used.

The constant marginal damage rates for first and second generation ECB, 0.055 and 0.028, are taken from Ostlie, Hutchison, and Hellmich (1997). Combined with the equilibrium populations, the implied average annual yield loss is 5.3 percent, which is 20 percent lower than the 6.4 percent reported in Calvin (1995).

The final parameter to specify is the length of the planning horizon for assessing the benefits and costs of resistance management. A fifteen-year planning horizon is used to conform to the 1998 EPA scientific advisory and ILSI/HESI (International Life Sciences Institute/Health and Environmental Sciences Institute) panel reports (EPA 1998b and ILSI/HESI 1999).

The model is implemented in C++ and solved using numerical optimization routines adopted from Press et al. (1992). It is important to note that the biological processes used to characterize resistance do not guarantee the satisfaction of second-order sufficiency conditions for a global optimum. Therefore, there is no guarantee that a numerical solution is globally optimal. Assuring a global optimum is computationally infeasible, so we use a range of starting values with the optimization routine to increase the robustness of the results.

3. RESULTS

The analytical characterization of the optimal dynamic path for refuge is generally not possible. Increasing refuge has a negative impact, a reduction in current production and increased ECB pressure in the future, and a positive one, the preservation of ECB susceptibility, which affords better control and reduced ECB pressure in the future. Adding additional structure and solving the model with parameter values found in the literature allows us to explore which of these countervailing effects tends to dominate and when.

Our results focus on four scenarios. We consider two alternative population models. The first assumes that ECB suppression is light, while the second assumes that suppression is heavy. We also consider two distinct salvage functions. The first assumes that resistance to the new technology is independent of the current technology because two novel toxins replace the existing toxin and there is no cross-resistance. The second assumes that resistance to the new technology is dependent on the current technology because the original toxin is supplemented with a novel toxin. It is important to note that while resistance to the new technology will be either dependent or independent, the value of the new technology always depends on the old one because the population of ECB when the new technology arrives depends on how the old technology is used. Combining the alternative population models with the alternative salvage functions yields the four scenarios.

Before interpreting the results, it is useful to summarize the optimal dynamic path for refuge, resistance, and ECB for each scenario, while highlighting important similarities and differences. Figure 1 reports the optimal dynamic refuge. The first interesting result is the consistent pattern for all scenarios. In the initial period, the optimal refuge is relatively low. It increases sharply in the second period, before a series of more moderate increases. Eventually, the optimal refuge begins to decrease, typically at an increasing rate. While this pattern is similar for all scenarios, there are notable differences. The pattern is more exaggerated with heavy suppression. With heavy suppression, the optimal refuge does not depend on whether the new technology is independent or dependent. When suppression is light, on the other hand, it is optimal to have more refuge if the new technology is dependent. This difference becomes more pronounced as the introduction of the new technology nears.

Figure 2 illustrates how the characteristics of the optimal time path for resistance differ substantially depending on whether suppression is light or heavy. When suppression is light, the optimal level of resistance for the original toxin increases at an increasing rate. The rate of increase is faster when resistance to the new technology is independent. But even when the new technology is independent it is not optimal to fully exhaust susceptibility. On the other hand, when suppression is heavy, the optimal evolution of

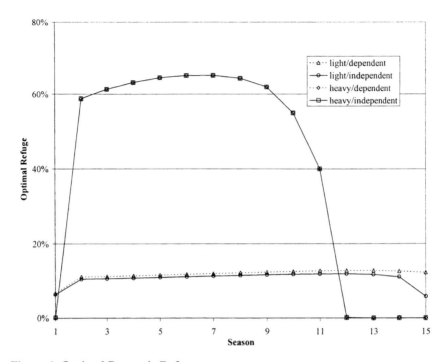

Figure 1. Optimal Dynamic Refuge

resistance is sigmoidal. Initially, resistance increases at an increasing rate. Later, it increases at a decreasing rate until susceptibility is fully exhausted. With heavy suppression, the evolution of resistance is not affected by whether resistance to the new technology is dependent or independent.

As with the optimal dynamic refuge, the optimal dynamic ECB population (Figure 3) for each scenario follows a similar pattern. The population rapidly declines in the first two periods. It then levels off and begins to increase. The increase is more pronounced as the introduction of the new technology nears. Despite these similarities there are several notable differences. First, populations are substantially lower (by two to three orders of magnitude) with heavy suppression. Also, with heavy suppression it takes longer for the population to recover and the type of new technology does not matter. When suppression is light, the population immediately begins to recover and the type of new technology does matter. The optimal population is always lower when the new technology is independent.

The general pattern of the optimal dynamic refuge in Figure 1 is due to the fact that increasing refuge reduces the current value of production and tends to increase pest pressure in the future, but it also increases susceptibility, which allows for better control and reduces future pest pressure. Since

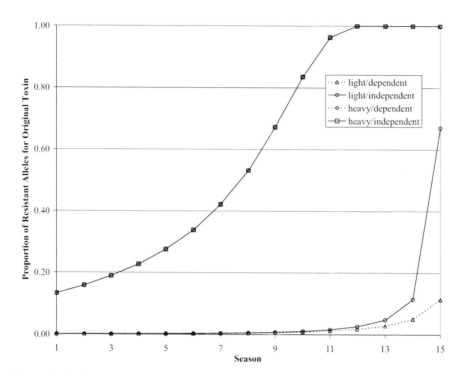

Figure 2. Optimal Dynamic Resistance

the starting value for ECB is the carrying capacity, the initial level of ECB pressure is high and the marginal cost of refuge in terms of reduced yield is high relative to the marginal benefit of managing resistance. Figure 3 shows that the initial emphasis on pest control reduces ECB substantially. Once there are few pests left to control, the marginal cost of refuge decreases relative to the marginal benefit of managing resistance, and more refuge is optimal. As the pest population begins to recover, the marginal cost of refuge increases once again. Additionally, as the arrival of the new technology nears, the value of susceptibility diminishes, particularly when the new technology is independent. With the marginal cost of refuge increasing and the marginal benefit of refuge declining, less refuge is again optimal.

Resistance management has different characteristics according to the resilience of the pest population. Susceptibility is more valuable with light than with heavy suppression. Figure 4 illustrates why by showing how fast the pest population recovers to carrying capacity after a four order of magnitude reduction. It takes three years for the population to exceed one ECB per plant and seven years to return to carrying capacity with light suppression. With heavy suppression, it takes fourteen years to exceed one ECB per plant and twenty years to return to carrying capacity. Economically

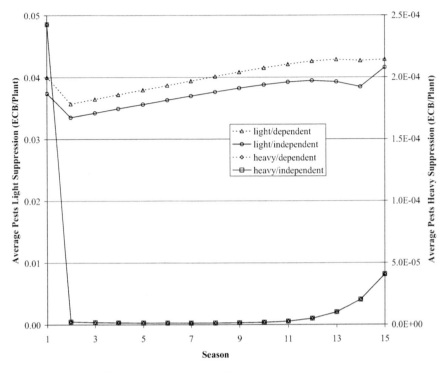

Figure 3. Optimal Dynamic Pest Population

important levels of pest population can return in just three years after the evolution of resistance with light suppression, while it takes over ten years with heavy suppression. When suppression is heavy, it is optimal to do no resistance management or to carry out intensive resistance management depending on whether there are a large number of ECBs to control. When suppression is light, the best strategy is to do some resistance management all of the time, but not as much as when suppression is heavy because there will always be more ECB to control and a higher marginal cost. With heavy suppression, the nature of the backstop is irrelevant because the pest population is essentially a nonrenewable resource in the time frame of analysis. Figure 2 shows that it is optimal to exhaust susceptibility with heavy suppression even when the new technology is dependent.[4] When suppression is heavy, the dependence of the new technology on the current toxin does not matter because it is not optimal to maintain susceptibility until the new technology arrives. Since ECB can be brought to near extinction, it

---

[4] Sensitivity analysis (available on request) indicates that this result is robust even if the delay in the new technology is substantially shorter or the discount rate is much less.

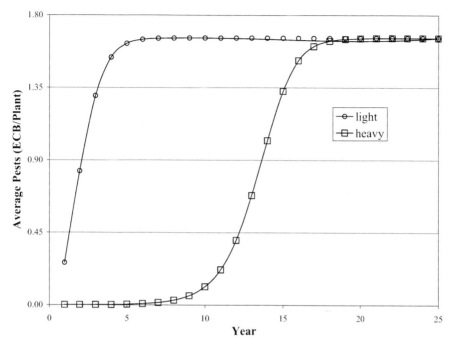

Figure 4. Comparison of the Recovery Rate of Pests for "Light" and "Heavy" Suppression Models

is optimal to do so by exhausting susceptibility regardless of the characteristics of the backstop.

On the other hand, if suppression is light (the more realistic case, according to entomologists), the nature of the backstop does matter in the terminal part of the life cycle of the single toxin technology. Since the pest population recovers quickly, the burden of maintaining susceptibility can be concentrated in the last part of the life span of the old technology. The relevance of the characteristics of the backstop for the late use of the old technology is even more evident in the optimal level of resistance. As Figure 2 illustrates, the optimal level of susceptibility when the backstop incorporates the previously used toxin is six times higher than if the backstop is independent. However, it is not optimal to fully exhaust susceptibility when suppression is light even if the technology is independent.[5] This result is justified by the biological constraints on exhaustion and the fact that the value of susceptibility is inextricably linked to controlling ECB. To fully

---

[5] Sensitivity analysis (available on request) shows that this result is robust for much larger discount rates and if the new technology is delayed much longer. However, with a long enough delay and a high enough discount rate it does become optimal to exhaust susceptibility even when suppression is light (e.g., 50 years and a 20 percent interest rate).

exhaust susceptibility, less refuge must be planted over a period of time. Planting less refuge over time imposes an implicit cost because resistance evolves sooner, thereby increasing pest pressure and reducing the value of production. When the cost of resistance is high, it is not optimal to fully exhaust susceptibility by planting less refuge over time. This extraction cost is higher when there are multiple generations of pest in a season because any resistance that develops during the first generation reduces control in subsequent generations and it is not possible to offset increased resistance by adjusting refuge during the season. Therefore, it is optimal to plant refuge even in the season before the introduction of a new technology. Sensitivity analysis shows that with a single generation of pests per season, it is not optimal to plant refuge in the season before the introduction of an independent new technology, but it may still not be optimal to exhaust susceptibility because of the implicit extraction cost.[6]

Figure 1 shows that over time the proportion of refuge that maximizes the long-run value of production changes in response to changes in the value of susceptibility and ECB control. Therefore, holding the proportion of refuge static over time will reduce the value of production. To understand the cost of using a second-best static refuge for resistance management, Table 4 reports the annualized net present value of production for the optimal dynamic refuge, optimal static refuge, and if Bt corn is never introduced. It also reports the optimal size of a static refuge.

Table 4 shows that the annualized value of Bt corn in all our scenarios is about $7.00 an acre, which represents just over a 6 percent increase in the value of production. What is more interesting is the difference in the value of production between the optimal dynamic and static refuge for all four scenarios. With light suppression, the dynamic refuge increases the annualized value of production by about $0.01 an acre when compared to the optimal

Table 4. Dynamic Versus Static Optima

Salvage Function	Suppression	Dynamic Optimum Value of Production	Static Optimum Value of Production	Refuge	Without Bt Corn Value of Production
		$/Acre		Percent	$/Acre
Independent	light	$120.34	$120.33	10.6	$113.36
	heavy	$120.50	$120.50	0.2	$113.36
Dependent	light	$120.32	$120.31	11.5	$113.36
	heavy	$120.50	$120.50	0.2	$113.36

---

[6] This sensitivity analysis is available on request.

static refuge, regardless of whether the new technology is "dependent" or "independent." This difference represents less than 0.1 percent of the value of production and less than 0.25 percent of the value of Bt corn. With heavy suppression, the difference is essentially zero.

Optimally varying refuge over time provides few benefits when compared to a second-best static refuge regardless of whether suppression is heavy or light and the new technology is dependent or independent of the current technology. This result is due to the effectiveness with which Bt corn controls ECB. When suppression is heavy, the effectiveness of Bt corn allows the immediate and near eradication of ECB. This is accomplished by planting almost no refuge in the first year. After that, how much refuge is planted has little effect on the value of production because the ECB is not able to reestablish itself and cause appreciable damage before the new technology arrives. When suppression is light, planting a modicum of refuge until the new technology arrives maintains resistance to levels that are low enough for Bt corn to still provide greater than 98 percent control. Comparing the optimal dynamic and static refuge reveals that there is little difference in the two strategies, with the exception of the initial period and the period right before the introduction of the new technology. This and the high level of control (greater than 98 percent) explains the small difference in the value of production.

4. CONCLUSIONS

Bt corn is a valuable new tool for controlling the European corn borer. This value will be diminished if the European corn borer (ECB) resistance to Bt emerges. Therefore, the U.S. Environmental Protection Agency has mandated insect resistance management guidelines based on farmers planting a proportion of their corn acreage to refuge—corn that does not use Bt for pest control. Refuge slows the proliferation of resistance by making more susceptible ECB available to mate with resistant ECB. So far, models used to guide EPA policy have focused on static recommendations and have not considered how the introduction of new technologies affects the value of resistance management. We explore how varying refuge optimally over time can increase the value of resistance management. We also consider how refuge requirements should account for pest population dynamics and the introduction of new technologies.

The results of our analysis show that varying refuge does improve the benefits of resistance management by accounting for the increased scarcity and diminished control as resistance develops. These opposing effects make it optimal to require less refuge when Bt corn is first introduced, more refuge once pests are better controlled and resistance starts to emerge, and less refuge as the introduction of a new technology nears. However, the improvement offered by optimally varying refuge is modest when compared to an optimal static refuge.

The optimal strategy for managing resistance is very sensitive to the population dynamics of the pest. If heavy suppression is feasible and the value of production is the primary objective, eradication type strategies that use little or no refuge until there is substantial resistance and a measurable loss of control will tend to be optimal. If heavy suppression is not possible, then a relatively consistent source of refuge tends to be optimal until a new technology is introduced.

We find that the effect of introducing a new technology on the optimal refuge depends on the population dynamics of ECB. If the ECB recovers slowly, it is optimal to fully exhaust pest susceptibility regardless of the type of new technology being introduced. If the ECB recovers rapidly, the type of technology introduced impacts the optimal refuge. If the new technology depends on susceptibility to the old, relatively more refuge should be planted over time. When the ECB is buoyant, it is not typically optimal to exhaust susceptibility regardless of the backstop because the evolution of resistance is biologically constrained and the value of susceptibility is inextricably linked to the value of pest control. These two factors impose an implicit extraction cost that tends to exceed the value of exhaustion.

The results of this analysis have important policy and research implications. Survey data suggests that farmers were confused by the changes in refuge recommendations that took place before the EPA mandated resistance management for Bt corn in 2000.[7] This data indicates that there is an implicit administrative cost associated with varying the size of refuge that could be avoided with a static policy. In light of these potential costs and the modest increase in the value of agricultural production provided by optimally varying refuge in response to scarcity and diminished control, a static policy that avoids these costs may be preferable.

Understanding the population dynamics of the ECB is important for resistance management, yet we are unaware of research currently underway to fill this information gap. Given the lack of information on the dynamics of ECB populations and other important parameters (e.g., the initial resistance frequency and dominance of resistance), if new information reveals that current assumptions are unfounded, adjustments to refuge in response to this new information could be valuable. How and when refuge requirements should adjust to new information is an important question for future research.

---

[7] Survey results provided by the Agricultural Biotechnology Stewardship Technical Committee and compiled by Marketing Horizons, Inc., in 2000 show that 26 percent of respondents thought a minimum of 5–15 percent refuge was mandated by the U.S. Environmental Protection Agency, while 39 percent indicated that they did not know the minimum amount of mandated refuge. A summary of survey results is available from the authors on request.

The optimal dynamic refuge we explore assumes that there is perfect control of the amount of refuge planted, but this is not the case. The EPA mandates refuge requirements and growers choose whether or not to meet or exceed those requirements. Our model suggests that Bt corn may substantially reduce ECB populations. If growers pay extra to plant Bt corn, there may be substantial incentives to discontinue or limit its use after a few seasons. The rapid adoption of Bt corn slowed in 2000. There are several explanations for this result, one of which is the low number of ECBs experienced across much of the Midwest since the 1997 growing season. Grower adoption and de-adoption of Bt corn and compliance with refuge requirements will have a substantial impact on the efficacy of EPA policy. New models integrating the complexities of pest biology and human behavior will provide the EPA with more reliable information and improve resistance management policy.

Our model focuses on optimizing agricultural productivity, while ignoring conventional pesticides. In the region we model, conventional pesticides are seldom used to control ECB because of high cost and poor efficacy. There are, however, regions and crops (e.g., cotton in the southern United States) where conventional pesticides are more important. In these regions, the EPA is also concerned about reducing the use of these pesticides because it believes they are more hazardous to the environment and human health. Therefore, a useful extension of the model could include an evaluation of conventional pesticides and the objective of reducing their use.

REFERENCES

Alstad, D.N., and D.A. Andow. 1995. "Managing the Evolution of Insect Resistance to Transgenic Plants." *Science* 268(5219): 1984–1996.

_____. 1996. "Evolution of Insect Resistance to *Bacillus Thuringiensis*-Transformed Plants." *Science* 273(5280): 1413.

Calvin, D.D. 1995. "Economic Benefits of Transgenic Corn Hybrids for European Corn Borer Management in the United States: A Report to the Monsanto Company." Department of Entomology, Pennsylvania State University, University Park, PA.

Clark, J.S., and G.A. Carlson. 1990. "Testing for Common Versus Private Property: The Case of Pesticide Resistance." *Journal of Environmental Economics and Management* 19(1): 45–60.

EPA [see U.S. Environmental Protection Agency].

Gould, F. 1998. "Sustainability of Transgenic Insecticidal Cultivars: Integrating Pest Genetics and Ecology." *Annual Review of Entomology* 43(1): 701–726.

Hartl, D.L. 1988. *A Primer of Population Genetics* (2nd ed.). Sunderland, MA: Sinauer and Associates, Inc.

Hueth, D., and U. Regev. 1974. "Optimal Agricultural Pest Management with Increasing Pest Resistance." *American Journal of Agricultural Economics* 56(3): 543–552.

Hurley T.M., B.A. Babcock, and R.L. Hellmich. 2001. "Bt crops and Insect Resistance: An

Economic Assessment of Refuges." *Journal of Agricultural and Resource Economics* 26(1): 176–194.

Hurley T.M., S. Secchi, B.A. Babcock, and R.L. Hellmich. 1999. "Managing the Risk of European Corn Borer Resistance to Transgenic Corn: An Assessment of Refuge Recommendations." Staff Report No. 99-SR88, Center for Agricultural and Rural Development, Iowa State University, Ames, IA.

ILSI/HESI. 1999. *An Evaluation of Insect Resistance Management in Bt Field Corn: A Science Based Framework for Risk Assessment and Risk management.* Washington, D.C.: ILSI Press.

Livingston M.J., G.A. Carlson, and P.L. Fackler. 2000. "BT Cotton Refuge Policy." Paper presented at the American Agricultural Economics Association Annual Meeting, July 30–August 2, in Tampa, Florida. Available online at http://agecon.lib.umn.edu/aaea00/sp00li01.pdf (accessed August 18, 2000).

Mason, C.E., M.E. Rice, D.D. Calvin, J.W. Van Duyn, W.B. Showers, W.D. Hutchison, J.F. Witkowski, R.A. Higgins, D.W. Onstad, and G.P. Dively. 1996. *European Corn Borer Ecology and Management.* North Central Regional Extension Publication No. 327, Iowa State University, Ames IA.

Onstad, D.W., and F. Gould. 1998a. "Do Dynamics of Crop Maturation and Herbivorous Insect Life Cycle Influence the Risk of Adaptation to Toxins in Transgenic Host Plants?" *Environmental Entomology* 27(3): 515–522.

_____. 1998b. "Modeling the Dynamics of Adaptation to Transgenic Maize by European Corn Borer (Lepidoptera: Pyralidae)." *Journal of Economic Entomology* 91(3): 585–593.

Ostlie, K.R., W.D. Hutchison, and R.L. Hellmich. 1997. "Bt Corn and the European Corn Borer." NCR Publication No. 602, University of Minnesota, St. Paul, MN.

Press, W.H., S.A. Teukolsky, W.T. Vetterling, and B.P. Flannery. 1992. *Numerical Recipes in C: The Art of Scientific Computing* (2nd ed.). New York: Cambridge University Press.

Regev, U., A.P. Gutierrez, and G. Feder. 1976. "Pests as a Common Property Resource: A Case Study of Alfalfa Weevil Control." *American Journal of Agricultural Economics* 58(2): 186–197.

Regev, U., H. Shalit, and A.P. Gutierrrez. 1983. "On the Optimal Allocation of Pesticides with Increasing Resistance: The Case of the Alfalfa Weevil." *Journal of Environmental Economics and Management* 10(1): 86–100.

Roush, R., and G. Osmond. 1996. "Managing Resistance to Transgenic Crops." In N. Carozzi and M. Koziel, eds., *Advances in Insect Control: The Role of Transgenic Plants.* London: Taylor and Francis.

U.S. Department of Agriculture. 2004. "Adoption of Genetically Engineered Crops in the U.S." Economic Research Service, U.S. Department of Agriculture, Washington, D.C. Available online at http://www.ers.usda.gov/Data/BiotechCrops/ (accessed April 8, 2005).

U.S. Environmental Protection Agency. 1998a. *The Environmental Protection Agency's White Paper on Bt Plant-pesticide Resistance Management.* Washington, D.C.: U.S. Environmental Protection Agency.

_____. 1998b. Sub-panel on *Bacillus thuringiensis* (Bt) Plant-Pesticide and Resistance Management, FIFRA Scientific Advisory Panel, February 9–10 (Docket No. OPP 00231).

_____. 2005. "Current and Previously Registered Section 3 PIP Registrations." Available online at http://www.epa.gov/pesticides/biopesticides/pips/pip_list.htm (accessed April 11, 2005).

APPENDIX

R_g is a 1×4 vector of the proportion of each type of gamete at the beginning of generation g: $[R^{r|r}_g, R^{r|s}_g, R^{s|r}_g, R^{s|s}_g]$, while $\rho_{\gamma i} = [\rho^{rr}_{\gamma i}, \rho^{ss}_{\gamma i}, \rho^{rs}_{\gamma i}]$ is the survival rate of resistant and susceptible homozygotes and heterozygotes for gene γ on crop i where $i = 0$ for the Bt crop and 1 for the refuge crop. It is also useful to define

$$
P_{ig} = (R_g'R_g) \times
\begin{bmatrix}
\rho^{rr}_{ai}\rho^{rr}_{bi} & \rho^{rr}_{ai}\rho^{rs}_{bi} & \rho^{rs}_{ai}\rho^{rr}_{bi} & \rho^{rs}_{ai}\rho^{rs}_{bi} \\
\rho^{rr}_{ai}\rho^{rs}_{bi} & \rho^{rr}_{ai}\rho^{ss}_{bi} & \rho^{rs}_{ai}\rho^{rs}_{bi} & \rho^{rs}_{ai}\rho^{ss}_{bi} \\
\rho^{rs}_{ai}\rho^{rr}_{bi} & \rho^{rs}_{ai}\rho^{rs}_{bi} & \rho^{ss}_{ai}\rho^{rr}_{bi} & \rho^{ss}_{ai}\rho^{rs}_{bi} \\
\rho^{rs}_{ai}\rho^{rs}_{bi} & \rho^{rs}_{ai}\rho^{ss}_{bi} & \rho^{ss}_{ai}\rho^{rs}_{bi} & \rho^{ss}_{ai}\rho^{ss}_{bi}
\end{bmatrix}
\tag{A.1}
$$

where $\times$ indicates multiplication by element. The net survival rate on the ith crop in generation g is $\rho_{ig} = I_4'P_{ig}I_4$, where I_4 is a 1×4 identity vector. The net survival rate in generation g and season t is $\rho_g = (1 - \phi_t)\rho_{ig} + \phi_t\rho_{ig}$. Let $P_g = [(1 - \phi_t)\rho_{0g} P_{0g} + \phi_t\rho_{1g} P_{1g}] / \rho_g$. Extending the Hardy-Weinberg model with random mating (see Hartl 1988), the evolution of resistance is characterized as

$$
R^{x|y}_{g+1} = P^{x|y\,x|y}_g + P^{x|y\,x|y'}_g + P^{x|y\,x'|y}_g + 0.5P^{x'|y'\,x|y}_g + 0.5P^{x|y'\,x'|y}_g
\tag{A.2}
$$

for all x, x', and y, and $y' \in \{r, s\}$, $x \neq x'$, and $y \neq y'$ where $P^{z\,z'}_g$ represents the z row and z' column of P_g.

Chapter 26

FARMER DEMAND FOR CORN ROOTWORM Bt CORN: DO INSECT RESISTANCE MANAGEMENT GUIDELINES REALLY MATTER?

Ines Langrock and Terrance M. Hurley
University of Minnesota

Abstract: We explore the potential for using the contingent valuation method to characterize the sensitivity of farmer demand for corn rootworm Bt corn for alternative insect resistance management (IRM) requirements. With a better understanding of farmer sensitivity to IRM, the U.S. Environmental Protection Agency (EPA) may be able to refine its IRM policy to improve farmer compliance, while reducing its regulatory burden and the likelihood of Bt resistance. We find that farmer demand is sensitive to IRM requirements, but not always in the anticipated direction and magnitude.

Key words: Bt corn, insect resistance management, corn rootworm, European corn borer

1. INTRODUCTION

Bt corn contains a gene from the soil bacterium *Bacillus thuringiensis*. For varieties first commercialized in 1996, this gene instructs the corn plant to produce proteins that are toxic primarily to the European corn borer (ECB) and southwestern corn borer, but that also have some activity on the corn earworm and cornstalk borer (U.S. Department of Agriculture, Economic Research Service 2001). Annual losses attributed to ECB prior to the release of Bt corn were estimated at $1 billion (Mason et al. 1996), and the near complete control offered by Bt corn has resulted in rapid and widespread adoption. Corn is planted on about 80 million acres annually in the United States, with the Bt share increasing to over 20 percent in 2002 (Minnesota

The authors would like to thank Ken Ostlie for his help in designing the survey instrument. The authors also gratefully acknowledge the financial support of the U.S. Department of Agriculture's Initiative for Future Agriculture and Food Systems (IFAFS) program.

Agricultural Statistics Service 2002). In 2003, a new corn rootworm (CRW) Bt corn was approved for commercial use in the United States by the U.S. Environmental Protection Agency (EPA).

Reductions in pest damage, conventional insecticide use, and farmer risk are believed to be major benefits of Bt varieties (EPA 2001). However, the widespread adoption of Bt corn also carries the risk of insect resistance, which poses an obstacle to sustainability (Alstad and Andow 1995). Empirical evidence suggests that the development of resistance is a real threat, although Bt corn resistance has not yet been documented in the field (e.g., Hama, Suzuki, and Tanaka 1992; Tabashnik et al. 1992).

The threat of insect resistance to Bt led the EPA to require farmers to follow an insect resistance management (IRM) plan. The EPA (1998, p. 1) expresses the agency's objectives for this plan: "pesticide resistance management is likely to benefit the American public by reducing the total pesticide burden on the environment, and by reducing the overall human and environmental exposure to pesticides." It also illuminates the important tradeoffs and constraints that concern the EPA: "It is the desire of the EPA that this focus on pesticide resistance management not overly burden the regulated community, jeopardize the registration of reduced risk pesticides, or exclude conventional pesticides or other control practices which can contribute to the further adoption of integrated pest management (IPM)."

The EPA believes it can accomplish its IRM objectives using a high-dose/structured-refuge strategy. For a high-dose refuse, Bt corn is designed to control all but the most resistant insects. To implement a structured refuge strategy, farmers plant some non-Bt corn in close proximity to their Bt corn. The high-dose/structured-refuge strategy works by supporting a population of susceptible insects to randomly mate with resistant insects, which will produce primarily Bt-susceptible progeny.

The effectiveness of the EPA's IRM plan will depend on farmer compliance. However, the opportunity costs of not planting Bt corn may give farmers an incentive not to comply. The requirements are also difficult to enforce because Bt corn is not easily distinguished from non-Bt corn. Furthermore, the EPA has made enforcement the responsibility of the companies that sell Bt corn, which may not be incentive-compatible. In accordance with EPA requirements, the Agricultural Biotechnology Stewardship Technical Committee (ABSTC) (2002) commissioned annual grower surveys. The results of these surveys indicate about 10 to 15 percent noncompliance with different aspects of the EPA's IRM requirements. Jaffe (2003a, 2003b) suggests noncompliance rates for the refuge size requirement as high as 20 percent. These results suggest that noncompliance could threaten the success of the EPA's current IRM plan, which has led the EPA to ask whether existing IRM plans can be amended to reduce compliance costs, improve compliance rates, and increase the likelihood of IRM success.

The purpose of this research is to explore the potential for using the contingent valuation (CV) method to test the sensitivity of farmer demand for CRW Bt corn to changes in EPA IRM requirements. The successful characterization of the sensitivity of farmer demand to EPA policy will help answer questions regarding farmer compliance with the regulatory burden of alternative IRM requirements. To accomplish this objective, we surveyed Minnesota farmers to ask if they would plant a Bt corn variety with a specific set of characteristics if it were available for the next growing season. Specifically, the Bt varieties that were described to farmers varied by spectrum of control, export market approval, IRM requirements, and price. Using a probit model, the distribution of farmers' willingness to pay (WTP) for Bt corn is characterized in terms of these factors. We then test whether these factors have a significant influence on the WTP and explore the reasonableness of the WTP estimates.

2. MATERIALS AND METHODS

2.1. Survey Sample

The survey was conducted between April and June 2002 following procedures outlined in Dillman (2000). The survey was mailed to 2,000 Minnesota corn farmers that produced a minimum of $1,000 worth of farm commodities. The sample of farmers was randomly and confidentially drawn from the Minnesota Agricultural Statistic Service's farmer database. The initial response rate was about 45 percent. After eliminating respondents who were no longer farming, explicitly refused to complete the survey, or did not complete the relevant portions of the survey, 630 responses remained, or 31.5 percent of the original sample.

2.2. Survey Design

The survey consisted of four sections. In the first, general information about the farmer and farm was queried (e.g., farm size, years of farming, and previous experience with genetically altered crops). The second section asked about the farmer's experience with insects, insect management, and current EPA regulations for planting Bt corn. The third section is the most relevant part for this research and will be described in further detail below. It focused on eliciting the value of a new Bt corn pest management program. The concluding section asked for typical demographic information.

The third section of the survey set up and asked a referendum type contingent valuation (CV) question (see Appendix for an example). The CV method is widely used to elicit preferences for non-market goods by asking what people would be willing to pay for the provision of these goods. It circumvents the absence of markets by presenting consumers with a hypothetical market in which they have the opportunity to buy the good in question (Mitchell and Carson 1989, pp. 2–3). The design of this survey follows the same principles in that a hypothetical market for new Bt corn hybrids was proposed. These hypothetical hybrids varied in terms of their spectrum of control, export market approval, IRM requirements, and price.

Four distinct spectrums of control were considered within two different dimensions. All hybrids were described as controlling CRW populations and damage to corn. Some were described as providing high CRW population control (i.e., 95 percent), while others were described as providing moderate population control (i.e., 75 percent). All were described as providing high CRW damage control. In addition to providing CRW control, some were described as providing high population and damage control for ECB. Table 1 provides the precise survey descriptions.

Whether the hybrid controlled CRW or CRW and ECB was varied to explore scope effects. It was also varied because both types of hybrids were headed for commercialization at the time of the survey. We varied the efficacy of CRW population control but not damage control to test for the publicness of CRW control (Clark and Carlson 1990). While damage control represents an immediate private benefit to farmers, population control represents a future benefit that may be private or public depending on the mobility of the insect. Since adult CRW beetles are quite mobile, we expected the population control benefits to be small because they are more public than private. From a more practical standpoint, exploring this type of

Table 1. Hybrid Product Descriptions

Product	Description
1	For example, the hybrid eliminates more than 95 percent of CRW and reduces lodging and yield loss due to CRW by more than 95 percent.
2	For example, the hybrid eliminates more than 75 percent of CRW and reduces lodging and yield loss due to CRW by more than 95 percent.
3	For example, the hybrid eliminates more than 95 percent of CRW and reduces lodging and yield loss due to CRW by more than 95 percent. It also eliminates more than 95 percent of ECB and reduces stalk breakage, eardrop, and yield loss due to ECB by more than 95 percent.
4	For example, the hybrid eliminates more than 75 percent of CRW and reduces lodging and yield loss due to CRW by more than 95 percent. It also eliminates more than 95 percent of ECB and reduces stalk breakage, eardrop, and yield loss due to ECB by more than 95 percent.

variation was reasonable because one company reported high CRW population and damage control for the product it was commercializing at the time of the survey, while another company reported moderate population and high damage control for the product it was commercializing.

Whether the produce of Bt corn could or could not be sold in major export markets was varied from one survey to the next to investigate the importance of farmer access to export markets. Farmer access to export markets was expected to increase a farmer's willingness to pay, but only if the farmer typically sells to these markets. The practical importance of this distinction continues because not all Bt corn hybrids are approved for sale in markets like Europe and Japan.

The surveys were varied in terms of whether or not farmers would be subject to EPA IRM guidelines if they purchased the new hybrid. Furthermore, if they were subject to IRM guidelines, the types of guidelines varied in terms of refuge size, refuge configuration, and the availability of the option to treat refuge with non-Bt insecticides. The alternative minimum refuge size requirements included 10, 20, 30, 40, and 50 percent of a farmer's Bt corn acreage. The alternative refuge configurations included nine combinations of four configurations that were either being allowed by or proposed to the EPA at the time of the survey: Bt and non-Bt seed mixes; multiple within-field refuge strips; within-field refuge blocks; and separate refuge fields within one-half mile of Bt corn fields. The nine specific combinations are described in Table 2. The alternative characterizations for refuge insecticide treatments were more complicated because sensible descriptions depended on the product type and configuration combinations. Table 3 summarizes the alternative descriptions associated with the different product types, configuration combinations, and treatment options. These refuge insecticide treatment options were sensible because the EPA had switched from not permitting them to permitting them while additional research was conducted to better understand the effect of the requirement on IRM success.

The additional cost of Bt seed corn proposed to farmers varied from $5 to $40 per acre of Bt corn in $1 increments. At the time of the survey, ECB Bt corn was selling with a price premium of about $8 to $10 per acre, with adoption rates in Minnesota around 30 percent. Therefore, a product with both ECB and CRW control could sell for more than $10 an acre. However, with crop rotations as a reasonable substitute for CRW Bt corn, many farmers might not be willing to pay as much as for ECB Bt corn even though they reported comparable damage from both pests.

In summary, each survey was assigned to one of four product descriptions, one of two market access conditions, one of two IRM regulatory conditions, one of five refuge size descriptions, one of nine refuge configuration descriptions, one of four refuge treatment descriptions, and one of 36 Bt seed corn price premiums. Figures 1 and 2 illustrate how these characteristics were randomly stratified across surveys. First, 200 surveys were

Table 2. Refuge Configuration Combinations

Configuration Combinations	Description
1	Planting refuge corn in a seed mix with your Bt corn
2	Planting refuge corn in a block in the same field as your Bt corn
3	Planting refuge corn in multiple strips in the same field as your Bt corn
4	Planting refuge corn in multiple strips in the same field as your Bt corn or in a seed mix with your Bt corn
5	Planting refuge corn in a block or multiple strips in the same field as your Bt corn
6	Planting refuge corn in a block or multiple strips in the same field as your Bt corn or in a seed mix with your Bt corn
7	Planting refuge corn in a separate field within one-half mile of your Bt corn
8	Planting refuge corn in a separate field within one-half mile of your Bt corn or in a block in the same field as your Bt corn
9	Planting refuge corn in a separate field within one-half mile of your Bt corn or in a block or multiple strips in the same field as your Bt corn

assigned to the stratification where farmers were told they would not be required to meet the described IRM guidelines if they planted the hypothetical hybrid (see Figure 1). For these surveys, the configuration combination for the IRM example description was randomly assigned. The remaining 1,800 surveys were assigned to the stratification where farmers were told they would have to meet the described IRM guidelines. These 1,800 surveys were then divided into nine groups of 200, with each group assigned a different refuge configuration combination. Figure 2 shows the distributions for the remaining characteristics within each of these stratifications. These values were randomly assigned to each survey.

2.3. Statistical Analysis

Farmers answered either "yes" or "no" to whether they would have been willing to plant the hypothetical hybrid in the current growing season if it were available. Therefore, a limited dependent variable model is most appropriate for interpreting farmer responses. We chose the probit model, while assuming the willingness to pay for Bt corn was log-linear in the product and regulatory characteristics:

Table 3. Refuge Insecticide Treatment Descriptions by Product, Configuration Combinations, and Availability of the Refuge Treatment Option

Product	Configuration Combinations	Availability of Treatment Option	Description
1 & 2	1–9	No	Using insecticides on your refuge corn to control CRW is not permitted
3 & 4	1,3, & 4	No	Using insecticides on your refuge corn to control CRW is not permitted
3 & 4	2, & 5–9	No	Using insecticides on your refuge corn to control CRW or ECB is not permitted
1 & 2	1–9	Yes	Using insecticides other than Bt microbial sprays on your refuge corn to control CRW only when economic thresholds are reached (as those recommended by local or regional professionals, such as Extension agents or crop consultants)
3 & 4	1,3, & 4	Yes	Using insecticides other than Bt microbial sprays on your refuge corn to control CRW only when economic thresholds are reached (as those recommended by local or regional professionals, such as Extension agents or crop consultants)
3 & 4	1,3, & 4	Yes	Using insecticides other than Bt microbial sprays on your refuge corn to control CRW only when economic thresholds are reached (as those recommended by local or regional professionals, such as Extension agents or crop consultants)
3 & 4	2, & 5–9	Yes	Using insecticides other than Bt microbial sprays on your refuge corn to control CRW or ECB only when economic thresholds are reached (as those recommended by local or regional professionals, such as Extension agents or crop consultants)

$$\ln(WTP) \quad = \sum_{j=1}^{4}\beta_{P_j}\ Product\ j + \beta_{EMA}\ Export\ Market\ Approval + IRM\ Required \quad (1)$$

$$(\beta_{IRM} + \beta_{RS}\ Refuge\ Size + \beta_{RT}\ Refuge\ Treatment + \beta_{SM}\ Seed\ Mix$$

$$+\ \beta_{MS}\ Multiple\ Strips + \beta_{BL}\ Blocks + \beta_{SF}\ Separate\ Field) + e$$

$$= X\beta + e,$$

where the subscripted βs are parameters; *Product j* is a dummy variable equal to 1 if the product described was product j; *Export Market Approval* is a dummy variable equal to 1 if the product was described as being approved for sale in major export markets; *IRM Required* is a dummy variable equal to 1 if the farmer was told IRM requirements were applicable; *Refuge Size* is the size of the described refuge in terms of the percentage of corn acreage; *Refuge Treatment* is a dummy variable that was equal to 1 if refuge insecticide treatments were described as being permitted; *Seed Mix* is a dummy variable equal to 1 if seed mixes were permitted for refuge; *Multiple Strips* is a dummy variable equal to 1 if multiple non-Bt strips were permitted for refuge; *Blocks* is a dummy variable equal to 1 if within field non-Bt blocks were permitted for refuge; *Separate Field* is a dummy variable equal to 1 if separate non-Bt fields were permitted for refuge as long as they were within one-half mile of Bt fields; and *e* is a normally distributed random error with mean zero and variance σ^2.

Using this characterization of the WTP, the probability that the *i*th farmer says "yes" to the referendum question is the probability that the farmer's WTP equals or exceeds the proposed price premium P_i: $\Pr(WTP_i \geq P_i) = \Pr(e_i \geq \ln(P_i) - X_i\beta) = 1 - \Phi(\sigma^{-1}\ln(P_i) - X_i\alpha)$, where $\Phi(\cdot)$ is the cumulative standard normal distribution and $\alpha = \beta\sigma^{-1}$. With this specification, the parameters represented by σ^{-1} and α can be estimated directly using the probit routine provided in a variety of different software packages.

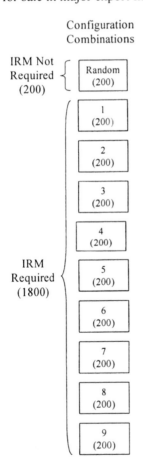

Figure 1. Stratifications by IRM Requirement and Configuration Combinations

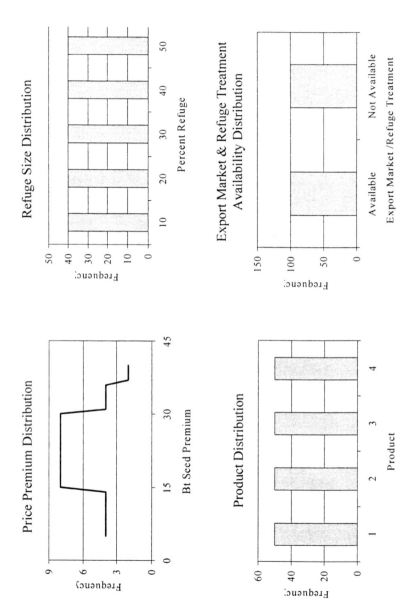

Figure 2. Distribution of Bt Corn and IRM Characteristics Within Each Stratification

It should be noted that this specification assumes that farmers actually valued Bt corn, which might not be the case given the controversy surrounding its introduction. Farmers who responded with "no" were also asked to select reasons for their negative response from a list or to provide their own reason (see Appendix). Out of the 630 farmers who responded to the WTP question, 526 answered "no." Out of these 526, fifty-nine or about 11 percent indicated that they would never plant Bt corn, which suggested that they placed no positive value on the product. Therefore, these individuals were not used to estimate the probit model. This left 571 observations to estimate the model. One in five of these respondents answered with a "yes."

2.4. Hypothesis Tests

The primary purpose of the survey was to estimate the sensitivity of a farmer's WTP to alternative EPA IRM requirements. However, we were also interested in how different control spectrums affected a farmer's WTP. To explore these issues, one unrestricted and 15 restricted versions of equation (1) were estimated.

Four alternative specifications for the effect of the EPA's IRM policy regime on a farmer's WTP were explored. The first, denoted by row (a) in Table 4, assumed that the WTP did not depend on any IRM requirements: $\beta_{IRM} = \beta_{RS} = \beta_{RT} = \beta_{SM} = \beta_{MS} = \beta_{BL} = \beta_{SF} = 0$. The second, denoted by row (b), assumed that having IRM requirements mattered, but what those requirements were did not: $\beta_{RS} = \beta_{RT} = \beta_{SM} = \beta_{MS} = \beta_{BL} = \beta_{SF} = 0$. The third, denoted by row (c), assumed that having IRM requirements mattered in terms of the refuge size and treatment option, but not in terms of the configuration requirements: $\beta_{SM} = \beta_{MS} = \beta_{BL} = \beta_{SF} = 0$. The fourth, denoted by row (d), assumed that all facets of the IRM requirements mattered, which placed no restrictions on the IRM-related parameters in equation (1).

Four specifications of the effect of the control spectrum on a farmer's WTP were explored. The first, denoted by column (i) in Table 4, assumed that the WTP did not depend on the control spectrum: $\beta_{P1} = \beta_{P2} = \beta_{P3} = \beta_{P4}$. The second, denoted by column (ii), assumed that the WTP depended on the difference in ECB control, but not on the difference in CRW population control: $\beta_{P1} = \beta_{P2}$ and $\beta_{P3} = \beta_{P4}$. The third, denoted by column (iii), assumed that the WTP depended on the difference in CRW population control, but not on the difference in ECB control: $\beta_{P1} = \beta_{P3}$ and $\beta_{P2} = \beta_{P4}$. The fourth, denoted by column (iv), assumed that the WTP depended on the difference in CRW population control and the difference in ECB control, which placed no restrictions on the control spectrum related parameters in equation (1).

The 16 alternative specifications of the model that were estimated are nested row- and column-wise, which allows for comparisons based on the likelihood ratio statistic. The statistic is asymptotically distributed χ^2 with the degrees of freedom equal to the number of parameter restrictions.

Table 4. Maximized Log-Likelihood for Alternative Models and Likelihood Ratio Statistic (χ^2) Model Comparisons

IRM Characteristics	Control Spectrum Characteristics				χ^2(d.f.) Model Comparisons		
	(i) No Spectrum Effect	(ii) ECB Control Effect	(iii) CRW Control Effect	(iv) CRW & ECB Control Effect	(i) vs. (ii)	(i) vs. (iii)	(i) vs. (iv)
(a) No IRM	-253.42 (3)	-251.37 (4)	-253.36 (4)	-250.53 (6)	4.09[b] (1)	0.11 (1)	5.77 (3)
(b) IRM Required	-252.72 (4)	-250.64 (5)	-252.66 (5)	-249.81 (7)	4.15[b] (1)	0.13 (1)	5.82 (3)
(c) IRM Required, Size, & Treatment	-247.60 (6)	-245.22 (7)	-247.60 (7)	-244.51 (9)	4.77[b] (1)	0.01 (1)	6.19 (3)
(d) IRM Required, Size, Treatment, & Configuration	-245.46 (10)	-243.03 (11)	-245.46 (11)	-242.42 (13)	4.86[b] (1)	0.00 (1)	6.07 (3)
χ^2(d.f.) Model Comparisons							
(a) vs. (b)	15.91[b] (7)	16.69[b] (7)	15.81[b] (7)	16.22[b] (7)			
(a) vs. (c)	14.52[b] (6)	15.23[b] (6)	14.39[b] (6)	14.78[b] (6)			
(a) vs. (d)	4.29 (4)	4.38 (4)	4.28 (4)	4.17 (4)			

[a] Significant at 1 percent for single tail test.
[b] Significant at 5 percent for single tail test.
[c] Significant at 10 percent for single tail test.

3. RESULTS

3.1. Overview of Respondents

The average respondent to the survey was male, 52 years old with at least a high school degree, and farmed about 500 acres. The majority (56 percent) worked off-farm, with 39 percent producing livestock as well as corn. The expected price per bushel of corn was about $2.00, with an expected yield of 137 bushels per acre.

Respondents confirmed the importance of corn rootworm (CRW) and European corn borer pests, with 55.4 and 83.9 percent reporting noticeable damage from each in the past five years. On average, farmers expected losses attributable to ECB and CRW of around 17 bushels per acre for either pest if no control measures were employed. The most important factors in deciding when and how to control these pests were costs followed by yield and harvest time.

The majority of farmers did not indicate having experience with insect resistance. Reported compliance rates with current ECB Bt corn requirements were relatively high, 72 percent, but lower than the results reported by ABSTC (2002). More than 80 percent expressed concern for ECB resistance to Bt corn.

3.2. Hypothesis Tests

Table 4 reports the maximized log-likelihood (number of parameters estimated) for the 16 combinations of the hypotheses regarding how IRM characteristics [rows (a)–(d)] and the proposed hybrid's control spectrum [columns (i)–(iv)] influence a farmer's willingness to pay (WTP). The right-hand side of the table reports χ^2-likelihood ratio statistics (degrees of freedom) for row-wise comparisons of the models in column (ii)–(iv) to the model in column (i). The bottom half of the table reports χ^2-likelihood ratio statistics (degrees of freedom) for column-wise comparisons of the models in row (a)–(c) to the model in row (d).

The comparisons reported in Table 4 offer two immediate results. First, a farmer's WTP for Bt corn appears to be sensitive to whether the product controls only CRW or both CRW and ECB. It does not appear to be sensitive to the level of CRW population control offered by the hybrid. As hypothesized, population control appears to be a public good, while damage control appears private. Second, some but not all IRM requirements influence a farmer's WTP for Bt corn. In particular, a farmer's WTP appears sensitive to refuge size and to whether refuge can be treated with non-Bt pesticides, but not to configuration requirements.

Table 5 reports the coefficient estimates (t-statistics) for the unrestricted model from equation (1) and the model with the CRW population control restriction and refuge configuration restrictions—row (c) and column (ii) in Table 4.

For the preferred model, the coefficient estimate for Products 3 and 4 exceed the estimate for Products 1 and 2, which implies that farmers value a product that controls CRW and ECB more than a product that controls only CRW. This result is as expected and statistically significant. The coefficient for export market approval is positive, which indicates that a farmer's WTP is higher when there is export market access for Bt corn. The coefficient for refuge size is negative, which indicates that a farmer's WTP is lower when the refuge size requirement is higher or stricter. While these two coefficient estimates have intuitive signs, their lack of statistical significance should be noted. The coefficient for required refuge is positive, which indicates that farmers value the EPA's IRM regulations. The coefficient for refuge treatment is negative, which indicates that farmers prefer not having the option to treat refuge with non-Bt insecticides. Both these coefficients have unexpected signs and are statistically significant.

3.3. Implications

Table 6 reports estimates of the mean, median, and standard deviation of the WTP for Bt corn based on the preferred model. These WTP statistics are reported for CRW control Bt corn without any IRM requirements and based on current IRM requirements (a 20 percent, treatable refuge in either blocks, multiple strips, or adjacent fields). The same statistics are reported for CRW and ECB Bt corn. Finally, the table reports estimates of the percentage increase in the willingness to pay associated with having export market approval, an IRM requirement, an increase of one percent in the refuge size requirement, and the option to treat refuge with non-Bt insecticides. Approximated standard errors for all these estimates are reported in parentheses.

The first notable result from Table 6 is the skew in the distribution of the WTP, with the mean WTP more than twice the median. The mean estimates for the WTP under current IRM regulations seem reasonable. For CRW Bt corn, our estimate of $8.59 is within about one standard deviation of the estimate reported in Alston et al. (2003). Our estimate of the added benefit of export market approval also seems reasonable, though still not significant. As noted earlier, our estimates imply that farmers actually prefer having IRM regulations and prefer not having the option to treat refuge. Table 6 shows how large these estimated effects are, while again calling into question the intuitive appeal of these results. Figure 3 shows the estimated demand curve in terms of the percentage of farmers adopting Bt corn at different price premiums. The adoption rates implied by a $10 per acre technology fee are between 20 and 30 percent, which is also within reason considering that adoption rates for ECB Bt corn at the time of the survey were about 20 percent with an \$8 to \$10 per acre price premium.

Table 5. Parameter Estimates (t-Statistic) for Unrestricted and Preferred Models

	Model	
Parameter	Unrestricted	Preferred
Product 1	0.91^b	0.78^b
	(2.11)	(1.87)
Product 2	0.77^b	0.78^b
	(1.70)	(1.87)
Product 3	1.06^a	1.06^a
	(2.50)	(2.56)
Product 4	1.21^a	1.06^a
	(2.73)	(2.56)
Export Market Approval	0.06	0.086
	(0.45)	(0.66)
IRM Required	0.54^c	0.61^b
	(1.62)	(2.23)
Refuge Size	-0.0055	-0.0057
	(1.09)	(1.20)
Refuge Treatment	-0.39^a	-0.41^a
	(2.66)	(2.88)
Seed Mix	0.18	
	(1.00)	
Multiple Strips	-0.23^c	
	(1.51)	
Block	0.221^c	
	(1.40)	
Separate Field	-0.02	
	(0.11)	
ln(Price Premium)	0.742^a	0.72^a
	(5.62)	(5.64)
Maximized Log-Likelihood	-242.42	-245.22
Estimated Parameters	13	7

[a] Significant at 1 percent for single tail test.
[b] Significant at 5 percent for single tail test.
[c] Significant at 10 percent for single tail test.

Table 6. WTP Distribution Estimates and Marginal Effects (Standard Deviations)

	CRW Control	CRW & ECB Control
No IRM	*$/Acre of Bt Corn*	
Mean	7.61	11.24
	(2.60)	(3.80)
Median	2.93	4.32
	(1.27)	(1.65)
Standard Deviation	18.26	26.95
	(10.42)	(16.39)
Current IRM		
Mean	8.59	12.68
	(1.86)	(2.73)
Median	3.30	4.88
	(1.22)	(1.51)
Standard Deviation	20.60	30.41
	(10.00)	(16.14)
	% Increase In WTP	
Export Market Approval	11.91	
	(18.13)	
IRM Required	84.16	
	(38.03)	
Refuge Size	-0.79	
	(0.65)	
Refuge Treatment	-56.38	
	(21.90)	

4. CONCLUSIONS

Existing guidelines for planting CRW Bt corn require a non-Bt corn refuge of 20 percent for corn grown in the Midwest. This refuge can be planted in a separate field if it is adjacent to the Bt field or it can be planted in a block or multiple strips within the Bt field. Non-Bt insecticides can be used for the supplemental control of insects as long as this control is based on accepted economic thresholds. These IRM requirements are designed to promote the sustainable use of Bt corn and reduce the use of more hazardous insecticides to the benefit of environmental and human health. However, for these guidelines to be effective, farmers must comply with them. There is increasing evidence that a significant proportion of farmers are not complying, which has prompted the EPA to look for ways to improve compliance.

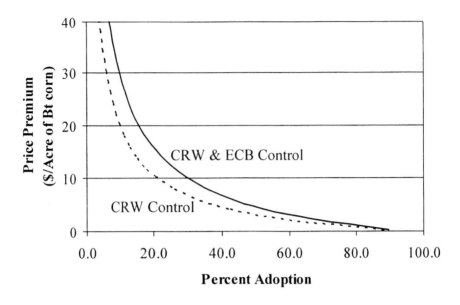

Figure 3. Estimated Demand for CRW and CRW and ECB Control Bt Corn Under Current IRM Guidelines and Without Export Market Approval

This chapter evaluates the utility of using the CV method to estimate how different components of the EPA's IRM requirements affect the value of Bt corn to farmers, which is directly related to compliance costs. The successful characterization of IRM compliance costs makes it possible to explore how regulatory refinements can reduce these costs and improve compliance without substantially threatening the primary policy objectives. While CV methods are commonly used, concerns about the validity and consistency of the method persist.

Our attempts to estimate the sensitivity of farmer demand for Bt corn to IRM guidelines using the CV method met with mixed results. We were able to estimate IRM effects that are intuitive in direction, reasonable in magnitude, but not statistically significant. We were able to estimate intuitive and statistically significant differences in a farmer's willingness to pay for Bt corn based on an increased spectrum of control: from CRW to CRW and ECB control. The magnitude and direction of the effect of export market approval is reasonable, but not statistically significant. Results that failed to meet our expectations include the significant and positive effect of IRM requirements on the willingness to pay, and the significant negative effect of having the option to treat refuge corn with non-Bt insecticides.

There are a range of possible explanations for the limited statistical significance of some anticipated results and strong statistical significance of some unanticipated results. For example, many farmers in our survey fed livestock with their produce, which weakens the general importance of export market approval. Alternatively, some farmers may have mistaken the refuge treatment option as a requirement instead, which would make Bt corn less attractive. Therefore, we believe there is more to be learned from further examination of our data and the collection of new CV data to better understand how farmer behavioral responses will affect the success of environmental regulatory policies that are becoming increasingly common.

REFERENCES

ABSTC [see Agricultural Biotechnology Stewardship Technical Committee].

Agricultural Biotechnology Stewardship Technical Committee (ABSTC). 2002. "Insect Resistance Management Grower Survey for Bt Field Corn: 2002 Growing Season." Available online at http://www.ncga.com/biotechnology/pdfs/IRM_exec_summary.pdf (accessed March 14, 2003).

Alstad, D.N., and D.A. Andow. 1995. "Managing the Evolution of Insect Resistance to Transgenic Plants." *Science* 268(5219): 1894–1896.

Alston, J.M., J. Hyde, M.C. Marra, and P.D. Mitchell. 2003. "An Ex Ante Analysis of the Benefits from the Adoption of Corn Rootworm Resistant Transgenic Corn Technology." *AgBioForum* 5(3): 71–84.

Clark, J.S., and G.A. Carlson. 1990. "Testing for Common versus Private Property: The Case of Pesticide Resistance." *Journal of Environmental Economics and Management* 19(1): 45–60.

Dillman, D.A. 2000. *Mail and Internet Surveys: The Tailored Design Method*. New York: Wiley.

EPA [see U.S. Environmental Protection Agency].

Hama, H., K. Suzuki, and H. Tanaka. 1992. "Inheritance and Stability of Resistance to *Bacillus thuringiensis* Formulations of the Diamondback Moth, *Plutella xylostella* (Linnaeus) (Lepidoptera: Yponomeutidae)." *Applied Entomology and Zoology* 27(3): 355–362.

Jaffe, G. 2003a. "Planting Trouble: Are Farmers Squandering Bt Corn Technology?" Center for Science in the Public Interest, Washington, D.C. Available online at http://cspinet.org/biotech/reports.html (accessed April 27, 2002).

———. 2003b. "Planting Trouble Update." Center for Science in the Public Interest, Washington, D.C. Available online at http://cspinet.org/biotech/reports.html (accessed May 10, 2002).

Mason, C.E., M.E. Rice, D.D. Calvin, J.W. Van Duyn, W.B. Showers, W.D. Hutchison, J.F. Witkowski, R.A. Higgins, D.W. Onstad, and G.P. Dively. 1996. "European Corn Borer: Ecology and Management." Publication No. 327, North Central Regional Extension Service, Iowa State University, Ames.

Minnesota Agricultural Statistics Service. 2002. "Minnesota AG News June Acreage." Minnesota Agricultural Statistics Service, Minnesota Department of Agriculture, St. Paul, MN.

Mitchell, R.C., and R.T. Carson. 1989. "Using Surveys to Value Public Goods: The Contingent Valuation Method." Resources for the Future, Washington D.C.

Tabashnik, B.E., J.M. Schwartz, N. Finson, and M.W. Johnson. 1992. "Inheritance of Resistance to *Bacillus thuringiensis* in Diamondback Moth (Lepidotera: Plutellidae)." *Journal of Economic Entomology* 85(4): 1046–1055.

U.S. Department of Agriculture, Economic Research Service. 2001. "Farm-Level Effects of Adopting Genetically Engineered Crops." Economic Issues in Agricultural Biotechnology Series, No. AIB 762. USDA/ERS, Washington, D.C.

U.S. Environmental Protection Agency. 1998. "The Environmental Protection Agency's White Paper on *Bt* Plant-Pesticide Resistance Management." Office of Pesticide Programs, U.S. Environmental Protection Agency, Washington, D.C. Available online at http://www.epa.gov/fedrgstr/EPA-PEST/1998/January/Day-14/paper.pdf (accessed October 28, 2002).

_____. 2001. "Bt Plant-Incorporated Protectants." Biopesticides Registration Action Document, U.S. Environmental Protection Agency, Washington, D.C. Available online at http://www.epa.gov/pesticides/biopesticides/pips/bt_brad.htm (accessed September 12, 2002).

APPENDIX

The following is an exact description of the third part of the survey for one combination of the product and IRM characteristics. Items that varied in this description from one survey to the next are in square brackets:

Please tell us about the value of a new program for managing insects:

New Bt corn hybrids are genetically engineered to control the corn rootworm (CRW). Some also control the European corn borer (ECB). The U.S. Environmental Protection Agency (EPA) is reviewing these new hybrids for registration and commercial sale to farmers. [For example, one hybrid eliminates more than 95 percent of CRW and reduces lodging and yield loss due to CRW by more than 95 percent. It also eliminates more than 95 percent of ECB and reduces stalk breakage, eardrop, and yield loss due to ECB by more than 95 percent.]

To reduce the chance of ECB resistance to Bt corn, EPA guidelines currently request farmers to plant non-Bt corn hybrids for refuge. The guidelines specify how much refuge corn to plant, where to plant refuge corn, and when to use insecticides on refuge corn. Similar guidelines are being considered for the new Bt corn hybrids. For example, the guidelines for the new hybrid might include:

- Planting at least [40] percent of your total corn acreage to non-Bt corn for refuge.
- [Planting refuge corn in a seed mix with your Bt corn.]
- [Using insecticides other than Bt microbial sprays on your refuge corn to control CRW only when economic thresholds are reached (as those recommended by local or regional professionals, such as Extension agents or crop consultants).]

D1. Suppose the example of a new Bt corn hybrid described above:

- was registered by the EPA for commercial sale to farmers,
- was the same as the non-Bt corn hybrids you commonly plant except for its insect control benefits (for example, it has the same maturity, yield potential, and herbicide tolerance),
- [was approved for marketing in the U.S. and all major corn export markets, and]
- [could be planted only if you follow all of the guidelines described above.]

Would you have planted this new hybrid in 2002 if it were available and its seed costs were [$5] per acre higher than the non-Bt hybrids you commonly plant?

(Please ✓ your answer)

❒ Yes   ❒ No

D2. If you would not have planted this hybrid in 2002, please tell us why not.

(Please ✓ all that apply)

❒   Would cost too much to plant it.
❒   Would want to first wait and see how it did in University performance trials or on a neighbor's farm.
❒   Would never plant a Bt corn hybrid.
❒   Would be concerned about being able to sell it.
❒   Would be concerned about getting a lower price for it.
❒   Would be concerned about possible environmental or safety issues.
❒   Would worry about having to keep it separate from my non-Bt corn.
❒   Other (Please describe): _____

D3. What is the most important reason why you would not have planted this new hybrid in 2002? (Please circle your response in question D2.)

Chapter 27

ADVERSE SELECTION, MORAL HAZARD, AND GROWER COMPLIANCE WITH Bt CORN REFUGE

Paul D. Mitchell[*] and Terrance M. Hurley[†]
University of Wisconsin [*] *and University of Minnesota* [†]

Abstract: We develop a principal-agent model of grower compliance with Bt corn refuge requirements for managing insect resistance to the Bt toxin. The model endogenizes the technology price, audit rate, and fine imposed on non-complying growers when grower willingness to pay for Bt corn and compliance effort is private information. Empirical analysis finds that practical application requires capping fine revenue. With such a program, the company raises the technology price and achieves complete compliance. The net welfare change (relative to competitive pricing) due to reducing company revenue and restricting technology access remains beyond the scope of this analysis.

Key words: asymmetric information, Compliance Assurance Program, corn rootworm, resistance management

1. INTRODUCTION

Bt corn is a popular term used to describe corn engineered to contain genetic material from the bacterium *Bacillus thuringiensis* (Bt). Bt corn produces proteins toxic to insects such as the European corn borer or corn rootworm, pests each estimated to cost growers over $1 billion annually in yield loss and control costs (Mason et al. 1996, Metcalf 1986). Bt corn offers excellent control of the targeted pests, which has resulted in its rapid adoption in the United States since commercial introduction in 1996. Over 25 percent of all corn acreage in the United States was planted to Bt corn for lepidopteran pests in 2003 (U.S. Department of Agriculture 2003). Bt corn for corn rootworm was commercially available in 2003 and 2004, but separate adoption data are unavailable.

Accompanying the excitement over the benefits of Bt corn is worry that pests will develop resistance to Bt corn. Current resistance management for

Bt corn is based on a high-dose refuge strategy, which requires growers to plant non-Bt corn as a refuge (Ostlie, Hutchison, and Hellmich 1997). This refuge generates pests not exposed to Bt corn that mate with any resistant pests emerging from nearby Bt corn. The goal is to produce an overwhelming number of susceptible pests for every resistant pest (at least 500:1) to slow the spread of resistance genes and prolong the efficacy of Bt. The refuge requirement has two primary components—a size requirement that depends on the region (20 percent or 50 percent of corn acres must be refuge) and a placement requirement (within one-half mile). If growers do not plant the required refuge, resistance will evolve more quickly (Hurley, Babcock, and Hellmich 2001). At this time, no viable field population of an insect pest resistant to Bt has been detected, but at least seven laboratory colonies of three insect species have developed resistance to Bt proteins (Tabashnik et al. 2003).

Surveys show that most Bt corn growers comply with refuge requirements, though significant noncompliance exists. An industry-sponsored survey of growers planting at least 200 acres of Bt corn in 2003 found that 92 percent of growers met the refuge size requirement and 93 percent met the placement requirement (National Corn Growers Association 2003). Jaffe (2003a, 2003b), using 2002 acreage data from USDA's National Agricultural Statistics Service (USDA-NASS), found that an average of 26 percent of farms planting Bt corn in 10 Corn Belt states violated the refuge size requirement, with an average of 15 percent not planting any refuge. The range for state-specific results around these averages was 11 percent to 46 percent of farms planting Bt corn violating the size requirement, with 9 percent to 38 percent not planting any refuge.

As a result of compliance concerns, the U.S. Environmental Protection Agency (EPA) required Bt corn registrants to develop a specific program to monitor and encourage refuge compliance. Together these companies announced the Compliance Assurance Program in 2002 with EPA approval (Agricultural Biotechnology Stewardship Technical Committee). Among the program's components, growers who do not comply with the refuge requirement for two consecutive years will be denied access to Bt corn by all companies. Company representatives visited farms in 2003, and noncomplying farms were guaranteed a visit during 2004. Summary statistics from these farm visits and revisits in 2004 have not been made public, and so the program's effectiveness at ensuring compliance remains to be established. The issue is whether the program's punishment is enforceable, since the registrants have licensed many seed companies to sell Bt corn, and whether the ban is an effective deterrent.

This chapter conceptually and empirically evaluates a fine program to ensure compliance. The program as evaluated randomly selects Bt corn growers for compliance audits, and non-complying growers pay a fine. To derive the optimal audit rate and fine, we develop a principal-agent model with a Bt corn registrant (a seed company) as the principal and growers as the agents. We first develop a conceptual model, and then specify an empirical model for Bt corn for corn rootworm to evaluate the practical application of the compliance program.

2. CONCEPTUAL MODEL

We develop a principal-agent model of the Bt corn refuge compliance problem with asymmetric information concerning a grower's willingness to pay for Bt corn (though the willingness-to-pay distribution among growers is common knowledge) and a grower's refuge compliance effort, which becomes known to the company if the grower is audited. Hidden information concerning grower willingness to pay creates adverse selection, and the potential for hidden action concerning compliance effort creates moral hazard. Currently, growers who purchase Bt corn must sign or renew an agreement confirming their awareness of refuge requirements. With the hypothetical fine program, a portion of these growers is randomly selected for an audit. Auditors visit selected growers, determine their compliance status, and impose fines for noncompliance. The company chooses the audit rate, fine, and technology fee to maximize expected net returns from sales, collected fines, and monitoring costs.

2.1. Grower Returns, Participation, and Incentive Compatibility

Random yield for conventional corn is y, while random yield for Bt corn is $y(1 + \lambda)$, where λ is the random net proportional yield gain for Bt corn relative to conventional corn. As a result, per acre returns for a grower planting conventional corn are $\pi_{cv} = py - K$, where p is the price of corn and K is the non-random production cost. Per acre returns for a non-complying grower who plants all Bt corn without paying extra for it are $\pi_{bt} = py(1 + \lambda) - K$. We focus only on the size requirement for refuge, so that growers choose the proportion ϕ of their corn to plant as non-Bt corn refuge. The required proportion of refuge is $\phi = \phi_r$, so returns for a complying grower are $\pi_{cp} = \phi_r \pi_{cv} + (1 - \phi_r)\pi_{bt}$. Per acre returns when the grower complies and pays an additional per acre cost for Bt corn are $\pi_{cp} - (1 - \phi_r)T$, where T is the non-random price of Bt corn usually identified with the technology fee. Following Mitchell et al. (2002) and Hurley, Mitchell, and Rice (2004), we assume that y and λ are independent.

Growers maximize expected utility of per acre profit, knowing the cost of production K and price p and the distributions of conventional yield y and proportional yield gain λ. The company also knows K and p, these distributions, and the expected utility-maximizing behavior of growers. However, a grower's maximum per acre willingness to pay W for Bt corn is private information known only to the grower, implicitly defined by

$$E[U(\pi_{cp} - W)] = E[U(\pi_{cv})], \tag{1}$$

where $U(\cdot)$ is the grower's utility function. From the company's perspective, W is a random variable with known distribution function $G(W)$ and density function $g(W)$ describing its distribution among growers.

A grower's compliance effort is ϕ, the proportion of corn acres planted as non-Bt corn refuge. Because this effort ϕ is private information known only to the grower, this potential for hidden action creates a classic moral hazard problem. Though a grower's choice of compliance effort ϕ is continuous, we follow the applied moral hazard literature (Laffont and Martimort 2002, p. 200) and focus on two cases. Grower effort is either high—that is, the grower complies and plants the required refuge ($\phi = \phi_r$)—or grower effort is low—that is, the grower plants no refuge ($\phi = 0$).

To address the moral hazard problem, the company's compliance program uses random field inspections to observe compliance effort, and then punishes non-complying growers. Let α be the probability that the company audits a grower for compliance and let F be the per acre fine for noncomplying growers. The probability that a grower must pay the fine depends on the grower's compliance effort ϕ. A complying grower pays no fine, while a non-complying grower pays the fine F with probability α, and pays nothing with probability $(1 - \alpha)$.

Given these definitions, a complying grower will buy Bt corn when a fine program is used if expected utility when complying equals or exceeds expected utility for conventional corn:

$$E[U(\pi_{cp} - (1 - \phi_r)T)] \geq E[U(\pi_{cv})]. \tag{2}$$

Condition (2) is the participation constraint for a company designing a fine program. Growers for whom the constraint is not satisfied will not buy Bt corn, and the company must account for this when choosing the price (technology fee) T.

With a fine program, a grower will comply if expected utility when complying equals or exceeds expected utility when planting all Bt corn and paying a fine F with probability α:

$$E[U(\pi_{cp} - (1 - \phi_r)T)] \geq (1 - \alpha)E[U(\pi_{bt} - T)] + \alpha E[U(\pi_{bt} - T - F)]. \tag{3}$$

Condition (3) is the incentive compatibility constraint for a company designing the fine program. Growers who buy Bt corn and for whom the constraint is not satisfied will not comply with the refuge requirement, but will plant all

Bt corn and pay the fine if audited. The company must account for this when choosing the technology fee T, the fine F, and audit rate α.

2.2. Constraint Reformulation

The participation and incentive compatibility constraints are reformulated as the probabilities that a grower buys Bt corn (participation) and complies (incentive compatibility), where the probabilities depend on the technology fee T, the audit probability α, and the fine F. Alternatively, for a population of growers, these participation and incentive compatibility probabilities can be interpreted as the proportion of the grower population buying Bt corn (participating) and complying (incentive compatibility).

PROPOSITION 1. *If grower utility is continuous and strictly increases in income, the participation condition can be expressed as $W \geq (1 - \phi_r)T$ and the probability a grower participates is $\beta = 1 - G((1 - \phi_r)T)$, which does not depend on the audit rate α or the fine F and is non-increasing in the technology fee T.*

Proof. Since grower utility is continuous and strictly increases in income, use the willingness to pay defined by equation (1) to reformulate participation condition (2) as $W \geq (1 - \phi_r)T$. Using $G(W)$, the distribution function for the willingness to pay W, express $W \geq (1 - \phi_r)T$ as $\beta = \Pr(W \geq (1 - \phi_r)T) = 1 - G((1 - \phi_r)T)$, which does not depend on α or F and is non-increasing in T, since $\partial\beta/\partial T = -g((1 - \phi_r)T)(1 - \phi_r) \leq 0$, where $g(\cdot) \geq 0$ is the density function for $G(\cdot)$.

Proposition 1 implies that the decision to purchase Bt corn depends on whether or not a grower's willingness to pay exceeds the price of Bt corn. That participation is non-increasing in the technology fee T is not surprising, since T is the price of Bt corn. Furthermore, Proposition 1 implies that the compliance program (α or F) does not affect participation, regardless of the grower utility function and willingness-to-pay distribution.

Incentive compatibility condition (3) cannot be expressed explicitly in terms of W without further assumptions concerning grower utility. Nevertheless, a condition similar to the reformulated participation condition in Proposition 1 exists.

PROPOSITION 2. *If grower utility is continuous and strictly increases in income, the incentive compatibility condition can be expressed as $W \geq Z(\alpha, F, T)$, where $Z(\cdot)$ is a function depending on the grower utility function. The probability that a grower complies is $v = 1 - G(Z(\alpha, F, T))$, which is non-*

decreasing in the audit rate α *and in the fine F and is non-decreasing (non-increasing) in the technology fee T if*

$$(1 - \alpha)E[U'(\pi_{bt} - T)] + \alpha E[U'(\pi_{bt} - T - F)]$$
$$- (1 - \phi_r)E[U'(\pi_{bt} - (1 - \phi_r)T - \phi_r py\lambda)] > (<) 0.$$

Proof. See Appendix.

That both the audit probability and the fine increase the probability of compliance is not surprising, since both increase the cost of cheating without affecting the cost of compliance. The technology fee's ambiguous effect on the compliance probability occurs because the technology fee affects grower utility whether the grower complies or cheats, but which it affects more cannot be determined without further assumptions concerning grower utility. The reported condition is the difference in the grower's marginal expected utility when the technology fee increases. The first two terms are the marginal utility if the grower cheats, and the last term is the marginal utility if the grower complies. The condition is positive (negative) if marginal utility for cheating is greater (less) than for complying.

Propositions 1 and 2 each define a lower bound on W such that any grower with W above this bound will either buy Bt corn or comply. If Proposition 1 defines a lower bound on W that is less than the lower bound defined by Proposition 2, then $\beta > v$; otherwise, $v \geq \beta$. Hence, Propositions 1 and 2 imply a condition indicating whether $\beta > v$ or $v \geq \beta$:

$$(1 - \phi_r)T < Z(\alpha, F, T). \tag{4}$$

If condition (4) holds, then $\beta > v$ and the participation and incentive compatibility constraints both bind for a company designing a compliance program. If condition (4) does not hold, then $v \geq \beta$ and only the participation constraint binds. Figure 1 illustrates these bounds for both cases, plus labels both the optimal behavior for growers with W in each range and the probabilities defined by Propositions 1 and 2.

The top plot in Figure 1 assumes that condition (4) holds, implying that a range for W exists for which some growers buy Bt corn and do not comply. In this case, $\beta > v$ and the probability that a grower buys Bt corn and complies is v, while the probability that a grower buys Bt corn and does not comply is $\beta - v$. The bottom plot assumes that condition (4) does not hold, so that all growers buying Bt corn comply. In this case, $v > \beta$ and the probability that a grower buys Bt corn and complies is β, while the probability that a grower buys Bt corn and does not comply is zero.

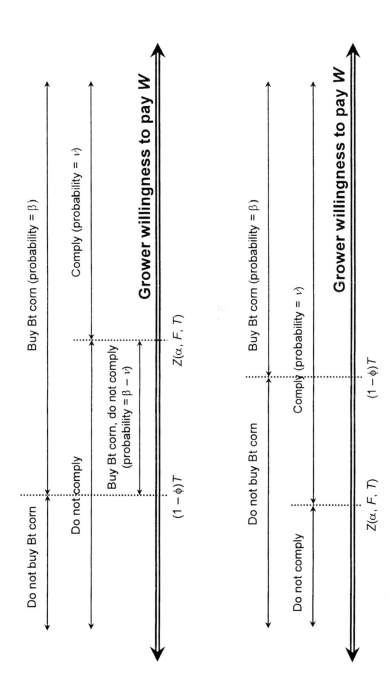

Figure 1. Illustration of the Relationship Between the Lower Bounds on Grower Willingness to Pay for Bt Corn Defined by the Participation and Incentive Compatibility Constraints and Associated Optimal Grower Behavior

2.3. Company Objective

The company maximizes net expected per acre returns from selling Bt corn to each grower. Relative to selling only conventional corn, when selling Bt corn using a fine program, the company earns additional revenue from charging more for Bt corn and from collecting fines. Additional costs include any extra costs for producing Bt corn and the cost of monitoring compliance. The applicable company objective depends on whether $\beta > v$. If condition (4) holds, then $\beta > v$ and the company's additional net expected per acre return is

$$V = (\beta - v\phi_r)(T - c) + (\beta - v)\alpha F - \beta k(\alpha), \tag{5}$$

which is the sum of expected net revenue from sales and collected fines, minus the expected costs of monitoring.

To focus on the participation and compliance issues the company faces, we assume a constant marginal cost c for each acre of Bt corn the company sells, so that company's net per acre revenue is $T - c$. The probability that any given grower buys Bt corn and complies is v, in which case the company's net per acre revenue is $(1 - \phi_r)(T - c)$, since the complying grower plants only the proportion $(1 - \phi_r)$ of fields in Bt corn. The probability that any given grower buys Bt corn and does not comply is $\beta - v$, in which case net per acre revenue is $(T - c)$. Hence, the expected net per acre revenue is $(\beta - v\phi_r)(T - c)$, the first term in equation (5).

For each grower, the probability that the company collects a fine is $(\beta - v)\alpha$, since with probability $(\beta - v)$ the grower does not comply and with probability α the grower is audited for compliance. Hence the expected per acre fine collected by the company is $(\beta - v)\alpha F$, the second term in equation (5). The company's per acre cost of monitoring compliance is $k(\alpha)$, where $k'(\cdot) > 0$ and $k''(\cdot) > 0$. The probability that a given grower will buy Bt corn is β, so the company's expected monitoring cost is $\beta k(\alpha)$, the third term in equation (5).

Because $G(W)$ describes the distribution of willingness to pay among growers and because the actual realization for any given grower is unknown to the company, β and v have two interpretations. Thus β is not only the probability that any given grower purchases Bt corn, but also the respective proportion of the grower population that purchases Bt corn. Similarly, v is not only the probability that any given grower complies, but also the proportion of the grower population that complies.

If condition (4) does not hold for the α, F, and T that maximize equation (5), then $v \geq \beta$, and the company's additional net expected per acre return is instead

$$V = \beta(1 - \phi_r)(T - c) - \beta k(\alpha). \tag{6}$$

When $v \geq \beta$, all growers who buy Bt corn also comply, so that the probability that a grower buys Bt corn and complies is β. Thus expected net sales revenue is $\beta(1 - \phi_r)(T - c)$, since the grower buys Bt corn with probability β and pays an additional $(1 - \phi_r)T$ because Bt corn is planted on the proportion $(1 - \phi_r)$ of all corn acres. The company must still monitor compliance and credibly threaten to impose fines on non-complying growers; otherwise, the incentive compatibility condition and the compliance probability v will change. Thus the expected cost of monitoring is $\beta k(\alpha)$, though no fines are collected, since all growers comply.

Because the probability $\beta = 1 - G((1 - \phi_r)T)$, the first-order condition for maximizing equation (6) with respect to α reduces to $k'(\alpha) = 0$, which defines the optimal α independent of grower utility and the grower willingness-to-pay distribution. The first-order condition for maximizing equation (6) with respect to T can be expressed as $\beta - g((1 - \phi_r)T)[(1 - \phi_r)(T - c) - k(\alpha)]$ $= 0$, which defines the optimal T independent of grower utility. Because no fines are collected, equation (6) does not depend on the fine F. However, to ensure that the incentive compatibility condition holds, the company must still conduct compliance audits and credibly threaten to impose fines. Condition (4) and the α and T that maximize equation (6) define a lower bound for this fine F.

Equation (5) defines a concave function for company net returns when grower participation exceeds compliance, implying that some growers do not comply (the top plot in Figure 1). Equation (6) defines a concave function for company net returns when grower compliance equals participation, implying that all growers buying Bt corn comply (the bottom plot in Figure 1). The company maximizes the upper envelope of these two functions, since the technology fee, audit rate, and fine can be chosen so that condition (4) is satisfied or not. Depending on the specific parameters (e.g., ϕ_r, c, $G(W)$, $E[\lambda]$), the upper envelope can have a variety of shapes. Most interesting for optimization, the two functions can intersect so that the upper envelope has two peaks, for which the relative magnitude of each depends on parameter values.

Figure 2 illustrates the company's upper envelope with a risk-neutral grower when no compliance program is used. Equations (5) and (6) still describe the company's objective, except that monitoring costs and fine revenue are zero ($k(\alpha) = \alpha = F = 0$) and the company's only choice variable is the technology fee T. The left curve is for equation (5) and the right curve is for equation (6). For the two plots, only the company's marginal cost c changes. The top plot uses a c less than for the bottom plot. For the top plot,

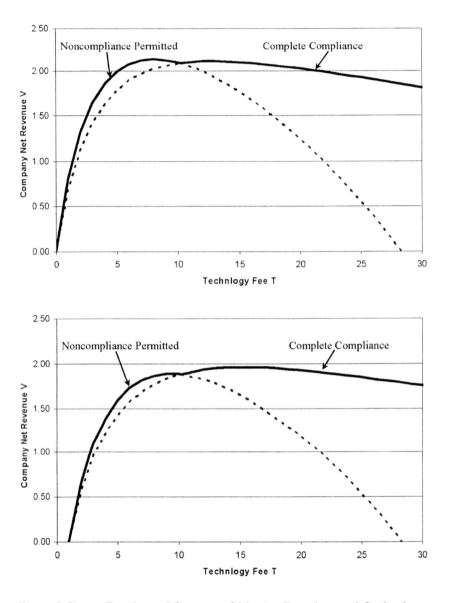

Figure 2. Upper Envelope of Company Objective Functions and Optimal Regime Shift from Permitting Noncompliance to Complete Compliance Due to Parameter Change

the optimal technology fee is $T = 8.06$, the peak of the first curve from equation (5). For the bottom plot, the optimal technology fee is $T = 15.20$, the peak of the second curve. As c increases, the magnitude of the first peak decreases relative to the second peak, so that the optimal technology fee jumps suddenly from the first to the second peak. Hence as c increases, eventually a critical c is reached at which the company's optimal response jumps from offering a low technology fee and permitting some non-compliance to using a higher technology fee that generates complete compliance. Other parameter values can generate other situations, such as peaks of equal height, merging of the peaks, and complete envelopment of one curve by the other. Optimization must take these possibilities into account.

2.4. Special Cases

This section examines two cases for grower utility that are used for empirical analysis—risk neutrality and constant absolute risk aversion. Corollary 1 demonstrates that for these cases, definitive results for the effect of the technology fee on grower compliance exist. Proposition 3 shows that with grower risk neutrality, the company's optimization problem separates so that the optimal audit rate α is determined independent of the technology fee and fine.

Corollary 1. *For a risk-neutral grower, the incentive compatibility condition is $W \geq E[py\lambda] - \phi_r T - \alpha F$. For a grower with constant absolute risk aversion, the incentive compatibility condition is $W \geq -\phi_r T + \{\ln(E[\exp(-R\pi_{cv})]) - \ln(E[\exp(-R\pi_{bt})]) - \ln(1 - \alpha + \alpha\exp(RF))\}/R$, where R is the coefficient of absolute risk aversion. For both cases, the probability that a grower complies is non-decreasing in the audit rate α, the fine F, and the technology fee T.*

Proof. For a risk-neutral grower, equation (1) implies that $E[\pi_{cp}] = W + E[\pi_{cv}]$ and incentive compatibility condition (3) becomes $E[\pi_{cp} - (1 - \phi_r)T] \geq (1 - \alpha)E[\pi_{bt} - T] + \alpha E[\pi_{bt} - T - F]$, which simplifies to $E[\pi_{cp}] + \phi_r T \geq E[\pi_{bt}] - \alpha F$. Substitute $E[\pi_{cp}] = W + E[\pi_{cv}]$ into this expression and rearrange to obtain $W \geq E[\pi_{bt}] - E[\pi_{cv}] - \phi_r T - \alpha F$. Lastly, because $\pi_{bt} = \pi_{cv} + py\lambda$, then $E[\pi_{bt}] - E[\pi_{cv}] = E[py\lambda]$. Thus, the incentive compatibility condition for a risk-neutral grower is $W \geq E[py\lambda] - \phi_r T - \alpha F$.

With constant absolute risk aversion, grower utility is $U(\pi) = 1 - \exp(-R\pi)$, where $R > 0$ is the coefficient of absolute risk aversion. Equation (1) becomes $E[1-\exp(-R\pi_{cp})\exp(RW)]=E[1-\exp(-R\pi_{cv})]$, which implies that $E[\exp(-R\pi_{cp})] = E[\exp(-R\pi_{cv})]/\exp(RW)$. Condition (3) becomes

$$E[1 - \exp(-R\pi_{cp})\exp(R(1 - \phi_r)T)] \geq (1 - \alpha)E[1 - \exp(-\pi_{bt})\exp(RT)]$$
$$+ \alpha E[1 - \exp(-R\pi_{bt})\exp(RT)\exp(RF)].$$

Rearranging gives $E[\exp(-R\pi_{cp})]\exp(R(1 - \phi_r)T) \leq E[\exp(-R\pi_{bt})]\exp(RT)$ (1–

$\alpha + \alpha \exp(RF)$). Substituting $E[\exp(-R\pi_{cp})] = E[\exp(-R\pi_{cv})]/\exp(RW)$ into this expression and rearranging gives the incentive compatibility condition with constant absolute risk aversion:

$$W \geq -\phi_r T + \{\ln(E[\exp(-R\pi_{cv})]) - \ln(E[\exp(-R\pi_{bt})]) - \ln(1 - \alpha + \alpha \exp(RF))\}/R.$$

The compliance probability is $v = 1 - G(Z(\alpha, L, T))$, so that $\partial v/\partial T = -g(\cdot)\partial Z/\partial T$, where $g(\cdot) \geq 0$ is the density function for $G(\cdot)$. For both special cases, previously derived expressions for $Z(\alpha, L, T)$ imply that $\partial Z/\partial T = -\phi_r > 0$. Hence, $\partial v/\partial T = g(\cdot)\phi_r \geq 0$ and, thus, the compliance probability is non-decreasing the technology fee T. Proposition 1 demonstrated that the compliance probability is non-decreasing in the audit rate α and in the fine F.

PROPOSITION 3. *For a risk-neutral grower, $k'(\alpha) = 0$ defines the optimal audit rate α regardless of the distribution of grower willingness to pay.*

Proof. See Appendix.

With grower risk neutrality, the optimization process simplifies, since equations (5) and (6) need only be optimized with respect to F and T, treating α as a parameter defined by $k'(\alpha) = 0$. No comparable result exists for constant absolute risk aversion.

3. EMPIRICAL ANALYSIS

We empirically parameterize this model to assess the practicality of this compliance program. We focus on Rock County in southern Wisconsin and the recently commercialized Bt corn for corn rootworm. Corn rootworms typically cause damage only in continuous corn, but the western corn rootworm soybean variant has spread into the region, so that corn-soybean growers regularly suffer economic damage in rotated corn (Levine, Oloumi-Sadeghi, and Fisher 1992; Cullen 2004). We specify stochastic models for conventional corn yield and the proportional yield benefit of Bt corn. Grower survey data from Minnesota are used to estimate the willingness-to-pay distribution. The Minnesota data should be comparable to southern Wisconsin, since south western Minnesota corn-soybean growers also regularly experience economic damage in rotated corn due to northern corn rootworm extended diapause (Krysan, Jackson, and Lew 1984).

3.1. Grower Returns

Consistent with the conceptual model, per acre returns for a grower planting conventional corn are $\pi_{cv} = py - K$, where p = \$2.25/bu is the price of corn and K = \$200/acre is the production cost. Both p and K are non-random to focus on yield risk and stochastic returns from planting Bt corn. Per acre returns for a non-complying grower who plants all Bt corn without paying extra for it are $\pi_{bt} = py(1 + \lambda) - K$, where λ is random proportional yield benefit for Bt corn relative to conventional corn treated with a soil insecticide.

Conventional yield y has a beta distribution, a common assumption for crop yields [Goodwin and Ker (2002) review several papers]. The linear trend yield for Rock County, Wisconsin, in 2006, estimated using least squares and USDA-NASS (U.S. Department of Agriculture 2005) yield data for 1990–2004, was 149.8 bu/acre. The yield coefficient of variation is 30 percent, which is consistent with farm-level yield data and other studies (Coble, Heifner, and Zuniga 2000). Minimum yield is 0 bu/acre and maximum yield is 239.7 bu/acre, the mean plus two standard deviations (Babcock, Hart, and Hayes 2004).

The proportional yield benefit for Bt corn, λ, is random because corn rootworm damage determining the benefit is random. However, the magnitude of this benefit varies by location and is difficult to estimate (Mitchell, Gray, and Steffey 2004). To focus on the technology pricing and compliance issue and not the technical details of developing a pest damage model, we specify a simple model for the proportional yield benefit of Bt corn and use sensitivity analysis. Specifically, we assume that λ has a beta distribution with a minimum of 0 and a maximum of 1. For the mean, we use 0.03, 0.05, and 0.07, i.e., on average Bt corn increases yields by 3 percent, 5 percent, and 7 percent. As Mitchell, Gray, and Steffey (2004) report in their Table 4 for the western corn rootworm soybean variant in Illinois using their composed error model, λ has a coefficient of variation of 1.0. Finally, following Mitchell et al. (2002) and Hurley, Mitchell, and Rice (2004), y and λ are independent.

3.2. Grower Preferences, Compliance Effort, and Willingness to Pay

We assume that growers are risk-neutral or have constant absolute risk aversion. Following Babcock, Choi, and Feinerman (1993), the coefficient of absolute risk aversion was chosen so that the implied risk premium is 25 percent of the standard deviation of returns for a moderately risk averse grower. For the specified model, this value is 0.005174. Grower compliance effort is either low or high. With low effort, the grower plants refuge $\phi = 0$.

With high effort, the grower plants $\phi = \phi_r = 0.20$, the minimum amount of refuge required under current regulations in the Midwest.

The grower willingness-to-pay distribution is estimated using data from a 2002 survey of growers in Minnesota. The survey included a referendum question asking growers if they would buy a hypothetical Bt corn product for corn rootworm at a given price. Because the price was randomly assigned to surveys, the willingness to pay can be estimated. The final model for grower willingness to pay is a lognormal distribution with a mean of $8.59/acre and a standard deviation of $20.60/acre. See Langrock and Hurley (2006) for additional information.

3.3. Company Costs and Returns

Proposition 3 indicates that with risk-neutral growers, the company chooses an optimal audit rate that minimizes monitoring costs. Hence, a monitoring cost specification with a unique minimum at some reasonable value for the audit rate α is needed. Ranges for audit rates or associated costs for the current Compliance Assurance Program were not available. As a result, we use a quadratic function for the company's monitoring costs with a minimum at $\alpha = 0.04$, as 4 percent seemed a reasonable baseline audit rate. The function is calibrated to match the cost for certifying hybrid seedcorn. Certification by the Iowa Crop Improvement Association (ICIA) involves five field visits by trained personnel to map the field, check isolation distances, confirm synchronization of pollination and silking, and ensure a sufficiently low number of rogue plants (Schmitz 2004). Besides an initial fee, the ICIA charges $6/acre for this certification (Schmitz 2004), which implies an average of $1.20/acre per visit. To complete the quadratic equation specification, we assume that the cost increases 25 percent if the inspection rate doubles. Hence, the cost of monitoring is $k(\alpha) = 1.5 - 15\alpha + 187.5\alpha^2$. Finally, c, the company's marginal cost of producing Bt corn, is first set at 0 and then increased for sensitivity analysis.

We examine three different regulatory programs to determine the company's optimal compliance program for each. As a benchmark, we examine the optimal technology fee when no compliance program is used. This case requires setting α and F to zero, dropping monitoring costs from the company objective, and then maximizing with respect to T. We also examine the company's optimal compliance program if the company keeps collected fines, but only up to its cost of monitoring; any additional fine revenue is retained by the government or some other third party. Thus the fine revenue term $(\beta - v)\alpha F$ in equation (5) is capped at the monitoring cost $\beta k(\alpha)$. Finally, we assume that the company retains all collected fines, as in equation (5).

An optimal solution α, F, and T conceptually exists for the company, but an analytical solution is generally intractable. Assumptions can simplify optimization, as Proposition 3 shows. Indeed, with a risk-neutral grower and a uniform distribution for grower willingness to pay, an analytical solution is tractable. However, numerical methods are needed for the model as specified here. We used a Monte Carlo approach with a grid search to find the company's optimal audit rate α, fine F, and technology fee T for the reported results.

3.4. Results and Sensitivity Analysis

Table 1 reports results when no compliance program is used, assuming different levels of grower risk aversion, expected benefits, and company marginal cost. The optimal technology fee ranges about $8/acre to $26/acre depending on assumptions. The current technology fee in southern Wisconsin is around $20/acre, which is consistent with these levels of expected benefits and a marginal cost ranging from $3/acre to $6/acre. Participation rates range 9 percent to 32 percent, with compliance less than participation in many cases. The ratio $(\beta - v)/\beta$ is the proportion of growers buying Bt corn who do not comply. For the results in Table 1, this proportion ranges from 22 percent to 65 percent for the cases with complete compliance, which is comparable to the results that Jaffe (2003a, 2003b) reports.

Risk aversion has a different effect on the technology fee depending on whether risk aversion shifts the company from maximizing equation (5) with noncompliance permitted to maximizing equation (6) with complete compliance. If adding risk aversion causes this shift, the optimal technology fee increases, as with an expected benefit of 3 percent and a marginal cost of $c = 0$. If adding risk aversion does not shift the optimal regime, then risk aversion decreases the optimal technology fee, as with an expected benefit of 5 percent and a marginal cost of $c = 0$.

Increasing the marginal cost c increases the optimal technology fee both at the margin and by shifting regimes. First, at the margin, increasing c increases T, as for the risk-neutral case with expected benefits of 5 percent and 7 percent. Also, increasing c increases T if it shifts the optimal regime, as for the risk-averse case with expected benefits of 5 percent and 7 percent. Though not reported, in all cases explored, increasing c eventually shifts the optimal regime.

Company net returns range from about $1.50/acre to $2.50/acre depending on assumptions. The company's marginal cost c has the largest effect—increasing the marginal cost $3/acre decreases the net returns around 20 percent to 25 percent. The expected benefit $E[\lambda]$ also has a sizable effect. For the reported cases, increasing the expected benefit from 3 percent to 5 percent or from 5 percent to 7 percent increases net returns less than 10 percent.

Table 1. Optimal Program Parameters with No Compliance Program

Grower risk preferences	Expected benefit E[λ]	Marginal cost c	Technology fee T	Audit rate α	Fine F	Participation rate β	Compliance rate v	Company revenue V
Risk neutral	3.0%	0	8.05	--	--	31.5%	24.6%	2.14
		3	19.70	--	--	12.9%	12.9%	1.73
	5.0%	0	9.60	--	--	27.1%	13.7%	2.34
		3	14.06	--	--	18.8%	14.7%	1.75
	7.0%	0	10.53	--	--	24.9%	8.7%	2.44
		3	15.58	--	--	16.9%	9.3%	1.89
		6	19.73	--	--	12.9%	9.8%	1.50
Risk averse[a]	3.0%	0	12.82	--	--	20.6%	20.6%	2.12
		3	19.70	--	--	12.9%	12.9%	1.73
	5.0%	01	8.92	--	--	28.9%	18.0%	2.26
		3	19.70	--	--	12.9%	12.9%	1.73
	7.0%	0	9.87	--	--	26.4%	12.1%	2.37
		3	14.49	--	--	18.2%	13.0%	1.79
		6	26.02	--	--	9.2%	9.2%	1.47

[a] Coefficient of absolute risk aversion $R = 0.005174$, implying a risk premium about 25 percent of the standard deviation of returns.

Adding risk aversion has a small (< 5 percent) negative effect on net revenue.

With the same assumptions as Table 1, Table 2 reports results when fine revenue is capped at the cost of monitoring, and Table 3 reports results with no fine revenue cap. The company must conduct compliance audits in both cases, even if complete compliance occurs. Thus, the optimal audit rate gravitates to $\alpha = 4$ percent, since it minimizes monitoring costs.

Results in Table 2 indicate that capping fine revenue generally shifts the optimal regime to a higher technology fee with complete compliance. Exceptions occur for risk-averse growers with high expected benefits and low company marginal cost. In these cases, the optimal technology fee is low (< $10/acre), the optimal fine is about $50/acre to $70/acre, and the optimal audit rate increases to around 5 percent. The audit rates and fines seem reasonable, but the technology fees seem rather low, less than half the current technology fee.

For cases with complete compliance, the optimal technology fee is independent of grower risk preferences and the expected benefit. The company maximizes equation (6) with respect to T (α is set at 4 percent to minimize monitoring costs), and Proposition 1 demonstrates that β depends only on $G(W)$, ϕ_r, and T, and that $U(\cdot)$ and $E[\lambda]$ have no effect. Hence the optimal T in Table 2 is 16.36 when $c = 0$, regardless of grower risk preferences and the expected benefit.

For the case of complete compliance, Tables 1 and 2 show that a compliance program with a revenue cap increases the technology fee (12 percent to 103 percent) but eliminates noncompliance. Higher technology fees reduce participation by 15 percent to 50 percent (2 to 16 percentage points). Company net returns decrease 2 percent to 22 percent due to added monitoring costs and reduced sales. Results for these cases indicate that a compliance program with a fine revenue cap can eliminate noncompliance at the expense of reduced company net returns and lower sales to growers.

Table 2 shows that a compliance program with a fine revenue cap in some cases (high expected benefits with low company marginal cost) creates incentives for the company to substantially decrease the technology fee to maximize sales revenue, which increases both participation and noncompliance, and to then use fine revenue to pay for the cost of monitoring. This scheme does not work for the other parameter assumptions in Table 2 because with risk neutrality and/or lower expected benefit, the lower technology fee does not create sufficient noncompliance incentives to generate enough fine revenue to pay monitoring costs. Also, as the company's marginal cost increases, the optimal technology fee must increase, so that eventually the higher technology fee decreases noncompliance incentives and not enough fine revenue can be generated to pay monitoring costs.

Table 2. Optimal Program Parameters with a Fine-Based Compliance Program That Caps Fine Revenue at the Cost of Monitoring

Grower risk preferences	Expected benefit $E[\lambda]$	Marginal cost c	Technology fee T	Audit rate α	Fine F	Participation rate β	Compliance rate v	Company revenue V
Risk neutral	3.0%	0	16.36	4.0%	0	16.0%	16.0%	1.90
		3	22.91	4.0%	0	10.8%	10.8%	1.59
	5.0%	0	16.36	4.0%	0	16.0%	16.0%	1.90
		3	22.91	4.0%	0	10.8%	10.8%	1.59
	7.0%	0	16.36	4.0%	0	16.0%	16.0%	1.90
		3	22.91	4.0%	0	10.8%	10.8%	1.59
		6	29.04	4.0%	0	7.9%	7.9%	1.36
Risk averse[a]	3.0%	0	16.36	4.0%	0	16.0%	16.0%	1.90
		3	22.91	4.0%	0	10.8%	10.8%	1.59
	5.0%	0	5.85	4.8%	62.18	40.1%	23.8%	2.07
		3	22.91	4.0%	0	10.8%	10.8%	1.59
	7.0%	0	8.79	4.7%	51.91	29.3%	14.8%	2.31
		3	10.01	5.0%	67.14	26.1%	16.6%	1.60
		6	29.04	4.0%	0	7.9%	7.9%	1.36

[a] Coefficient of absolute risk aversion $R = 0.005174$, implying a risk premium about 25 percent of the standard deviation of returns.

Table 3. Optimal Program Parameters with a Fine-Based Compliance Program With No Cap on Fine Revenue

Grower risk preferences	Expected benefit E[λ]	Marginal cost c	Technology fee T	Audit rate α	Fine F	Participation rate β	Compliance rate v	Company revenue V
Risk neutral	3.0%	0	16.36	4.0%	0	16.0%	16.0%	1.90
		3	22.91	4.0%	0	10.8%	10.8%	1.59
	5.0%	0	0.10	4.0%	293.84	99.6%	37.3%	6.22
		3	0.14	4.0%	296.43	99.3%	37.9%	3.46
	7.0%	0	0.07	4.0%	431.05	99.8%	31.4%	10.67
		3	0.07	4.0%	433.74	99.8%	31.8%	9.36
		6	0.10	4.0%	436.10	99.6%	32.2%	5.07
Risk averse[a]	3.0%	0	16.36	4.0%	0	16.0%	16.0%	1.90
		3	22.91	4.0%	0	10.8%	10.8%	1.59
	5.0%	0	0.25	7.2%	93.94	97.9%	38.6%	2.88
		3	22.91	4.0%	0	10.8%	10.8%	1.59
	7.0%	0	0.14	8.7%	109.41	99.3%	32.6%	4.87
		3	0.30	8.6%	110.69	97.1%	32.9%	2.12
		6	29.04	4.0%	0	7.9%	7.9%	1.36

[a] Coefficient of absolute risk aversion $R = 0.005174$, implying a risk premium about 25 percent of the standard deviation of returns.

Table 3 shows that with no fine revenue cap, the company may find it optimal to raise the technology fee so that complete compliance will occur, especially when expected benefits are low. Also, as the company's marginal cost increases, eventually the optimal technology fee increases, and complete compliance occurs when the optimal regime shifts. Though not shown, this shift occurs at higher marginal cost for the risk-neutral case (by $c = 6$ for an expected benefit of 5 percent, by $c = 11$ for an expected benefit of 7 percent). After this shift, since the technology fee is independent of risk preferences and the expected benefit, technology fees are the same as in Table 2.

The interesting result in Table 3 is that without a fine revenue cap, the company can use fines as a revenue source. The company uses a very low technology fee to obtain almost 100 percent participation and high noncompliance, and then fines become a lucrative revenue source. For the noncompliance cases, participation exceeds 97 percent and the proportion of growers buying Bt corn who do not comply, $(\beta - v)/\beta$, ranges from 61 percent to 69 percent. However, once the marginal cost c becomes sufficiently large, fine revenue is insufficient to pay the cost of giving Bt corn away, and the company shifts to the regime with a high technology fee and complete compliance.

For the cases in Table 3 with positive fines, the optimal audit rates for risk-averse growers are about twice the cost-minimizing rate of $\alpha = 4$ percent, and the fines seem reasonable. Company net revenue is larger than in Table 2, slightly more than doubling when the expected benefit is 7 percent, but much less for the other risk-averse cases. For risk-neutral growers, the optimal audit rate is $\alpha = 4$ percent, as demonstrated by Proposition 3. Given the lower audit rate and the lack of grower aversion to income risk from potential fines, the optimal fines become substantial, ranging from about $300/acre to over $400/acre, or about three times the amount of the fines for risk-averse growers. As a result, company net revenue for these risk-neutral cases are much larger than in Table 2—over two times to almost six times larger, depending on assumptions.

4. SUMMARY AND CONCLUSION

We developed a model of a company selling Bt corn in the presence of asymmetric information concerning grower willingness to pay for Bt corn and compliance with refuge requirements that prolong the efficacy of Bt corn. The company maximizes expected net revenue from selling Bt corn, collecting fines, and paying monitoring costs. The participation and incentive compatibility constraints for the principal-agent model were reformulated as probabilities using the grower willingness-to-pay distribution, and then the effects of the technology fee, audit rate, and fine on participation and compliance were characterized.

The conceptual model was empirically parameterized for corn rootworm Bt corn for Rock County in southern Wisconsin. With no compliance program, the optimal technology fees are similar in magnitude to those currently charged. In addition, the associated participation rates and compliance rates are similar to those observed for lepidopteran Bt corn.

When the company implements a fine-based compliance program that caps fine revenue at the cost of monitoring, the company's optimal technology fee typically increases, which decreases participation and increases compliance until complete compliance occurs. The optimal fine is zero and the optimal audit rate is 4 percent (the calibrated rate that minimizes monitoring costs). Interestingly, this program is operationally similar to the Compliance Assurance Program (CAP) that seed companies recently implemented for lepidopteran Bt corn. For the CAP, companies pay the cost of inspecting farmers for compliance, but collect no fines. Based on results in Table 2, the optimal company response to the CAP may encourage compliance, regardless of the penalties the CAP imposes.

When expected benefits are high and company marginal cost low, even with a cap on fine revenue, the company may find a lower technology fee optimal, implying high participation and low compliance. Hence, companies may find it optimal, even when using a fine-based compliance program, to permit some noncompliance. Empirically, we find that this possibility is even more likely if no cap on fine revenue is imposed. Indeed, when expected benefits are high and company marginal cost low, it can become optimal for the company to essentially give Bt corn away, so that participation approaches 100 percent, but compliance remains relatively low. The company then audits and imposes large fines, and the company's primary source of revenue is collected fines. As the company's marginal cost increases, eventually this sort of program becomes less profitable and the company switches to charging technology fees sufficiently high for complete compliance to occur.

In general, these results indicate that an inspection–fine compliance program could work, but care must be taken to ensure that the program design does not create perverse incentives for the company to encourage substantial noncompliance. Empirical results indicate that capping fine revenue at the cost of monitoring is sufficient in most cases.

Compliance programs that generate complete grower compliance at the cost of reducing company revenue and restricting grower access to the technology imply welfare losses that are missing from this analysis. Incorporating such welfare losses would greatly improve this economic analysis. Indeed, it is conceivable that, when no compliance program is used, the grower welfare gain from increased access to the technology due to lower technology fees and noncompliance could completely offset the grower welfare loss from the monopolist restricting access to the technology. If so, imposing a compliance program would imply a net loss in grower welfare relative to the competitive market equilibrium. Hence, results reported here should be considered preliminary, and final conclusions should be withheld until the completion of a more comprehensive welfare-based analysis.

REFERENCES

Agricultural Biotechnology Stewardship Technical Committee. 2002. "Insect Resistance Management Compliance Assurance Program: Backgrounder." Available online at http://www. ncga.com/public_policy/PDF/CAPbackgrounder.pdf (accessed March 8, 2005).

Babcock, B.A., E.K. Choi, and E. Feinerman. 1993. "Risk and Probability Premiums for CARA Utility Functions." *Journal of Agricultural and Resource Economics* 18(1): 17–24.

Babcock, B.A., C.E. Hart, and D.J. Hayes. 2004. "Actuarial Fairness of Crop Insurance Rates with Constant Rate Relativities." *American Journal of Agricultural Economics* 86(3): 563–575.

Chiang, A. 1984. *Fundamental Methods of Mathematical Economics* (3rd edition). New York: McGraw-Hill Publishing Company.

Coble, K., R. Heifner, and M. Zuniga. 2000. "Implications of Crop Yield and Revenue Insurance for Producer Hedging." *Journal of Agricultural and Resource Economics* 25(2): 432–452.

Cullen, E. 2004. "Southeast Wisconsin Variant Western Corn Rootworm Trapping Results 2004." *Wisconsin Crop Manager* 11(28): 200–204. Available online at http://ipcm.wisc. edu/wcm/ (accessed March 8, 2005).

Goodwin, B.K., and A.P. Ker. 2002. "Modeling Price and Yield Risk." In R.E. Just and R.D. Pope, eds., *A Comprehensive Assessment of the Role of Risk in U. S. Agriculture*. Boston: Kluwer Academic Press.

Hurley, T.M., B.A. Babcock, and R.L. Hellmich. 2001. "Bt Corn and Insect Resistance: An Economic Assessment of Refuges." *Journal of Agricultural and Resource Economics* 26(1): 176–194.

Hurley, T.M., P.D. Mitchell, and M.E. Rice. 2004. "Risk and the Value of Bt Corn." *American Journal of Agricultural Economics* 86(2): 345–358.

Jaffe, G. 2003a. "Planting Trouble: Are Farmers Squandering Bt Corn Technology?" Center for Science in the Public Interest, Washington, D.C. Available online at http://cspinet.org/ new/pdf/planting_trouble_update1.pdf (accessed March 8, 2005).

_____. 2003b. "Planting Trouble Update." Center for Science in the Public Interest, Washington, D.C. Available online at http://cspinet.org/new/pdf/bt_corn_report.pdf (accessed March 8, 2005).

Krysan, J.L., J.J. Jackson, and A.C. Lew. 1984. "Field Termination of Egg Diapause in *Diabrotica* with New Evidence of Extended Diapause in D. barberi (Coleoptera: Chrysomelidae)." *Environmental Entomology* 13(6): 1237–1240.

Laffont, J.-J., and D. Martimort. 2002. *The Theory of Incentives*. Princeton, NJ: Princeton University Press.

Langrock, I., and T.M. Hurley. 2006. "Farmer Demand for Corn Rootworm Bt Corn: Do Insect Resistance Management Guidelines Really Matter?" In R.E. Just, J.M. Alston, and D. Zilberman, eds., *Regulating Agricultural Biotechnology: Economics and Policy.* New York: Springer.

Levine, E., H. Oloumi-Sadeghi, and J.R. Fisher. 1992. "Discovery of Multiyear Diapause in Illinois and South Dakota Northern Corn Rootworm (Coleoptera: Chrysomelidae) Eggs and Incidence of the Prolonged Diapause Trait in Illinois." *Journal of Economic Entomology* 85(1): 262–267.

Mason, C.E., M.E. Rice, D.D. Calvin, J.W. Van Duyn, W.B. Showers, W.D. Hutchison, J.F. Witkowski, R.A. Higgins, D.W. Onstad, and G.P. Dively. 1996. "European Corn Borer Ecology and Management." North Central Regional Extension Publication No. 327, Iowa State University, Ames, IA.

Metcalf, R.L. 1986. "Forward." In J.L. Krysan and T.A. Miller, eds., *Methods for the Study of Pest Diabrotica.* New York: Springer-Verlag.

Mitchell, P.D., M.E. Gray, and K.L. Steffey. 2004. "A Composed Error Model for Estimating Pest-Damage Functions and the Impact of the Western Corn Rootworm Soybean Variant in Illinois." *American Journal of Agricultural Economics* 86(2): 332–344.

Mitchell, P.D., T.M. Hurley, B.A. Babcock, and R.L. Hellmich. 2002. "Insuring the Stewardship of Bt Corn: A Carrot Versus a Stick." *Journal of Agricultural and Resource Economics* 27(2): 390–405.

National Corn Growers Association. 2003. "Survey Shows Corn Growers Good Stewards of Bt Technology." *News Direct from the Stalk* (November 13). Available online at http://www.ncga.com/news/notd/2003/november/111303.htm (accessed March 8, 2005).

Ostlie, K.R., W.D. Hutchison, and R.L. Hellmich. 1997. "Bt Corn and European Corn Borer: Long-Term Success Through Resistance Management." North Central Region Extension Publication 602, University of Minnesota, St. Paul, MN.

Schmitz, D. 2004. Iowa Crop Improvement Association, Ames, IA. Personal communication.

Tabashnik, B.E., Y. Carriere, T.J. Dennehy, S. Morin, M.S. Sisterson, R.T. Roush, A.M. Shelton, and J.-Z. Zhao. 2003. "Insect Resistance to Transgenic Bt Crops: Lessons from the Laboratory and Field." *Journal of Economic Entomology* 96(4): 1031–1038.

U.S. Department of Agriculture. 2003. "Crop Production—Acreage—Supplement (PCP-BB)." National Agricultural Statistics Service, U.S. Department of Agriculture, Washington, D.C. Available online at http://usda.mannlib.cornell.edu/reports/nassr/field/pcp-bba/ (accessed March 8, 2005).

_____. 2005. "Published Estimates Data Base On Line." National Agricultural Statistics Service, U.S. Department of Agriculture, Washington, D.C. Available online at www.nass. usda.gov:81/ipedb/ (accessed March 8, 2005).

APPENDIX

Proof of Proposition 2. Rearrange equation (1) and condition (3) and add expressions to obtain:

$$H(W, \alpha, F, T) = E[U(\pi_{cp} - (1 - \phi_r)T)] - (1 - \alpha)E[U(\pi_{bt} - T)] \qquad (A1)$$

$$- \alpha E[U(\pi_{bt} - T - F)] + E[U(\pi_{cp} - W)] - E[U(\pi_{cv})] \geq 0.$$

The Implicit Function Theorem (Chiang 1984, pp. 205–208) implies that condition (A1) as an equality defines a continuous implicit function $W = Z(\alpha, F, T)$ in the neighborhood of a point $(W_0, \alpha_0, F_0, T_0)$ satisfying equation (A1) and that it has continuous partial derivatives if $H(\cdot)$ has continuous partial derivatives with respect to W, α, F, and T, and if, at the point $(W_0, \alpha_0, F_0, T_0)$, the partial derivative with respect to W is not zero. The partial derivatives of (A1) are:

$$\partial H \big/ \partial W = E[U'(\pi_{cp} - W)], \tag{A2}$$

$$\partial H \big/ \partial \alpha = E[U(\pi_{bt} - T)] - E[U(\pi_{bt} - T - F)], \tag{A3}$$

$$\partial H \big/ \partial F = \alpha E[U'(\pi_{bt} - T - F)], \tag{A4}$$

$$\partial H \big/ \partial T = (1 - \alpha)E[U'(\pi_{bt} - T)] + \alpha E[U'(\pi_{bt} - T - F)] \tag{A5}$$
$$- (1 - \phi_r)E[U'(\pi_{bt} - (1 - \phi_r)T - \phi_r py\lambda)].$$

The last term in (A5) follows because $\pi_{cp} = \pi_{bt} - \phi_r py\lambda$. If utility is continuous and strictly increases in income, then (A2)–(A5) are continuous and $\partial H/\partial W > 0$, so that $W = Z(\alpha, F, T)$ exists. The Implicit Function Theorem implies that

$$\partial W \big/ \partial x = \partial Z \big/ \partial x = - \frac{\partial H \big/ \partial x}{\partial H \big/ \partial W}.$$

Since $v = 1 - G(Z(\alpha, F, T))$, $\partial v/\partial x = -g(Z(\cdot))\partial Z/\partial x$ for all $x \in \{\alpha, F, T\}$, where $g(\cdot) \geq 0$ is the density function for $G(\cdot)$. Thus $\partial v/\partial x$ has the opposite sign of $\partial Z/\partial x$. Since $\partial H/\partial W > 0$, $\partial Z/\partial x$ has the opposite sign of $\partial H/\partial x$, and so $\partial v/\partial x$ has the same sign as $\partial H/\partial x$. If utility strictly increases in income, (A4) implies $\partial H/\partial F > 0$, so that $\partial v/\partial F > 0$, and, if in addition $F > 0$, (A3) implies $\partial H/\partial \alpha > 0$, so that $\partial v/\partial \alpha > 0$. The sign of $\partial v/\partial T$ is the same as the sign of (A5), the condition reported in the proposition, and has an ambiguous sign since all three terms in (A5) are positive.

Proof of Proposition 3. By Proposition 1, $\partial \beta/\partial \alpha = \partial \beta/\partial F = 0$. By Corollary 1, $Z(\cdot) = E[py\lambda] - \phi_r T - \alpha F$, so that $\partial v/\partial \alpha = Fg(\cdot)$ and $\partial v/\partial F = \alpha g(\cdot)$. The first-order condition for maximizing equation (5) with respect to F is $\alpha(\beta -$

$v) - \partial v/\partial F[\phi_r(T - c) + \alpha F - 2(\beta - v)\theta_n M] = 0$. Using $\partial v/\partial F = \alpha g(\cdot)$ and rearranging gives

$$\phi_r(T - c) + \alpha F - 2(\beta - v)\theta_n M = (\beta - v)/g(\cdot). \tag{A6}$$

The first-order condition for maximizing equation (5) with respect to α is

$$F(\beta - v) - \partial v/\partial \alpha \, [\phi_r(T - c) + \alpha F - 2(\beta - v)\theta_n M] - \beta k'(\alpha) = 0. \tag{A7}$$

Substituting in both $\partial v/\partial \alpha = Fg(\cdot)$ and (A6) for the term in square brackets and simplifying gives $-\beta k'(\alpha) = 0$, which has the rejected trivial solution $\beta = 0$ and the reported solution, which is independent of the distribution $G(W)$: $k'(\alpha) = 0$. If α, F, and T from maximizing equation (5) imply $(1 - \phi_r)T \geq Z(\alpha, F, T)$, equation (6) is the applicable objective, and the first-order condition with respect to α is $-\beta k'(\alpha) = 0$. Thus in both cases, the optimal α is defined by $k'(\alpha) = 0$, which does not depend on the grower utility function or willingness-to-pay distribution.

Chapter 28

DAMAGE FROM SECONDARY PESTS AND THE NEED FOR REFUGE IN CHINA

Shenghui Wang, David R. Just, and Per Pinstrup-Andersen
Cornell University

Abstract: Because Bt technology targets only bollworm populations, secondary pest populations have slowly eroded the benefits of Bt technology in China. Stochastic dominance tests based on primary household data from 1999–2001 and 2004 in China provide strong evidence that secondary pests have completely eroded all benefits from Bt cotton cultivation. Refuge, while currently used to prevent Bt resistance in the United States, could be used to control secondary pest populations by increasing their exposure to lethal toxins. We show that such a strategy would be profitable in China.

Key words: Bt cotton, secondary pest, refuge, pest control

1. INTRODUCTION

Since its introduction in 1996, adoption of Bt cotton has been rapid and widespread. Often touted as being particularly suited to conditions in developing countries, Bt cotton is cotton that has been genetically modified to produce toxins that are lethal to leaf eating bollworm, the most prevalent pest affecting cotton. Thus, Bt cotton should not require the same expenditures on pesticide as conventional cotton. In fact, Bt cotton has been shown to be extremely effective in fighting bollworm infestation in developing countries. For example, Qaim and Zilberman (2003) find that Indian farmers were able to save 39 percent on pesticide expenditures by using Bt cotton.

Adoption of Bt cotton has been widespread in China, where 3.7 million hectares of Bt cotton were grown in 2004 (Huesing and English 2004), with

We are grateful to Jikun Huang and the staff of the Center for Chinese Agricultural Policy (CCAP) for collaboration on data collection. Also, this paper has benefited greatly from the comments of Jianzhou Zhao at Cornell University and Andrejus Parfionovas at Utah State University.

an average of 4 to 6 million hectares of cotton (conventional or otherwise) grown in total since 1984. The adoption rate in China increased from 1 percent in 1997 (when the technology was first made available) to over 45 percent in 2001 (Huang et al. 2002, James 2002) and 65 percent in 2004. Farmers were enthusiastic about the new technology, with farmers who used Bt cotton overwhelmingly choosing to use Bt on all their cotton lands. Thus, farms in China have not employed the refuge (land planted in conventional cotton) as the U.S. Environmental Protection Agency (EPA) recommends to prevent resistance to the toxins produced by Bt cotton. The toxins produced by Bt cotton are narrowly focused on bollworm pests, affecting only a very few pests as opposed to traditional pesticides which employ *broad spectrum* toxins. For the years 2000 and 2001, Bt cotton was associated with a 55 percent reduction in pesticide for the average Chinese farm (Pray et al. 2002), thus limiting pest exposure to broad spectrum toxins. By limiting the prevalence of broad spectrum toxins, farmers may have unintentionally created a safe haven for other pests not affected by Bt technology.

While Bt has been shown to be effective in controlling bollworm infestation, to date little is known about the potential effects of Bt use on other potential pests. In this chapter, we use farm survey data from China to illustrate a growing problem of secondary pest infestation in China due to Bt cotton adoption. Our data show that secondary pests are now so prevalent that they completely erode the net profits of Bt cotton. This infestation has appeared only after several years of Bt use, and is cited by many farmers as a major challenge to the profitability of Bt adoption. Using a simulation, we show that use of refuge can increase profitability by reducing pesticide expenditures due to secondary pests. In the following section we review the relevant literature on Bt effectiveness and the economic modeling of secondary pest infestations. We will then present survey data and analysis of the size of the effect of the secondary pest infestation. We will use this data to inform our simulation of farm profitability under various levels of refuge use.

2. Bt TECHNOLOGY AND SECONDARY PESTS

Several researchers have measured the effectiveness of various Bt crops and the potential to increase profits. Many of these studies have targeted the potential for increasing profits in developing countries. If Bt technology is truly a more efficient mechanism to combat pest infestation, it could provide a way for farmers in developing countries to produce more with fewer resources, reducing the dual problems of poverty and hunger. Qaim and Zilberman (2003) examine the effectiveness of Bt cotton in India, finding a 39 percent reduction in the use of pesticide. Similar results have been found in Argentina (Qaim and de Janvry 2003), Mexico (Traxler et al. 2003), and

South Africa (Bennett et al. 2003), with reductions of 47 percent, 77 percent, and 58 percent, respectively. This reduction in pesticide translates into cost savings and increased profits, so long as the price difference between Bt and conventional seeds do not totally offset the savings.

The impacts of Bt cotton adoption on production in China is well documented. Several Bt varieties were approved by the Chinese Biosafety Committee in 1997, and subsequently became available for commercial sale. Based on data from the first seasons of Bt adoption, Huang et al. (2002) and Huang et al. (2003) find that farmers were able to reduce the average number of pesticide applications from 20.0 to only 6.6. This translates to a reduction in pesticide use by 43.3 kilograms per hectare in 1999, or a 71 percent decrease in total pesticide use (Huang et al. 2003). Pray et al. (2002), using data from 2000 and 2001 find an average reduction in pesticide use of 35.6 kilograms per hectare, or a 55 percent reduction. However, as with studies conducted in many other countries, research on the profitability of Bt technology in China has focused only on the first few seasons of adoption. Some of the more prominent studies have focused only on a single growing season. Given initial populations that are negligible, infestations from secondary pests, such as mirids, may take several seasons to develop. Under this scenario, currently available studies of Bt production would not provide evidence of growing secondary pest problems.

Researchers have often noted the importance of managing multiple pests simultaneously for farm profitability (e.g., Getz and Gutierrez 1982, Feder and Regev 1975, Boggess, Cardell, and Burfield 1985, Harper and Zilberman 1989). Nonetheless, few have specifically examined the interaction between the control of one pest and the impacts this control may have on the population of a secondary pest. Harper and Zilberman (1989) outline the potential problems of managing several interrelated pests, noting that the use of chemical pesticides may produce what they call a *pest externality*. A pest externality occurs when the chemicals used to target one pest inadvertently increase the concentration of and damage from a second pest. If a farmer applies a narrow spectrum pesticide that can kill only the primary pest, the secondary pest may grow in population if (i) the primary pest is a natural predator of the secondary pest, or (ii) the primary and secondary pest are in competition for food (i.e., the crop). Harper and Zilberman suggest that farmers may not understand the nature of the relationship between primary and secondary pests. Thus farmers changing their pest control strategy may unintentionally induce secondary pest damage.

If the farmer does not recognize the problem until the secondary pest has a significant foothold, the secondary pest infestation may be more costly to control than the primary pest infestations under the previous technology. Thus, Harper and Zilberman (1989) suggest that ignoring the secondary pest when adopting the new strategy can lead to a prolonged recurrence of crop damage that may be more difficult to control than primary pest infestations. Harper (1991) developed a dynamic model of multi-pest management, explicitly modeling the predator-prey relationship of a primary and secondary pest, illustrating the importance of timing in combating secondary pest infestations. Getz and Gutierrez (1982) suggest that farmers making pesticide decisions based on primary pests, largely ignorant of the existence and interaction with secondary pests, are responsible for "worldwide elevation of certain species from relatively innocuous to highly destructive levels" (p. 447).

Within China, entomologists have recognized the potential for pest externalities due to the use of Bt cotton. Wu, Li, and Guo (2002) and Wu and Guo (2005) conducted field experiments designed to measure the impact of Bt use on secondary pest populations in northern China. Their field experiments consisted of two treatments: Bt cotton grown without pesticide and conventional cotton grown with pesticides designed to control bollworm infestations. Both studies find that the secondary pest is significantly more densely populated in Bt plots than in the conventional plots. Thus broad spectrum pesticides may be less effective in controlling bollworm, but more effective in controlling mirids and other pests. Wu concludes that in the Bt plots, secondary pests have become the primary pests and could be responsible for significant damage to the expanding region of Bt cotton cultivation, if no further action is taken.

3. EMPIRICAL RESULTS

As in many developing nations, in China comprehensive production data is not collected by the government or industry. Thus, we designed and conducted a household survey in November 2004, with the cooperation of the Center for Chinese Agricultural Policy (CCAP), Beijing, located in the Chinese Academy of Science (CAS). The survey was conducted in 5 provinces: Anhui, Hebei, Henan, Hubei, and Shangdong. These provinces were chosen as they are each major centers of cotton production. The survey was administered to 481 farmers, each interviewed for about two hours. Farmers were selected using a stratified sampling method, with provinces and counties carefully selected to allow us to compare the performance of Bt and conventional cotton farms. After selecting counties, villages and farmers within the villages were selected at random proportional to farm populations within the village and county. The sample contains 20 villages within 10 counties of 5 provinces. A similar survey had been conducted previously by CCAP in the years 1999, 2000, and 2001, when 283, 407, and 306 farmers were interviewed, respectively. This data structure allows us to analyze how Bt cotton has performed relative to conventional cotton over several growing seasons.

Data on pesticide expenditures were collected separately for pesticides targeting primary pests and pesticides targeting secondary pests.

Wang, Just, and Pinstrup-Andersen (2005) (henceforth WJP) analyze the data by constructing cumulative densities for total pesticide expenditure, net revenue, and primary and secondary pesticide expenditures in 2004. Cumulative densities were constructed using 10,000 boot-strapped samples to find average values of the cumulative distribution function for each value. These cumulative densities were then used to conduct tests for first-order stochastic dominance. While first-order stochastic dominance can tell us that a significant change has occurred, it is also important to note the size of the change. In order to address the magnitude, we report the quantile values of the cumulative density functions and the standard errors for various measures of net revenue and expenditures in Tables 1 through 8. Standard errors were calculated using quantile regression, also known as least absolute deviation estimation (see Koenker and Bassett 1982 and Rogers 1993 for descriptions).

For the early years of adoption (2000 and 2001), WJP find that net revenue for Bt farmers dominates that of non-Bt farmers. This trend, however, reverses itself in 2004, with the net revenue of conventional cotton farmers dominating that of their Bt colleagues. Table 1 displays the 20th, 40th, 60th, and 80th quantiles of net revenue for Bt and conventional cotton farmers in 1999. Here we find that while Bt net revenues exceed conventional cotton net revues for the majority of the distribution, the revenues of the most profitable non-Bt farms exceed the profits of the most profitable Bt farmers (measured at the 80th percentile). There are several reasons to discount this result. Bt technology was first made commercially available only in 1997, and thus Bt technology was still in the earliest stages of diffusion in China. Secondly, in 1999 China was just at the end of a wave of economic change—loosening the government controls on production and exchange. Finally, we note that the standard errors in Table 1 suggest that the differences in net revenues at the 80 percent quantile are not significant at any reasonable level. Tables 2 and 3 tell a different story for the years 2000 and 2001. In the year 2000, net revenues for Bt farmers were significantly larger than those of their conventional counterparts at each of our selected quantiles. In fact, the difference is more pronounced at the top end of the distribution, with around $700/ha more in net revenue. While the differences are substantially less in 2001 (about $100/ha), we continue to find that net revenues were significantly higher for Bt farmers than conventional cotton farmers at each chosen quantile. Note that net revenues for Bt farmers decline significantly in 2001, in fact returning to levels observed in 1999. By 2004, we find that non-Bt net revenues are in fact significantly greater than for Bt farmers for all selected quantiles. By comparing Tables 3 and 4, it can be seen that while conventional farmers experienced net revenue declines of about $200/ha

Table 1. Net Revenue (US $/ha) in 1999

Quantile	Bt Farmers		Non-Bt Farmers	
	Estimated Value	Standard Deviation	Estimated Value	Standard Deviation
20%	1147.982	14.369	900.824	67.924
40%	1291.353	17.150	1145.199	69.639
60%	1430.149	16.190	1369.645	65.267
80%	1567.024	13.925	1594.349	56.835

Table 2. Net Revenue (US $/ha) in 2000

Quantile	Bt Farmers		Non-Bt Farmers	
	Estimated Value	Standard Deviation	Estimated Value	Standard Deviation
20%	1192.401	35.142	760.294	28.282
40%	1525.204	24.108	923.226	30.072
60%	1741.947	19.992	1073.394	27.895
80%	1944.689	15.739	1230.337	29.402

Table 3. Net Revenue (US $/ha) in 2001

Quantile	Bt Farmers		Non-Bt Farmers	
	Estimated Value	Standard Deviation	Estimated Value	Standard Deviation
20%	1018.978	14.955	933.236	18.727
40%	1182.898	14.538	1062.648	19.629
60%	1332.796	13.800	1200.542	24.295
80%	1514.555	20.205	1364.327	27.753

Table 4. Net Revenue (US $/ha) in 2004

Quantile	Bt Farmers		Non-Bt Farmers	
	Estimated Value	Standard Deviation	Estimated Value	Standard Deviation
20%	408.4001	34.357	480.567	29.771
40%	713.0021	32.170	838.836	23.955
60%	961.4159	28.943	1096.89	19.755
80%	1208.208	29.157	1350.86	19.839

relative to 2001 (with no decline on the high end of the distribution), Bt farmers experienced significant declines of about $400/ha at each level, and more pronounced declines on the low end of the distribution. In the following, we argue that much of this decline was due to increased expenditures on secondary pests.

WJP find that conventional cotton farmers' total expenditure on pesticides targeting secondary pests in 2004 dominates that of Bt farmers (meaning conventional farmers expended less on pesticides targeting pests other than bollworm). Table 5 presents the quantile estimates and standard deviations. Here we can see that the difference is stark at each level. The median amount spent by Bt farmers per hectare is comparable to the amount spent by the 80th percentile conventional cotton farmer. The amount spent by the 20th percentile Bt farmer is close to the median amount spent by conventional cotton farmers. Bt farmers spend significantly more than conventional farmers at each chosen quantile. Thus, once a farmer begins growing Bt we suggest that expenses on secondary pests will increase significantly (increasing by between 127 percent and 470 percent).

More insight can be gained by comparing the amount of pesticide used to combat secondary pests in 2001 and 2004. To show how pesticide use evolved over time for those growing Bt cotton, WJP compare the amount of pesticide used on secondary pests in 2001 to the amount used on secondary pests in 2004 for Bt cotton farmers. Here, WJP find that pesticide usage by Bt farmers for secondary pests in 2004 stochastically dominates that of Bt farmers in 2001, meaning that Bt farmers now use more pesticide to control secondary pests than was needed in 2001. In Table 6 we report the quantiles of the amount of pesticide used by Bt farmers in 2001 and 2004 to combat secondary pests. The table displays an increase in pesticide use on secondary pests by between 310 percent and 460 percent, with more pronounced differences on the higher end of the distribution. Again, these differences are all significant, showing that Bt farmers severely increased their use of pesticides to fight secondary pests. This is consistent with many of the open-ended descriptions given our interviewers, sighting a spike in prevalence of secondary pests. Thus secondary pests have become a substantial problem where they were nearly non-existent during early Bt adoption.

As is expected, WJP find that pesticide expenditures in 2004 on primary pests by conventional cotton farmers first-order stochastically dominates expenditures by Bt cotton farmers, meaning conventional farmers must spend more money on pesticide to fight bollworm infestations. Thus if there are problems with *resistance* due to the lack of refuge, they are not apparent in the expenditure data. Table 7 displays the quantiles for pesticide expenditures on bollworm in 2004 for Bt and non-Bt farmers. Bt farmers use significantly less pesticide on bollworm at each selected quantile, reducing expenditure by 49.5 percent to 83 percent relative to conventional cotton farmers.

Table 5. Pesticide Expenditure (US $/ha) on Secondary Pests in 2004

	Bt Farmers		Non-Bt Farmers	
Quantile	Estimated Value	Standard Deviation	Estimated Value	Standard Deviation
20%	2.192	0.280	0.384	0.310
40%	6.041	0.616	1.060	0.672
60%	15.136	1.237	3.339	1.913
80%	33.238	2.609	14.642	2.145

Table 6. Amount of Pesticides (kg/hectare) Used on Secondary Pests by Bt Farmers

	Bt Farmers 2004		Bt Farmers 2001	
Quantile	Estimated Value	Standard Deviation	Estimated Value	Standard Deviation
20%	1.675	0.094	0.408	0.037
40%	3.970	0.221	0.940	0.059
60%	7.770	0.423	1.622	0.070
80%	15.498	1.078	2.840	0.174

Table 7. Pesticide Expenditure (US $/hectare) on Primary Pest Bollworm in 2004

	Bt Farmers 2004		Non-Bt Farmers	
Quantile	Estimated Value	Standard Deviation	Estimated Value	Standard Deviation
20%	3.737	0.235	21.470	1.667
40%	9.849	0.678	39.356	2.217
60%	23.625	1.890	58.058	2.894
80%	52.898	3.341	104.701	8.111

The magnitude of the differences in expenditure on bollworm range from $17.73/ha to $51.80/ha. This compares with an increase in expenditure on secondary pests between $1.81/ha and $18.60/ha. At first blush, it appears that the change in secondary pest expenditures is minute as compared to the savings on primary pest expenditures. However, this is not the case. This difference in secondary pest expenditure represents the difference for only a small subset of secondary pests appearing on the survey. Table 8 displays the quantiles of total pesticide expenditures for both Bt and non-Bt

Table 8. Total Pesticide Expenditure (US $/hectare) in 2004

Quantile	Bt Farmers		Non-Bt Farmers	
	Estimated Value	Standard Deviation	Estimated Value	Standard Deviation
20%	62.504	1.899	65.942	2.975
40%	99.261	2.099	96.081	3.803
60%	128.728	3.057	129.252	5.577
80%	183.671	3.917	184.449	10.374

farmers in 2004. Notably, there is no significant difference (at any reasonable level) between pesticide expenditure for Bt and conventional cotton farmers for any of the quantiles we have chosen (with the difference that is closest to significant at the 20th percentile with a p-value of 0.18). The quantiles are nearly identical, and WJP find the distributions to be nearly identical. The difference in expenditures range from $0.52/ha to $3.44/ha. The difference at the 40th percentile favors conventional cotton growers.

Table 8 excludes the cost of seed, and hence the added costs of Bt seed. Clearly, where Bt seeds may cost anywhere from 2 to 3 times as much as their conventional counterparts, Bt cotton is no longer profitable relative to conventional seed. Our evidence suggests that farmers using Bt have been overrun by secondary pests. In fact a large portion of the Bt farmers interviewed in 2004 cite a growing prevalence of secondary pests as the greatest challenge to their profitability.

4. SIMULATING REFUGE

Bt cotton that requires the spraying of substantially extra amounts of pesticide to control secondary pests may result in suboptimal profits, if the level of pesticide used could have controlled the primary pest absent Bt technology. When secondary pests are eroding the benefits of Bt, the primary question becomes how farmers can effectively use Bt without inducing secondary pest infestation. One potential solution is the use of refuge. Refuge sprayed with broad spectrum toxins present in pesticides should increase the exposure of secondary pests to lethal toxins (toxins lethal to both primary and secondary pests) and reduce the threat of the secondary pest before they proliferate to a damaging level of concentration.

In the case illustrated by our data, using Bt requires that a pesticide targeting a secondary pest be applied, while conventional cotton can be treated only with pesticide targeting the primary pest. Suppose that total expenditures on pest control are given by

$$y = \phi\left(w^1 p_{BT}^1 + w^2 p_{BT}^2 + k\right) + \left(1 - \phi\right)\left(w^1 p_R^1 + w^2 p_R^2\right), \qquad (1)$$

where y is expenditure, ϕ is the proportion of land devoted to Bt cotton production, w^1 and w^2 are the price of pesticide targeting the primary pest and the secondary pest respectively, p_{BT}^1 and p_R^1 the amounts of pesticide applied per hectare targeting the primary pest for the Bt cotton and refuge lands respectively, p_{BT}^2 and p_R^2 are the amounts of pesticide applied per hectare targeting the secondary pest for the Bt cotton and refuge lands respectively, and k is the price differential per hectare between Bt and conventional seed. Suppose that the farmer wishes to minimize expenditure on pesticide subject to some level of crop damage, $z(\phi, p_{BT}^1, p_{BT}^2, p_R^1, p_R^2) = \bar{z}$. The data show that $p_R^1(\phi = 0) > p_{BT}^1(\phi = 1)$, $p_R^2(\phi = 0) < p_{BT}^2(\phi = 1)$, and $k > w^1(p_R^1 (\phi = 0) - p_{BT}^1 (\phi = 1)) + w^2 (p_R^2 (\phi = 0) - p_{BT}^2 (\phi = 1))$. Introducing refuge will decrease expenditures if

$$k + w^1 \left(p_{BT}^1 - p_R^1\right) + w^2 \left(p_{BT}^2 - p_R^2\right) + \left(w^1 \frac{\partial p_{BT}^1}{\partial \phi} + w^2 \frac{\partial p_{BT}^2}{\partial \phi} \right) > 0,$$

where all pesticide levels are evaluated at the constraint imposed by $\phi = 1$. This must be the case if the interaction between ϕ and the pesticide concentrations required on both refuge and Bt lands are minimal. Alternatively, we could think of maintaining the levels of pesticide on refuge given by $\phi = 0$, and the levels of pesticide on Bt cotton given by $\phi = 1$. In this case, reducing ϕ should reduce expenditures (at least by k per hectare) but may lead to greater crop damage if either the primary or secondary pest populations exceed a critical level. While we cannot calculate the resulting pest damage for the various possible levels of ϕ, we can use our data to calculate the cumulative density for expenditures on pesticide supposing farmers maintain pesticide levels on both conventional and Bt cotton lands [see Reichelderfer and Bender (1979) for a comprehensive discussion of pest control simulations]. This appears in Table 9.

The table reflects substantial savings on pesticide for farms with 20 percent, 40 percent, 60 percent, and 80 percent refuge. Some of this savings may be diminished by increased pest damage due to the interaction of pest populations in the refuge and Bt plots. This may happen if, for example, a refuge plot located directly next to a Bt plot faces added pressure from the secondary pest population. In this case the savings on pesticide displayed in Table 9 translate into real savings only if the added pesticide necessary to reduce damage to levels observed in our data does not exceed the savings we have calculated. This added expense on pesticide should be the smallest for

Table 9. Simulated Total Pesticide Expenditure (US $/hectare) for Various Levels of Refuge in 2004

	80% Refuge		60% Refuge		40% Refuge		20% Refuge		Non-Bt	
Quantile	Expense	Std. Dev.	Expense	Std. Dev.	Expense	Std. Dev.	Expense	Std. Dev.	Expense	Std. Dev.
20%	24.241	0.996	29.770	1.123	34.837	1.345	39.478	1.566	65.942	2.975
40%	42.658	1.683	49.772	1.655	56.983	1.888	63.820	1.944	96.081	3.803
60%	67.908	2.587	75.416	2.591	83.516	2.656	91.891	2.814	129.252	5.577
80%	106.563	4.712	114.197	4.963	123.903	4.644	134.399	4.335	184.449	10.374

the smaller refuge farms. The middle range of farms should save an average of between $32.26/ha and $39.10/ha, equal to approximately a third of their total pesticide expenditure. Thus, the increase in required pesticide would have to be substantial in order to eclipse the potential savings. If the refuge is large enough, the secondary pest population should be held to negligible levels, as was the case prior to introduction of Bt technology. Thus we believe savings should be realized even for farms with a large majority of refuge.

5. CONCLUSION

While agricultural economists have long recognized the importance of pest interactions and potentially hidden secondary pests in farmer decisions, the issue has been largely ignored in the recent rash of studies on Bt technology [with the exception of Livingston, Carlson, and Fackler (2004)]. In reality, the near complete reliance on narrow spectrum toxins, like those produced by Bt cotton, provides a perfect opportunity for secondary pests to increase their concentrations and their damage to crops. Our data show clear evidence of an increasingly devastating secondary pest infestation among Chinese cotton farmers. This increased secondary pest activity appears to be largely isolated to farms growing Bt cotton. Further, the increase in pesticide expenditures on pests other than bollworm (the primary pest) completely offsets the savings in pesticide expenditures on bollworm realized by Bt cotton growers in China.

Refuge has been recommended primarily to combat the possibility of pest resistance to Bt. Many have suggested that the lack of refuge in developing countries (and in the United States) may not be detrimental, particularly if there is some intra-regional diversity in adoption of Bt and non-Bt varieties. In fact, the evidence of resistance has been sparse. In this chapter and in WJP we provide evidence that the lack of refuge may not be so innocuous. Refuge may be an essential check on secondary pest populations for which Bt technology is ineffective. This is a stark illustration of the need for technology-specific training and education in developing countries, even for seemingly simple technologies such as Bt seed. Without such education, farmers left to their own devices may opt to reject new technologies that, when used properly, promise to improve food supplies and farm profitability in areas where such improvements are badly needed.

REFERENCES

Bennett, R., S. Morse, and Y. Ismael. 2003. "The Benefits of Bt Cotton to Small-Scale Producers in Developing Countries: The Case of South Africa." Paper presented at the Seventh ICABR (International Consortium on Agricultural Biotechnology Research) International Conference on Public Goods and Public Policy for Agricultural Biotechnology, June 29–July 3, in Ravello, Italy.

Boggess, M.G., D.J. Cardell, and C.S. Barfield. 1985. "A Bioeconomic Simulation Approach to Multi-species Insect Management." *Southern Journal of Agricultural Economics* 17(2): 43–55.

Harper, C.R. 1991. "Predator-Prey Systems in Pest Management." *Agricultural and Resource Economics Review* 20(1): 15–23.

Harper, C., and D. Zilberman. 1989. "Pest Externalities from Agricultural Inputs." *American Journal of Agricultural Economics* 71(3): 692–702.

Feder, G., and U. Regev. 1975. "Biological Interactions and Environmental Effects in the Economics of Pest Control." *Journal of Environmental Economics and Management* 2(2): 75–91.

Getz, W.M., and A.P. Gutierrez. 1982. "A Perspective on Systems Analysis in Crop Production and Insect Pest Management." *Annual Review of Entomology* 27(January): 447–466.

Huang, J.K., R. Hu, C. Fan, C.E. Pray, and S. Rozelle. 2002. "Bt Cotton Benefits, Costs and Impacts in China." *AgBioForum* 5(4): 153–166.

Huang, J.K., R. Hu, C. Pray, F. Qiao, and S. Rozelle. 2003. "Biotechnology as an Alternative to Chemical Pesticides: A Case Study of Bt Cotton in China." *Agricultural Economics* 29(1): 55–67.

Huesing, J., and L. English. 2004. "The Impact of Bt Crops on the Developing World." *AgBioForum* 7(1&2): 84–95.

James, C. 2002. "Global Status of Commercialized Transgenic Crops: 2002." Volume 27 of *ISAAA Briefs*, ISAAA (International Service for the Acquisition of Agri-Biotech Applications) *AmeriCenter*, Cornell University, Ithaca, NY.

Koenker, R., and G. Bassett. 1982. "Robust Tests for Heteroscedasticity Based on Regression Quantiles." *Econometrica* 50(1): 43–61.

Livingston, M., G. Carlson, and P. Fackler. 2004. "Managing Resistance Evolution in Two Pests to Two Toxins with Refugia." *American Journal of Agricultural Economics* 86(1): 1–13.

Pray, C.E., J. Huang, R. Hu, and S. Rozelle. 2002. "Five Years of Bt Cotton in China: The Benefits Continue." *The Plant Journal* 31(4): 423–430.

Qaim, M., and A. de Janvry. 2003. "Genetically Modified Crops, Corporate Pricing Strategies, and Farmers' Adoption: The Case of Bt Cotton in Argentina." *American Journal of Agricultural Economics* 85(4): 814–828.

Qaim, M., and D. Zilberman. 2003. "Yield Effects of Genetically Modified Crops in Developing Countries." *Science* 299(5608): 900–902.

Reichelderfer, K.H., and E.E. Bender. 1979. "Application of a Simulative Approach to Evaluating Methods for the Control of Agricultural Pests." *American Journal of Agricultural Economics* 61(2): 258–267.

Rogers, W.H. 1993. "sg11.2: Calculation of Quantile Regression Standard Errors." *Stata Technical Bulletin* 13(May): 18–19.

Traxler, G., S. Godloy-Zvila, J. Falck-Zepeda, and J. Espinoza-Arellan. 2003. "Transgenic Cotton in Mexico: Economic and Environmental Impacts." In N. Kalaitzandonakes, ed., *Economic and Environmental Impacts of First Generation Biotechnologies.* New York: Kluwer Academic Press.

Wang, S., D.R. Just, and P. Pinstrup-Andersen. 2005. "Tarnishing Silver Bullets: Bt Technology Adoption, Bounded Rationality and the Outbreak of Secondary Pest Infestations in China." Working paper, Department of Applied Economics and Management, Cornell University, Ithaca, NY.

Wu, K.M., and Y.Y. Guo. 2005. "The Evolution of Cotton Pest Management Practices in China." *Annual Review of Entomology* 50(January): 31–52.

Wu, K.M., H.F. Li, and Y. Guo. 2002. "Seasonal Abundance of the Mirids, Lygus lucorum and Adelphocoris spp. (Hemiptera: Miridae) on Bt Cotton in Northern China." *Crop Protection* 21(10): 997–1002.

Chapter 29

REGULATION OF BIOTECHNOLOGY FOR FIELD CROPS

Richard K. Perrin
University of Nebraska, Lincoln

Abstract: Biotechnology innovations in field crops are jointly regulated by the Animal and Plant Health Inspection Service (APHIS) of the U.S. Department of Agriculture, the U.S. Environmental Protection Agency, and the Food and Drug Administration, with the objectives of ensuring safety of the environment and human health. The coordination mechanism has been criticized, but seems likely to persist, barring some dramatic safety failure. Economic analysis has established costs of alternative refugia requirements for insect-resistance technologies, but has to date contributed little else to determining the nature and extent of regulatory oversight.

Key words: agricultural biotechnology, regulation, APHIS

1. INTRODUCTION

There are no U.S. laws specifically establishing the regulation of crop biotechnology. It is a stepchild, along with other agricultural biotechnologies, attended to by three parents.

The Animal and Plant Health Inspection Service (APHIS) of the U.S. Department of Agriculture has primary custody under its authority under the Plant Protection Act to regulate agricultural pests. The U.S. Environmental Protection Agency (EPA) exercises oversight because it has authority to regulate pesticides under the Federal Insecticide, Fungicide and Rodenticide Act (FIFRA) and toxic substances under the Toxic Substances Control Act (TOSCA). Third, the Food and Drug Administration (FDA) bases its claim to parenthood on its responsibility for food safety under the Federal Food, Drug and Cosmetic Service Act (FD&C) and for pharmaceuticals under the Federal Health Act (FHA). Under the National Environmental Policy Act (NEPA), all three agencies must conduct environmental assessments for major actions.

This shared parenthood was a deliberate decision after discussions in the 1980s led to the rejection of a new law and new agency for agricultural biotechnology, in favor of the "Coordinated Framework for Regulation of Biotechnology Products." This framework was established in 1986 by the Office of Science and Technology Policy (51 Fed. Reg. 23302). The coordinated approach is still being debated, however, as indicated by the concerns and proposals collated in a recent report by the Pew Initiative on Food and Biotechnology (2004).

Transgenic field crops are of two types for regulatory purposes. Those that produce grain for food and feed use are subject to a simpler regulatory process than the more exotic pharmaceutical-producing plants. Brief summaries of these current regulatory procedures follow.

2. THE REGULATORY PROCESS FOR TRANSGENIC FIELD CROPS WITH TRADITIONAL COMPOSITION

Details on the regulatory procedures of the three agencies can be obtained from their websites.[1] Recent studies that provide more detail and evaluation of these procedures include Nelson, Babinard, and Josling (2001), Belson (2002), National Research Council (2002), and the Pew Initiative on Food and Biotechnology (2004).

2.1. Animal and Plant Health Inspection Service

The Animal and Plant Health Inspection Service defines all transgenic crop organisms as having the potential to be "pests," and they are therefore designated as "regulated articles." A permit from APHIS is required to conduct a field test of any transgenic crop. Plants with insect or herbicide resistance, or enhanced grain qualities for food or feed use (often referred to as first- and second-generation transgenics), are becoming well understood and thus demand less stringent regulatory oversight than the more exotic third-generation transgenics with pharmaceutical or industrial traits.

A request for a field trial permit ("permit for introduction") must contain extensive descriptions of the organism, and of the testing, monitoring, disposal, and reporting procedures to be followed, and it must disclose any evidence of unusual or harmful aspects of the plant. APHIS provides guidelines for testing and reporting (minimum isolation requirements, etc.), but applica-

---

[1] USDA Animal and Plant Health Inspection Service, Biotechnology Regulatory Services (http://www.aphis.usda.gov/brs/index.html); FDA Center for Food Safety and Applied Nutrition (http://www.cfsan.fda.gov/~lrd/biotechm.html#reg); EPA Office of Pollution Prevention & Toxics, Biotechnology Program (http://www.epa.gov/opptintr/biotech/); and EPA Office of Pesticide Programs (http://www.epa.gov/pesticides/biopesticides/).

tions are evaluated on a case-by-case basis, with the review process requiring several months. Once the permit is issued, APHIS conducts field inspections and monitors reports to verify compliance.

In lieu of requesting a permit for a field test, a developer may use a streamlined "notification process" if the plant meets six well-defined standards that establish that it involves substances that are reasonably well understood and safe. The notification requires essentially the same information as the permit request, but APHIS will acknowledge within a month whether the permit is granted. The majority of recent field test permissions have been granted under the notification process.

APHIS contacts regulatory agencies in the states where tests are conducted, to ascertain their approval prior to issuing the permit for testing.

Once field tests are completed, over perhaps 2–3 years, a request for release is submitted (technically a "petition for determination of nonregulated status"), based on the results of the field tests and other information submitted by the petitioner. The review of this petition may take a number of months. If the review determines that the plant will not become a pest nor pose a significant risk to the environment, APHIS grants the release and exercises no further regulatory oversight, although the determination can be reversed if new information so warrants.

2.2. U.S. Environmental Protection Agency

Field trials for plants with "plant-incorporated protectants" require an Experimental Use Permit (EUP) from EPA if the test exceeds 10 acres. If the review of test data and other information submitted by the applicant indicates no unreasonable adverse health, occupational, or environmental risks, EPA clearance for commercialization is granted by registering the plant. Because EPA is responsible for pesticide residues in crop products, it must either establish a tolerance for the pesticide substance or provide an exemption from tolerances, as it has done for Bt toxins. While the EPA did register StarLink™ corn for feed uses but not for food uses, it has announced it will no longer grant such split registrations.

The EPA may require resistance-management refugia or other planting restrictions as a condition of registration, but it has no direct authority to enforce compliance on farmers. Instead, it has required, as a condition of registration, that registrants establish a compliance assurance program that demands farmer compliance as a condition of the sale of seed.

2.3. Food and Drug Administration

The FDA does not issue permits, but rather exercises its regulatory authority ex-post by taking legal action against marketers of unsafe or mislabeled substances. To preclude this threat, FDA encourages voluntary consultation during the development and testing of plant biotechnology substances. The FDA asks for a considerable amount of data related to toxicology, allergenicity, nutritional content, etc., and offers guidelines for tests to generate the data. These consultations end when the FDA states that it has no further questions about the developer's position that the material is not significantly different from its traditional counterpart and safe for human health. The FDA does not require labeling of a product simply because genetic engineering was involved, but it offers guidelines about labeling products regarding the use of genetic engineering. However, if the nutritional or allergenic content of the product differs significantly from its traditional counterpart, labeling is required to indicate those contents.

3. THE REGULATORY PROCESS FOR TRANSGENIC FIELD CROPS WITH PHARMACEUTICAL OR INDUSTRIAL TRAITS

Field crops such as corn and soybeans are among the plants that have been engineered to produce pharmaceutical proteins, vaccines, or industrial compounds ("third generation" biotechnology). The regulatory process for these plants is fundamentally the same as for other biotechnology field crops, but it is rapidly evolving. Isolation and containment requirements are much more demanding than for other crops, with larger perimeter fallow zones, one-mile distances from other corn fields in the case of corn, management of all residuals, etc. Dedicated equipment, training programs for workers, and specified shipping containers are also required, and compliance inspections are more frequent.

In August 2003, APHIS announced that the notification process for field testing would no longer be available for plants engineered to produce pharmaceutical or industrial products. In January 2004, APHIS proposed a risk-tiered regulatory process that would establish different regulatory processes for organisms with three levels of risks.[2] The most flexible processes would be for products with low risks owing to previous experience, while the most stringent and lengthy processes would apply to plants produced for pharmaceutical or industrial traits, products with more unknown risks. Organisms with new agronomic traits or nutritional qualities would be subject to an intermediate risk tier. Mechanisms would be developed to allow the developer of pharmaceutical/industrial plants to produce and process the crops under confinement with governmental oversight, rather than using the approval process for unconfined deregulation of tier one and tier two risk levels. These proposals have not yet been implemented.

---

[2] Federal Register 69:15, pp. 3271–3272, January 23, 2004.

4. ISSUES

Regulatory issues related to transgenic field crops are not different in principle from those for other forms of agricultural biotechnology. But field crops are widely grown, and the potential for contamination of their commodity marketing chains with transgenic materials is a huge economic threat.

This risk was made evident by contamination events in 2000 and 2002. Traces of the insecticidal Cry9C protein from StarLink™ corn were found in food products in 2000, even though Starlink™ had been originally registered in 1998 only for feed use. Growers had signed agreements on segregation and containment measures for these crops, but it is not clear whether the contamination occurred through pollen drift or non-compliance. Contaminated seed stocks were removed from the market and contaminated lots of corn were directed into livestock uses to the extent possible. No harmful health or environmental effects have been demonstrated to have resulted from this episode.

In 2002, two cases resulted from non-compliance by ProdiGene in growing pharmaceutical-producing corn in 2001. In both cases, APHIS inspectors discovered volunteer corn plants in soybean crops grown the year after the corn crop. In the minor case, APHIS required that over a hundred acres of nearby corn fields be destroyed. In the other case, the soybean crop was harvested before all corn plants were removed, and a half-million bushels of soybeans in the elevator had to be destroyed.

These contamination experiences caused hundreds of millions of dollars of cleanup losses to the developing companies and to USDA, but no apparent harm to consumers or the environment. Because of these events, field crop producers and their associations are intensely concerned about losing markets due to contamination, and have therefore been strong supporters of proposed changes that result in more stringent regulatory oversight in field crops.

4.1. Risk and Costs

Regulation policy can reduce biotechnology risks but at the cost of increased regulatory and compliance efforts and the opportunity costs of valuable products that may not be brought to market, though they might have been found to be safe. How many tests and of what kinds are enough? How do we measure the risk inherent in previously unknown products? What is the optimal expenditure on inspection and compliance measures? What measures for containment and segregation should be required? These are all legitimate policy questions but it is impossible to imagine definitive answers.

The response of U.S. regulatory agencies to this dilemma has been to approve substances that meet such rather vague criteria as "no unreasonable adverse risks to the environment," "no significant risks under CFR 340," and "not significantly different from traditional counterparts." [One critic, Consumers' Union, prefers an equally vague standard of "reasonable certainty of no harm" (Hanson 2001).] Procedures required by current regulations have evolved as the agencies and product developers have acquired experience with potential risks and the effectiveness of various testing and compliance procedures. APHIS has established a multi-stakeholder advisory committee to help guide continued evolution of these rules.

The contribution of economic analysis to these issues has been largely confined to the estimation of costs of alternative refugia requirements for Bt corn and cotton (Hurley, Babcock, and Hellmich 1997, 2001; Hyde et al. 1999; Hyde 2000; Laxminarayan and Simpson 2002; Livingston, Carlson, and Fackler 2004; Secchi et al. 2006) Economic analysis also has the potential to improve regulatory procedures by using principles of mechanism design to compare the outcomes from alternative rules for the implicit game that is being played by regulators, developers, and producers (Giannakas and Kaplan 2005, Mitchell and Hurley 2006). Some additional study has been made of costs of segregation of grains in the marketing channels, but at this point this is not a regulatory issue *per se*.

4.2. Structure of Regulatory Agencies

The Pew Initiative and others cited in the Pew Initiative's report have been critical of the coordination problems inherent in having three agencies exercising joint oversight under the coordinated framework developed two decades ago. Concerns include the lack of clear legal authority, the possibility of some products slipping through the regulatory cracks, the potential for confusion and delay required by coordination, and the lack of clear "post-market" authority and responsibility for monitoring approved products. While some critics support new legislation creating a new agency, the Pew report suggests a non-legislated "single door" approach by which super coordination responsibilities would be assigned to one of the current agencies or to the Office of Science and Technology Policy. Defenders of the current system note that it has so far been successful in protecting human health and environmental safety, and that the continuing evolution of regulations under the coordinated framework is progressing satisfactorily, as compared with the costs and uncertainties associated with a radically altered structure.

Many other substances are simultaneously subject to regulatory oversight by two or more federal agencies, so agricultural biotechnology is not unique in that regard. APHIS, EPA, and FDA have been reasonably responsive to critics by establishing more open procedures and by modifying procedures substantially. Unless some dramatic case of contamination should occur, it appears that regulation of crop biotechnology will continue to evolve under the coordinated framework.

REFERENCES

Belson, N.A. 2000. "U.S. Regulation of Agricultural Biotechnology: An Overview." *AgBio Forum* 3(4): 268–280. Available online at http://www.agbioforum.org.

Giannakas, K., and J. Kaplan. 2005. "Policy Design and Conservation Compliance on Highly Erodible Lands." *Land Economics* 81(1): 20–33.

Hanson, M. 2001. "Make Sure It's Safe and Label It." In G. Nelson, ed., *Genetically Modified Organisms in Agriculture: Economics and Politics*. New York: Academic Press.

Hurley, T.M., B.A. Babcock, and R.L. Hellmich. 1997. "Biotechnology and Pest Resistance: An Economic Assessment of Refuges." Working Paper No. 97-WP 183, Center for Agricultural and Rural Development, Iowa State University.

_____. 2001. "Bt Corn and Insect Resistance: An Economic Assessment of Refuges." *Journal of Agricultural and Resource Economics* 26(1): 176–194.

Hyde, J. 2000. "The Economics of Within-Field Bt Corn Refuge." *AgBioForum* 3(1): 63–68.

Hyde, J., M.A. Martin, P.V. Preckel, C.L. Dobbins, and C.R. Edwards. 1999. "The Economics of Refuge Design for Bt Corn." Selected paper presented at the annual meetings of the American Agricultural Economics Association, Nashville, TN.

Laxminarayan, R., and R.D. Simpson. 2002. "Refuge Strategies for Managing Pest Resistance in Transgenic Agriculture." *Environmental and Resource Economics* 22(4): 521–536.

Livingston, M.J., G.A. Carlson, and P.L. Fackler. 2004. "Managing Resistance Evolution in Two Pests to Two Toxins with Refugia." *American Journal of Agricultural Economics* 86(1): 1–13.

Mitchell, P.D., and T.M. Hurley. 2006. "Adverse Selection, Moral Hazard, and Grower Compliance with Bt Corn Refuge." In R.E. Just, J.M. Alston, and D. Zilberman, eds., *Regulating Agricultural Biotechnology: Economics and Policy*. New York: Springer.

National Research Council. 2002. *Environmental Effects of Transgenic Plants: The Scope and Adequacy of Regulation*. Washington, D.C.: National Academy Press.

Nelson, G.C., J. Babinard, and T. Josling. 2001. "The Domestic and Regional Regulatory Environment." In G. Nelson, ed., *Genetically Modified Organisms in Agriculture: Economics and Politics*. New York: Academic Press.

Pew Initiative on Food and Biotechnology. 2004. *Issues in the Regulation of Genetically Engineered Plants and Animals*. Pew Initiative on Food and Biotechnology, Washington, D.C. (available online at http://www.pewagbiotech.org).

Secchi, S., T.M. Hurley, B.A. Babcock, and R.L. Hellmich. 2006. "Managing European Corn Borer Resistance to Bt Corn with Dynamic Refuges." In R.E. Just, J.M. Alston, and D. Zilberman, eds., *Regulating Agricultural Biotechnology: Economics and Policy*. New York: Springer.

Chapter 30

REGULATION OF TRANSGENIC CROPS INTENDED FOR PHARMACEUTICAL AND INDUSTRIAL USES

Gregory D. Graff
University of California, Berkeley

Abstract: The most interesting case in regulation of non-food agricultural biotechnology products is the manufacture of novel pharmaceutical and industrial products. The current regulatory situation in the United States has arisen out of a political economy driven largely by the food industry's risk exposure. Yet, the resulting "zero tolerance" regulatory situation appears inefficient and unstable: set up for another incident such as befell Aventis and ProdiGene. The regulatory situation could be resolved by taking dissemination and susceptibility characteristics into account in setting containment requirements, including threshold allowances in the case of minor breaches of containment. A workable regulatory regime will need to be driven by major agricultural states' political interests, along with the biotechnology and pharmaceutical industries, and mindful of the interests of the food industry. Countries intent on developing their domestic biotechnology industry might attract investment by providing a workable regulatory regime that ensures lower risk at lower cost.

Key words: agricultural biotechnology, regulation, plant-made pharmaceuticals (PMPs), plant-made industrial products (PMIPs), molecular farming

1. INTRODUCTION

The products of agriculture find a wide diversity of applications beyond those of just human food uses—including animal feed, fiber, ornamentals, medicines, and a range of industrial uses. When biotechnology is employed in the development of crops for such non-food uses, the technology employed, as well as the resulting genetically modified varieties, are typically subject to the same regulatory requirements as crops intended for food uses. In some cases, non-food crops invoke fewer regulatory requirements, given lower levels of concern over human health impacts when the product is not eaten. Non-food uses of genetically modified *food* crops already in development include animal feed (high-lysine corn, for improved animal nutrition and reduced nitrogenous waste), fuel feedstocks (corn with high levels of fermentable content, for ethanol production), paper products (potato with pure amylose starch, for paper finishing), as well as soaps and cosmetics (such as low-linolenic canola oil). Non-food applications of genetically modified *non-food* crops in development or already on the market include animal forage (low-lignin alfalfa, for improved digestibility by animals), fiber (*Bt* cotton), paper products (low-lignin aspen, for easier pulping), landscaping (herbicide-tolerant turfgrass), and ornamentals (novel colored carnations). However, one sub-class of agriculturally based non-food biotechnologies does pose a unique and challenging regulatory situation: those genetically engineered to express novel proteins and other biological products intended for pharmaceutical and industrial uses.

2. CROPS AND THE BUSINESS OF BIOMANUFACTURING

Crops engineered to produce plant-made pharmaceuticals (PMPs) and plant-made industrial products (PMIPs)—also known as "molecular farming"—are virtually all still in research or development. A few have progressed to the stage of applying for regulatory approval. Only one product has been commercialized to date. It is a recombinant plant-made trypsin, an enzyme intended primarily for laboratory uses. A recent survey (bio-era 2004a) found 80 different transgenic plants expressing novel compounds for pharmaceutical or industrial use that have been reported in the scientific literature or field trial registrations of the U.S. Department of Agriculture's Animal and Plant Health Inspection Service (APHIS) (Table 1), and more have certainly been reported since that survey. Pharmaceutical products expressed in plant hosts include vaccines, monoclonal antibodies, and other therapeutic proteins such as blood factors, anticoagulants, human hormones, interferons, interleukins, and therapeutic enzymes. Industrial products consist mostly of industrial, food processing, and analytic enzymes—such as trypsin, amylase, laccase, and GUS—but also include a few other protein- and carbohydrate-based compounds, such as brazzein (a protein sweetener), biopolymers, and a biological liquid crystal.

Demand for biopharmaceutical and bioindustrial products is already significant and is expected to grow rapidly in the next decade (bio-era 2004a). The market for therapeutic proteins, consisting largely of biologics on patent, is estimated to be around $30 billion per year, with an expected surge in demand that will result from the 500 or so new biologics currently in the regulatory phase at the FDA, including over 70 monoclonal antibodies which require particularly large production volumes. Other sources of demand for biomanufacturing capacity will undoubtedly come from the

Table 1. Numbers of Plant-Made Pharmaceuticals (PMPs) and Plant-Made Industrial Products (PMIPs) in Research and Development

Class of compound	Number of candidates
Pharmaceuticals	
Vaccines	12
Monoclonal antibodies	13
Other therapeutic proteins	33
blood factors and anticoagulants	
hormones	
interferons and interleukins	
therapeutic enzymes	
Industrial Products	
Industrial and analytic enzymes	17
(amylase, trypsin, laccase, GUS, etc.)	
Others	5
(avidin, elasin, brazzein, biopolymers, etc.)	
Total	**80**

Source: bio-era (2004a).

emergence of biogenerics, as current biologics are already beginning to go off patent in Europe, the United States, and throughout the rest of the world. Supplies for public health programs (including vaccinations and biological treatments) in low income countries may also require high-volume, low-cost biomanufacturing capacity. The market for industrial enzymes is smaller, currently estimated at around $2 billion per year, and consists of an upper tier of proprietary specialty, analytic, or "designer" enzymes, and a lower tier of bulk commodity enzymes (bio-era 2004a).

The biomanufacturing business is characterized by a set of incumbent technologies that have been essentially in place for almost two decades, plus a competitive fringe made up of an increasing array of alternative technologies seeking to enter the market. The incumbent technologies—which today provide the vast majority of material that make up the $32 billion in sales of therapeutic proteins and industrial enzymes—include protein expression by transgenic *E. coli*, yeast, and mammalian (hamster) cell lines housed in closed-vessel fermentation facilities. Alternative technologies vying to enter the market, in addition to crop plants, include transgenic livestock, insect larvae, mosses, and algae, to name a few. Ultimately there is likely room in the market for a wide taxonomic range of host organisms, with each specializing in the production of those biological materials for which it enjoys some technical advantage that translates into real economic advantage.

Crop-based biomanufacturing technology—as an alternative that is not yet widely used—is situated within a somewhat peculiar division of labor in the R&D-commercialization processes. First, the discovery of a bioactive compound is usually accomplished and patented at a large pharmaceutical or chemical firm, at a biotech firm that specializes in "discovery" or at a university that then licenses it exclusively to a firm. A biomanufacturing technology is chosen early in the process and the biomanufacturing firm becomes involved in development of the product and its production processes up through the regulatory phase, since FDA approval of biologics includes approval of the method of manufacture as an integral part of approval of the product. However, after FDA approval is achieved, the relationship reverts to something that more closely resembles that of a contract manufacturer. The real value thereafter lies in (the patents on) the active compound. Returns on investment are much more a function of monopoly pricing power than they are on costs. The real costs associated with the product are those sunk in achieving FDA approvals. Any costs incurred in ongoing biosafety compliance in the course of manufacture of the active compound will continue to be a relatively small portion of total costs. Even total ongoing manufacturing costs are themselves not a major variable in the profit function, with safety, flexibility, and reliability—to maximize the window of time in which monopoly rents can be collected—likely far more important than achieving economies of scale in manufacture. Indeed, given realistic assumptions about the amount of protein demand that will be likely for crop-based biomanufacturing, estimated acreage requirements will be insignificant in agricultural terms, no more than 25,000 acres through the next decade (bio-era 2004b). A company representative from one of the firms engaged in crop biomanufacturing technology made it very clear at a recent meeting when he emphasized, "This is not agriculture. Let's be clear about that. This is a pharmaceutical manufacturing process that happens to use a crop plant in one of its steps."

3. UNCERTAINTY AND THE PERCEPTION OF RISK-BENEFIT TRADEOFFS ARE DRIVING THE POLITICAL ECONOMY OF REGULATION

The nature of the primary risks introduced by this technology are straightforward: the testing and manufacture of pharmaceutically active and "industrial grade" (i.e., non food grade) products in crop systems can directly introduce novel, biologically active compounds into the environment or the food supply that—while valuable and beneficial when confined to their proper applications—may harm wildlife, agricultural productivity, or human health if inadvertently contacted or consumed in sufficient quantities.

This is a classic situation of the introduction of a new negative external-ity, but it is complicated in at least two important ways. First, there is significant uncertainty over the actual probability and extent of damages that could arise from an incident or accident. This uncertainty will be resolved in time by scientific study and by experience of practitioners and regulators. Meanwhile, the greatest "damage" caused by an incident or accident will likely be public reactions based on perceptions of "something going wrong" rather than anyone actually being injured. Second, there is significant hetero-geneity in how each different product would behave as a contaminant in the event of an incident or accident, in terms of introduction, dissemination, and susceptibility (Figure 1)—the three components of a typical biophysical risk function as discussed by Lichtenberg (2006). The overall risk of contamina-tion can be described as a function of (i) the probability of introduction of the compound into the environment or food supply, (ii) the rate of dissemi-nation of that compound through the environment or food supply, and (iii) the degree of susceptibility of humans or wildlife to the likely resulting levels of the compound eventually encountered in the environment or food supply. Or, to summarize,

$$R_j(x) = I_j(x_i) \, D_j(x_d) \, S_j(x_s)$$

is the overall risk for product j given regulatory actions $x = (x_i, x_d, x_s)$ de-signed to mitigate or account for each of these three components of risk. The x_i are containment requirements to minimize $\mathbf{I}$, the risk of introduction of the biologic into the environment or the food supply. The x_d are regulations to mitigate the risk of dissemination, $\mathbf{D}$, such as cleanup or product recalls. The x_s are regulations to limit the risk of susceptibility, $\mathbf{S}$, among human or wildlife populations, such as setting acceptable threshold levels of exposure.

However, such objective measures of the overall risk are not, as of yet, effectively addressed in the debate over how to regulate this crop biomanu-facturing technology. Rather, broad categorical perceptions of the risk/bene-fit tradeoffs are prevailing within each of the affected sectors of the econ-omy, including consumers, the food industry, agriculture, the pharmaceutical and chemical industries, and the current agricultural biotechnology industry. These perceptions in turn are determining the sectors' respective positions

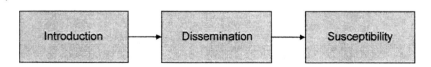

Figure 1. The Three Components of Biophysical Risk

within the political economic process unfolding over the regulation of plant-made pharmaceuticals and industrial products (Stewart and McLean 2004).

3.1. Consumers

Consumers express significant "fear and dread" factors over what they perceive as an "unnatural" or "uncontrollable" risk in the food supply or the wider environment. Such anxieties are neither new nor particular to this technology, but are similar to those that have arisen around the introduction of radical technological breakthroughs throughout history. On the other hand, some consumers groups seeking affordable prescription drugs as well as some patient advocacy groups have voiced anticipation of tangible benefit from crop biomanufacturing in the form of new, less expensive drugs. While there is no product in the marketplace yet, priors are currently being formed. The prototypical consumer-voter may be considered "highly impressionable" at this point in time. A "contamination" incident or accident could invoke a decided consumer response, including immediate avoidance of products associated with the incident and—in the longer run—broad but shallow manifestations of loss of confidence, softening of demand, and greater risk aversion toward the associated products and technologies. For example, many have attributed the long-term response of European consumers to food biotechnology in general as a reaction to several related but separate food safety incidents in the previous decade, including BSE (mad dow disease), dioxin contaminations, and foot-and-mouth disease.

3.2. The Food Industry

The food industry is highly averse to any incidents or accidents involving PMP contamination and the ensuing consumer response, which the industry feels it would have to face. At the same time, while supportive of technology that promises to deliver lower-priced drugs to consumers, the food industry sees little in the way of direct benefits to itself from the technology as it is currently being developed. The immediate aftermath of an incident or accident would impose costs on the food industry through the resulting demand shock, product recalls, legal bills, and product liabilities, as well as expenditure of political capital to manage the fallout from such an incident. Longer-term shifts might then be seen in the level or volatility of demand for what may become viewed as "tainted" classes of food product. Food companies that market directly to the final consumer are exposed *de facto* to liability for the safety of their product, regardless of the source of the contaminant. They are understandably reluctant to welcome a technology over which they have little control as a source of risk and where they see little benefit from its main output of therapeutic proteins and industrial enzymes.

3.3. Agriculture

Agricultural producers seem to be divided into two camps in terms of their perceptions of the technology. Most mainstream producers and those characterized as relatively late adopters of novel or alternative technologies tend to be quite concerned about the possible impacts of the technology on their own conventional agriculture, including its longer-term impacts on commodity perceptions and prices. On the other hand, progressive producers and early adopters of technologies appear to be (if anything, overly) optimistic about the value-added opportunities to be found in crop biomanufacturing. As already noted, the actual economic opportunity for growers is likely to be quite limited as volume requirements are likely to be quite small by agricultural standards (bio-era 2004b), and the owners of key intellectual property and biomanufacturing know-how are likely to maintain tight control over the full value chain, even keeping the growing operations in-house, or at the least contracting out to just a handful of highly specialized growers for the field production step. This has not, however, dissuaded agricultural states from trying to promote the opportunity for their farmers to cultivate pharmaceutical proteins.

3.4. Pharmaceutical, Biotech, and Chemical Industries

Most pharmaceutical, biotechnology, and chemical firms—the potential direct commercial customers of this new technology—still equate biomanufacturing with the mainstream incumbent technologies of closed vessel cell-culture fermentation systems using *E. coli*, yeast, or mammalian cell lines. Industry tends to view crop biomanufacturing as a yet unproven emerging option, with its unresolved regulatory issues as a decided negative. Those who have considered the technology, in fact, express concerns over the idea of open field production: not because the pharmaceutical compound is feared to contaminate food supplies, but rather because conditions typical to food supplies might contaminate the pharmaceutical. Numerous questions are raised about maintaining quality manufacturing standards in an open system with so many uncontrollable variables such as temperature, moisture, post-harvest protein degradation, microbial contamination, and more.

3.5. Agricultural Biotech Industry

Finally, the agricultural biotechnology innovators who are actively developing and advancing the crop biomanufacturing technology are mostly small entrepreneurial firms, such as ProdiGene (College Station, Texas), Epicyte (San Diego, California), Meristem Therapeutics (Clermont-Ferrand, France), SemBioSys (Calgary, Canada), and Ventria Biosciences (Sacramento, California). The major corporate seed and agrochemical firms typically associated with agricultural biotechnology have, to a large extent, exited the crop biomanufacturing arena. Pioneer divested its programs and assets with the spin out of ProdiGene in 1996. Monsanto decided to shut down its division of Monsanto Protein Technologies more recently, in October 2003. Dow appears to be the only major agricultural inputs firm left with serious involvement. A number of reasons can be cited for such exit. The technology may be considered to be too early stage and uncertain to invest in out of the corporate R&D budget, better left to venture capital until some of the inherent uncertainty has resolved itself. These corporations may view the target markets as simply too small: essentially niche or specialty agricultural products. Also, major agbiotech firms may be intentionally distancing their investments in multibillion-dollar agronomic traits from any possibly damaging association by the public with the risks perceived to be inherent in PMPs and PMIPs. Or, perhaps the agbiotech majors are in a position where it is more important to them to appease their downstream corporate customers in the food industry, many of whom have been forced into uncomfortably defensive positions in many countries with the introduction of biotech food crops.

4. CURRENT STATE OF U.S. REGULATIONS

Currently, very little has been commercialized, and thus the technology is still almost solely regulated by the Biotechnology Regulatory Service (BRS) of USDA-APHIS under requirements for release of transgenic organisms. The APHIS system consists of two tracks. Permits are the standard mode of regulating the field trial, release, or interstate transport of transgenic organisms. A quicker, more streamlined option for transgenic organisms on a list of standard traits and species involves merely giving notification to APHIS about the trial, release, or transport. PMPs and PMIPs, however, are not eligible for the relatively less intensive notification track: they must in every case be permitted. In addition, permits governing the field trials of PMPs or PMIPs carry new requirements under a March 2003 revisions of rules,[1] which may be summarized into three categories: (i) field test site selection (including perimeter distance requirements), (ii) equipment and facilities (including dedicated equipment for the duration of a field trial, storage in dedicated facilities, and special cleaning procedures at the end of a field trial), and (iii) field test procedures (governing seed cleaning and drying, personnel training, as well as multiple APHIS site inspections during and after a field trial).

---

[1] Federal Register Notice 68 FR 11337, March 10, 2003 (http://www.aphis.usda.gov/brs/pdf/7cfr.pdf),

Under APHIS rules, transgenics that have been thoroughly tested and that are ready for commercialization can apply to become "deregulated" and thus can be grown without the need for a permit or notification. However, PMPs and PMIPs are not at all eligible for eventual deregulation under APHIS rules. Thus, without the option of notification track or deregulation, commercial production will presumably have to be conducted under the system of field release permits and its attendant requirements. The entire regulatory strategy at APHIS is thus one of categorical containment. Indeed, the protein may even be generally regarded as safe (GRAS) and completely innocuous, but if deemed "novel" in the context of the host plant and thus categorized as a PMP or PMIP, it would fall under the containment requirements. However, APHIS's regulatory procedures for PMPs and PMIPs are currently undergoing a thorough review.[2]

Since PMPs and PMIPs are not likely to be considered plant-incorporated pesticides, the U.S. Environmental Protection Agency (EPA) becomes involved only indirectly via the mechanism of APHIS permitting, reviewing environmental assessments written for a PMP or PMIP by APHIS. The EPA is also involved to some extent in the overall review of APHIS regulations, which involves an environmental impact study under the requirements of the National Environmental Policy Act of 1969 (NEPA).

The FDA as well as the USDA's Bureau of Veterinary Medicine (BVM) regulate the products and processes of plant-based biomanufacturing through new drug and new biologic applications,[3] primarily to ensure the efficacy and safety of the resulting therapeutics or diagnostics for their prescribed indications. In the process they do not rule on the safety of these novel biological materials for unintended, adventitious presence. Under the logic and intent of FDA regulations, these fall into something of a gray zone. For pharmaceutical candidates that have failed the approvals process, there is an explicit prohibition against selling them for that pharmaceutical use, but even such a failure may not necessarily preclude sale and use of the same compound for other non-pharmaceutical (non-regulated) uses. The FDA Office of Food Additive Safety does conduct consultations for crops that are close to being commercialized to review "relevant safety, nutritional, or other regulatory issues regarding the bioengineered food."[4] The FDA is

---

[2] See U.S. Department of Agriculture's press release "USDA Announces First Steps to Update Biotechnology Regulations" (January 22, 2004) (http://www.usda.gov/Newsroom/0033.04.html).

[3] "Guidance for Industry: Drugs, Biologics, and Medical Devices Derived from Bioengineered Plants for Use in Humans and Animals," the U.S. Department of Agriculture and the U.S. Food and Drug Administration, September 2002 (http://www.fda.gov/cber/gdlns/bio[-]plant.pdf).

[4] See list of completed consultations on bioengineered foods at http://www.cfsan.fda.gov/~lrd/biocon.html.

considering procedures to conduct similar consultations significantly earlier in the development process, providing in essence a "pre-screening" of the safety of non-pesticidal proteins expressed in crops and intended for food use.[5] It is not clear to what extent both the early and the pre-commercialization consultations encompass those transgenic proteins expressed in crops but *not* intended for food use. Beyond scrutinizing specific applications for new drug, biologics, and food additives, it is of course improbable if not impossible for the FDA to pass judgment on the enormous diversity of naturally occurring proteins and other compounds that may be encountered in the human diet, other than to state that they are generally regarded as safe (GRAS).

The food industry would generally prefer involving the FDA in the regulation of the unintended/adventitious presence of PMPs and PMIPs in the food supply.[6] Indeed, it would be a natural role of the agency to determine many of the questions central to the safety of PMPs and PMIPs, including a particular protein's status as novel to the food supply, whether it is expressed in bioactive form or merely as an innocuous precursor, its survival through standard handling and processing procedures, and the levels of susceptibility (including digestability and allergenicity) towards it in the population. However, until questions of any such proposed FDA food safety standards can be worked out and implemented in collaboration with USDA-APHIS, the food industry is taking the position that APHIS should enforce a strict policy of zero percent introduction of any transgenic deemed a "PMP" or "PMIP" into the conventional food supply.

There is, however, an inefficiency and instability inherent in requiring such complete containment. Simply, the regulation does not take into account the dissemination and susceptibility components of the standard risk function, the $D_j(x_d) \cdot S_j(x_s)$, nor the fact that they will vary greatly over the various products j. Compliance by a firm or a farmer with containment regulations, $x_j(c)$, thus requires costly interventions to drive the level of introductions to zero, $I_j(x_j) \rightarrow 0$, even in those cases where the likelihood of dissemination and level of susceptibility is already at or near zero, $D_j(x_d) \cdot S_j(x_s) = 0$, and thus the overall objective risk is already at or near zero. Compliance thus requires treating any breach in the containment system as a full-blown contamination event, even in cases when it may be essentially harmless.

---

[5] "Guidance for Industry: Recommendations for the Early Food Safety Evaluation of New Non-Pesticidal Proteins Produced by New Plant Varieties Intended for Food Use," U.S. Food and Drug Administration, November 2004 (http://www.cfsan.fda.gov/~dms/bioprgui.html).

[6] See Grocery Manufacturers of America (GMA) press releases "GMA Comments on USDA Bio-Pharma Permit Regulations" (May 9, 2003) and "GMA Statement on FDA's Early Food Safety Evaluations for Biotech Foods" (November 19, 2004) (http://www.gmabrands.com/publicpolicy/biotechnology.cfm).

It follows that a rational compliance strategy on the part of firms and farmers involves minimizing the chances of any breach in any containment system by evenly allocating resources and efforts across all field trials (and, eventually, production plots), which means over-allocating resources and effort to contain low risk PMPs or PMIPs, and under-allocating resources and effort to contain the objectively higher risk products. Differentiating the handling of different PMPs and PMIPs requires of course collection of additional information about the risks involved, much of which is, however, certainly already available and could be employed already to rationalize the current system.

For instance, some of the compounds being produced in plants, such as trypsin, are already in the food supply and generally regarded as safe: trypsin is an enzyme commonly present in meat products at levels comparable to or higher than any that would be consumed if a trypsin-producing corn were to escape containment. Under a legal interpretation of current regulations, even the adventitious presence of a single volunteer trypsin-producing corn plant in a field of food-grade corn could trigger a federal violation, a total recall of all co-mingled corn, and significant fees and damages, even if eating a hot dog would expose a person to more trypsin than eating transgenic trypsin-producing corn would.

While perhaps nothing more than an excessive exercise of the "precautionary principle," such regulatory requirements serve to place the technology in a precarious position when it comes to compliance and, thereby, acts as a disincentive for further investment in the development and improvement of the technology. The zero-tolerance rule, by not providing reasonable yet safe allowances for negligible residues, greatly increases the chances of another crisis like the one that followed the enforcement by APHIS against ProdiGene in late 2002 when traces of a vaccine-producing corn were found in a shipment of conventional food grade soybeans.[7] The risk of introduction of at least some PMP seed, gene, or gene product into the wider environment or food supply is—while highly unlikely—nonetheless not zero under any level of containment efforts. This situation could, in several ways, be compared to the "split registration" which led to the StarLink incident involving Aventis CropScience in late 2000, wherein a product was approved for one use (animal consumption) but not another (human consumption), a regulatory situation with which it was very difficult if not impossible to comply in practice, leading to an eventual breach, even though in retrospect it appears that no human or animal health was in objective danger.

---

[7] See USDA press release "USDA Announces Actions Regarding Plant Protection Act Violations Involving Prodigene Inc.," December 6, 2002 (http://www.usda.gov/news/releases/2002/12/0498.htm).

5. FOUR WAYS FORWARD

As the development of both incumbent and alternative biomanufacturing technology is racing ahead, if crop biomanufacturing is to remain viable it needs to maintain a healthy trajectory in terms of investment in R&D for the future. There are compelling opportunities for crop biomanufacturing to supply therapeutics and other biological materials in ways that significantly improve in terms of cost, flexibility, and environmental impact. There is also the possibility that crop biomanufacturing can bring high-tech manufacturing opportunities into rural areas traditionally dependent on just agriculture. However, there is real risk that the technological trajectory along which crop biomanufacturing is developing will be stalled and diverted by the lack of a coherent, sound, and cost-effective regulatory framework. Under these less than optimal conditions, however, there do appear to be some basic steps forward for both regulators and the new industry.

Take dissemination and susceptibility characteristics into account in setting containment requirements, including threshold allowances in the case of minor breaches of containment. Initially this will need to be done on a case-by-case basis, just as in any new area of regulatory intervention. However, in short order, a framework should be set up that effectively classifies PMPs and PMIPs into several levels or categories of risk. (See Figure 2 for a hypothetical example.) Precedents can be found in National Institutes of Health (NIH) biohazard guidelines (Risk Groups 1 through 4) and EPA toxicity ratings (toxin categories 1 through 4). Such an approach would help to systematize containment requirements according to objective risk, reduce regulatory uncertainty, and allow for a more effective allocation of effort and resources, both by producers and by regulators, easing up unnecessary precautions and penalties on low risk products and focusing greater care onto higher risk products. Setting safe and reasonable tolerance thresholds would in effect "decriminalize" minor breaches of containment in low risk products when they are objectively harmless, and would play an important role in reassuring the public of the safety of many of these products.

One approach that may facilitate the development and implementation of such a framework would be for firms developing PMPs and PMIPs to submit the biologic as a novel food additive for review by the FDA's Office of Food Additive Safety, as has been common practice with other transgenic proteins expressed in crops even when they are not themselves directly intended as a food ingredient.[8] A challenge to developing such a framework for transgenic crops is, of course, the need for further deepening of coordination between

---

[8] A list of completed reviews of transgenic crops by the FDA Center for Food Safety and Applied Nutrition's Office of Food Additive Safety is online at http://www.cfsan.fda.gov/~lrd/biocon.html.

		Level of exposure, dosage-response risk			
		Negligible	Low	Moderate	High
Grade of intended use	Food grade	0	n/a	n/a	n/a
	Novel food additives/ dietary supplements	0	1	n/a	n/a
	Over-the-counter drugs	0	1	2	n/a
	Prescription drugs	0	1	2	3
	Industrial or research (non-food, non-therapeutic)	0	1	2	3

Figure 2. Hypothetical Risk Levels 0–3 for Plant-Made Pharmaceutical and Industrial Products

USDA, EPA, and FDA, but it would be a worthwhile undertaking for the existing coordinated framework among the three agencies.[9]

Achieve containment in a cost-effective, scalable manner. There are several ways this may be pursued, and it will be easier once the risk levels of PMPs and PMIPs are more systematically differentiated and classified. Likely the simplest approach is to utilize geography and to locate the growing of higher risk PMPs in regions that would minimize the chance of cross-contamination with food supplies or sensitive environments. Such a strategy was proposed by the Biotechnology Industry Organization (BIO) in 2002 in the form of voluntary safeguards by its member firms not to grow PMP corn in the Corn Belt. However, BIO reversed its position in response to political pressure arising in the Midwestern farm states.[10] Yet, economic logic will likely prevail over political logic in the long run, and given the relatively

---

[9] See http://usbiotechreg.nbii.gov/.

[10] See Fox (2003) and press releases from Senator Tom Harkin, "Harkin Urges Review of Biotech Crop Ban," November 1, 2002 (http://harkin.senate.gov/press/print-release.cfm?id= 188229), and Senator Chuck Grassley, "Grassley Receives New Position Statement from Biotechnology Industry Organization: Letter Outlines BIO's Commitment to Iowa," December 3, 2002 (http://grassley.senate.gov/releases/2002/p02r12-03.htm).

small scale of operations needed, cultivating PMPs in isolated areas such as irrigated desert valleys or remote islands would decrease cost because it would minimize risk.

A second approach is to build redundancy into technical and managerial systems for containment. Putting two complementary low-tech containment measures in place can in many cases achieve higher safety levels than a single sophisticated, high-tech solution, and at lower cost.

Third, it may make sense to utilize economies of scope by clustering crop biomanufacturing operations in a given region. Again, given the relatively small scale of agricultural productivity required, as well as the highly specialized nature of the business, the associated containment and tracking infrastructures, and the value of accumulating human capital expertise, costs can be expected to decline as a function of experience and sharing of physical and human capital. In the medium to long run, innovation can also be expected to increase more rapidly in the midst of such a crop biomanufacturing cluster (Porter 1998).

Finally, costs of containment will certainly fall as a result of investment in innovation by firms, by states interested in building up a crop biomanufacturing industry, and by the USDA. Areas of particular value for improving containment at lower cost would include diagnostics to monitor presence of PMPs and PMIPs, IT-based inventory control systems, and genetic mechanisms for fertility control of host plants.

Outsource the growing of PMPs/PMIPs. The adoption of a more stable regulatory framework along with measures to reduce containment costs can and likely will be pursued by a number of governments—both U.S. state governments and other national governments. Thus, it may be a reasonable commercial decision to outsource or relocate PMP growing operations to jurisdictions with the most favorable regulatory environment, as part of a strategy to minimize risk and cost. Indeed, some governments may make regulatory and infrastructure commitments to create an environment for the pursuit of PMP technology as part of a broader strategy to attract and promote a local biotechnology industry.

Pursue an activist agenda. Finally, the technology will become truly viable only as it wins public affirmation and the trust and confidence of the food industry. Targeting products that demonstrate the unique cost, product safety, and flexibility profile of crop biomanufacturing may be particularly valuable investments for the nascent industry to make at this juncture. These might include contracts for ultra-low-cost production of vaccines for global public health initiatives in low income countries or the supply of products with clear environmental benefits.

6. CONCLUSIONS

It is clear why the interesting case in regulation of non-food products is the manufacture in food crops of novel pharmaceutical and industrial products. The current regulatory situation in the United States has arisen out of a political economy largely driven by the food industry's risk exposure. Major political will is lacking on the part of major agbiotech players such as Monsanto and DuPont, as they likely do not see significant value in providing what is in essence contract manufacturing. They have in fact actively distanced themselves from the technology and its attendant risks. The resulting "zero tolerance" regulatory situation is inefficient and unstable—set up for another incident such as befell Aventis with its StarLink corn or ProdiGene with its vaccine-producing corn. The regulatory situation in the United States could be resolved by a relatively straightforward integration of the transmission and susceptibility components of risk (and thus input from the EPA and FDA) into containment requirements (administered by USDA). Such a workable regulatory regime will likely need to be driven by major agricultural states' political interests, with possible backing from the biotechnology and pharmaceutical industries, yet it will likely remain subject to veto by the interests of the food industry. However, the story may not, in the end, be played out in the United States. Other countries intent on developing their biotechnology industry may wake up to the relative ease with which they might attract investment and excel in this technology by providing a regulatory regime that ensures low risk at relatively low cost.

REFERENCES

bio-era. 2004a. "Crop Biomanufacturing, Part 1: The Economic Opportunity." Bio Economic Research Associates (bio-era), Cambridge, MA (36pp).

———. 2004b. "Crop Biomanufacturing, Part 2: Implications for the Farm Sector." Bio Economic Research Associates (bio-era), Cambridge, MA (14pp).

Fox, J.L. 2003. "Puzzling Industry Response to ProdiGene Fiasco." *Nature Biotechnology* 21(1): 3–4.

Lichtenberg, E. 2006. "Regulation of Technology in the Context of Risk Generation." In R.E. Just, J.M. Alston, and D. Zilberman, eds., *Regulating Agricultural Biotechnology: Economics and Policy*. New York: Springer.

Porter, M.E. 1998. "Clusters and the New Economics of Competition." *Harvard Business Review* 76(6): 77–90.

Stewart, P.A., and W. McLean. 2004. "Fear and Hope over the Third Generation of Agricultural Biotechnology: Analysis of Public Response in the *Federal Register*." *AgBioForum* 7(3): Article 5.

Chapter 31

REGULATION OF BIOTECHNOLOGY FOR FORESTRY PRODUCTS

Roger A. Sedjo
Resources for the Future

Abstract: Genetically improved trees are playing a greater role in meeting the world's industrial wood requirements. Genetically engineered trees, trees created by asexual means, offer potential to provide higher-quality and lower-priced industrial wood as well as environmental benefits. However, there may also be environmental risks. A regulatory system exists in the United States for assessing the safety and environmental impacts of transgenics, including trees. This chapter discusses some aspects of that regulatory system as it applies to transgenic trees.

Key words: trees, biotechnology, genetic engineering, forests, wood

1. INTRODUCTION

Humans have been striving to improve their ability to survive by making modifications and innovations in food and fiber for countless millennia. Hunting and gathering gradually gave way to herding and primitive farming, and crop seeds from one region were transported to other areas. As Bradshaw (1999) points out, the domestication of a small number of plants, particularly wheat, rice, and maize, is among the most significant accomplishments in the human era. Modern civilization would be impossible without this innovation. Common features associated with plant domestication include high yields, large seeds, soft seed coats, nonshattering seed heads that prevent seed dispersal and thus facilitate harvesting, and a flowering time that is determined by planting date rather than by natural day length.

Forestry today is following a pattern very similar to that established over the past several millennia in agriculture but has only recently begun a serious transition from gathering (harvesting natural forests) to cropping (planting and managing) trees (Table 1). Although fruit trees and ornamentals

Table 1. Transitions in Technology and Forest Management

Type	Period
Wild forests	10,000 B.C.E. (or earlier) to present
Managed forests	100 B.C.E. to present
Early planted forests	circa 1800–1900
Development of principles of inheritance by Gregor Mendel	mid- to late 1800s
Development of commercial hybrids of annual crops	1930s
Planted, intensively managed forests	1960 to present
Planted, superior trees from traditional breeding techniques	1970 to present
Planted, superior trees from clones	1990 to present
Field trials of genetically engineered trees	1990 to present
Commercial plantings of genetically engineered trees	2005? to future

have been domesticated for hundreds if not thousands of years, timber species have been partly domesticated for wood production only in the past half-century (El-Kassaby 2003). Early on, tree planting benefits were often derived from the superior growth productivity observed in introduced exotics that thrived in new environments (Table 2). Although exotics have not played a major role in U.S. forest production, their success abroad—in South America and Oceania, for example—provided an impetus for improving domestic trees to increase domestic productivity and competitiveness (Sedjo 1999, 2004a, 2004c).

Only after trees were being planted for commercial purposes did it pay to undertake the investments necessary for serious efforts to increase their productivity. Once a decision is made to invest in planting, a reasonable follow-up is to plant highly productive species that have the characteristics to improve financial returns. Today, improved or superior trees are increasingly being planted and are in fact the norm in the United States. In many other countries, emphasis is on improving exotic species that have become the dominant commercially planted trees.

Associated with planting improved genetic stock has been an increase in management intensity. Contemporary planted forests are treated like agricultural crops. Sites are prepared, seedlings planted, vegetative control undertaken, and fertilizer applied where needed; trees are thinned to promote growth and finally harvested, after which the cycle is repeated.[1] Plantation

---

[1] Selection harvesting is practiced for some forest types and tree species in which the rotation cycle is more complex.

Table 2. Worldwide Timber Yields

Site	Yield (m3/ha/yr)	Rotation (years)
Temperate and boreal indigenous softwood forests		
Canada average	1.0	--
British Columbia	1.5-5.3	--
Sweden average	3.3	--
Finland	2.5	60
Russia	1.0-2.9	--
Siberia	1.0-1.4	70
Softwood plantations, exotics		
Britain (Sitka spruce)	14	40
South Africa (pine spp.)	10-25	20-35
New Zealand (Monterey pine)	18-30	20-40
East Africa (pine spp.)	25-45	20-30
Brazil (pine spp.)	15-35	15-35
Chile (Monterey pine)	20-30	15-35
Tropical hardwoods		
Malayan dipterocarp forest	up to 17	--
Mixed tropical high forest	0.5-7.0	--
Teak plantations	14	40-60

Source: Clapp (1993).

forestry has already demonstrated the potential not only to dominate industrial wood production, but also to help protect and conserve much of the natural forest and the environmental and ecosystem services it provides (Sohngen, Mendelsohn, and Sedjo 1999).

Today, plantation forests have become an important source of timber, accounting for about one-third of the harvested industrial wood by the end of the twentieth century (Food and Agriculture Organization 2001). This contrasts dramatically with the situation just 50 years ago, when planted forests accounted for a negligible portion of the world's industrial wood harvest. Although forest biotechnology, including genetic engineering, is still in its infancy, much of the biotechnology already developed for agriculture has direct applications in forestry. A logical next step would be the introduction of transgenic—that is, genetically engineered—trees, again following the agricultural cropping model. To begin this next step, appropriate new technology must be developed and become financially viable, the transgenic trees must pass the hurdle of deregulation, and there must be some degree of public acceptance (Sedjo 2004b).

2. TRANSGENIC PLANT REGULATION IN THE UNITED STATES

A genetically engineered plant—that is, one that involves the insertion of genes using a nonsexual approach—is considered a transgenic and, under U.S. law, is automatically a "regulated article" subject to a set of regulatory constraints. In the United States, the Animal and Plant Health Inspection Service (APHIS) of the U.S. Department of Agriculture has been regulating plant biotechnology since 1987. For certain genetic modifications, the U.S. Environmental Protection Agency (EPA) also has a regulatory role, by virtue of its responsibilities for the regulation of toxics and pesticides (under the Toxic Substances Control Act of 1976 and the Federal Insecticide, Fungicide, and Rodenticide Act of 1996) and overall environmental safety (under the National Environmental Policy Act of 1969).

APHIS administers regulations for most genetically engineered plant organisms, which are initially classified as regulated articles. The basic APHIS deregulation process for trees is identical to that for other plants, including annual crops, although the time frame for field tests and other investigations may vary. As with crops, the tree developer must obtain a permit from APHIS to field-test, provide the results of the field tests to APHIS, and support any petition for deregulation with various other types of information, such as scientific literature and statistical test results.

In January 2004, USDA (2004) announced its intention "to update and strengthen its biotechnology regulation for the importation, interstate movement and environmental release of certain genetically engineered (GE) organisms." As part of the overall process, APHIS was to prepare an environmental impact statement (EIS) on its biotechnology regulations. The EIS would provide a broader coverage of risks and benefits as part of an updated protocol for trees (Cordts 2004).

Since revisions to biotechnology regulation are in process, this chapter focuses on the current APHIS system and considers the prospective changes that appear likely, on the basis of the announced goals.

3. FOREST BIOTECHNOLOGY

Biotechnology has been defined as comprising five major categories: markers, propagation and multiplication, functional genomics, marker-aided selection and breeding, and genetic modification (El-Kassaby 2003). This chapter focuses largely on the genetic modification, but propagation and multiplication are enabling technologies for genetic modification, and hence we begin with these traditional techniques.

Traditional breeding involves genetic modification of a plant through various sexual processes and procedures, as well as asexual reproduction through vegetative cloning. Over the past 30 years, considerable improvements have been made in forest stock utilizing traditional breeding approaches.

3.1. Selection and Breeding Orchards

Tree improvement has usually relied on traditional breeding techniques, such as the selection of superior trees for volume increases and stem (trunk) straightness, and on the grafting of these traits into breeding orchards and producing seed orchards. When breeding orchards begin to flower, pollination of selections is artificially controlled, seeds are collected, progeny tests are established, and the best offspring are selected for the next cycle of breeding. By identifying and selecting for desired traits, breeders can choose a set of traits that will improve wood and fiber characteristics, improve tree form, improve growth, and provide other desired characteristics. The potential improvements that can be expected from the various traditional breeding approaches are presented in Table 3.

Table 3. Gains from Various Traditional Breeding Approaches: Loblolly Pine

Technique	Effect (increase in yields)
Orchard mix, open pollination, first generation	8 %
Family block, best mothers	11 %
Mass pollination, full sibling (control for both male and female)	21 %

Source: Westvaco Corporation (1997).

3.2. Cloning and Vegetative Reproduction

Vegetative reproduction comprises a broad range of techniques for manipulating plant tissue that ultimately allow for vegetative reproduction of the whole plant. The vegetatively regenerated plant is a clone having the same genetic composition as the original plant. The simple form of vegetative propagation involves planting cuttings from a plant, such as a branch or root. Fences consisting of live trees, common in much of the tropics, are created in this fashion. Vegetative propagation has been practiced for centuries with many plants, including grapes, potatoes, and many deciduous trees.

The development of cloning techniques in forestry is important for a number of reasons. First, the approach allows the propagation of large numbers of seedlings of superior trees. That is, a clone is a vehicle for mass production. Second, for genetic engineering of trees, the clone provides a vehicle through which desired foreign or artificial genes can be transferred and thus provides an excellent platform for the application of genetic engineering. Cloning can be viewed as an enabling technology that will facilitate the transgenic transformation of conifer trees.

The clonal approach has the advantage of capturing all the desired genetic superiority of the donor plant because the process relies on mitosis cell division, which does not impart any gene segregation. The rooted plantlets can be planted en masse and the beneficial traits of the single tree duplicated in each new tree.

This approach has commonly been used for poplar, eucalyptus, and other deciduous trees. Vegetative propagation, however, has not been an effective technique for most conifers—a biological family in which vegetative propagation is extremely rare.

Though often referred to as biotechnology, even sophisticated cloning techniques are not considered genetic engineering, however, since genes are not transferred asexually. Clonal activities are not typically regulated and do not trigger any review for deregulation.

4. POTENTIAL BENEFITS OF TRANSGENIC TREES

A plant that is altered through the insertion of a gene using a nonsexual approach is considered a bioengineered plant and defined as a transgenic. This approach provides an enhanced ability to transfer genes that could be transferable using traditional approaches, as well as to transfer genes across species—something that could not be accomplished by traditional approaches.

High-productivity plantation forestry, with control from seedling to harvest, has created the preconditions necessary to financially justify tree improvement through both traditional and transgenic breeding. The financial and economic benefits expected from transgenic trees would come from increased productivity, increased quality, and lower costs (Table 4).

Table 4. Forest Tree Traits That Can Be Improved Through Biotechnology

Silviculture	Adaptability	Wood
growth rate	drought tolerance	wood density
nutrient uptake	cold tolerance	lignin reduction
crown and stem form	fungal resistance	lignin extraction
flowering time	insect resistance	juvenile fiber
herbicide resistance	salt tolerance	branching

Source: Context Consulting provided information on potential innovations and their likely cost implications based on the best judgment of a panel of experts.

Despite the wide range of transgenic possibilities for trees, in very recent years a primary target of interest for genetic engineering in trees has been lignin and cellulose modification. Lignin, which imparts rigidity to the tree, makes wood fiber difficult to process into paper. Attempts are under way to reduce lignin content or vary the lignin-to-cellulose composition for easier lignin removal—something of great interest to the pulp and paper industry. Some work suggests that a reduction of lignin content can be accompanied by an absolute increase in the volume of cellulose (Fladung 2004).

Another potential of genetic engineering is creating trees resistant to the application of glyphosate, an herbicide used to control weeds commonly used in some genetically engineered agricultural crops. Inserting herbicide-resistant genes into soybeans, for example, has been highly successful. The herbicide is usually applied before planting and again shortly after planting, thus reducing weed competition early in the crop's growth cycle. There is evidence that for some crops, such as cotton, the total herbicide use is reduced.

A further potential application to trees is the Bt *(Bacillus thuringiensis)* gene, which, used in agriculture, has been very effective in protecting the host plant against certain insect pests (Tiang et al. 1994). However, the deregulation process in the United States would require tree breeders to obtain authorization from EPA, which has responsibility for pesticides and toxics, as well as APHIS, and therefore this innovation has a lower priority within the tree development community.

The prospective economic benefits from biotechnology in trees are expected to result from the complementary and coordinated development of both traditional breeding and genetic engineering. A superior tree developed through traditional breeding would provide the basic tree into which the desired genes would be asexually introduced. The introduction of the desired traits into the already-superior tree gives great promise of allowing the expression of those traits, thereby increasing productivity, increasing product quality, and expanding the range and types of land and climatic conditions under which productive forests can thrive.

Table 5 provides possible financial gains and operating costs estimated to be associated with tree biotechnology innovations. The first of these innovations—a low-cost, pine cloning technique—is a biotechnological but not transgenic innovation.

5. CONCERNS ABOUT TRANSGENICS

The rationale for regulating transgenics arises from two types of concerns: health and safety, and environmental. The question of health and safety usually involves whether ingesting a transgenic plant will have any deleterious

Table 5. Possible Gains from Future Biotech Innovations

Innovation	Benefit	Operating costs
Superior pine clone	20 percent yield increase after 20 years	$40 per acre or 15–20 percent increase
Wood density gene	Improved lumber strength	None
Herbicide tolerance gene in eucalyptus (Brazil)	Reduced herbicide and weeding costs, potentially saving $350 or 45 percent per hectare	None
Improved fiber characteristics	Reduced pulping digester cost, potentially saving $10 per m^3	None
Reduced amount of juvenile wood	Increased usable wood, raising value $15 per m^3	None
Reduced lignin content	Reduced pulping costs, potentially saving $15 per m^3	None

Source: Context Consulting. The actual net benefits experienced by the tree planter will depend on the pricing strategy used by the gene developer and the portion of the savings to be captured by the developer and passed on to the grower.

effects on human or animal health. For forestry this is not a major issue, since wood ingestion is a minor use of trees, confined to fillers in certain food products. Rather, the major concerns relate to possible ecological damage that might be associated with the release of a transgenic tree into the environment (for an example see Mullin and Bertrand 1998). One concern is that a transgenic tree may directly become a type of invasive. Botkin (2001) has likened a transgenic to the introduction of exotics, some of which have become invasive.[2]

A fundamental concern in tree genetic engineering is the question of gene escape, or "gene flow"—that is, whether the transferred gene can escape to a wild relative, thereby increasing the fitness of that relative and disrupting the existing ecosystem (DiFazio et al. 1999). One principal concern in forestry is that the exotic gene in a transgenic may be passed from the plantation forest to trees in adjacent stands. For example, anxiety has been expressed over the risk of transgenic forest tree invasiveness at the interface of private forests and public lands (Williams 2004). The concern is twofold. As with agricultural crops, one concern is that the transfer of a gene from transgenic to nontransgenic crops could disqualify the "tainted" crop

---

[2] However, other ecologists have argued that the risks associated with transgenics are generally lower and more predictable than for an exotic because the plant has only a couple of introduced genes, the general expression of which is known. Thus problems associated with transgenics should be easier to identify than the effects of exotics with large numbers of unknown genes.

from nontransgenic status and hence preclude it from sale in certain markets—for example, the European Union. This effect might apply to tree nurseries as well. A second major concern involves the possibilities of gene escape into a natural forest.

The general issue with gene flow is "flow versus fate." Given that gene flow will occur, under what circumstances will it be deleterious to a planted forest or to the natural environment, and what approaches would contain or mitigate these effects?

6. REGULATORY CHALLENGE

The challenge for regulators is limiting gene escape—a particularly complex issue with species that have close relatives in the surrounding natural environment. One approach is formal containment, which is often required in trial research plots. Other techniques are under consideration or development, including ways to minimize or eliminate the tree's ability to transfer genes by delaying or preventing (terminating) flowering, thereby promoting actual or de facto sterile trees (see Meilan et al. 2004, Kellison 2004).[3] This approach, a common area of bioengineering tree research, is likely to be a precondition, at least in the United States, of the deregulation of transgenic trees. In agriculture, sterilization has been contentious and related to property rights issues: it would prevent growers from using improved seed for the next planting without payment to the developer. In forestry, with its long growth periods, the property rights aspect is unlikely to be important, since the technology is likely to be obsolete before reproduction could be undertaken.

Another method for containing gene flow is selecting locations where there are no compatible wild or weedy relatives in the natural environment. An advantage of planting an exotic species with altered genes is that close relatives are usually absent, making the probability of gene transfer nonexistent. For example, pines and eucalyptus that are commonly planted in South America are not indigenous. So the problem of a gene transfer from a planted exotic to a native tree is nonexistent (DiFazio et al. 1999).

7. U.S. REGULATORY FRAMEWORK

A consistent principle of health and environmental law in the United States is that products introduced into commerce should either be safe or present no

---

[3] Terminator genes are some used in annual crops to protect the property right embodied in the transgenic seed by preventing seed of the transgenic from being used for future planting.

unreasonable risk to humans or the environment. How this principle is applied varies, depending on which law applies, which agency has jurisdiction, and the social perception of risk.

Products of biotechnology do not always fit comfortably within the lines the law has drawn, which is based on the historical function and intended use of products. The relationship and coordination of the various authorities is governed by the policy statements contained in the 1986 Coordinated Framework for the Regulation of Biotechnology (51 Fed. Reg. 23302) and the 1992 Policy on Planned Introductions of Biotechnology Products into the Environment (57 Fed. Reg. 6753). The acts were designed to provide for a coordinated regulatory approach to be adopted by federal agencies. Products of biotechnology are regulated according to their intended use, with some products being regulated under more than one agency.

Three main agencies are involved in regulating transgenics:

- The Food and Drug Administration (FDA) of the U.S. Department of Agriculture is concerned with food safety.
- The Animal and Plant Health Inspection Service (APHIS) of the U.S. Department of Agriculture, under the Plant Protection Act (especially Title 7 U.S.C. Sections 7701 *et seq.*), has authority to determine whether a gene-altered plant, crop, or tree is likely to be a plant pest that could harm U.S. agriculture.
- The U.S. Environmental Protection Agency requires that federal agencies publicly address any impact of their activities that may significantly affect the environment and also has responsibilities in the regulation of toxics and pesticides (under the Toxic Substances Control Act of 1976 and the Federal Insecticide, Fungicide, and Rodenticide Act of 1996).

Associated with the 1986 framework are two laws identified as containing requirements applicable to all agencies reviewing biotechnology products. The National Environmental Policy Act of 1969 (NEPA) preceded the creation of EPA and is now administered by that agency. The Plant Protection Act of 2000 supersedes the Federal Plant Pest Act of 1957 and broadens the definition of a noxious weed to include plants (previously the definition was limited to nonnative plants). It is under this broader definition that APHIS regulates genetically engineered organisms. The Plant Protection Act is generally applied to all genetically engineered plants, including trees (Bryson et al. 2001, Bryson, Quarles, and Mannix 2004).

The Plant Protection Act consolidated and enhanced the Department of Agriculture's authority to prohibit or restrict the importation, entry, exportation, or movement in interstate commerce of any plant, plant product, biological control organism, noxious weed, article, or means of conveyance if the secretary determines that the prohibition or restriction is necessary to prevent introduction into the United States or the dissemination of a plant pest or noxious weed within the United States (7 U.S.C. 7712(a)).

8. THE DEREGULATION PROCESS

The assessment for the deregulation of transgenic plants centers on determining the health, safety, and environmental implications of the modified plant. The risk criterion is that the new varieties must be as safe to use as are varieties modified by traditional breeding techniques, which do not require deregulation.

A regulated article is defined as "any organism which has been altered or produced through genetic engineering of the donor organism, recipient organism, or vector, or vector agent belonging to any genera or taxa designated in the regulation (provision 340.2) and meeting the definition of a plant pest or any organism or product which APHIS determines or has reason to believe is a plant pest (7 CEF 340.1)." Regulated status has been applied to most of the genetically engineered plants that have been developed to date (Bryson et al. 2001).

8.1. Permitting

The deregulation approach used by APHIS can be briefly summarized as follows: permitting, notification, and petition. A *permit* must be obtained from APHIS for the importation, interstate movement, or release of a regulated article, such as a transgenic plant, into the environment. When the developer undertakes the field testing required for deregulation, APHIS must receive *notification*. Upon completion of field testing and other relevant procedures, the developer may submit a *petition* for deregulation to APHIS. The petition details the field test results and provides a comprehensive literature review together with any other relevant information and experience. APHIS, using a scientific committee, then makes a determination of whether to deregulate. Once a "determination of nonregulation status" is made, the product and its offspring no longer require APHIS authorization for transport, release, or communication in the United States. In the petition process, APHIS generally works cooperatively with developers. Petitions are seldom rejected outright but are not uncommonly returned for being incomplete or providing insufficient information.

Although the overall assessment by APHIS includes a consideration of the potential effects on the "wider" environment to ensure that any environmental impacts are not likely to be significant, broader environmental considerations are mandated under the National Environmental Policy Act (Title

42, U.S.C. sections 4321 *et seq.*). In addition, the EPA is directly involved in the deregulation process for any transgenic plant that has pesticidal or toxic properties under the Toxic Substances Control Act and the Federal Insecticide, Fungicide, and Rodenticide Act, as well as for overall environmental safety under NEPA.

8.2. Notification of Field Testing

Field tests for most genetically engineered plant varieties are usually undertaken by the developer and occur under controlled conditions.[4] Consistent with the basic criterion, they are designed to demonstrate that the transgenic variety is as safe to use as traditional varieties.

Under the Plant Protection Act, APHIS issues field-test permits for new plants that have the potential to create pest problems in domestic agriculture. This could apply to plants, plant products, and other articles developed through biotechnological processes if such plants, plant products, or articles present a risk of plant pest introduction, spread, or establishment. The APHIS regulation specifically applicable to genetically modified organisms was first promulgated in 1987 and controls the introduction of a class of organisms referred to as "regulated articles" (Bryson et al. 2001).

To obtain a permit for field testing, a plant breeder must provide detailed information, including scientific details relating to the development and identity of the regulated article, the purposes for introduction of the regulated article, and the procedures, processes, and safeguards that will be employed to prevent escape and dissemination of the regulated article.

The plant breeder may field-test a plant that meets the eligibility criteria by simply submitting a notification letter to APHIS and by meeting certain performance standards. The eligibility criteria require, among other things, that the genetic material be "stably integrated" in the plant genome, that the function of the genetic material is known and its expression does not result in plant disease, that it does not produce an infectious entity or will not be toxic to nontarget organisms, and that it has not been modified to contain certain genetic materials from animal or human pathogens. The performance standards also include controls on shipment, storage, planting, identification, and conduct and termination of the field trial.

8.3. Petitioning

If testing demonstrates that the organism is not a plant risk, an APHIS assessment will consider data and information showing that, with the excep-

---

[4] The first field release of a transgenic forest tree was in Belgium in 1988; however, today most forest tree trials are in the United States. The first U.S. field trial was in 1993 (Beardmore, forthcoming).

tion of the deliberately introduced gene, the genetically engineered line is the same as a nonengineered parental line with respect to a suite of traits. If there is sufficient familiarity with the introduced trait, the recipient plant, and its environment, APHIS can determine with a high degree of confidence that the engineered plant meets the criterion of being no more likely to become a plant pest than a traditionally bred plant. Once a determination of nonregulated status is made, the new plant variety may be developed further through traditional breeding. It may be produced, marketed, distributed, and grown without any other special oversight by APHIS. Nonregulated status permits unencumbered commercialization.

9. RISK ASSESSMENT

A major issue for regulators is whether the regulation should apply on the basis of the genetic modification process or on the basis of attributes of the plant that may pose risks. The formal U.S. decision criterion is that the product present "no significant or unreasonable adverse risks." Thus, some "reasonable risk" is allowed.[5]

Although some biologists have argued that regulation would better be applied to plants on the basis of the plant attributes, regulation tends to be applied simply on the basis of the genetic engineering process.

Environmental assessments require the following steps: (i) identifying hazards, (ii) assessing actual risks that may arise from a hazard, (iii) determining how risk can be managed and whether to proceed with the proposed action, and (iv) comparing the risks with those posed by actions involving comparable organisms.

In conducting risk assessments, APHIS begins with consideration of the existing knowledge base and the traditional procedures that are used in developing any new crop variety. This baseline enables APHIS to identify hazards and then determine whether the risk posed differs significantly from well-known risks. This process, referred to as "familiarity," is based on the assumption that the types of safety issues raised by genetically engineered plants are no different from those for traditional breeding when similar traits are being conferred. However, the magnitude of a particular risk may differ. The extensive experience gained from traditional plant breeding provides useful information in establishing parallel risk associations for newly developed crops.

For plants, familiarity takes account of knowledge and experience with the following:

  - the particular crop, including its flowering and reproductive charac-

---

[5] By contrast, the EU regulatory system has been characterized as allowing no risk (Pachico 2003).

teristics, ecological requirements, and past breeding experience;
- the agricultural and surrounding environment of the trial site;
- the specific traits transferred to the plant;
- the results of previous research;
- the scale-up of the plant crop varieties developed by more traditional techniques;
- the scale-up of other plant lines developed by the same technique;
- the presence of related and sexually compatible plants in the surrounding natural environment and knowledge of the potential for gene transfer between the plant and its relatives; and
- interactions among the crop plant, the environment, and the trait (OECD 1993).

Familiarity can range from very high to very low. The standard has been that for a genetically engineered crop to be commercialized in the United States, there must be a high degree of familiarity (Bryson et al. 2001).

Major modifications or hazards that have been identified for which risks are assessed include the following:

- plant pathogenic potential of the transgenic—for example, the ability to harm other plants;
- the potential to negatively affect handling, processing, and storage of commodities containing the genetically engineered plant;
- changes in cultivation that might accompany adoption of the transgenic;
- potential harm to nontarget organisms;
- changes in the potential of the genetically engineered plant to become a weed;
- the potential to affect weediness of sexually compatible plants; and
- potential impacts on biodiversity.

10. REGULATORY CONCERNS SPECIFIC TO TREES

Forest trees—largely undomesticated species with long lifespans—present special challenges. APHIS has noted that field trials of many species of trees can be safely performed over several years under the notification procedures, since trees do not become sexually mature for a considerable and well-established period. Moreover, tree species can be effectively isolated from wild populations by the appropriate choice of test location, by the use of physical methods for confinement of pollen, or through the application of various sterility approaches.

Nevertheless, APHIS has acknowledged that long-term vigilance is required. Because field tests involving trees may be of several years' duration and could involve unexpected exposures of nontarget organisms, continual adherence to performance standards must be maintained. Moreover, procedures used to ensure reproductive confinement during the first years of a field trial may not be adequate later in the trial. For that reason, APHIS requires that all field trials under notification for more than one year be renewed annually (Bryson et al. 2001).

10.1. Bioconfinement

To address potential problems, some transgenic plants will require bioconfinement during field testing. The National Academy of Sciences (2004) points out that many crops pose little hazard because the traits that make them useful to humans also reduce their ability to establish feral populations in either agricultural or nonagricultural habitats. However, if some transgenics should confer the ability to overcome factors that limit wild populations, significant invasive problems could ensue. Solutions include rendering plants sterile and confining pollen-mediated gene flow, but the efficacy of some approaches is untested.

10.2. Delayed or Prevented Flowering

Since environmental concerns about trees often involve issues related to the transfer of pollen into similar species in the natural environment, the delay (beyond the harvest) or prevention of flowering offers a means of addressing this concern. Sterility is particularly attractive for trees: the lack of fertile pollen or seeds prevents transgenic escape. Furthermore, since the stem is usually the important output of the forest tree, the loss of the seed is not significant.[6] However, the degree of sterility may vary by plant and by environmental condition and often may not be complete. Also, there are concerns about maintaining long-term stability of the sterility trait.

10.3. Conditional Release

Another outstanding issue is the question of conditional release. Under the current system, a plant is either regulated or nonregulated. Conditional release would enable the developers to answer questions related to the development and use of a transgenic tree prior to its unconditional release and thus allow movement toward unconditional release in a step-by-step process. Restrictive conditions could be altered as more information becomes available. Ac-

---

[6] Given the longevity of a tree and the tree improvements through time, the genetic improvements embodied in the seed at maturity are likely to be obsolete.

cording to the USDA announcement and press report of January 2004, revisions in the area of conditional release are under consideration (USDA 2004).

11. APHIS PERFORMANCE

The number of transgenic plants, including trees, being field-tested has increased dramatically in recent years. AHPIS now reviews about 1,000 applications for field-testing transgenics annually. In no instance has any biotech plant approved for field testing by USDA created an environmental hazard or exhibited any unpredictable or unusual behavior compared with similar crops modified using traditional breeding methods (U.S. House of Representatives 2000). To date most of the field tests have been agricultural crops. As of 2000, only 124 field tests of genetically altered trees had been authorized (McLean and Charest 2000), including transgenic spruce, pine, poplar, walnut, citrus, cherry, apple, pear, plum, papaya, and persimmon.

APHIS has overseen several thousand field trials and has received around 100 petitions for deregulations of genetically modified crops (National Research Council 2002). Only 61 transgenics, representing 13 species, have been deregulated over the past 20 years. Only one tree has achieved deregulated status.

Trees can be classified as orchard, ornamental, and timber. From 1987 to 2001, timber species were involved in only 1.2 percent of the total number of field tests, for both agricultural and forest crops, and 91 percent of those occurred in the latest reported period (1997–2001). A total of 90 timber tree field tests, representing four tree genera, were undertaken between 1987 and 2001; the poplar genus was the subject of well over one-half of the trials. About 57 percent of the trees tested are timber species.

The United States accounts for an estimated 61 percent of worldwide tree trials. Other countries undertaking field trials include Australia, Canada, Chile, France, Italy, Japan, New Zealand, and South Africa.

Despite the increase in field-testing in recent years, however, only one petition for deregulation of a tree has been submitted and granted—the fruit tree papaya. Orchards in Hawaii were experiencing severe disease problems, and a genetic modification was developed to impart disease resistance. This transgenic papaya was deregulated and is now in widespread use in Hawaii. No other trees of any type appear ready for imminent deregulation: APHIS has received no petitions for the deregulation of a transgenic timber tree. Worldwide, there are no documented transgenic timber trees that have been commercially released, although there are rumors that transgenic trees are being planted commercially in China.[7]

---

[7] At a November 2004 meeting of the Food and Agriculture Organization (FAO) Panel on Forest Genetic Resources, a principal research scientist of the Chinese Institute of Forestry

12. CONCLUSIONS

Genetically improved trees are playing a greater role in meeting the world's industrial wood requirements. Genetically engineered trees offer potential to provide higher-quality and lower-priced industrial wood as well as environmental benefits. However, there may also be environmental risks. A regulatory system exists in the United States for assessing the safety and environmental impacts of transgenics, including trees.

Currently, the regulatory system is being assessed for possible updating to allow for scientific advances in biotechnology and to address plants with unique features, such as tree longevity.

REFERENCES

Beardmore, T.L. Forthcoming. "Where Are We Today with Regard to Genetically Modified Trees?" In S. MacRae, ed., *Forestry Biotechnology Task Force State of Knowledge Report*. International Union of Forest Research Organizations, Vienna, Austria.

Botkin, D. 2001. "Comments on the Environmental Effects of Transgenic Trees." Paper presented at the Conference of Biotechnology in Forestry, Stevenson, WA, July 22–27.

Bradshaw, T. 1999. "A Blueprint for Forest Tree Domestication." College of Forest Resources, University of Washington, Seattle (available online at http://poplar2.cfr.wash[-]ington.edu/toby/treedom.htm).

Bryson, N.S., R.C. Davis Jr., P. Katz, R. Mannix, and W.M. Cohen. 2001. *Environmental Law Biotechnology Deskbook*. Environmental Law Institute, Washington, D.C.

Bryson, N.S., S.P. Quarles, and R. Mannix. 2004. "Have You Got a License for That Tree?" In S.H. Strauss and H.D. Bradshaw, eds., *Forest Biotechnologies: Technical Capabilities, Ecological Questions, and Social Issues in Genetic Engineering of Plantation Trees*. Resources for the Future, Washington, D.C.

Clapp, R.A.F. 1993. "The Forest and the End of the World: The Transition from Old-Growth to Plantation Forestry in Chile." Ph.D. dissertation, University of California, Berkeley.

Cordts, J. 2004. Project Director, U.S. Department of Agriculture's Animal and Plant Health Inspection Service (APHIS), Riverdale, MD. Personal communication (July 19).

DiFazio, S.P., S. Leonardi, S. Cheng, and S.H. Strauss. 1999. "Assessing Potential Risks of Transgenic Escape from Fiber Plantations." In P.W. Lutman, ed., *Gene Flow and Agriculture: Relevance for Transgenic Crops*. British Crop Protection Council Symposium Proceedings 72: 171–176 (Farham, UK).

El-Kassaby, Y.A. 2003. "Feasibility and Proposed Outline of a Global Review of Forest Biotechnology." Working paper prepared for the Food and Agriculture Organization (FAO) Panel of Experts on Forest Gene Resources. FAO Forest Resources Division, Forestry Department, Rome, Italy (available online at http://www.fao.org/forestry/foris/PFGR/en/html/FORGEN-2003-6-E.htm).

---

reported on the establishment of close to 300 ha. of transgenic poplar in China (El-Kassaby 2004).

_____. 2004. University of British Columbia, Vancouver. Personal communication (January 20).

Fladung, M. 2004. "GMO and Environmental Concerns." In S. MacRae, ed., *Forestry Biotechnology Task Force State of Knowledge Report*. International Union of Forest Research Organizations, Vienna, Austria.

Food and Agriculture Organization. 2001. *The State of the World's Forests*. Food and Agriculture Organization, Rome, Italy.

Kellison, R.C. 2004. "The Future of Forest Biotechnology." In S. MacRae, ed., *Forestry Biotechnology Task Force State of Knowledge Report*. International Union of Forest Research Organizations, Vienna, Austria.

McLean, M.A., and P.J. Charest. 2000. "The Regulation of Transgenic Trees in North America." *Silvae Genetica* 49(6): 233–239.

Meilan, R., D. Ellis, G. Pilate, A.M. Brunner, and J. Skinner. 2004. "Accomplishments and Challenges in Genetic Engineering of Forest Trees." In S.H. Strauss and H.D. Bradshaw, eds., *Forest Biotechnologies: Technical Capabilities, Ecological Questions, and Social Issues in Genetic Engineering of Plantation Trees*. Resources for the Future, Washington, D.C.

Mullin, T.J., and S. Bertrand. 1998. "Environmental Release of Transgenic Trees in Canada—Potential Benefits and Assessment of Biosafety." *The Forestry Chronicle* 74(2): 203–219.

National Academy of Sciences. 2004. "Biological Confinement of Genetically Engineered Organisms." Available online at http://www.nap.edu/books/0309090857/html/.

National Research Council. 2002. *Environmental Effects of Transgenic Plants: The Hope and Adequacy of Regulation*. Washington, D.C.: National Academy Press.

OECD (Organization for Economic Cooperation and Development). 1993. "Field Releases of Transgenic Plants, 1986–1992: An Analysis." OECD, Paris.

Pachico, D. 2003. "Regulation of Transgenic Crops: An International Comparison." Paper delivered at the Seventh International Consortium on Agriculture Biotechnology Research's International Conference on Public Goods and Public Policy for Agriculture Biotechnology, Ravello, Italy, June 29–July 3 (available online at http://www.economia.uni[-]roma2.it/conferenze/icabr2003/).

Sedjo, R.A. 1999. "The Potential of High-Yield Plantation Forestry for Meeting Timber Needs." *New Forests* 17: 339–359.

_____. 2004a. "Potential for Biotechnology Application in Plantations Forestry." In C. Walter and M. Carson, eds., *Plantation Forest Biotechnology for the 21st Century*. Trivandrum, India: Research Signpost.

_____. 2004b. "Transgenic Trees: Implementation and Outcome of the Plant Protection Act." Discussion Paper No. 04-10, Resources for the Future, Washington, D.C.

_____. 2004c. "Biotech and Planted Trees: Some Economic and Regulatory Issues." *AgBioForum* 6(3): 29–35.

Sohngen, B., R. Mendelsohn, and R. Sedjo. 1999. "Forest Management, Conservation, and Global Timber Markets." *American Journal of Agricultural Economics* 81(1): 1–13.

Tiang, Y., T. Li, K. Mang, Y. Han, L. Li, X. Wang, M. Lu, L. Dai, Y. Han, J. Yan, and D.W. Gabriel. 1994. "Insect Tolerance of Transgenic *Populus nigra* Plants Transformed with *Bacillus thuringienis* Toxin Gene." *Chinese Journal of Biotechnology* 9: 219–227.

U.S. Department of Agriculture. 2004. News release 0033.04. Available online at http://www. usda.gov/Newsroom/0033.04.html (accessed March 5, 2004).

U.S. House of Representatives. 2000. "Seeds of Opportunity: An Assessment of the Benefits, Safety and Oversight of Plant Genomics and Agriculture Biotechnology." Subcommittee on Basic Research, Committee of Science, Committee Print 106-B (April 13), Washington, D.C.

Westvaco Corporation. 1997. Conversation with researchers in spring 1997. Westvaco Corporation, Summerville, SC.

Williams, C.G. 2004. "Genetically Modified Pines at the Interface of Private and Public Lands: A Case Study Approach." Paper presented at the Forest Service, U.S. Department of Agriculture, Washington, D.C. (January 12).

Chapter 32

REGULATION OF BIOTECHNOLOGY FOR SPECIALTY CROPS

Kent J. Bradford,[*] Julian M. Alston,[*] and Nicholas Kalaitzandonakes[†]
University of California, Davis,[*] and University of Missouri-Columbia[†]

Abstract: While crops improved using biotechnology (recombinant DNA methods) have been widely adopted in soybeans, cotton, maize, and canola, only a few varieties of horticultural or specialty crops have been commercialized. Numerous traits developed through biotechnology would be valuable for specialty crops. However, commericalization of these traits is limited by the diversity of species involved, multiple niche markets, small production windows per cultivar, requirements of processors, distributors and retailers, and access to intellectual property required for developing transgenic varieties. Regulatory requirements for biotech crops, particularly the separate regulation of each transgenic event, are also uniquely burdensome for specialty crops. Targeted assistance with the regulatory process, analogous to the IR-4 program for the registration of agricultural chemicals for minor crops, is recommended as a way to encourage the commercialization of biotech specialty crops. Continuing development of biotech specialty crops in China, India, and other countries may eventually open international markets to these products.

Key words: biotechnology, specialty crops, vegetables, fruits, ornamental plants, regulation

1. BIOTECHNOLOGY IN HORTICULTURE

Biotechnology has transformed the production systems of major field crops, including soybeans, corn, cotton, and canola. Since their first large-scale introduction in 1996, the global area planted to crops developed using recombinant DNA techniques (i.e., biotech crops) grew to 200 million acres by 2004 (James 2004). Almost 60 percent of those acres were in the United States, where biotech varieties represented 85 percent of the soybeans, 76 percent of the cotton, and 45 percent of the corn acreage (National Agricultural Statistics Service 2004). This high rate of adoption reflects the farmer benefits associated with these crops, which have been almost exclusively targeted toward providing resistance to herbicides (72 percent), insects (19 percent), or both (9 percent) (James 2004). Such benefits come from increased yields, lower risks, reduced use of chemical pesticides, savings in management, labor, and capital equipment, and gains from reduced tillage and other modified production practices (Kalaitzandonakes 2003).

While biotechnology has been adopted rapidly in certain field crops, it has had limited commercial success in horticultural or specialty crops, including vegetables, fruits, nuts, nursery and floral crops, mushrooms, seeds, and other specialty agricultural commodities. While these are sometimes referred to as "small market" crops, collectively they represent over $50 billion in farm income each year, or almost half of the crop value produced in the United States (Table 1). In addition, they account for 45 percent of U.S. agricultural import value and 24 percent of U.S. agricultural export value (Jerardo 2004). Remarkably, specialty crops are produced on less than 3 percent of total agricultural crop acreage in the United States (Table 1). Thus, they represent very high value commodities, but are minor markets for most agricultural inputs, including seeds, agricultural chemicals, fertilizers, equipment, etc.

The first biotech crop to reach the market was the Flavr Savr™ tomato, engineered to remain more firm during shipping. Sweet corn, potato, squash, and papaya varieties engineered to resist insects and viruses, and carnations exhibiting novel colors, have been approved for commercial use and marketed. However, papaya is currently the only horticultural crop for which

Table 1. Farm Cash Receipts and Crop Area for Horticultural and Field Crops, United States, 2003

Crop	Farm Cash Receipts ($ billions)	Share (percent)	Crop Area (million acres)	Share (percent)
All crops	106.2	100.0	409.2	100.0
Horticultural crops	50.0	47.1	10.6	2.6
Vegetables	16.0	15.1		
Nursery and greenhouse	15.2	14.3		
Fruits and melons	11.5	10.8		
Tree nuts	2.4	2.3		
Other (mushrooms, seeds)	4.9	4.6		
Field crops	56.1	52.9	398.6	97.4
Feed crops (including corn)	24.3	22.9		
Oil crops (including soybeans)	17.3	16.3		
Food grains (wheat, rice)	8.0	7.5		
Cotton	5.0	4.7		
Tobacco	1.7	1.5		

Cash receipts compiled by A. Jerardo, Economic Research Service, U.S. Department of Agriculture (http://www.ers.usda.gov/briefing/farmincome/data/cr_t3.htm). Acreage data from National Agricultural Statistical Service (www.nass.usda.gov).

transgenic varieties have achieved a significant market share (about 70 percent of the Hawaiian crop shipped to the U.S. mainland is transgenic). Development of papaya cultivars resistant to the papaya ringspot virus has allowed the recovery of that industry in Hawaii after devastation from the disease in the early 1990s (Gonsalves 2004). Insect-resistant sweet corn and potatoes were initially adopted by growers and dramatically reduced the use of pesticides in those crops, yet market resistance has largely prevented their widespread production, and the transgenic potato cultivars are no longer commercially available. Only limited amounts of insect-resistant sweet corn and virus-resistant squash are currently marketed in the United States, and with the exception of papaya and a few flowers engineered for novel color (www.florigene.com), biotech fruit or ornamental crops are not produced commercially.

Virtual absence in the market does not mean that biotech traits would not be useful for horticultural crops. Producers of horticultural crops would benefit from the herbicide-tolerance and insect-resistance traits that have been successful in field crops, and would particularly benefit from development of genetic resistances to viral, fungal, and bacterial diseases (Clark, Klee, and Dandekar 2004; Gianessi 2004; Gianessi et al. 2002; Gianessi, Sankula, and Reigner 2003). The needs are arguably greater in the horticultural crops, since the options for chemical controls for weeds, insects, and diseases are more limited due to the restricted markets they represent. In addition, as many perennial crops are vegetatively propagated, highly heterozygous, and slow to reach sexual maturity, the ability to introduce a single advantageous trait into an elite cultivar without the necessity for extensive backcrossing would be extremely useful. In tree and vine crops, modifications could be made in rootstocks while leaving the scions that produce the marketable product essentially unchanged (Driver, Castillón, and Dandekar 2004). Genetic enhancements that increase production yields, improve ease of harvesting, extend storage life, and reduce costs of processing would also be of significant value. Longer-lasting flowers, slower-growing grass, and novel ornamentals now in development would be attractive to the floriculture and landscape industries.

Despite demonstration of diverse traits that would be desired by growers and consumers of horticultural crops, what once was a significant pipeline of research and development in horticulture, as evidenced by the number of U.S. field trials of biotech horticultural crops, has recently dwindled to only a handful of trials in fruits and vegetables, with turf grasses (e.g., creeping bentgrass) and tree species (e.g., poplar) accounting for the majority of continuing field trials of ornamentals (Figure 1). In 1999, 374 field test notifications or permits were filed for biotech horticultural crops. Of these filings, 116 were from 16 public institutions and 258 were from 16 private companies. In 2003, the total number of notifications or permits had fallen to 97, of

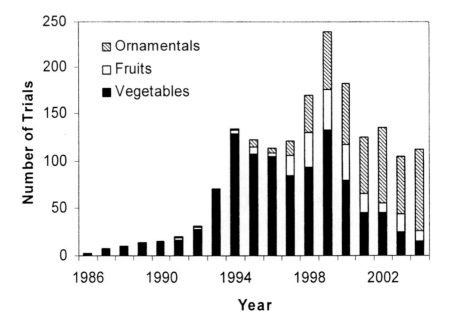

Figure 1. Field Trials of Biotech Vegetables, Fruits, and Ornamentals Conducted Under Permits in the United States, 1986 to 2004

Notes: Data for vegetables do not include potatoes or sweet corn. Data for ornamentals include tree species.

Source: U.S. Department of Agriculture's Information Systems for Biotechnology (http://www.nbiap.vt. edu/cfdocs/fieldtests1.cfm).

which 51 were from 17 public institutions and only 46 were from 8 private companies (Information Systems for Biotechnology 2004). This contrasts with continued research in biotech corn, cotton, and soybean, for which 506 permits or notifications were recorded in 1999 versus 520 for these three crops in 2003. No biotech horticultural variety has been deregulated since 1999, and only one since 1997 (National Biological Information Infrastructure 2004).

As a consequence of the disappointing past commercial results, current market outlook, and particularly regulatory costs and uncertainty, many horticultural seed and nursery companies have reduced their investments in research involving genetic engineering. After briefly surveying some of the unique challenges faced by biotech specialty crops, we will focus specifically on those associated with governmental regulations.

2. HURDLES FOR COMMERCIALIZATION OF BIOTECH SPECIALTY CROPS

An array of technical and market factors combine to create hurdles for the utilization of biotechnology in horticultural crops, many of which have been discussed in detail previously (Alston 2004; Alston, Bradford, and Kalaitzandonakes 2006; Bradford and Alston 2004a). Some of the most important of these are summarized below (Bradford and Alston 2004b).

Species and cultivar diversity. In contrast to the few species and moderate number of varieties of agronomic crops that are each grown on millions of acres each year, fruit, vegetable, and ornamental crops comprise hundreds of species and thousands of cultivars each produced on a relatively small acreage. Engineering a trait into a specific crop and cultivar may require considerable research and development, even for traits that have been previously utilized in other species. This diversity inherently delays development of biotech products for horticultural crops and increases costs relative to potential returns.

Multiple niche markets. Horticultural markets are highly segmented into a multitude of niches by location, season, consumer preferences, and other factors. Satisfying these diverse markets requires many cultivars within each species that may vary for resistance to pests and diseases, dates of maturity, seasonal adaptation, color, shape, taste, and other attributes. Thus, introducing a particular trait into a horticultural species likely requires its introduction into multiple cultivars to achieve market success. This generally necessitates extensive backcross programs using a single initial transformed line, which can delay advancement in overall breeding programs. This situation is already evident in soybean, cotton, and corn breeding, and is exacerbated by the diversity of cultivars required for multi-season and multi-location production in many horticultural crops.

Small production/market windows per cultivar. The diversity of production environments for horticultural crops also means that any single cultivar can command only a relatively small fraction of the total market for that crop. The potential crop area (and sales) of a given cultivar is therefore limited, and the returns on a biotech trait solely from seed or propagation materials may be insufficient to justify the investment in development. For example, seed sales of a single cultivar of lettuce, worth $150,000 to $250,000 during its typical 5-year market lifetime (Redenbaugh and McHughen 2004), cannot cover the additional costs associated with developing and marketing a biotech trait on a cultivar-by-cultivar basis. For perennial crops such as fruit or nut trees, the value of a novel biotech trait extends over many years after the initial sale of propagative materials, and mechanisms for value capture by the developer have not been established in the market.

Requirements of processors. Many vegetables and fruits are processed (frozen or canned), and some biotech traits, such as high viscosity in tomatoes or insect resistance in sweet corn, would be highly beneficial for processors. However, processors must balance these benefits against risking their overall market position and brand name, given the potential for protests, pickets, or boycotts from anti-biotech activists. In addition, many processed products are marketed internationally, potentially requiring segregation or labeling of processed products to comply with multiple regulatory jurisdictions and requirements.

Requirements of distributors and retailers. Distribution and retailing of horticultural products is increasingly global and concentrated. Only 30 firms are estimated to account for 10 percent of global grocery sales, and 30 percent of Walmart's $259 billion in global 2003 sales were in groceries (Cook 2004). Like processors, these large distribution firms operate internationally and must meet diverse local requirements for importation or labeling of biotech-derived products. Labeling, traceability, liability, and global markets are the major concerns for marketers handling biotech products. Without reasonable thresholds for adventitious presence of biotech DNA or protein, the risk is high with little benefit to the distributor.

Benefits throughout the marketing chain. While the biotech products to date have been targeted primarily to growers (i.e., input traits), products having clear benefits to the consumer (i.e., output traits) are needed to develop demand in the marketing chain. These may include increased nutritional content, better flavor, or longer storage life. As these products will likely require a premium price to compensate for the additional tracking and identity preservation that is needed, a very high quality product will have to be delivered to provide benefits throughout the marketing chain (Cook 2004).

Access to intellectual property and enabling technologies. The enabling technologies required to develop and market a biotech cultivar have been patented or licensed by a limited number of large corporations. In some cases, the costs of licensing patented technologies from diverse sources may be too great to be economical for a small-market crop. New licensing structures for intellectual properties developed in universities and public research institutions may be particularly helpful for small-revenue crops (as well as for developing country applications) (Atkinson et al. 2003, Graff et al. 2004).

3. REGULATION OF BIOTECH SPECIALTY CROPS

While all of the factors noted above contribute to the difficulties encountered in commercializing biotech specialty crops, the unique regulatory status of biotech crops in general is the underlying cause of most of the problems. Clearly, food and environmental safety must be established prior to commercialization of biotech cultivars. However, the same is true of crops developed using traditional breeding methods that include the use of techniques such as mutagenesis, wide crosses with other species, or protoplast fusion, which also have the potential for unexpected consequences. Current regulations consider cultivars developed using recombinant DNA as a distinct category from those developed using other genetic technologies, markedly increasing the costs of development, testing, marketing, and stewardship (Redenbaugh and McHughen 2004). A number of reports have recently examined the regulatory process for biotech crops (e.g., National Research Council 2004; Pew Initiative on Food and Biotechnology 2004, 2005), so we will focus only on those aspects that particularly affect specialty crops.

Event-specific regulation. The regulation of biotech crops has come to be based on specific transgenic "events," meaning a single insertion of a transgene into the host genome. Since the site of integration of the transgene cannot currently be controlled, different events generally represent different chromosomal locations of the inserted gene. A gene may insert into an existing gene, possibly disrupting or altering expression of that endogenous gene. Different insertion events also may express the transgene to different extents or in different patterns within the plant (Landsmann, van der Hoeven, and Dietz-Pfeilstter 1996). Thus, regulators have focused on specific events and require each event to be separately evaluated, even when the transgene and its product have been demonstrated to be safe. The concern is that unexpected consequences may occur associated with different insertion events, and therefore each event requires separate regulatory approval. However, all breeding methods, particularly mutagenesis, also have the potential for such random genetic effects, and yet only crops developed using recombinant DNA techniques are subjected to this additional regulatory scrutiny. Furthermore, while it is true that unexpected consequences *could* occur from different insertion events (National Research Council 2004), experience with over 2200 cultivars developed using mutagenesis as a method for generating variation has not resulted in unexpected adverse consequences for safety (van Harten 1998). Since mutagenesis creates random changes throughout the genome, there has been ample opportunity for adverse unexpected consequences to occur, yet normal phenotypic selection procedures that are routine in cultivar development have thus far prevented any such occurrence in marketed crops. Thus, extensive experience with diverse methods of genome modification, along with expanding knowledge of the variable and dynamic nature of plant genomes, indicates that when combined with routine phenotypic screening, as is done in all breeding and selection programs, regulating individual events of an approved trait does not contribute significantly to ensuring human or environmental safety (Bradford et al. 2005).

However, event-specific regulation does create major additional hurdles for biotech specialty crops. As described above, multiple varieties are required to match diverse growing seasons, locations, and markets. Thus, a desirable trait needs to be introduced into a range of cultivars having different environmental adaptations and product attributes for it to be successful in the market. As many as 60 different cultivars of crisphead lettuce alone may be grown in California throughout the year to match different growing location requirements, not to mention the dozens of romaine, leafy, and other specialty lettuce types now widely marketed. A new trait, no matter how desirable, would need to be available in a range of these cultivars to achieve market success. Event-specific regulatory requirements and the costs thereof lead to the necessity of deregulating a single transgenic event in one cultivar and then introducing it into other cultivars by backcrossing. While backcrossing can be efficient when DNA markers are employed, those marker systems are generally not available for specialty crops. As a result, it is a laborious and time-consuming process to move a single transgenic event into all of the varieties of a given crop that are required for successful market introduction. As a consequence, these varieties developed in backcrossing programs inevitably lag behind the improved varieties that use forward breeding approaches. A tragic example is the delay in release of Golden Rice, which was engineered to produce beta-carotene to help alleviate vitamin A deficiency. Instead of transforming the genes for beta-carotene production directly into a range of local varieties, the cost of meeting regulatory requirements made it necessary to select a single event in a single rice variety, get it through the deregulation process, then use backcrossing to transfer the trait to locally adapted varieties in the countries where it would be grown (Hoa et al. 2003). In addition to delaying release of new varieties, these policies actually reduce biological diversity by forcing backcrossing of a single event from one host genotype rather than encouraging multiple transformation events in diverse genetic backgrounds.

Perennial crops, such as vines, fruits, and nuts, face even greater challenges from event-specific regulation. These crops have long generation times and are generally genetically heterozygous, making it essentially impossible to use backcrossing to transfer valuable transgenic traits to additional cultivars. It will be necessary and desirable to transform each cultivar directly in order to retain all of the original horticultural and product quality traits. While U.S. regulations allow extension of nonregulated status to additional events within a species using the same gene(s) with somewhat reduced data requirements (www.aphis.usda.gov/brs/gaddbeg.html), each variety/trait combination requires a separate deregulation process under European Union and most foreign regulations, even for traits that have been demonstrated to be safe in that species. Similarly, multiple accessions of forest or landscape trees adapted to different ecological zones would each need to be engineered to provide accessions that are adapted to the diverse environments they will occupy for many years. The requirement for complete deregulation data packages for each new event-variety-provenance combination discourages biological diversity and creates financial and practical hurdles to commercialization.

Regardless of these adverse consequences of event-specific regulation for the variety development and commercialization process, the biggest casualty of this regulatory approach has been the marketing of biotech commodities. As national or international jurisdictions independently approve or disapprove individual events, one variety may be legal and a second one a "contaminant" simply by virtue of where the same transgene has been incorporated into the genome. For example, Syngenta (2005) announced in March 2005 that some of its corn varieties thought to contain a particular event conferring insect resistance (Bt11) actually contained a closely related event of the same gene (Bt10). This admission will likely result in fines from regulatory agencies and market impacts, yet the difference between the two events is genetically trivial and without consequence for health or safety. This regulatory focus on specific events has led to a burgeoning infrastructure of product channeling, identity preservation, commodity testing, and auditing for commodity crops. However, relatively little market premium has been generated for non-biotech commodities (other than in conjunction with organic certification) as there is no relationship to true risk, hazard, or product quality. For specialty crops, which are often traded internationally, the maze of regulatory jurisdictions that could require segregation and labeling of individual varieties of each commodity creates a major disincentive to market biotech crops, regardless of how desirable they may be to growers or consumers. As this event-specific approach to regulation becomes entrenched into international agreements such as the Cartegena Protocol, marketing of biotech products will continue to be saddled with market barriers that have no foundation in science or safety. A Public Research and Regulation Foundation has recently been formed to provide input from public sector scientists into the international biotechnology regulatory process (see www.pubresreg.org). Elimination of event-specific regulation has been advocated in general previously (Bradford et al. 2005), but is particularly critical for specialty crops.

Specialty crops regulatory initiative. As illustrated in Table 1, specialty crops produce almost as much product value as commodity field crops on less than 3 percent of the area devoted to field crops. As such, the markets for agrichemicals and other farm inputs are in aggregate much smaller for specialty crops. It is seldom economical for developers to produce new agrichemical products specifically for the specialty market. This difficulty in attracting commercial research and development of chemical pest controls for small-market crops has long been recognized and led to the establishment of the U.S. Department of Agriculture's IR-4 program, which supports research required to gain registration or special limited-use permits for agrichemicals for those crops (Holm and Kunkel 2004; http://ir4.rutgers.edu). A similar program is in place for the development of "specialty drugs" (the Food and Drug Administration's Orphan Drug Program: www.fda.gov/orphan/). These organizations have been instrumental in extending products developed for large markets or at public expense to smaller markets for the public good.

Recently, calls have come for a similar program to be established to assist both public and private developers of biotech specialty crops in gaining regulatory approval for those crops (Goldner, Thro, and Radin 2005, Pew Initiative on Food and Biotechnology 2005). Navigating the regulatory process and obtaining approval from as many as three federal agencies is a significant challenge in commercializing a new biotechnology-derived specialty crop. While multinational companies employ specialists to conduct these tasks for large acreage commodity crops, this is often out of the question for public institutions or small companies developing specialty crops. Some have challenged whether a separate regulatory scheme for biotech crops that requires such an investment of resources is even necessary or beneficial (Miller and Conko 2004), but given the current regulatory structure, a program is needed that could assist developers of biotech specialty crops to conduct the tests required to obtain regulatory approvals. While still in the formative stages, such an organization is envisioned to be a partner with public sector (universities and government research institutes) and private sector developers of specialty crops. A primary function of the program would be to assist in generating required regulatory safety data and in obtaining regulatory clearances for a biotech product. For example, if animal feeding or product composition studies are required, these could be performed by standardized laboratories specializing in these areas supported or sanctioned by the program. Providing supplementary funding for collecting such regulatory data, as is done through the IR-4 program, must be a high priority. The envisioned program would evaluate potential products on the basis of need (as measured by demand from the specialty crops community, including growers, farming organizations, the U.S. Department of Agriculture, and university Extension specialists) and potential public benefit (broad societal value, particularly to the environment, agriculture, consumers, or health). Once the program had prioritized a potential biotech product for support, it would work cooperatively with the various entities to identify the data needed for regulatory compliance, to generate those data, to submit safety and environmental data to regulatory agencies, and to work with those agencies to obtain necessary approvals. Such a program would lower the barriers to development of new biotechnology-derived specialty crops of interest to consumers and growers, and would provide returns to society on the public investment in biotechnology and genomics research.

4. CONCLUDING COMMENTS

While the adoption of biotech field crops is spreading around the globe, biotech horticultural products are struggling to emerge into the marketplace. There is no shortage of targets and applications, many of which have already been demonstrated or even marketed on a limited basis. However, it will be difficult for additional biotech traits that provide primarily producer benefits (input traits) to break into the horticultural market. Horticultural biotech products will need to appeal to consumers to create demand and provide economic benefits to the processor and distributor segments of the supply chain as well to the producers. Nutritionally improved horticultural products that appeal to consumers could meet this need; however, testing requirements to obtain regulatory approval for nutritionally enhanced products could also be higher than for current products that do not substantially alter composition, exacerbating the regulatory hurdle for specialty crops. Nonetheless, such consumer-oriented products may be necessary to open horticultural markets to biotech before benefits from input traits can be realized.

Horticultural crops represent gross market value roughly equivalent to that of the field crops, but are produced on less than 3 percent of the acreage (Table 1). Thus, the potential for economic returns through sales of seeds or propagation materials is more limited for horticultural crops. Development costs are further increased and potential returns per variety are decreased by the need for multiple varieties adapted to diverse locations in order to provide season-long production. When these determinants of market potential are combined with high transaction costs for accessing intellectual property rights and for obtaining regulatory approvals, the economic environment is not favorable for private investment in new horticultural biotech products. Public-sector pooling of enabling technologies (Atkinson et al. 2003, Graff et al. 2004) and governmental support for programs to assist in meeting regulatory testing requirements, analogous to those supporting the registration of agrichemicals for minor crops, could lower these hurdles. The lack of harmony in international regulations for sale and labeling of biotech food products will continue to discourage introduction of biotech specialty crops.

While this chapter has adopted a U.S. perspective, there are prospects for significant developments in horticultural biotechnology to occur outside the United States. Most of the crops considered important in developing and tropical countries would be classified as specialty crops in the United States, and a number of countries are actively pursuing biotech enhancements for those crops. Cohen (2005) identified 201 genetic transformation events in 45 crops under investigation by public researchers in 15 developing countries. In fact, China may already have taken the lead from the United States in horticultural biotechnology (Huang and Rozelle 2004). China has been rapidly expanding its horticultural industry and has a large and successful agricultural research system with rising strength in plant biotechnology. China's large domestic market and rapidly expanding horticultural sector, combined with receptive domestic attitudes to the development and adoption of biotech products, provide a comparatively favorable setting for horticultural biotech products. India and Brazil may be on the threshold of a similar expansion of biotech activity in horticulture, and Cuba has developed a number of products that are currently stymied by export market resistance (Runge and Ryan 2004). Scientific and commercial progress in China and other countries could change the global environment for horticultural biotechnology by bringing publicly developed products to market and pushing for their acceptance in international markets.

Some additional positive developments could improve the economic environment for horticultural biotechnology. Public institutions and foundations are collaborating through the Public Intellectual Property Resource for Agriculture (PIPRA), the Center for the Application of Molecular Biology in International Agriculture (CAMBIA), the African Agricultural Technology Foundation (AATT), and other organizations to lower the intellectual property and regulatory hurdles for international agriculture and specialty crops (Delmer et al. 2003). The expanding adoption of biotech field crops is stimulating movement toward more uniform regulatory and biosafety protocols around the world, albeit slowly, and the European Union is gradually beginning to relax its moratorium on approvals of biotech crops. Nutritionally enhanced "foods for health," such as soybean and canola crops with more healthful oils (Ursin 2003), are being developed in field crops, and, if accepted by consumers, these products could open the door for acceptance of similar products in horticultural commodities. A few ornamental biotech products are in the market, and additional ones may face lower hurdles for acceptance since they are not consumed as food. A potentially significant event is the purchase of Seminis Vegetable Seeds by Monsanto in March 2005. Seminis represents approximately 40 percent of the global vegetable seed market and was active in researching and developing biotech traits before reducing its investment in transgenic breeding methods in recent years. It will be of interest to see whether Monsanto resumes research on and development of biotech products for the vegetable market.

Thus, although the timeline for a significant economic impact of biotechnology on horticulture will likely be pushed back from earlier predictions, continued research, coupled with sensible, scientifically based regulatory policies and targeted assistance programs, will create products desired by producers and consumers and an economic environment in which they can be supplied.

REFERENCES

Alston, J.M. 2004. "Horticultural Biotechnology Faces Significant Economic and Market Barriers." *California Agriculture* 58(2): 80–88.

Alston, J.M, K.J. Bradford, and N. Kalaitzandonakes. 2006. "The Economics of Horticultural Biotechnology." *Journal of Crop Improvement* 16 (in press).

Atkinson, R.C., R.N. Beachy, G. Conway, F.A. Cordova, M.A. Fox, K.A. Holbrook, D.F. Klessig, R.L. McCormick, P.M. McPherson, H.R. Rawlings III, R. Rapson, L.N. Vanderhoef, J.D. Wiley, and C.E. Young. 2003. "Public Sector Collaboration for Agricultural IP Management." *Science* 301(5630): 174–175.

Bradford, K.J., and J.M. Alston. 2004a. "Diversity of Horticultural Biotech Crops Contributes to Market Hurdles." *California Agriculture* 58(2): 84–85.

_____. 2004b. "Horticultural Biotechnology: Challenges for Commercial Development." *Chronica Horticulturae* 44(4): 4–8.

Bradford, K.J., A. Van Deynze, N. Gutterson, W. Parrott, and S.H. Strauss. 2005. "Regulating Transgenic Crops Sensibly: Lessons from Plant Breeding, Biotechnology and Genomics." *Nature Biotechnology* 23(4): 439–444.

Clark, D., H. Klee, and A. Dandekar. 2004. "Despite Benefits, Commercialization of Transgenic Horticultural Crops Lags." *California Agriculture* 58(2): 89–98.

Cohen, J.I. 2005. "Poorer Nations Turn to Publicly Developed GM Crops." *Nature Biotechnology* 23(1): 27–33.

Cook, R.L. 2004. "Transgenic Produce Slow to Enter Evolving Global Marketplace." *California Agriculture* 58(2): 82–83.

Delmer, D.P., C. Nottenburg, G.D. Graff, and A.B. Bennett. 2003. "Intellectual Property Resources for International Development in Agriculture." *Plant Physiology* 133(4): 1666–1670.

Driver, J., J. Castillón, and A. Dandekar. 2004. "Transgenic Trap Crops and Rootstocks Show Potential." *California Agriculture* 58(2): 96–97.

Gianessi, L. 2004. "Biotechnology Expands Pest-Management Options for Horticulture." *California Agriculture* 58(2): 94–95.

Gianessi, L., S. Sankula, and N. Reigner. 2003. *Plant Biotechnology: Potential Impact for Improving Pest Management in European Agriculture. A Summary of Nine Case Studies.* Washington, D.C.: National Center for Food and Agricultural Policy. Available online at www.ncfap.org (accessed May 18, 2005).

Gianessi L.P., C.S. Silvers, S. Sankula S, and J.E. Carpenter. 2002. *Plant Biotechnology: Current and Potential Impact for Improving Pest Management in U.S. Agriculture—An Analysis of 40 Case Studies.* Washington, D.C.: National Center for Food and Agricultural Policy. Available online at www.ncfap.org (accessed May 18, 2005).

Goldner, W., A.M. Thro, and J.W. Radin (eds.). 2005. "Public Research and the Regulatory Review of Small-market (Specialty) Biotechnology-Derived Crops Workshop." Animal and Plant Health Inspection Service (APHIS), U.S. Department of Agriculture, Riverdale, Maryland. Available online at www.csrees.usda.gov/nea/biotech/pdf/small_mkt.pdf (accessed May 18, 2005).

Gonsalves, D. 2004. "Virus-Resistant Transgenic Papaya Helps Save Hawaiian Industry." *California Agriculture* 58(2): 92–93.

Graff, G., B. Wright, A. Bennett, and D. Zilberman. 2004. "Access to Intellectual Property Is a Major Obstacle to Developing Transgenic Horticultural Crops." *California Agriculture* 58(2): 120–126.

Hoa, T.T.C., S. Al-Babili, I. Potrykus, and P. Beyer. 2003. "Golden Indica and Japonica Rice Lines Amenable to Deregulation." *Plant Physiology* 133(1): 161–169.

Holm, R.E., and D. Kunkel. 2004. "IR-4 Project Targets Specialty Crops." *California Agriculture* 58(2): 110–112.

Huang, J., and S. Rozelle. 2004. "China Aggressively Pursuing Horticulture and Plant Biotechnology." *California Agriculture* 58(2): 112–113.

Information Systems for Biotechnology. 2004. "Field Test Releases in the U.S." Virginia Tech, Blacksburg, VA. Available online at www.isb.vt.edu/cfdocs/fieldtests1.cfm (accessed May 18, 2005).

James, C. 2004. "Preview: Global Status of Commercialized Biotech/GM Crops: 2004." ISAAA Briefs No. 32, International Service for the Acquisition of Agri-biotech Applications, Cornell University, Ithaca, NY. Available online at www.isaaa.org (accessed May 18, 2005).

Jerardo, A. 2004. Economist, Economic Research Service, U.S. Department of Agriculture. Personal communication.

Kalaitzandonakes, N. (ed.). 2003. *Economic and Environmental Impacts of Agbiotech: A Global Perspective.* New York: Kluwer-Plenum Academic Publishers.

Landsmann, J., C. van der Hoeven, and A. Dietz-Pfeilstter. 1996. "Variability of Organ-Specific Expression of Reporter Genes in Transgenic Plants." In E.R. Schmidt and T. Hankeln, eds., *Transgenic Organisms and Biosafety.* Berlin: Springer-Verlag.

Miller, H.I., and G. Conko. 2004. *The Frankenfood Myth: How Protest and Politics Threaten the Biotech Revolution.* Westport, CT: Praeger Publishers.

National Agricultural Statistics Service. 2004. "Statistical Information." U.S. Department of Agriculture, NASS, Washington, D.C. Available online at www.usda.gov/nass/ (accessed May 18, 2005).

National Biological Information Infrastructure. 2004. "U.S. Database of Completed Regulatory Agency Reviews." United States Regulatory Agencies Unified Biotechnology Website, http://usbiotechreg.nbii.gov (accessed May 18, 2005).

National Research Council. 2004. *Safety of Genetically Engineered Foods: Approaches to Assessing Unintended Health Effects.* Washington, D.C.: National Academy Press. Available online at http://books.nap.edu/catalog/10977.html (accessed May 18, 2005).

Pew Initiative on Food and Biotechnology. 2004. *Issues in the Regulation of Genetically Engineered Plants and Animals.* Washington, D.C.: The Pew Initiative on Food and Biotechnology. Available online at www.pewagbiotech.org/research/regulation (accessed May 18, 2005).

_____. 2005. *Impacts of Biotech Regulation on Small Business and University Research: Possible Barriers and Potential Solutions.* Washington, D.C.: The Pew Initiative on Food and Biotechnology. Available online at http://pewagbiotech.org/events/0602/proceedings.pdf (accessed May 18, 2005).

Redenbaugh, K., and A. McHughen. 2004. "Regulatory Challenges Reduce Opportunities for Horticultural Biotechnology." *California Agriculture* 58(2): 106–115.

Runge, C.F., and B. Ryan. 2004. "The Global Diffusion of Plant Biotechnology: International Adoption and Research in 2004." University of Minnesota, St. Paul, MN. Available online at www.apec.umn.edu/faculty/frunge/globalbiotech04.pdf (accessed May 18, 2005).

Syngenta. 2005. "Following Syngenta-initiated Investigation of Unintended Corn Release, EPA and USDA Conclude Existing Food Safety Clearance Applies, No Human Health or Environmental Concerns." Syngenta International AG, Basel, Switzerland. Available online at www.syngenta.com/en/downloads/Final_Syngenta_March_21_Statement_e.pdf (accessed May 18, 2005).

Ursin, V.M. 2003. "Modification of Plant Lipids for Human Health: Development of Functional Land-Based Omega-3 Fatty Acids." *Journal of Nutrition* 133(12): 4271–4274.

van Harten, A.M. 1998. *Mutation Breeding. Theory and Practical Applications.* Cambridge: Cambridge University Press.

Conclusions

WHAT HAVE WE LEARNED, AND WHERE DO WE GO FROM HERE?

Chapter 33

WHAT HAVE WE LEARNED, AND WHERE DO WE GO FROM HERE?

Julian M. Alston,[*] Richard E. Just,[†] and David Zilberman[‡]
University of California, Davis,[*] University of Maryland,[†] and University of California, Berkeley[‡]

Abstract: This chapter revisits the themes and issues explored in the book, drawing upon the content of its chapters. The main points are summarized, key results are synthesized, conclusions are drawn, and implications for policy and for further work by economists are suggested.

Key words: agricultural biotechnologies, causes and consequences of technological regulation, overview and synthesis, conclusion

1. INTRODUCTION

Agricultural biotechnology is engulfed in controversy. The discovery of the molecular structure of DNA in the 1950s generated new opportunities in the life sciences. This and subsequent innovations enabled the application of genetic engineering to produce agricultural crop varieties that were first introduced commercially in 1995. During the ten years since then, more than 1 billion acres have been planted with biotech crop varieties, mostly in North and South America, as well as in China and India (see James 2004 for details). Yet, the technology is virtually banned in Europe, and its adoption in Africa and much of the developing world has been minimal. While biotechnology is now prevalent in the production of corn, soybeans, and cotton, it has not been significantly adopted in the production of either of the major food grains, wheat and rice, nor in fruits and vegetables.

Various concerns have been expressed about biotech crops, often based on perceptions for which there may be little or no scientific foundation, including claims that agricultural biotechnology (i) is mostly beneficial to producers rather than consumers, (ii) addresses the problems of the developed world rather than the developing world, (iii) is controlled by a small number of corporations that make it inaccessible for the development of technologies helping the poor, (iv) involves risks to human health and food safety that outweigh its potential benefits, (v) is a source of environmental risk that must be regulated rigorously, and (vi) is worse in these various regards than the conventional technologies that it is designed to replace. Such claims can be addressed and informed by empirical economic analysis.

Much research has been done addressing various aspects of these issues. One body of literature investigates the performance and profitability of biotechnology relative to traditional technologies. Significant research has addressed issues of intellectual property rights and access to biotechnologies. Also, much ongoing research is investigating consumer and market acceptance of biotech crop products. However, the *regulation* of agricultural biotechnologies and its implications have not been the subject of much research. This book begins with a rallying call by Berwald, Matten, and Widawksy (2006) for economic research that will support the regulation of biotechnology by aligning testable hypotheses with issues that arise in regulatory decision making, testing those hypotheses with models that are empirically tractable and robust based on feasible and verifiable data, and communicating those results effectively to non-economists. This volume documents and extends what is known about the economics of regulation of agricultural biotechnology and lays a foundation to support the regulation of agricultural biotechnology. It also provides an overview of some related issues that are essential to assessing the future of biotechnology.

One major theme that emerges from the chapters in this volume is that regulation of biotechnology is (i) valuable, but (ii) expensive, and (iii) sometimes exploitive and predatory. Regulation is valuable to the extent that it prevents the introduction of technologies that involve net social costs. It is also valuable because the introduction of some socially beneficial new biotechnologies might not be socially acceptable in the absence of the government safeguards and endorsements that are provided implicitly through the regulatory process. Even if the technology is completely safe and is an improvement over existing technologies, it has certain dimensions that alarm some members of the public. Regulatory activity allows these concerns to be overcome. But regulation is expensive, both in terms of its direct use of resources and in terms of its market implications. And as demonstrated within this volume, it can create opportunities for predatory behavior.

A major challenge is to develop an optimal regulatory framework that maximizes social welfare and responds to concerns and fears. A key element of this regulatory framework is reasonable quantification of benefits and costs of agricultural biotechnologies. Prior to quantification, the categories of costs and benefits presented by agricultural biotechnologies must be identified and methods to evaluate each of them must be determined. Several chapters in this volume present such frameworks for evaluating standard costs and benefits. The chapter by Felicia Wu (2006), for example, suggests that, in addition to the standard impact on yields and costs, some biotech crop varieties may actually reduce the danger of other health threats—in the case of her chapter, from mycotoxins.

A second important aspect of research is to identify the costs of regulation. Several authors have provided arguments as to why regulation (and its attendant costs) may be excessive in some cases and inadequate in others. Thus, the study of costs is a high priority. As argued by Kalaitzandonakes, Alston, and Bradford (2006), who present preliminary estimates of the costs of compliance in the United States, understanding the cost of regulation and the obstacles to adoption will provide a starting point for analysis aimed at streamlining the regulatory process.

The rather complex and diffuse nature of biotechnology regulation in the United States is outlined in Chapter 1 (Alston, Just, and Zilberman 2006). Regulations for agricultural biotechnologies differ greatly depending on exactly which statute is applicable, and which government agency administers the regulations. For example, the introduction of new crops, such as herbicide-tolerant or pest-resistant crops, is regulated by the U.S. Department of Agriculture's Animal and Plant Health Inspection Service (APHIS), while the U.S. Environmental Protection Agency (EPA) regulates pesticidal substances, including those used on herbicide-tolerant crops, only after experimentation. In contrast to APHIS, which imposes regulations only during introduction and experimentation, the EPA requires registration before commercial use and occasional further testing and re-registration throughout the product life cycle, and imposes specific limits on how pesticides may be used. Thus, both at the stage of product development and testing and after later commercialization, new biotech crops face different standards than chemical pesticides do, even though the new pest-resistant crops are much less environmentally toxic than the chemical pesticides they replace. By dividing regulations among multiple agencies, the importance of comparing social benefits and costs between such alternatives may be underemphasized. Nevertheless, given current entrenchment of the established political economy, major changes in this overall regulatory framework are unlikely (Perrin 2006).

The regulatory framework of biotechnology in other countries is similarly complex and is divided among multitudes of agencies. Biotechnology is based on new science, and its regulation is subject to many concerns and uncertainties. The regulatory framework is evolving in response to developments in the field and to new knowledge. Economics can and should play a role in the design of the regulatory process and in actual decision making. Understanding adoption behavior, profit and yield effects of biotechnologies, and the economic benefits and costs of biotechnologies in the context of financial, human, and environmental risk can be of great value to the ongoing regulatory effort. Such studies and research on the impact of the regulatory process and methodologies for regulatory decision making are needed to assess the potential for modification of the regulatory process.

2. DIMENSIONS OF REGULATION

The regulatory process must address a wide variety of issues, including biosafety, resistance, liability, and labeling, as well as decisions about the characteristics of the regulations, including whether to regulate ex ante or ex post, the specifics of registration requirements, and the "scope" of regulatory approval (whether to regulate traits, events, or varieties).

2.1. Biosafety

What is the impact of biotech crops on the environment and human health? Concerns have been expressed about the potential for unintended damage to valued species (e.g., the Monarch butterfly) and about genetic drift, which could result in super weeds and other undesirable organisms. The regulatory process for pesticides provides insights for addressing some biosafety issues associated with biotechnology, but not others.

A major need in assessing biosafety is consideration of the environmental opportunity costs of not using biotechnology. Biotechnology may cause some environmental damage, but the activities that it replaces may cause more or less damage. For example, the use of pest-resistant crops (such as Bt corn) reduces pesticide use, which provides an important human and environmental benefit, and higher-yielding biotech crop varieties may result in reduced dependence on land in agriculture, leading to less deforestation. Therefore, assessment is needed not only of the direct effects of biotech crops on human and environment health, but also of the overall net effects compared with the relevant alternative (which might involve heavy use of chemical pesticides), considering induced adjustment of other activities.

2.2. Resistance

Complete adoption of biotech crops that are pest-resistant or herbicide-tolerant may lead to the development of resistant insect pests or weeds. To avoid resistance build-up, under U.S. regulations for most biotech crops farmers are required to establish refuges. For example, a farmer growing a Bt corn crop is required to plant 20 percent of his acreage devoted to that crop to a non-Bt crop variety. The reason is that these refuges provide a breeding ground for Bt-vulnerable pests that will mate with resistant pests, and thereby, slow the development of pest resistance. While refuge regulations implemented thus far have been inflexible, Frisvold (2006) finds that more flexible alternatives can appreciably increase welfare. Secchi et al. (2006) also find that introducing dynamic refuge requirements may yield modest gains. However, evidence to date suggests that U.S. farmers' non-compliance with these requirements is widespread (Langrock and Hurley 2006, Mitchell and Hurley 2006). To date, the acreage of conventional varieties has been sufficient to serve as a refuge in most countries and local areas, so resistance has been a minor problem. Thus, little is know about how much a given pest's resistance may increase if refuge requirements continue to be ignored as rates of adoption of biotech crops increase. Furthermore, the Chinese example discussed by Wang, Just, and Pinstrup-Andersen (2006) points out that, with a high rate of adoption in the absence of refuges, a secondary pest previously controlled by the primary pest can became a more expensive problem than the primary pest had been.

2.3. Liability

One concern about agricultural biotechnology is the potential that biotech crop products can become commingled with non-biotech products in the food chain as a result of pollen drift across farm boundaries if not by deliberate or unintentional mixing of products beyond the farm gate. These risks of adventitious presence of genetically modified (GM) crop products in otherwise non-GM products give rise to a demand for the development of standards of product quality and systems of liability. The resulting institutions differ among and within countries and are evolving over time. When many farmers in a given neighborhood grow biotech crops, assigning individual liability for the adventitious presence of GM crop products in other farmers' fields or in grain supplies may be difficult, much as in the case of non-point source pollution. In these cases, mechanisms such as collective punishment as well as due-care standards can be used to improve efficiency (Smyth, Phillips, and Kerr 2006).

The potential for coexistence of biotech crops with conventional or organic crops turns significantly on the assignment of these rights and obligations. Countries differ in their approach to this issue. In the European Union (EU), for instance, an organic grower may lose his organic certification as a result of the adventitious presence of biotech crop material, and neighboring growers of biotech crop varieties may be held liable for the costs, whereas in the United States adventitious presence does not have such dire consequences.

2.4. Labeling

Several dimensions of labeling policy have emerged.

- When should the government require labeling? Should labeling be required to warn against hazards reflecting the "best scientific knowledge"? Or should labeling be motivated by perceptions of consumer concerns about substances such as GM products? Scientific knowledge enters the political debate but may or may not receive much emphasis compared with the influence on political action by groups such as Greenpeace and distorted reporting in the media, which itself can focus more on public opinion than on scientific knowledge.

- Should labels be limited only to seeds and feeds, or should labels also be required for derivative products like oil or meat? The answer may depend on whether the guide to labeling requirements is based on perceptions of consumer preferences or on the best scientific knowledge.

- Should labeling be mandatory or voluntary? An alternative to regulation is a commercial commodity grading system including a labeling policy supported by a well-developed set of contracts that are enforceable by the courts. In most cases, such a system can be used to inform people about the presence of GM crop products. But such a system carries substantial costs that are ultimately borne by consumers. If consumers are adequately informed about the costs of labeling, their preference for such a system may be altered (Huffman and Rousu 2006, McCluskey, Grimsrud, and Wahl 2006). With a voluntary system, consumers willing to pay for products with specific labeling can select them, while other consumers need not bear the cost of a labeling system. Nevertheless, the cost of a regulatory agency that ensures truthfulness in labeling will likely be borne by taxpayers.

- Labeling poses risks for sellers and buyers. For a buyer, the risk is accepting a shipment of a product that is labeled as GM-free but that exceeds tolerances (the maximum amount of biotech material that it can contain and still qualify as GM-free under the prevailing definitions) and should be rejected. For a seller, it is having a shipment of a product rejected that should be accepted because it is within tolerance. These issues are complicated by adventitious commingling that can occur at locations within the production and marketing system. More generally, there is the issue of whether "GM-free" is a meaningful grade.

- Segregation costs imposed on producers of non-GM crops are socially costly and have the potential to convert what would otherwise be a social welfare-improving technology into a social welfare-decreasing one. Hence, the size of segregation costs is important to the GM technology debate (Moschini and Lapan 2006).

2.5. Ex Ante Versus Ex Post Regulation

To what extent does the regulatory process aim to provide a risk-free, final outcome, and to what extent is some learning and adjustment allowed? The experience with pesticides suggests that while regulation aims to minimize negative side effects, the screening is still not perfect, and occasionally chemicals have to be recalled or banned. Regulation by APHIS under the Plant Protection Act (PPA) focuses heavily upon prevention at the stage of pre-market testing but largely ignores post-commercialization monitoring. This contrasts sharply with substances regulated by the EPA under the Federal Insecticide, Fungicide, and Rodenticide Act, where strict standards limit how a pesticide can be used and the crops to which it can be applied. In addition, the EPA requires periodical data call-ins from registrants of particular pesticides as standards are tightened, including occasional re-evaluations triggered by Congressional amendments. Once a product is in the market and demonstrates a negative effect, the regulation of its continued use takes into account economic benefits even though risk considerations are paramount. The regulatory process thus has an element of continuous learning and reassessment. Given the high degree of uncertainty regarding biotech crops, these features should be included in biotechnology regulation as well, but because most such technologies are introduced in the form of crops under the PPA, such regulation is lacking.

2.6. Specification of Registration Requirements

Testing before registration is costly, and specifications present major decisions for the regulator. Any new variety may have a large range of economic and environmental impacts. The depth (degree of detail of investigation of each aspect) and breadth (the range of issues investigated) determine the costs of regulation and its reliability. With substances regulated under FIFRA, for example, the battery of tests that must be conducted to ensure environmental safety (toxicity, environmental fate, teratology, oncogenicity) as well as efficacy can cost tens of millions of dollars.

2.7. Should Traits, Events, or Varieties Be Registered?

The cost of regulation depends on what is being regulated and registered. A technology represented by a trait may be inserted in several varieties of crops (which constitutes several events) and then inserted into other varieties through back crossing. Obviously, regulation of traits may be much cheaper, while regulation of every variety may be redundant and expensive, especially when the insertion occurs through back crossing.

2.8. Registration Requirements and National Boundaries

Each national government establishes its own registration requirements. Many tests may be satisfied by the studies required for registration in other countries but many typically are not. This causes repetition of tests when the same or different companies register the same or similar products in different countries, or redundancy of effort when different tests are required by different countries to satisfy similar scientific concerns. International uniformity of registration requirements could substantially reduce the waste from requiring similar tests by the same company. However, firms will not voluntarily share tests with other companies that wish to register a product, because that leads to increased competition and lower product prices. Mechanisms are needed that allow not only for sharing of results among national regulatory agencies but also for registration by competitors without duplication of tests or excessive (anticompetitive) compensation.

2.9. Retroactive Assessment of Cost Effectiveness of Regulations

Although GM-contract terms have undoubtedly been violated and contracts rejected, not even one human health liability claim has been awarded against GM products. This raises an issue of whether the government's regulatory program, even in the United States, has been too stringent. The performance of regulatory regimes should be reassessed from time to time based on experience, and the results should be used for modification and readjustment.

3. REGULATION IN A RISKY AND UNCERTAIN ENVIRONMENT

Some of the methodological issues that have to be addressed in designing the regulatory process include quantification of risk and the processes that generate it and the definition of optimal regulatory criteria that give appropriate consideration to the different kinds of risks that are involved in technological regulation.

3.1. Quantification of the Risk Generation Process

The environmental side effects of transgenic crops are outcomes of several processes with significant time and spatial dimensions that are subject to a high degree of variability. Quantification of these processes and estimation of their key parameters are needed for estimation of the impact of alternative regulatory decisions. Lichtenberg (2006) suggests that the risk of a transgenic crop can be approximated as the product of the rate of introduction of a transgenic crop, the rate of its dissemination over space and time, and the likelihood of susceptibility of vulnerable populations, which each depend on parameters to be estimated. Research to estimate these parameters is necessary to establish sound quantitative understanding of the risk of biotechnology and is an interdisciplinary challenge essential for a successful regulatory design.

3.2. Optimal Regulatory Criteria

Economic analysis suggests that an optimal approach to determining a regulatory process is to maximize expected net discounted social benefits. But the costs of the environmental side effects as well as the benefits of new varieties are not easy to estimate. Thus, alternative safety rules, which minimize cost to attain safety targets, have been introduced. Lichtenberg (2006) suggests a mechanism that minimizes cost subject to attaining a certain threshold of environmental safety. His approach takes into account the spread of environmental risks from the planting of seeds, to drift, to consumer use of the product. It also recognizes that there are different means to control risks. The experience of pesticide use suggests that regulation based on the understanding of an overall risk-generation process can result in a cost-effective outcome. The safety-first regulatory approach, which is the typical approach for regulating carcinogens, requires imposing more stringent requirements to reduce risk at the point of introduction in order to offset less stringent requirements and an implied increase in the other two components of risk. In this framework, the regulator must assess not only risk (on average) but also uncertainty about risk.

3.3. Regulation of Risky Activities

Genetic drift from biotech crops into other crops or wild reserves may generate undesirable effects both on humans and the environment. Faced with this risk, a key choice is whether to allow a certain technology such as a specific biotech crop variety to be used at all, much like pharmaceutical regulation determines whether to allow use of a specific drug. Development of criteria for risk regulation is critical. The standard approach taken in the United States, for example by the EPA and APHIS, is to evaluate the performance of the technology and, using classical statistics, decide within a high degree of statistical reliability whether a technology has a negative effect. These tests aim to limit Type I errors (rejecting the hypothesis of no adverse effect when there is none). However, such tests do not necessarily result in a small likelihood of Type II errors (failing to reject a hypothesis of no adverse effect when there is one). The difference between the EU perspective and the U.S. perspective on biotechnology risk can be viewed as the difference between limiting Type II errors and limiting Type I errors. The precautionary approach of the EU, as pointed out by Ervin and Welsh (2006), assumes that the biotechnology poses a certain risk and tests for rejection of that assumption. A major problem is that some potential risks may be difficult to identify, and the potential outcomes that cannot be ruled out may vary widely because of lack of information, which makes the precautionary approach much more stringent. Furthermore, as pointed out by Ervin and Welsh, inability to rule out highly adverse outcomes can prevent introduction of a new technology even when benefits are great. Therefore, an approach that balances Type I and Type II errors may be preferable. For example, a Bayesian approach that weights the net benefits of all possibilities with best available information may be preferable as a decision making criterion.

4. COST AND IMPACT OF BIOTECHNOLOGIES

The merit of introducing a biotech crop variety depends on both its benefits and costs relative to the alternatives. The agricultural biotechnology industry has chosen to develop and market input traits, including herbicide tolerance and pest resistance. Farmers outside of the European Union have been quite accepting of these traits. The relevant traits are active in plants from the beginning of growth and greatly reduce problems that arise with chemical pest control, including timing as affected by weather and premature washing away of chemicals from treated plants. The literature identifies several major categories of benefits associated with these biotech crops.

First, and perhaps most important, is enhanced pest control and lower costs of pest management, involving either reduced pesticide use, reduced pest damage, or both. Second, in some situations, biotech crops reduce the health risks faced by farmers and hired farm labor. For instance, Huang et al. (2005) found direct evidence of gains from improved health of farm workers associated with reduced exposure to pesticides when growing GM rice in China. Marra and Piggott (2006) report on estimates of farmers' perceived benefits from reduced exposure to chemical pesticides. Third, and particularly in the context of developing countries that lack effective alternative methods of pest-control, biotech crops may have higher yields (Qaim and Zilberman 2003). For instance, in India and South Africa Bt cotton has a 30–80 percent yield advantage over conventional varieties. Fourth, the use of GM crops may reduce environmental side effects relative to pesticides, though transgenic crops may also introduce new sources of environmental risks.

Some effects become known only after commercial introduction. For example, Bt technology yields human and animal health benefits from reducing mycotoxin levels that have been identified only recently (Wu 2006). Additionally, farmers have come to appreciate the non-pecuniary benefits of agricultural biotechnologies, including increased convenience, reduced risk, and reduced environmental side effects that are valued by farmers (Marra and Piggott 2006).

Both the risks and benefits of agricultural biotechnologies are subject to random effects. Several studies have shown that the costs and benefits of agricultural biotechnologies depend on a variety of factors and that heterogeneity means that impacts differ according to location and application.

The key for sound economic assessment of biotechnology is recognizing that the costs and benefits of new biotechnologies should be measured relative to existing technologies. Bt varieties in many cases will replace varieties that are grown with pesticides and, whatever environmental side effects they may have, may reduce the side effects associated with the technologies they are replacing. For example, one of the main benefits of Bt cotton already documented is reduction of health effects associated with pesticide use. Failure to consider the opportunity costs of regulation may prevent the introduction of varieties that enhance overall social welfare (Zilberman 2006).

The assessment of opportunity costs must also take into account that when new varieties increase yields, they may reduce farmland use. Thus, the extra environmental benefit from increased open space should also be taken into account. One of the major advantages of intensification of agriculture is that it has allowed dramatic increases in food demand to be met without proportional increases in utilized land. Biotechnology may contribute to this trend, which should also be considered in assessing its impacts.

The overall costs of new agricultural biotechnologies depend on consumer attitudes and whether their introduction has implicit costs to farmers that grow existing varieties. The introduction of supply-enhancing biotech crops improves overall welfare if costly means of market segregation and identity preservation activities are not required. But when biotech crop products are considered inferior by consumers, sufficiently high costs of segregation and identity preservation can lead to an overall welfare reduction even though the crops are cost-saving for farmers. To have a welfare gain, the relative gains in production efficiency and net environmental gains from reduced pesticide use have to be larger than the extra cost required for market segregation and identity preservation. Excessive regulatory constraints on the transfer, storage, and handling of GM crop products are also likely to reduce the overall benefits from the introduction of biotech crops. Similarly, if the introduction of biotech crops leads to the creation of new international trade barriers, then the net welfare consequences may be turned from a potential gain to a loss (Anderson 2006). The introduction of segregation and labeling requirements by the EU can be seen as a de facto trade barrier with consequences of this type.

5. COST OF REGULATION

The regulatory process is costly in many ways. It imposes direct added costs on the introduction of new varieties. The part of these expenses borne by private firms increases the initial investment needed to develop a new technology, which detracts from the incentive to develop and supply new technologies. One of the major remaining research challenges is quantification of both the public cost (borne by taxpayers) and private cost (borne initially by private companies and ultimately by consumers and producers of crop products) of regulation of new technologies in various countries and under various regulatory schemes.

Kalaitzandonakes, Alston, and Bradford (2006) document the regulatory process and costs of compliance from the point of view of a biotechnology company, and provide some preliminary estimates of the costs in terms of the expenses incurred and the delay in commercialization.

Pray et al. (2006) suggest that the cost of regulation varies across countries and is dependent on the identity of the regulated party. They find that an Indian company may be charged significantly less than an international company to register a new variety. Falck Zepeda and Cohen (2006) show that the costs of compliance and biosafety regulation in developing countries such as in Africa and Latin America are prohibitively high and may be much higher than a typical research budget required to develop an applicable crop. The high cost of regulation, in many cases, leads to illegal introduction of new varieties and much less control of the impacts.

Bradford, Alston, and Kalaitzandonakes (2006) argue that, when it comes to specialty crops, requiring separate registration of the same trait for each crop that includes it, rather then a one-time registration of the trait, causes a significant reduction of investment in biotechnologies and prevents the introduction of many valuable traits into specialty crops. Sedjo (2006) argues that the introduction of transgenic tree varieties in forestry production may be constrained by concerns about genetic drift. Given the large acreage and low labor intensity in forestry production, monitoring for possible side effects of biotechnologies will be a challenge.

Graff (2006) suggests that biotechnologies may have valuable applications in the production of crops used to produce non-food products. Also, when value per acre is high, production should be economical within contained environments. However, the food industry can be expected to be wary of introducing GM traits or food additives in food products because of the risk of consumer concerns and adverse media accounts. As a result, the future of biotechnologies may be brighter for non-food crops used to manufacture industrial products.

In some cases, concern about environmental side effects may lead to the concentration of some applications in overseas markets where the regulatory framework is more lenient. Relocating production to developing countries where biotechnologies are less restricted than in the United States could result in the loss of the U.S. advantage in many aspects of biotechnology. When it comes to the cost of regulatory constraints, informed regulation requires knowing the direct costs of regulation, the time delays it imposes, and the options that are lost by maintaining, investing in, or discouraging agricultural biotechnology capital.

The cost of regulation also has to take into account the dynamics of innovation. Early applications provide both new solutions to problems and also establish tools that can be used elsewhere for more advanced innovations. Excessive regulatory processes that constrain the application and utilization of the early applications of biotechnology may negatively affect its future. Biotechnology is in its early stages. Significant investment in research and development is needed for it to reach its potential. Heavy regulatory requirements with high uncertainty regarding the likelihood and timing of approval will make investment in new forms of biotechnology less profitable and harm the future of the technology.

6. POLITICAL ECONOMY

A recurring theme among the chapters in this book is the political economy of regulation which determines what is regulated and how and to what extent regulations are imposed. Regulations provide regulating agencies with considerable discretion for rule making, and in some cases provide important opportunities to influence benefits and costs. Some regulators may be predisposed to imposing a heavy set of regulations and then using their discretion to relax and modify them. Regulations also depend on geographic conditions and the strengths of the jurisdictional government authorities. This can result in significant redundancies in registration requirements depending on whether countries impose similar requirements and depending on whether state and local as well as national governments impose overlapping regulations.

The extent to which governments choose to regulate a product depends on the net benefits or costs borne by those governments and the individual parties affected by their actions. Some companies are very good at extracting rents through the regulatory process. Both companies and countries use the regulatory process for rent seeking. Carter and Gruère (2006) and Anderson (2006) argue that some regulations on biotech products are mechanisms that allow countries to achieve an advantage in trade, and some regulations may serve as trade barriers. Regulation of biotechnology is thus becoming a major disputed topic in the World Trade Organization (WTO) and is likely to become an important subject of international trade negotiations.

One plausible explanation for Europe's non-acceptance of biotech crop products is that Europe has little to gain and more to lose. Results such as those presented by Scatasta, Wesseler, and Demont (2006) provide quantitative evidence of the limited or negative benefits of biotech crop adoption. In addition, Europe exports large quantities of pesticides, which are likely to be curtailed by agricultural biotechnology adoption elsewhere. African countries, on the other hand, may be large beneficiaries because no alternatives to biotech crops are available to control some of the devastating diseases, and these countries may not have any feasible alternative mechanism for addressing food deficits and starvation (Evenson 2006). However, African countries have taken little initiative in biotechnology development. Brazil, China, and India, on the other hand, have their own capacities to develop biotech crop varieties. China in particular is developing a significant export market for biotechnologies. To some extent, the future of the biotechnologies is more about what China and India are developing than about what Europe does not allow.

The political process may also affect the adoption of agricultural biotechnologies through farm commodity programs (Gardner 2006). Farm programs in the United States cover eight program crops including cotton, corn, and soybeans. These programs have reduced price risk to farmers who raise program crops. As a consequence, farm programs may have encouraged biotechnology adoption just as for earlier mechanical and hybrid variety technologies. Government polices may also affect the future of the biotechnologies for various crops by allocation of research funding and policy modifications that reduce regulatory costs.

The political economy of regulation can also produce odd bedfellows. Sometimes both environmental groups and major biotechnology companies can favor certain stricter regulations. For example, multinational biotechnology companies may consider heavy regulation as a source of advantage because they have a relative advantage in dealing with the regulatory process. By comparison, a complicated regulatory process reduces the likelihood that innovations will be commercialized by small start-up firms. The size, capital base, and financial muscle of multinational firms gives them the ability to absorb product testing and extensive registration expenses, but this also discourages the development of products to meet the needs of smaller niche markets.

7. NONCOMPETITIVE BEHAVIOR AND THE FUTURE OF AGRICULTURAL BIOTECHNOLOGY

The agricultural biotechnology industry has become very concentrated in three large U.S. firms and three foreign firms. Intellectual property rights were created to bestow a limited monopoly position on the use of new intellectual property, thereby creating an income stream to support its development, just as patents provide for new physical inventions. But the patent duration is limited intentionally to balance incentives for innovation and development under monopoly pricing with the benefits from expansion of access to technology under post-patent competitive pricing. In areas of innovation where the pace is slow, a real possibility exists that going off patent is more important for social welfare than product innovation (Just 2006, Oehmke 2006). Policies that slow this process reduce social welfare.

Pesticide regulation provides insights about how biotechnology regulation might be undertaken given that some biotechnologies are regulated under the same statutes (Just 2006). Post-patent competition can lead to 20–50 percent price reductions for pesticides, even with only a 10–20 percent generic market penetration. The transfer of surplus from monopolistic developers (under patent protection) to farmers and consumers (upon generic entry) is enormous, typically amounting to several times the competitive profits that exist in the industry after generic entry. Thus, the generic firms that provide this competition receive only a fraction of the benefits of other market groups. As a result, manipulation by original entrants of regulatory provisions for sharing of regulatory testing costs between market developers and generic entrants under FIFRA can discourage or prevent generic entry. When this occurs, farmers and consumers do not receive the major share of benefits from innovation that they would receive otherwise under patent policy.

Montana-Alberta cross-border comparisons of pesticide prices confirm the existence of non-competitive pricing (Smith 2006). The substantial fixed cost required to comply with government regulations has thus tended to create an entry barrier for small firms and an advantage to larger firms, which has probably only increased with continuing rounds of mergers and acquisitions. This has been the pattern of the agricultural biotechnology industry just as it has been for the agricultural chemical industry. Furthermore, economies of scale in obtaining regulatory approval also appear to contribute to concentration in both industries. With interest group politics, larger firms tend to be more astute in making political contributions to promote or protect their interests (Heisey and Schimmelpfennig 2006). Thus, the current regulatory framework for biotechnologies in the United States prevents efficient functioning of markets and denies or delays post-patent farmer and consumer benefits.

8. CONSUMERS AND MARKET ACCEPTANCE

While regulation hampers the speed of introduction of many new biotech-nologies, and prohibits others, in many cases consumer and market accep-tance may be the main obstacle. Research on consumer and market accep-tance enables understanding more about what types of people like or dislike the technology, but it is harder to determine the causes of positive or nega-tive attitudes and to judge the strength of preferences, or how they would be reflected in an unregulated market environment.

In many cases consumers are not able to reveal their true preferences in the marketplace because retailers or food manufacturers have opted not to handle biotech crop products or because governments have adopted policies that keep them out of the market. In those instances, information about con-sumer preferences is limited to that derived from stated preference studies or from experimental studies using synthetic markets. Such research has re-vealed demographic patterns in attitudes towards biotech crops but little is known about the reasons for negative attitudes or the process by which negative attitudes may have evolved. Further, some economists have raised questions about the extent to which stated preferences, or even the results from experimental studies, accurately represent actual attitudes as they would be revealed in a real marketplace. In particular, studies have shown that the answers to hypothetical questions about attitudes are highly depend-ent on the phrasing of the questions. Thus, these results in the literature must be interpreted carefully with some degree of skepticism.

Consumers are heterogeneous with respect to age, education, income, and prior information about biotechnology, as well as their preferences for agricultural biotechnology. In a recent study, 43 percent of U.S. consumers, but only 30 percent of Norwegian consumers, were extremely or somewhat willing to consume foods that were produced using GM ingredients. When asked about the importance of price as factor when deciding whether or not to buy GM foods, 67 percent of U.S. consumers but only 36 percent of Nor-wegian consumers said that price was extremely or somewhat important. These differences in willingness to consider foods containing biotech crop products are consistent with other evidence in the literature about the U.S. versus European divide over GM products (for instance, see Lusk et al. 2005 as well as Huffman and Rousu 2006 and McCluskey, Grimsrud, and Wahl 2006).

A considerable amount of evidence has been accumulated indicating that consumers in Western developed countries would discount food products made using GM ingredients. Also, a small amount of evidence exists indi-cating that consumers would discount GM fish and animal products more heavily than plant products. However, the average premium for non-GM traits is quite small. Across a broad set of countries, consumer acceptance of

GM products has been shown to depend on cultural, religious, and political factors. Evidence also exists that Western consumers' resistance to GM products is related to information effects. For example, the probability that consumers will reject biotech food products is significantly higher if they have received anti-biotechnology information and significantly lower if they have received pro-biotechnology information. Introduction of third-party information almost exactly offsets the negative effects of anti-biotechnology information. Hence, some of the difference between acceptance and resistance in the United States versus Europe is undoubtedly the result of a different intensity of information flow.

Labeling for GM content is an important and controversial topic for consumers, farmers, and the agricultural biotech industry. Labeling would be part of an effective policy of segregating GM and non-GM products. In a recent U.S. and Norwegian study, 87 percent of U.S. consumers and 99 percent of Norwegian consumers indicated that GM labeling was extremely or somewhat important to them. This result, however, is interpreted as a preference when labeling is costless. When they were told that GM labeling might increase food costs by 5 percent, the percentage of consumers in the United States and Norway that indicated that labeling was extremely or somewhat important dropped by 44 percentage points in both cases (McCluskey, Grimsrud, and Wahl 2006). Hence, consumers' preferences for GM-labeling in Western developed countries are price-responsive.

9. A GLOBAL PERSPECTIVE

Agricultural biotechnology has an important international dimension, beyond the acceptance contrast between the United States and the EU. The greatest potential impact of biotechnologies may be in developing countries. Distinction between two tiers of nations is important for this discussion: the "up-and-coming" nations, including India, China, Brazil, South Africa, and Argentina; and the "poor nations," which are mostly in sub-Saharan Africa. Some African nations were largely missed by the Green Revolution and are ill-prepared to deal with the issues raised by agricultural biotechnologies. Rough estimates show that great potential exists for biotechnologies to solve hunger problems in Africa, but development has not taken place because of insignificant economic incentives for biotechnology developers (Evenson 2006). The six large multinational agricultural biotechnology firms make their profits by "renting" their GM traits, using contracts that prevent second-generation use of seed stock. Without a reliable system of enforcement, as is the case in most poor countries, profitability is not assured for these developers. Some potential exists for eliminating the contract enforcement problem associated with poor nations by incorporating a terminator gene that limits second-generation benefits, but biotechnology companies have opted not to introduce this gene in commercial applications in the face of opposition from environmentalists.

The EU's conservative policy on agricultural biotechnology, even though it imposes relatively little cost on the EU because current input traits are not well suited to EU crops, has discouraged biotechnology development worldwide. In part, the discouragement may be the result of adverse attention in the media. Although the EU can easily afford the losses that it incurs by going slowly, the same cannot be said for the poor countries of Africa, which are affected by the discouraging posture of the EU. Evenson (2006) estimates that only about $1 billion of the $8 billion in potential annual benefits from GM crops has been realized, with most of the unrealized potential remaining in poor countries.

China, India, and Brazil have adopted transgenic crops on large amounts of land and have developed the technical capacity to introduce new traits to their crop varieties and even to identify new traits. China has probably as many experimental plots as any country and is poised to become a major developer and user of agricultural biotechnology. The less restrictive regulatory environment in China may lead to faster development than in countries where biotechnologies were initially developed. But as China becomes a major producer of plant-made pharmaceuticals and plant-made industrial products, its exports will be constrained by more stringent regulation in the developed world, just as has been the case in agricultural chemical production. Thus, the utilization and development of innovative capacities in agricultural biotechnologies will become a source of growth and export earnings for some of the up-and-coming countries, while the lack of adoption of transgenic crops may increase the disadvantage of the poorest nations.

10. THE FUTURE OF AGRICULTURAL BIOTECHNOLOGY

Agricultural biotechnology promises to be a source of great benefits and value for humanity, but this technology is very new and still in its infancy. Only some of its basic features, potential, and risks are observable. At this early stage, which is crucial for the development of the industry, some basic tools, both technological and institutional, are emerging. They include enabling technologies (allowing early applications) and establishment of a regulatory framework that enables a global scientific, educational, and industrial infrastructure that will facilitate further development of the technology, finance investment in expansion of its capacity, and produce and market its product. The early applications of the technology have been those easiest to conceive and implement, taking advantage of existing knowledge, market demand, and institutional and financial capacity.

The early applications of agricultural biotechnology, which have been mostly for the control of pests and disease, establish the capacity of the technology to provide new tools that are easy to implement. They suggest huge potential demand for biotech crop varieties and demonstrate the fast rate at which they can be adopted. The early outcomes, especially given the high rates of adoption, are promising in terms of safety and control of side effects. But, they also indicate the potential for consumer concerns and market resistance to stymie the development and adoption of the technology. The sources of consumer concerns and market resistance must be identified and addressed if the ultimate potential benefits are to be realized. The introduction of some second-generation biotechnology products, which are being developed and tested now, will include traits that have more apparent consumer benefits by enhancing the nutritional content of foods or further reducing environmental footprints (Moschini and Lapan 2006). Education may also help consumers to realize the reduction in carcinogenic and other side effects of pesticides that are displaced. Agricultural economists and other professionals in the public sector can play a role in enhancing consumer acceptance of the technology by developing and communicating relevant and credible information about the technologies and the costs and benefits of regulation.

The evolution of agricultural biotechnology has drastically altered the structure of related seed and chemical input industries and has significant distributional implications. Emerging institutional innovations (such as intellectual property rights and initiatives to establish easier access to enabling technologies) and a prominent role of public research and venture start-ups in triggering new breakthrough innovations has provided hope that the bulk of benefits from biotechnology will not be captured in the end by only a few corporations. However, the typical evolution whereby discoveries forthcoming from public research and successful venture start-ups are bought up by a small number of large multinational firms before commercialization suggests that regulation may be necessary to ensure that the benefits of biotechnology are shared among many.

The regulation of biotechnologies thus faces a twofold challenge: (i) to balance the benefits from maintaining and building technological and commercial momentum against the benefits from protecting human health and environmental safety, and (ii) to balance incentives for private investment against the sharing and expansion of benefits through competition and open access. Lack of sufficient oversight may result in malfunctioning products and dangerous accidents that would create further objections and resistance to the development of the technology. Costly and time-consuming regulations will delay product introduction, reduce the range of new products, diminish their profitability, lower the rate of return on investment in agricultural biotechnology, slow the build-up of new capacity and human capital, and diminish research efforts to expand the knowledge base that serves to fuel further development of the technology. Heavy-handed regulation is likely to choke agricultural biotechnology and prevent harvesting of its massive potential benefits.

While market acceptance poses major obstacles on the path of agricultural biotechnology development, expensive and redundant regulatory requirements are also major causes for delays in the evolution of biotechnology and expansion of its application, especially in developing countries and specialty crops. They also lead to high concentration in the biotechnology industry and sometimes to illegal introduction of biotech crop varieties, which circumvents the regulatory process. Emphasis on imposing registration requirements at the appropriate level (by trait rather than event or variety) and for the appropriate location (broad agro-ecological zones rather than smaller political boundaries) will reduce regulatory costs and may also provide opportunities to intensify testing. Prior to regulatory reform, however, the costs of various approaches to regulation require quantification. This is a major immediate priority for agricultural economists and policy research.

REFERENCES

Anderson, K. 2006. "Interactions Between Trade Policies and GM Food Regulations." In R.E. Just, J.M. Alston, and D. Zilberman, eds., *Regulating Agricultural Biotechnology: Economics and Policy*. Amsterdam: Springer.

Berwald, D., S. Matten, and D. Widawsky. 2006. "Economic Analysis and Regulating Pesticide Biotechnology at the U.S. Environmental Protection Agency." In R.E. Just, J.M. Alston, and D. Zilberman, eds., *Regulating Agricultural Biotechnology: Economics and Policy*. Amsterdam: Springer.

Bradford, K.J., J.M Alston, and N. Kalaitzandonakes. 2006. "Regulation of Biotechnology for Specialty Crops." In R.E. Just, J.M. Alston, and D. Zilberman, eds., *Regulating Agricultural Biotechnology: Economics and Policy*. Amsterdam: Springer.

Carter, C.A., and G.P. Gruère. 2006. "International Approval and Labeling Regulations of Genetically Modified Food in Major Trading Countries." In R.E. Just, J.M. Alston, and D. Zilberman, eds., *Regulating Agricultural Biotechnology: Economics and Policy*. Amsterdam: Springer.

Ervin, D.E., and R. Welsh. 2006. "Environmental Effects of Genetically Modified Crops: Differentiated Risk Assessment and Management." In R.E. Just, J.M. Alston, and D. Zilberman, eds., *Regulating Agricultural Biotechnology: Economics and Policy*. Amsterdam: Springer.

Evenson, R.E. 2006. "Status of Agricultural Biotechnology: An International Perspective." In R.E. Just, J.M. Alston, and D. Zilberman, eds., *Regulating Agricultural Biotechnology: Economics and Policy*. Amsterdam: Springer.

Falck Zepeda, J., and J.I. Cohen. 2006. "Biosafety Regulation of Genetically Modified Orphan Crops in Developing Countries: A Way Forward." In R.E. Just, J.M. Alston, and D.

Zilberman, eds., *Regulating Agricultural Biotechnology: Economics and Policy*. Amsterdam: Springer.

Frisvold, G.B. 2006. "Bt Resistance Management: The Economics of Refuges." In R.E. Just, J.M. Alston, and D. Zilberman, eds., *Regulating Agricultural Biotechnology: Economics and Policy*. Amsterdam: Springer.

Gardner. B. 2006. "Regulation of Technology in the Context of U.S. Agricultural Policy." In R.E. Just, J.M. Alston, and D. Zilberman, eds., *Regulating Agricultural Biotechnology: Economics and Policy*. Amsterdam: Springer.

Graff, G.D. 2006. "Regulation of Transgenic Crops Intended for Pharmaceutical and Industrial Uses." In R.E. Just, J.M. Alston, and D. Zilberman, eds., *Regulating Agricultural Biotechnology: Economics and Policy*. Amsterdam: Springer.

Heisey, P., and D. Schimmelpfennig. 2006. "Regulation and the Structure of Biotechnology Industries." In R.E. Just, J.M. Alston, and D. Zilberman, eds., *Regulating Agricultural Biotechnology: Economics and Policy*. Amsterdam: Springer.

Huang, J., R. Hu, S. Rozelle, and C. Pray. 2005. "Insect-Resistant Rice in Farmers' Fields: Assessing Productivity and Health Effects in China." *Science* 308 (April 29): 688–690.

Huffman, W.E., and M. Rousu. 2006. "Consumers Attitudes and Market Resistance to Biotech Products." In R.E. Just, J.M. Alston, and D. Zilberman, eds., *Regulating Agricultural Biotechnology: Economics and Policy*. Amsterdam: Springer.

James, C. 2004. "Preview: Global Status of Commercialized Biotech/GM Crops: 2004." ISAAA Briefs No. 32. International Service for the Acquisition of Agri-Biotech Applications, Cornell University, Ithaca, NY. Available online at www.isaaa.org (accessed November 2005).

Just, R.E. 2006. "Anticompetitive Impacts of Laws That Regulate Commercial Use of Agricultural Biotechnologies in the United States." In R.E. Just, J.M. Alston, and D. Zilberman, eds., *Regulating Agricultural Biotechnology: Economics and Policy*. Amsterdam: Springer.

Kalaitzandonakes, N., J.M. Alston, and K.J. Bradford. 2006. "Compliance Costs for Regulatory Approval of New Biotech Crops." In R.E. Just, J.M. Alston, and D. Zilberman, eds., *Regulating Agricultural Biotechnology: Economics and Policy*. Amsterdam: Springer.

Langrock, I., and T.M. Hurley. 2006. "Farmer Demand for Corn Rootworm Bt Corn: Do Insect Resistance Management Guidelines Really Matter?" In R.E. Just, J.M. Alston, and D. Zilberman, eds., *Regulating Agricultural Biotechnology: Economics and Policy*. Amsterdam: Springer.

Lichtenberg, E. 2006. "Regulation of Technology in the Context of Risk Generation." In R.E. Just, J.M. Alston, and D. Zilberman, eds., *Regulating Agricultural Biotechnology: Economics and Policy*. Amsterdam: Springer.

Lusk, J.L., M. Jamal, L. Kurlander, M. Roucan, and L. Taulman. 2005. "A Meta-Analysis of GM Food Evaluation Studies." *Journal of Agricultural and Resource Economics* 30(1): 28–44.

Marra, M.C., and N.E. Piggott. 2006. "The Value of Non-Pecuniary Characteristics of Crop Biotechnologies: A New Look at the Evidence." In R.E. Just, J.M. Alston, and D. Zilberman, eds., *Regulating Agricultural Biotechnology: Economics and Policy*. Amsterdam: Springer.

McCluskey, J.J., K.M. Grimsrud, and T.I. Wahl. 2006. "Comparisons of Consumer Responses to Genetically Modified Foods in Asia, North America, and Europe." In R.E. Just, J.M.

Alston, and D. Zilberman, eds., *Regulating Agricultural Biotechnology: Economics and Policy*. Amsterdam: Springer.

Mitchell, P.D., and T.M. Hurley. 2006. "Adverse Selection, Moral Hazard, and Grower Compliance with Bt Corn Refuge." In R.E. Just, J.M. Alston, and D. Zilberman, eds., *Regulating Agricultural Biotechnology: Economics and Policy*. Amsterdam: Springer.

Moschini, G., and H. Lapan. 2006. "Labeling Regulations and Segregation of First- and Second-Generation GM Products: Innovation Incentives and Welfare Effects." In R.E. Just, J.M. Alston, and D. Zilberman, eds., *Regulating Agricultural Biotechnology: Economics and Policy*. Amsterdam: Springer.

Oehmke, J.F. 2006. "The Social Welfare Implications of Intellectual Property Protection: Imitation and Going Off Patent." In R.E. Just, J.M. Alston, and D. Zilberman, eds., *Regulating Agricultural Biotechnology: Economics and Policy*. Amsterdam: Springer.

Perrin, R.K. 2006. "Regulation of Biotechnology for Field Crops." In R.E. Just, J.M. Alston, and D. Zilberman, eds., *Regulating Agricultural Biotechnology: Economics and Policy*. Amsterdam: Springer.

Pray, C.E., J. Huang, R. Hu, Q. Wang, B. Ramaswami, and P. Bengali. 2006. "Benefits and Costs of Biosafety Regulation in India and China." In R.E. Just, J.M. Alston, and D. Zilberman, eds., *Regulating Agricultural Biotechnology: Economics and Policy*. Amsterdam: Springer.

Qaim, M., and D. Zilberman. 2003. "Yield Effects of Genetically Modified Crops in Developing Countries." *Science* 299 (February 7): 900–902.

Scatasta, S., J. Wesseler, and M. Demont. 2006. "Irreversibility, Uncertainty, and the Adoption of Transgenic Crops: Experiences from Applications to HT Sugar Beets, HT Corn, and Bt Corn." In R.E. Just, J.M. Alston, and D. Zilberman, eds., *Regulating Agricultural Biotechnology: Economics and Policy*. Amsterdam: Springer.

Secchi, S., T.M. Hurley, B.A. Babcock, and R.L. Hellmich. 2006. "Managing European Corn Borer Resistance to Bt Corn with Dynamic Refuges." In R.E. Just, J.M. Alston, and D. Zilberman, eds., *Regulating Agricultural Biotechnology: Economics and Policy*. Amsterdam: Springer.

Sedjo, R.A. 2006. "Regulation of Biotechnology for Forestry Products." In R.E. Just, J.M. Alston, and D. Zilberman, eds., *Regulating Agricultural Biotechnology: Economics and Policy*. Amsterdam: Springer.

Smith, V.H. 2006. "Regulation, Trade, and Market Power: Agricultural Chemical Markets and Incentives for Biotechnology." In R.E. Just, J.M. Alston, and D. Zilberman, eds., *Regulating Agricultural Biotechnology: Economics and Policy*. Amsterdam: Springer.

Smyth, S., P.W.B. Phillips, and W.A. Kerr. 2006. "Managing Liabilities Arising from Agricultural Biotechnology." In R.E. Just, J.M. Alston, and D. Zilberman, eds., *Regulating Agricultural Biotechnology: Economics and Policy*. Amsterdam: Springer.

Wang, S., D.R. Just, and P. Pinstrup-Andersen. 2006. "Damage from Secondary Pests and the Need for Refuge in China." In R.E. Just, J.M. Alston, and D. Zilberman, eds., *Regulating Agricultural Biotechnology: Economics and Policy*. Amsterdam: Springer.

Wu, F. 2006. "Bt Corn's Reduction of Mycotoxins: Regulatory Decisions and Public Opinion." In R.E. Just, J.M. Alston, and D. Zilberman, eds., *Regulating Agricultural Biotechnology: Economics and Policy*. Amsterdam: Springer.

Zilberman, D. 2006. "The Economics of Biotechnology Regulation." In R.E. Just, J.M. Alston, and D. Zilberman, eds., *Regulating Agricultural Biotechnology: Economics and Policy*. Amsterdam: Springer.

INDEX